Community Ecology: Pattern and Process

Community Ecology: Pattern and Process

EDITED BY

Jiro Kikkawa
Department of Zoology,
University of Queensland, Brisbane, Australia

Derek J. Anderson
School of Botany,
University of New South Wales, Sydney, Australia

BLACKWELL SCIENTIFIC PUBLICATIONS
MELBOURNE OXFORD LONDON
EDINBURGH BOSTON PALO ALTO

Editorial offices:
107 Barry Street, Carlton
Victoria 3053, Australia
Osney Mead, Oxford, OX2 OEL
8 John Street, London, WC1N 2ES
23 Ainslie Place, Edinburgh, EH3 6AJ
52 Beacon Street, Boston
Massachusetts 02108, USA
744 Cowper Street, Palo Alto
California 94301, USA

First published 1986

Typeset by Setrite Typesetters
Printed by Kyodo-Shing Loong Pte Ltd

DISTRIBUTORS

USA and Canada
Blackwell Scientific Publications Inc
PO Box 50009, Palo Alto
California 94303

Australia
Blackwell Scientific Publications Pty Ltd
107 Barry St, Carlton
Victoria 3053

Cataloguing in Publication Data

Community ecology.
Bibliography.
Includes index.
ISBN 0 86793 272 4 (PBK.).
ISBN 0 86793 264 3
1. Biotic communities. 2. Ecology.
I. Kikkawa, Jiro, 1929– .
II. Anderson, D.J. (Derek John), 1935–

574.5′24

Contents

List of Contributors

DEREK J. ANDERSON *School of Botany, University of New South Wales, Kensington, NSW 2033, Australia*

A. (TONY) D. BRADSHAW *Department of Botany, University of Liverpool, Liverpool L69 3BX, UK*

PETER D. DWYER *Department of Zoology, University of Queensland, St Lucia, Qld 4067, Australia*

W. (BILL) J. FREELAND *Conservation Commission of the Northern Territory, Winnellie, NT 5789, Australia*

PETER GREIG-SMITH *School of Plant Biology, University College of North Wales, Bangor LL57 2UW, UK*

DAVID W. GOODALL *CSIRO Division of Groundwater Research, Wembley, WA 6014, Australia*

JOHN C. HOLMES *Department of Zoology, University of Alberta, Edmonton, Alberta T6G 2E9, Canada*

JIRO KIKKAWA *Department of Zoology, University of Queensland, St Lucia, Qld 4067, Australia*

ROGER L. KITCHING *Department of Ecosystem Management, University of New England, Armidale, NSW 2351, Australia*

YASUSHI KURIHARA *Biological Institute, Faculty of Science, Tohoku University, Sendai 980, Japan*

JOHN H. LAWTON *Department of Biology, York University, Heslington, York YO1 5DD, UK*

MALCOLM MACGARVIN *Department of Biology, York University, Heslington, York YO1 5DD, UK*

A. MARTIN MORTIMER *Department of Botany, University of Liverpool, Liverpool L69 3BX, UK*

PETER W. PRICE *H.S. Colton Research Center, Museum of Northern Arizona, Flagstaff, AZ 86001, USA*

SCOTT ROBINSON *Illinois Natural History Survey, 607 East Peabody Drive, Champaign, Il 61820, USA*

THOMAS W. SCHOENER *Department of Zoology, University of California, Davis, CA 95616, USA*

JOHN TERBORGH *Department of Biology, Princeton University, Princeton, NJ 08540, USA*

A. (TONY) J. UNDERWOOD *School of Biological Sciences, University of Sydney, Sydney, NSW 2006, Australia*

Preface

The past quarter of a century has seen considerable advances in the methods of community ecology; new techniques of environmental measurement, community metabolism studies, complex data analysis and model building have developed and an array of new information has become available through organized research. As a result, many patterns and processes at the community level of organization have been described and the range of communities studied has been greatly expanded. For example, tropical communities, both marine and terrestrial, have never received so much attention in the past as they do now. Similarly, microbial and parasite communities are very much part of the living systems being discussed today. Critical examination of new facts in relation to established concepts, as well as developing concepts, has not yet found its way into textbooks. Our intention has been to seek authors who will take a fresh and critical look at major patterns and processes that must underlie all community organization. We have based our intention on our own idiosyncratic viewpoints concerning the need for integration and the need to resolve some ecological paradoxes (Anderson 1971; Kikkawa 1977). It seems to us that the authors represented in this book have accepted most graciously the role imposed upon them. One of them has even shown us that one aspect of humour — punning — can provide other and maybe unexpected intellectual insights.

This book will, in most areas of community ecology, focus on 'real world' systems as revealed by recent studies, evaluate existing theories and provoke new thoughts. It consists of five parts. The first is an introductory chapter which gives a historical perspective to the development of modern concepts necessary for the understanding of natural communities. Part 2 deals with an area in which our understanding of communities is particularly tied to methodologies developed in recent years. Part 3 presents not only structural but also functional patterns seen in the horizontal and vertical organization of communities. Part 4 is concerned with ecological processes which contribute to the organization of communities. Here the biotic interactions are emphasized and interpreted at community level of organization rather than as population processes. Part 5 examines evolutionary processes which contribute to the organization of communities. This is one area in which community ecology is struggling to establish methodologies and coherent theories. We have explored inductive, deductive and traditional methods to elucidate what are clearly evolutionary processes occurring in communities. Not least, we have included the patterns produced by human societies that have developed through the utilization of natural resources.

We are most grateful to the authors who have been so responsive, so co-operative and so patient in meeting our goal of producing a textbook which, while scientifically rigorous in addressing its major themes, simultaneously offers some new perspectives on community patterns and process.

Jiro Kikkawa
Derek J. Anderson

Acknowledgements

We are most grateful to Robert Campbell of Blackwell Scientific Publications for his encouragement, advice and sustained enthusiasm, without which this book would never have seen the light of day. Over a period of 5 years in which plans, drafts and manuscripts travelled satellite distances many colleagues in different countries provided us with constructive criticisms, Mark Robertson of Blackwell in Melbourne provided unfailing support, Fay Marsack retyped many a draft and Jenny Rushton redrew most diagrams. Drawings in Table 5.3 and Fig. 6.1 are due to Naoko Kikkawa and Fig. 14.13 to Lynn Pryor. The authors of respective chapters thank the following for discussion or critical comments on drafts: J.M. Cherrett, P.W. Greig-Smith, R.N. Hughes, M.A. Lock, W.W. Sanford (Chapter 2); A.M. Cameron, M.B. Dale, P.D. Dwyer, S.L. Pimm (Chapter 4); T. Case, C. Toft, M. Turelli (Chapter 6); J.G. Greenwood, K. Plowman (Chapter 7); M.J. Crawley for loan of book proofs (Chapter 8); A.O. Bush, A.E. Butterworth, T.M. Stock (Chapter 9); N.E. Pierce, C. Catterall, J. Covacevich, R. Floyd, J. Gibb, M. Zalucki (Chapter 10); J.H. Choat, J.H. Connell, P.G. Fairweather, the late R.A. Fisher, P. Jernakoff, K.A. McGuinness, P.A. Underwood (Chapter 11); C. Barbour, P.D. Dwyer, S. Freeland, D.H. Janzen, B. Kerans, P.W. Price, J. Runkle, D.R. Strong (Chapter 13); J. Lazarus, M. Minnegal (Chapter 15). We also wish to thank Joan Ratcliffe for typing references and index and Sandie Wilson and Brad Congdon for helping with proof reading.

Permission to reproduce copyright material was given by: Academic Press (Table 14.3, Fig. 14.9), Acta XVII Congressus Internationalis Ornithologici (Tables 5.5, 5.6), the American Association for the Advancement of Science (Figs 6.5a, b, 10.3, 10.4), American Ornithologists' Union (Figs 5.4, 5.5), the American Society for Limnology and Oceanography (Fig. 7.6), Annual Reviews Inc. (Table 5.4, Fig. 6.16), the Australian Museum (Table 10.4), the Cambridge University Press (Fig. 14.11), the Ecological Society of America (Figs 6.11, 6.12, 6.13), Elsevier Biomedical Press (Table 7.7), the Genetic Society of Great Britain (Fig. 14.6), the Harvard University Press (Table 5.2), the Linnean Society of London (Fig. 6.8), Princeton University Press (Fig. 6.4), Smithsonian Institution (Fig. 5.3), the Society for the Study of Evolution (Fig. 6.14), the Society of Systematic Zoology (Fig. 6.2), the University of Chicago Press (Fig. 14.4), the University of New England Publishing Unit (Fig. 7.3), the University of Texas Press (Fig. 14.13a), and the Zoological Society of London (Fig. 7.4).

Part 1
Introduction

1 Development of Concepts

Derek J. Anderson and Jiro Kikkawa

1.1 Introduction

It is one and a quarter centuries since Saint Hilaire wrote of "the study of the relations of the organism within the family and society in the aggregate and in the community", yet modern ecologists still equivocate in their use of the word 'community'. For some the term is still redolent of the early plant geographers attempting to classify the range of vegetation covering the earth's surface. For example, Mueller-Dombois and Ellenberg (1974) write: "...plant communities may be considered (as) subdivisions of a vegetation cover. Wherever the cover shows more or less obvious spatial changes, one may distinguish a different community". For others the term can be usefully imprecise as in Poore (1964): "...it is an ecological maid-of-all-work with undefined duties and responsibilities...", or definable conditionally as in Hairston (1964): "Animal communities may be considered organized if any property of a natural assemblage of species can be predicted".

That the concept of 'community' developed from a clearly appreciated social structure of plant and animal groupings in the wild seems clear from the distinction already made between autecology and synecology by the turn of the century (Schröter & Kirchner 1896, 1902). In 1933 the Committee on Nomenclature of the Ecological Society of America followed Warming's earlier description of communities as "ecological units of every degree" by defining 'community' as "a general term to designate sociological units of every degree from the simplest (as an unrooted mat of algae) to the most complex biocoenosis (as a multistoried rain forest)".

For our own purposes, particularly in structuring the contributions that make up this book, there seems much to be gained by adopting and endorsing Robert Whittaker's (1975) view that "much ecological understanding can be integrated also around the concepts of communities as assemblages of different species which interact with one another, and ecosystems as functional systems formed by communities and their environments". Assemblages of species interacting with one another, or at least sharing the same general resources, such as herbivores, parasites and microbial detritivores, all growing, reproducing and surviving in a locally 'social' and larger-scale environmental context, through time, contribute to the framework of community ecology. The constituents of that framework (Fig. 1.1) we now mostly take for granted, but the development of the component concerns of community ecology are themselves of some historical interest, not least because they illustrate the impact of 'context sensitivity'. Just as Charles Elton appreciated "...The Arctic made me see ecological processes in full swing..." (Elton & Miller 1954), Pearsall (1964) perceptively pointed out the more general truth that "...an ecologist, like his subject, owes much to the impact of environment".

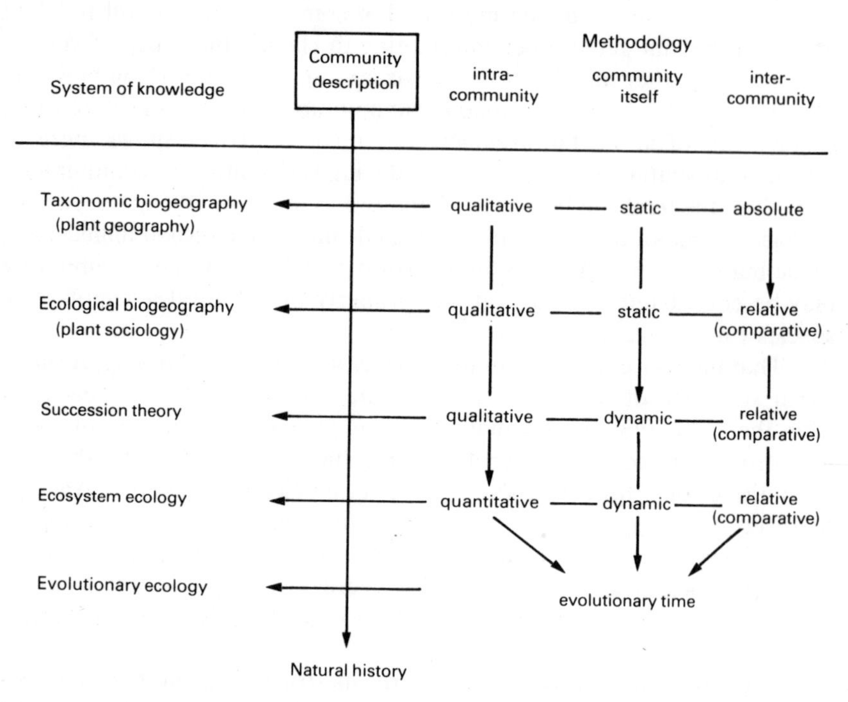

Fig. 1.1. Development of the concept of community in descriptive ecology. Groupings of organisms have been described variously as the concept of community developed from the qualitative–static–absolute entity to quantitative–dynamic–relative entity, signifying establishment of a new discipline or theory with each change in methodological treatment of the subject matter (inter-community, community itself and intra-community). Evolutionary time is yet another dimension leading to a distinct area of community ecology. Natural history remains central to community description throughout the development of community ecology. Other approaches, such as mathematical ecology, systems ecology and physiological ecology, are considered as additional means to establish systems of knowledge concerning biotic communities.

1.2 Natural history

Community ecology as a descriptive science began with the first attempts to group organisms. In this sense we may consider Aristotle's classification of animals according to habitat, Theophrastus's (372–288 BC) description of plant distribution in relation to altitude, or the more pragmatic natural history book of Plinius (23–79 AD) as rudimentary approaches to community ecology. As a system of knowledge, however, natural history did not emerge until the eighteenth century, when Réaumur described community life and other habits of insects in *Memoires pour servir à l'histoire des insectes* (1734–42) and Buffon founded adaptation ecology in his six-volume *Histoire Naturelle* (1780–1785).

1.3 Plant geography

It was in the nineteenth century that the substance of natural history was given a global perspective. From the widespread study of physiognomic types (size and form of vegetation) von Humboldt (1850) arrived at an explanation of the distribution of plant formations in relation to broad climatic belts. He defined the climatic belt in terms of the mean air temperature, which increased about 1°F with each decreasing degree in latitude. His environmental induction was the first attempt at establishing plant geography as a system of knowledge to be distinguished from descriptive natural history.

Plant geography flourished in the nineteenth century as terrestrial communities were grouped according to physiognomy (von Humboldt), life forms based on floristic responses (Grisebach, Drude), ecological responses (Henfrey, Warming, Raunkiaer) or vegetation structure (Kerner), and habitat types discriminated by temperature and humidity (de Candolle).

1.4 Plant sociology

In the first quarter of this century a group of European botanists based in Switzerland and southern France set out to achieve a coherent classification of plant communities, basing their strategy on the somewhat loosely assumed analogy that assemblages of species in a range of vegetation types could be treated equivalently with taxonomic discriminations based on assemblages of organisms with common genetic lineages. Today the analytical techniques employed by this 'Zürich–Montpellier' school of phytosociology led by J. Braun-Blanquet have been largely superseded by computer-assisted sorting strategies. However, the group left an important legacy, not only in terms of a comprehensive vegetation classification for much of Europe, but also by providing an impetus for other ecologists to further develop multi-objective approaches to vegetation classification *per se*, and more particularly to seek a clear understanding of the ecological relevance of 'sociality'.

The Z–M approach to community analysis was essentially floristic. Semiquantitative data recorded as cover-abundance indices were collected systematically from a range of vegetation samples (*relevés*), tabulated and sorted to achieve a display of discriminated communities known technically as associations. The key floristic discriminant was held to be the existence of certain diagnostic species, originally conceived as *faithful* but latterly designated as *character*-species (defining communities by their distributional modes) or *differential*-species (defining communities by their distributional limits).

These early attempts to define the limits of communities were a natural legacy from the newly appreciated global system of biomes and plant formations recognized by the well-travelled plant geographers of the nineteenth century. If the approach appeared anthropocentric and subjective, objectivity at least could be entertained by employing formalized statistical methods to divide regional vegetation into circumscribed communities. The ecologists who were either led or who later owed 'genealogical' allegiance to G.E. Du Rietz at Uppsala formed such a school. Dubbed by Braun-Blanquet *Die Herren Quadratica* for their insistence on the need for quadrat sampling, the International Botanical Congress held in 1910 determined that the communities defined by the Uppsala School were not identical to the Z–M model, and therefore had to be distinguished as *sociations* (Whittaker 1962). Distinct or not, the Uppsala group used coincidence tables to great effect in establishing communities based on the similarity of groups of samples taken within a regional area, and their insistence on the use of a statistical approach provided a considerable stimulus if not impetus for the more rigorously mathematical analysis of communities that was to follow in the second half of the century. Not least, the Uppsala School clearly recognized the existence and possible circumscription of animal communities, a recognition recorded for other reasons in North America by the publication in 1913 of Shelford's *Animal Communities in Temperate America*.

These early attempts to define and circumscribe plant and animal communities were followed by a myriad of classificatory approaches, particularly after the introduction of digital computers as standard items of hardware in ecological laboratories in the 1960s. Although much of the growth of computer-assisted, multivariate analysis had its origins in the fields of human behaviour and psychology, it had other, lateral roots in the controversy that stemmed from the zealous espousal of dynamic change — the doctrine that ecology could be encapsulated simply as succession and climax theory — emanating from North America, particularly through the pens of Cowles and Clements. The attention paid to succession as a cornerstone of community ecology has been at once both energizing and ennervating for at least three generations of ecologists; and among those generations have been not only believers and converts but also the occasional infidel (e.g. Egler 1942, 1951).

1.5 Ecological succession

Although many early ecological writings contain seeds of the notion that communities of plants and animals can change, and change more or less predictably through time, it was two young American ecologists who communicated this message to very considerable effect. The European and Scandinavian hierarchy of associations established by careful annotation and classification of seemingly static communities became instead a dynamic hierarchy of time-related phases in the eyes of the North American phytosociologists.

In 1899 H.S. Cowles published an account of his carefully documented observations and inferences on the series of communities, from dune grassland to beech–maple forest, developed extensively over the southern shores of Lake Michigan. Circumstantial as much of the evidence was, it seemed incontrovertible to Cowles and his students that the various vegetation phases which could be floristically and physiognomically distinguished — cottonwoods, pine forest, oak woods; *Typha* marsh, *Salix* shrubland and oak–hickory woodland — all led inexorably by successional stages to a climax beech–maple forest.

The ubiquity of both primary and particularly secondary successions was borne in on Clements too as he toured his native state while preparing material for a monograph, coauthored with Pound, *The Phytogeography of Nebraska.* With an intuitive conviction that all communities within a geographical region were but seral stages leading to the climatically-determined formation for that region, and an impressive advocacy of his viewpoint, Clements injected a considerable dynamic stimulant and a new terminological base into community ecology. If all successional pathways in the same climatic region led to a single common climax, then all such routes were surely both directional and self-determining; and self-determination suggested organismal qualities for communities. The hierarchical nature of the community — the whole being more than merely the sum of its constituent parts — served not only to feed developing concepts of holism (and the subsequent, even modern debates on the utility of an 'holistic' as opposed to a 'reductionist' approach to ecological science) but also confirmed an already widely accepted view that communities had 'emergent' properties. The manner in which successional vegetation obviously repaired itself after absorbing wounds incurred through disturbance promoted Clements' view of the climax as a 'superorganism'.

Although Clements promulgated his new-found generalization with great effect throughout the English-speaking world, many other field-experienced ecologists remained unconvinced of his monoclimax concept and adopted rather a polyclimax theory as a more realistic intellectual position, particularly in those parts of Europe where variation in topography and soils was sufficiently dramatic to produce markedly different types of vegetation, despite a seemingly uniform climate. Perhaps more importantly, some ecologists took issue with Clements' organismal view of community structure and functioning. Gleason in particular (Gleason 1917, 1926) argued forcibly that field evidence pointed to the existence of species assemblages achieved by random spread and establishment of individual plants along particular environmental gradients. It is in Gleason's writings that we see the concept of a vegetation continuum being developed, a concept of community 'organization' on a larger scale which was the underlying theme for much of the ordination studies conducted by Curtis and his colleagues (e.g. Curtis 1959) and by Whittaker through his approach to 'gradient analysis' (e.g. Whittaker 1952, 1956, 1965, 1967).

For a period, an intellectual (and sometimes fierce) dispute raged between ecologists committed to a Clementsian organismal view, who saw a range of discrete boundaries between communities and who thereby saw reason to classify them (e.g. Daubenmire 1947) and those who, like Curtis (e.g. Curtis 1959) saw favour in Gleason's 'individualistic' concept and thereby adopted ordination procedures as being more appropriate to large scale community description and ordering. Fortunately these methodological approaches have been disentangled from their associated philosophical bases (cf. D.J. Anderson 1965b) and classification, ordination and gradient analysis all form standard tools of trade now for preliminary analysis of community structure and community–environment relationships (Gauch 1982).

None of this is to suggest, of course, that successional theory is passé or useless within the ambit of modern community ecology. While many 'obvious' successions have been demonstrated to students on lake margins which in fact have no temporal connections whatever (Walker 1970), there are many secondary successions which provide an entirely predictable series of post-disturbance phases. Indeed, an understanding of successional sequences and an ability to maintain or initiate such sequences is not only a key to agricultural production but also the basis of many land reclamation and management programmes (e.g. Bradshaw & Chadwick 1980).

While much of the earlier debate about the organismal, quasi-organismal (Tansley 1935) or individualistic nature of vegetation has receded from centre stage, the existence of emergent community characteristics is tacitly accepted (if not always philosophically endorsed) in that modern community theory addresses such concepts as diversity, stability and resilience. When we speak of perturbation, we are implying the operation of a selective force on a (community) system — and it is the system, rather than its component individuals, which we expect to respond.

1.6 Ecosystematics

So far in this historical sketch, little reference has been made to the second part of Whittaker's definition "...and ecosystems as functional systems formed by communities and their environments". Given that environmental measurements of any kind were in their infancy, and given that ecological science was being developed primarily by biologists, perhaps it is not surprising that the term 'ecosystem' came into the literature only 50 years ago, via the pen of A.G. Tansley, by then the doyen of British ecologists. Dominant leader though he undoubtedly was, he saw some value in attempting to effect a compromise between the strongly developed Clementsian tradition on the one hand and the European/Scandinavian procedures on the other. In 1935 he wrote:

> "I have already given my reasons for rejecting the terms 'complex organism' and 'biotic community'...But the more fundamental conception is...the whole system...including not only the organism-complex, but also the whole complex of physical factors forming what we call the environment of the biome...It is the systems so formed which...are the basic units of nature...These ecosystems, as we may call them, are of the most various kinds and sizes..."

Ecology owes much to C.S. Elton for beginning to flesh-out this notion of a functioning ecosystem, and his book *Animal Ecology* published in 1927 was a landmark in developing that concept. As a young scientist he had had a strong desire to describe ecological groupings of species and was much inspired by Shelford's book when he first read it in 1920. He soon realized, however, that animal communities differed fundamentally from vegetation units in the transfer of energy, and he turned his attention to these and related pathways "forming a network of biotic connections between the populations of different species" (Elton & Miller 1954). In Elton's own estimation

> "Animals are organised into a complex society, as complex and fascinating to study as human society. At first sight we might despair of discovering any general principles regulating animal communities. But

careful study of simple communities shows that there are several principles which enable us to analyze an animal community into its parts, and in the light of which much of the apparent complication disappears. These principles are food-chains and the food-cycle; size of food; niches; (and) the pyramid of numbers."

The realization of the importance of food-chains in the economy of natural communities was fast gaining ground in the 1920s; Lotka (1925) had called such chains of species 'connected energy transformers' and Elton himself had seen and appreciated connected food chains and webs during his early period of field research in Spitzbergen. Elton also recognized the fact that progressively fewer animals occupied the upper levels of any chain and that a graphical representation of size class against number of individuals inevitably led to a representation of his 'pyramid of numbers' (Elton 1927).

Not surprisingly the pyramid of numbers is also a pyramid of species number. The feeding habit — together with all its metabolic and behavioural characteristics — of a species fits it for a particular role in a community; this was the Eltonian profession or 'niche' for a species, whose 'address' was the habitat it occupied. Elton's succinct definition of the niche was to prompt a great deal of later theoretical work on this concept, now a central concern of community ecology. As importantly, he was to focus ecologists' attention too on the importance of diversity in both natural and man-made ecosystems. Pearsall (1964) paid an appropriate compliment on the value of Elton's work when he wrote:

"I certainly owe it two things — a much more thorough understanding of the fluctuating nature of animal populations and hence of the *dynamic* character of ecosystems, and secondly, a fuller appreciation of the philosophy that a successful ecosystem possesses a high degree of diversity. This diversity is not only shown in the large number of constituent species and the numerous small habitat units which result particularly from topography and from the structure and pattern of the vegetation, but also in the *processes*, the numerous food-chains and energy relations which underlie the Eltonian concept of the pyramid of numbers."

If Pearsall felt scientifically indebted to Elton, it was clear that other ecologists saw a wise investment of research effort into the energetics of food cycles as a result of Elton's perceptions. Just as community ecology had seemed to centre around Clements' views on succession earlier in the century, now the ecosystem — and particularly the flow of energy and materials through it — became a natural focus for understanding how communities functioned. Tansley's (1935) phrase, "But the more fundamental conception is. . . the whole system (in the sense of physics). . .", was about to be explored with a vengeance, and H.T. Odum would see the wheel complete its circle in the publication of his book *Environment, Power and Society* (1971), wherein the seeds are sown not only for a definitive energetic analysis of 'conventional' ecosystems, but also of human 'belief' systems: a new holism via reductionism, writ large.

1.6.1 Ecosystems and energy flow

A little more than 40 years ago R.L. Lindeman, following collaboration with G.E. Hutchinson, published his paper "The trophic-dynamic aspect of ecology" (Lindeman 1942). That paper provided very considerable insight into what most ecologists would now take for granted:energy flows through a community, and the dissipation of energy within each trophic level, provides a coherent explanation for the Eltonian pyramid of numbers. Indeed it seemed plain that the Eltonian pyramid was in fact a necessary consequence of the second law of thermodynamics, and that empirical assessments of the ecological efficiencies (*Ef*) based on ingestion rate (I) between trophic levels t and $t-1$, i.e.

$$Ef = \frac{I_t}{I_{t-1}} \times 100,$$

of the order of 10%, were commonplace in a wide range of communities.

As data on energy and material flows were accumulated for a wide range of ecosystems, some ecologists were emboldened to make a range of generalizations about the bioenergetics in 'developing' ecosystems (Table 1.1), and wrote pointedly of such features as 'the strategy' of ecosystem development (E.P. Odum 1969, 1971). In Odum's words

Table 1.1. Trends in community attributes to be expected on the basis of successional development (after Odum 1969).

Community attribute	Early stages	Late stages
Total biomass	small	large
Gross productivity/biomass	high	low
Biomass supported/unit energy flow	low	high
Food chains	simple, mainly grazing	complex, mainly detrital
Species diversity	low	high
Pattern diversity	poorly organized	well organized
'Information'	small	large
Entropy	high	low
Stability	poor	good

> "The overall strategy...is directed towards achieving as large and diverse an organic structure as is possible...In many cases...biotic controls of grazing, population density, and nutrient cycling provide the chief negative feedback mechanisms that contribute to stability in the mature system...The intriguing question is: do mature ecosystems age as organisms do?"

From this it seems we have a community 'organism' once more, at least by analogy, and one that contains sufficient negative feedbacks for successional determinism to be apparent. Margalef contributed to this viewpoint in his 1963 paper "On certain unifying principles in ecology": increasing complexity of structure and information content goes hand in hand with successional development; information feedback to the developing system becomes a self-directing and self-organizing force.

While it is not our purpose in this introductory chapter to make critical judgments on the major methodological contributions to community ecology and theory, it is perhaps not unfair to say that the analogic hydraulic model used to considerable pedagogic effect by both E.P. and H.T. Odum is necessarily a gross reconstruction of the intricate array of food-webs, niches and diversity in the Eltonian world from which their models sprang. This is not to say their models have not been useful — the IBP Biome projects in a major sense were predicated on just such models — but the great promise held up for ecosystematics two decades ago has not been entirely realized, at least as a singularly coherent and embracing theory which underpins all the legitimate concerns of community ecology. In retrospect this is not surprising, since the modular concepts of ecosystems have rarely encompassed the field biologist's perception of the real world as a rather patchy place. But patches and gradients certainly exist, and it is within their framework that niches (and the diversity needed to fill those niches) are created and destroyed. Not surprisingly, then, patchiness, niche and species diversification have become in their own right important community-level constructs (see Section 1.8).

1.7 Biotic interactions

Elton's reasoned and indeed prophetic conviction, subsequently shared by many animal ecologists, that the controlling processes in animal populations were brought about primarily by biotic interactions in the species network of the community, led naturally to investigations of population fluctuations which caused disturbances within communities. The prophesy, retrospectively reconstructed by Elton, is itself informative:

> "All these highly abstruse theories (mathematical theories propounded by Lotka, Volterra, Thompson, Nicholson and others...and some pioneer laboratory model experiments being developed by Chapman and Gause) were really saying that limitation of numbers must be brought

> about primarily by biotic relationships. Such relationships can be reduced to five kinds: (1) intraspecific competition, (2) interspecific competition, (3) herbivore–plant relations, (4) predator–prey relations, and (5) parasite–host relations" (Elton & Miller 1954).

Initially some of these population processes were isolated from their community context and studied instead in controlled laboratory environments. For example, a series of experiments on competition between two ciliated protozoans by Gause (1934), the experimental work by Crombie (1946) and the analysis of competition between two species of flour beetle by Park (1948) laid the foundation for much of the theoretical discussion of this topic during the following decade. Similarly the early experimental work on prey–predator interactions by Gause (1934) underpinned much of the later experimental analysis of this process (e.g. Huffaker 1958), although such biotic interactions were also studied in the field (e.g. Leopold 1943, cf. Caughley 1970) or were analyzed on the basis of field-derived data (e.g. Elton 1942; Reynoldson 1964) or were even subjected to the newly burgeoning science of biological modelling (e.g. Holling 1959). The early experimental work on host–parasite interactions by Nicholson (1933) seemed to transform what had originally been a microbiological bailiwick into an important ecological consideration of the impact of evolutionary selection. Interestingly, similar evolutionary interests relating to joint selection between species (coevolution) (e.g. Breedlove & Ehrlich 1968) grew out of the early experimental work on plant–herbivore interactions (e.g. Itô 1960) a topic which in its own right has now grown to considerable dimensions (e.g. Crawley 1983).

Ecologists gradually began to relate their findings concerning biotic interactions to population phenomena which were responsible for the pattern and dynamics of communities. This volume is testimony to the ongoing struggle of ecologists to provide a further perspective on these fundamental interactions, which themselves are played out in the environmental theatre.

1.8 Patchiness, diversity and resource exploitation

That the world is made up of patchy environments was in one sense denied by the early phytosociologists, with their insistence on taking samples from 'representative' and 'homogeneous' areas of vegetation. This is all the more curious since they clearly recognized the patchiness of the plant populations they recorded, for their measures of 'sociality' were the first crude approaches to the now rigorously statistical descriptions of distributional patterns of both animals and plants (cf. Hutchinson 1953; Greig-Smith 1961, 1983).

Most ecologists now take for granted that as a rule environments are made up of mosaics (the 'harlequin' environment), with mosaic patches fluctuating in time. The notion was perhaps most cogently presented by A.S. Watt in 1947 when he addressed the British Ecological Society as its president, and since that time a plethora of ecologists have sought to understand how species have become 'packed' into patchy, unstable or stable environments.

Watt's insight clearly had an enormous impact on our present understanding of the way that niches are occupied by different species in time and space (Newman 1982). Across the Atlantic other minds were turning to the problem of niche characterization, and it was G.E. Hutchinson (1959) who provided the now distinctive definition of a niche as 'an *n*-dimensional hypervolume'. This Hutchinsonian niche included physical dimensions as well as the Eltonian 'profession', and therefore a central question of community ecology became 'How do species divide their world into niches?' With the evolutionary perspective, which is always implicit, the question also becomes central to evolutionary ecology: 'Why are there so many or so few species in a given community?' Or more explicitly, 'Do competitive processes lead to diversification of niches and specialization of species or survival of dominant species and extinction of others?'

Whether fortuitously or by a considerable prophetic insight, but certainly to great effect, Robert H. MacArthur in 1955 commenced the publication of a series of papers which explored the niche — and its filling, diversity. Maybe one of MacArthur's great fortes was in the choice of his group of organisms — birds — which at least in their feeding modes operated along a relatively simple dimension of the niche hypervolume. What if one took the environment occupied by birds, represented it as a line, and apportioned that line at random to match the phenomena of supposed Gausean exclusion and character divergence for a group of birds observed in nature? MacArthur was that rare scientist — a gifted field naturalist with an enormously inventive and theoretical turn of mind: the broken stick model was born, as was a whole dependent era of investigation and controversy over the nature of niche diversification, occupation and, not least, the concomitant concerns of stability and resilience. The engineering discipline of general systems analysis had been harnessed into modelling energy and information flow/content in ecosystems; agricultural and psychological biometrical methods had provided characterizations of species distributions; and surely applied mathematics could be pressed into service to predict the likely relative abundance of birds occupying distinctive niches in real environments!

The success of the broken stick model itself seems to oscillate; it seemed not to mimic effectively the intercommunity appearance of species commonness and rarity, which was so carefully matched to known statistical distributions by the engineer-turned biologist Preston (e.g.

Preston 1948), but it did match effectively some intracommunity data, particularly if 'equilibrium' species really did divide their total environments up into discrete niches.

There is no doubt that MacArthur's work provided an invaluable focus and incentive for the now vigorous attempts to encapsulate in simple mathematical terms the existence of closely packed species and the differentiation of the niches they occupied. His original model was tested against data collected from bird communities, within which feeding niches at least are usually readily discriminated. It might be that birds in any event are a special case: that on a simple resource spectrum like food availability they have little difficulty in rationalizing this niche dimension. For a botanist it seems not at all so easy to appreciate how terrestrial plants might so simply discriminate between carbon dioxide, water and nutrient availability, particularly since low-nutrient environments appear to promote rather than diminish species diversity (Tilman 1980).

Although relatively few studies of plant guilds have so far been conducted, it seems likely from limited field evidence (e.g. Werner & Platt 1976) that niche separation in well established vegetation is achieved by interspecific competitive outcomes. No doubt this in part is why competition (or interference) studies have rated rather highly in plant community process studies (Harper 1977) compared with equivalent studies of animal niche diversification: it is not difficult to demonstrate the measurable effects of interference in some plant populations, particularly if the density (of individuals) is sufficiently high, but it is much more difficult to do so in motile animal populations.

The problem of niche diversification — or rather the coexistence of a group of plant species along a particular, common resource spectrum — is in fact a variation on the problem of coexistence of genetically different strains within a population of a single species. Not only is the genetical structure of a population then a more important unit than a species in assessing niche occupation and discrimination (Harper 1977), it also points up a significant interaction which links animal and plant communities, that of predation by herbivores on autotrophic energy sources.

It is quite common for insect herbivores to attack plants in a frequency-dependent manner, and this pattern of interaction has been cited as an underlying mechanism for both frequency-dependent mortality of certain polymorphs caused by pests in crops (e.g. Scott *et al* 1980) and as an explication for the spectacular diversity of tree species in tropical rainforest (Janzen 1980b). The natural dominance hierarchy which may develop in some plant communities may well be modified dramatically by grazing herbivores, with equally dramatic consequences for the animals that might have exploited niches created by a more complex structure promoted by realized dominance (e.g. Watt 1957).

1.9 Epilogue

It will be apparent from this brief historical survey of community ecology that this book can only be an excursion into the complex realm of biotic community organization and process; it is certainly not an attempt to provide a definitive, nor even a particularly coherent, account of phenomena which will need to receive much more detailed treatment before the essence of 'community' is ultimately distilled out. Indeed we suspect that another generation of ecologists after us may well decide that several essences are necessary for a satisfactorily coherent explication of community biology.

Part 2
Limits of Communities

Three chapters in Part 2 examine the methodological framework of community ecology. The idea behind it is to bring the readers up to date with the current focus and terminology in community ecology. This is done by asking fundamental questions about biotic communities and then by analysing concepts in realistic terms.

Chapter 2 starts with the most general definition of the term community and then examines its implications in terms of organization. The questions asked are: To what extent can a community be considered organized? What are the community processes that produce patterns of distribution? Does chance play a part in the distribution of organisms at different scales? And how can the degree of organization be assessed?

Chapter 3 explores the area of interaction between communities and their environments (biotopes). The dynamic approach taken is deliberately thought-provoking in order to challenge the view that the biotope is the physical environment providing the necessary conditions for the community to persist in a given ecosystem. Both temporal and spatial patternings are considered to occur in the biotope as exemplified by forest succession and a pond ecosystem.

Chapter 4 concentrates on the properties of communities as they emerge from trophic and competitive interactions of constituent species. Complexity and diversity are related to stability of the community; it is argued that stability is an historical function of diversity. Some useful methods of comparing communities are then presented using classification and ordination as established means of pattern analysis.

Part 2
Limits of Communities

2 Chaos or Order — Organization

Peter Greig-Smith

2.1 Introduction

The term community is very generally applied to assemblages of plants and/or animals found living together with some degree of permanence. In this loose sense communities are readily recognized; the same combination of organisms does tend to recur on different sites and, given knowledge of the physical characteristics of the environment, geographical location and history of management, if any, of a site, it is in principle possible to predict what organisms will be found there. Recognition of communities in this way does not, however, imply that they are necessarily anything more than assemblages of organisms whose ranges of tolerances in relation to physical factors overlap, selected from the pool of organisms present within dispersal distance. Even if this essentially unorganized concept of the community were correct, it would still be useful to study the relationships of communities to habitat as an economical first approach to understanding the factors which determine the occurrence and performance of particular species or subspecific taxa, but there is always some degree of interaction between organisms within the community by which the occurrence and performance of a species is constrained by the other species present. No herbivores are completely generalist, capable of grazing any plant species and performing equally well whatever their food. Similarly no carnivores are completely generalist; at the very least there are mechanical and energetic limitations on the prey available to a predator. From such obvious and inevitable interactions there is a wide range to very subtle and finely tuned ones such as the relation between an internal animal parasite and its host.

The intensity of interactions and the degree to which they maintain the composition of the community represent a major aspect of community organization but there are other ways in which communities show a greater or lesser degree of organization. Almost without exception they show spatial heterogeneity or pattern in their composition. This is partly inherent in the site, resulting from spatial variation in the physical habitat, but the activities of both plants and animals may superimpose further heterogeneity. The degree to which the component organisms impose pattern is a further aspect of organization. Secondly, the degree to which the nutrient requirements of individuals in the community are met by internal cycling rather than by inputs (from rainfall, weathering of mineral nutrients, etc.) balancing losses (by leaching, export by animals, immobilization in peat, etc.) contributes to organization. Though there may be some uptake of mineral nutrients by larger animals from drinking water, or by microfauna by direct absorption from water, animals obtain their mineral nutrient requirements mainly from other organisms. The important factor is how far nutrients released by decomposer organisms are immediately taken up by plants.

2.2 Interactions

Although there is an increasing interest in plant–animal interactions, the great majority of community studies have been directed either to the plant component, with animals regarded as an influencing 'biotic factor', or to the animal component with plants regarded primarily as a habitat and food resource; the principal exception has been in studies of plankton communities. The effect has been to think of plant–animal interactions as one-way. Historically this division is understandable — ecologists have rarely had comparable botanical and zoological expertise — but it has resulted in the complexity of community interactions being underestimated. A further difficulty in obtaining an overall view is the differing importance of direct environmental control and interactions in determining the composition of the plant and animal components. Although the presence of large grazing animals has a profound effect on the plant component it is largely determined by environment, and plant–plant interactions are effective mainly by modifications of the immediate environment. The animal component, within broad limits, is less affected by the physical environment and is determined mainly by food and shelter, i.e. by interactions, though experience from biological control that a control organism will only be effective over a limited part of the range of the food or prey organism emphasizes that the animal component is not immune to relatively minor differences in the physical environment.

A further, equally understandable, limitation has been to concentrate attention on organisms of a limited range of size. This is most evident in animal studies, as the titles of reports often demonstrate, e.g. 'The bird communities of X', 'The insect community on plant Y', but studies of vegetation have often ignored non-vascular plants or dealt only with cryptogams, e.g. 'The bryophyte communities of Z'. Micro-organisms, understandably in view of the very different techniques involved, have rarely been considered in community studies in spite of the essential part they play in the decomposition of dead organic matter and in the transfer of energy to animals. Such studies, covering a segment only of the community, have contributed much to the understanding of the ecology of individual taxa or groups of taxa, but less to the understanding of community organization. The object of this section is to review briefly the types of interaction that may contribute to the overall organization of the community, emphasizing those where more information is most needed.

Terrestrial communities show varying degrees of structural complexity. Pattern, whether due to site characteristics or to the effects of organisms, is essentially two-dimensional and commonly relatively unordered. The plants of a community extend heterogeneity into a third dimension. This is an ordered heterogeneity, recognized as layering or stratification of the vegetation. Some degree of such stratification is found in all but the simplest vegetation, but is best expressed in forest. The layering is an expression of the different morphology of different species and does not in itself represent organization within the community, but it does result in an ordered heterogeneity of microclimate and of available surfaces, which imposes organization. While vascular epiphytes are commonly included in vegetation description, less attention has been paid to cryptogamic epiphytes as part of the community though there is a fairly extensive literature on, for example, epiphytic 'communities' of bryophytes and lichens. Particular epiphytic species are often confined to certain parts of a phorophyte ('host' plant), but it is not clear how far this is determined by microclimate and how far by differences in the physical characteristics of the substratum or the availability of leachates. That many species of bryophytes in wet tropical forest appear to be obligately epiphyllous suggests that the nature of the substratum is important.

Consideration of microclimatic modification by plants and variation in the substratum presented by different species leads on to the question whether and to what extent epiphytic species are limited to or are more abundant on particular phorophyte species. A certain correlation is evident; e.g. rough-barked trees tend to carry a greater epiphyte load than smooth-barked and epiphylls are more scarce on plants with a rapid turnover of leaves, but this does not imply a high degree of organization. Relationships between individual species of epiphyte and phorophyte would indicate much more precise organization. There is a considerable amount of conflicting anecdotal information on this but hard evidence is lacking; see, for example, Sanford's (1974) discussion of epiphytic orchids.

Similar considerations apply to the part played by the structural heterogeneity of the vegetation in ordering the distribution of animals within the community. The dependence of specialized invertebrate herbivores on their food plants is obvious, but many herbivores are further limited to particular parts of the food plant; i.e. they are leaf feeders, seed feeders, stem borers, etc. The extent to which less specialized herbivores are limited to parts of the community by physical conditions of the habitat appears to have been little considered; e.g. how far are molluscs limited to lower strata by their need for shelter or, for larger snails, by the diameter of stems up which they have to climb to reach leaves? One may further ask how far carnivores are ordered in their distribution within the community by physical characteristics of the vegetation, quite apart from the distribution of their prey or potential prey.

The remarks above relate to the aerial part of the community. Information on ordered structure beneath the ground surface is very sparse. Species of plants generally have characteristic rooting depths and patterns of root branching. Exudation from roots and fragmentation of root-cap and root-hair material contribute to the resources available in the soil and, more especially, to the micro-organisms of the rhizosphere, the layer of soil immediately adjacent to the root surface. The organisms of the rhizosphere, bacteria and fungi, and the ectotrophic mycorrhizal fungi associated with the roots of many species, play an important but incompletely understood part in the making available of mineral nutrients and in their uptake (e.g. Harley 1969, 1971; Nye & Tinker 1977). How far do these products of the roots of different species differ and play a part in determining the occurrence of different micro-organisms?

In aquatic communities there is ordered heterogeneity of physical characteristics imposed by the decrease in light intensity with depth, but in habitats shallow enough for rooted plants to occur layering results also from effects comparable to those in terrestrial communities.

Horticultural experience makes it clear that many if not most species of plants can grow in isolation under a wider range of physical conditions than they are found in mixed communities. This indicates that in communities an individual is subject to competition or better (Harper 1977) interference from other individuals both of the same and, more importantly in the present context, other species. In principle, the main mechanism of interference is obvious; one plant depletes the resources, of light, water or nutrients available to another. There is abundant experimental evidence from mixtures of two or a few species grown together in artificially uniform conditions and from the effect of weeds on the yield of crops that interference effects do occur, but little certain information is available on the total interactions among the species of even relatively species-poor communities, let alone species-rich ones.

The clearest evidence of the importance of interference is from the occurrence of 'dominance' of vegetation by a single species. If a plant species is the one in a community which most affects the presence or absence and performance of other species and is least affected by other species present, it is dominant in the most meaningful sense (Greig-Smith 1983). The term was originally applied simply to that species present in greatest amount, i.e. constituting the greatest proportion of plant biomass. Such a species is not necessarily dominant in this more restricted sense, though it is likely to be so because it has obtained the largest share of the resources available. It is not easy to confirm that a species dominant in the looser sense is also the most independent of others without extensive autecological investigation of at least all the more abundant species. If, however, a species varies in its potential for interference through its life cycle and is accompanied by different sets of species at different stages, then it is clearly dominant in the restricted sense. Watt (1947) demonstrated the occurrence of such mosaic patterns, determined by the dominant species, in a number of communities in the British Isles, ranging from forest to grass-heath, and they have been recognized in various parts of the world since.

Though achievement of dominance is largely by depleting the resources available to other species, other mechanisms may contribute. The production of substances by one species that are to some extent toxic to other species, allelopathy (Rice 1974), has been widely claimed from laboratory experiments to be important, but critical experiments are difficult to design (e.g. Harper 1977) and demonstration of their effectiveness in the field is very difficult. Indirect toxic effects may be more important. For example, Pigott (1970) suggested that the suppression of other species by the invading grass *Deschampsia flexuosa* was the result of increased solubility of manganese, to which the former were susceptible. The release of manganese was brought about by reducing conditions, caused by the nature of the humus produced by *Deschampsia*. It is not clear how widespread such indirect effects are.

The concept of dominance is more familiar in relation to plants, but comparable effects occur among animals. Ants in the canopy of rainforest form a mosaic pattern of the territories of different species, which have differential effects on the insect fauna (Leston 1973, 1978). More detailed studies in cocoa farms have made it clear that the ants exert dominance in a sense comparable to that in vegetation (e.g. Majer 1976a, b, c; Jutsum, *et al* 1981).

The importance of food-chain and food-web relationships between animals scarcely calls for comment and clearly leads to a degree of organization. Specialist predators depend on the presence of appropriate prey and all larger animals have specialized parasites for which the host provides not only food, but also, for internal parasites, their habitat. The increase of a prey species following the decrease or elimination of a predator may, by competition for resources of space or food, affect the abundance of other species, the predators of which are in turn affected, so that 'modules' of species whose abundances are correlated are formed (Paine 1980). Further, an animal species may affect others by modification of habitats without any feeding relationships being involved. "Who would guess that woodpeckers, by the making of nest-holes, opened the gates to twelve vertebrate and two invertebrate species, which would otherwise be absent or scarcer in the community?" (Southern 1970). How widespread and important are effects of this kind?

Some interactions between animals and plants are very evident. Animals are dependent directly or indirectly on plants for food and the differential consumption of different species of plants will be reflected in the composition of the vegetation. The result may be obvious, as in comparison of grazed and ungrazed grassland, but there can be little doubt that the differing palatability of different plant species [e.g. Grime *et al* (1968) on the snail *Cepaea nemoralis*] results in a balance between the particular set of herbivores present and the composition of the vegetation. More detailed information is needed. Some mechanical effects of animals are likewise obvious, e.g. the effects of trampling by larger animals (and by man), but less obvious effects are easily overlooked. For example, the characteristic vegetation of sea-bird colonies is generally attributed to the increased nutrient levels resulting from the deposition of bird faeces. This is certainly important but Sobey and Kenworthy (1979) have shown for herring gull (*Larus argentatus*) colonies the importance of the destructive effect on the vegetation of boundary clashes between neighbouring birds. The role of animals, especially insects, as pollinating agents can also play a part in controlling vegetation. Plants lacking means of vegetative propagation are dependent on pollination for their longer term survival in a community. If a plant species has only a limited range of potential pollinators, this implies a strong interaction between it and the animals present.

2.3 Pattern produced by organisms

Both plants and animals may initiate and maintain spatial heterogeneity in the community, producing some degree of organization within it. Any plant species that interferes markedly with other species will, if not randomly distributed within the community, impose pattern on the species it influences. Insofar as its pattern is environmentally determined this cannot be regarded as more than an extension of the sifting of the available species by the environment. If, however, the species spreads mainly vegetatively, or has markedly inefficient seed dispersal, patches would occur even in a completely uniform environment. Seed dispersal, however inefficient, may be expected eventually to spread the species throughout the available habitats, given sufficient time. Vegetative spread, on the other hand, commonly involves the older parts dying as the younger advance, producing bands of the species if the vegetative growth is unidirectional, or arcs or rings if the growth is radial, so that pattern is maintained.

Apart from the general effect of animals grazing on plants, which is liable to alter the balance between plant species even if there is little selection by the grazing animal, colonial animals may produce pattern in the vegetation and hence in the whole community. The effect of colonial animals is threefold, alteration of physical environment, often very obvious as on termite mounds, production of a gradient of grazing pressure with correlated differencc in vegetation, and the rather less immediately obvious transfer of nutrients from the surrounding area to the base or nest (Greig-Smith 1979). If the colonies are territorial, nests will tend to be regularly spaced, giving a further degree of organization (Glover *et al* 1964).

Both plants and animals impose pattern on organisms of smaller size by the nature of the organic matter they disperse.This varies in its 'quality' as food, i.e. its energy and nutrient content and digestibility, and in the ease with which it can be decomposed. Much of it is released in discrete packets, e.g. deposits of dung, individual fruits, fallen branches, etc. which give rise to patchiness of those organisms able to make use of them.

2.4 Cycling of nutrients

Two sources of mineral nutrients are available to an individual organism. One is from other organisms of the community, either living organisms (herbivory, parasitism, predation) or dead organisms (either directly or after release by decomposers). The other is from nutrients not previously available in the community under consideration, rainfall, weathering of minerals, the carrying in of particulate matter by wind or water and the entry of mobile animals which are then predated or die. Loss of potentially available nutrients may occur by leaching, mainly of nutrients released by decomposers but to some extent directly from living plant tissue, by immobilization *in situ* either chemically or by peat formation, by loss of particulate matter and by emigration of animals. The degree to which the nutrient requirements of members of a community are met internally is one appropriate criterion of organization but is difficult to determine.

There is a very extensive literature reporting on particular sections of nutrient cycles (e.g. Swift *et al* 1979). There has also been much work on the input of nutrients though often ignoring input from the decomposition of unweathered minerals, and on the loss of nutrients (mainly derived from estimates of concentration in drainage water); both have frequently ignored movement of nutrients by mobile animals. Such determinations indicate whether the input and outflow are in balance and, if they are in balance, provide an estimate of overall flow of nutrients through the system. If they are not in balance it does not necessarily indicate that the nutrient status of the community is changing; the amount of dead organic matter accumulated or undergoing decomposition may be increasing or decreasing. What is needed in the present context is a comparison of flow into and out of the system with cyclic flow within the system, perhaps ideally in the form of an estimate of the average number of passages from one organism to another, either directly or via decomposer activity, before an atom of a nutrient element is lost from the system, together with an estimate of the average retention time within an organism.

Such estimates are unlikely to be attainable, but some features may be indicative of cycling that is efficient in this sense. Longer food-chains imply more transitions before a nutrient atom is 'at risk' of loss from the system by leaching, the major factor in nutrient loss. Decomposition continues as long as temperature and water availability are adequate; a spread of periods of maximum growth of plant species over the growing season implies a continued uptake of nutrients and lower risk of loss from the system.

2.5 Chance

Having considered those features of communities which give them some degree of organization, it is important to consider evidence for lack of organization. How far are the composition of the community, the spatial arrangement of individuals of species of plants and sedentary animals and the movements of individuals of animal species a matter of chance? The answer to this question will depend partly on the scale at which it is asked. On a broad geographical scale the chance introduction of a species may sometimes have a marked effect. The widespread death of the American chestnut *Castanea dentata*, following the chance introduction of a fungal pathogen to which it was very susceptible, resulted not only in its virtual elimination as a component of deciduous forest in the United States, but also in a change in the balance of representation of other tree species (Korstian & Stickel 1927). The effect of the introduction of animals into communities from which they were previously absent, e.g. the rabbit *Oryctolagus cuniculus* into Australia, is often marked. These more spectacular consequences have generally been the result of accidental or deliberate introduction by people (Elton 1958). Chance effects of natural dispersal or evolutionary change are less easy to identify. The effects of delayed dispersal are only obvious when the species is gregarious, e.g. *Mora excelsa* in Trinidad (Beard 1946), estimated to have a rate of spread of *c.* 10 m in 50 years (Greig-Smith 1979). The great reduction in abundance of the marine angiosperm *Zostera marina* along the coasts of both the eastern United States and western Europe in the 1930s, attributed to a pathogen (Tutin 1942), is an example of a chance effect apparently not the result of human activity. Although at this geographic scale the arrival of a new species may have very large effects, the result is the establishment of a new set of interactions and the communities involved again show organization.

On an intermediate scale numerous analyses by ordination techniques of the overall composition of stands of vegetation within relatively limited areas indicate that composition generally correlates with features of the physical environment, i.e. that chance plays relatively little part. Comparable analyses of the animal component or total composition of communities are fewer, but do not conflict with this conclusion (Hughes & Thomas 1971). There is, however, some evidence that chance can be important. Webb *et al* (1972) have suggested from a study in Australian rainforest that different combinations of species may exist in equilibrium in any one habitat. If this is generally so, it is a matter of the chance of which species first established which determines the composition of the community, but that only certain combinations of species are stable is an indication of organization rather than the reverse. Where an animal habitat is limited by environmental discontinuity, the chance of the species that first establishes may determine the community that develops. Fager (1968) in a study of invertebrates in synthetic oak 'logs' found that individual species varied in their occurrence in different logs from single individuals to being the most numerous species present.

On smaller scales — what kind of plant occupies a particular point, which particular plant is predated by an animal — a considerable chance element is clearly involved. To say this does not imply that no causation can be recognized as determining which species has established at a particular point. When a site becomes vacant by the death or destruction of an individual plant, the ability of different species to establish on it will depend on such factors as the size and shape of the gap created, the availability of propagules, the time of year and the weather at the time and subsequently. Plant species have characteristic 'regeneration niches' (Grubb 1977) which determine the likelihood of their establishing on a vacant site with stated characteristics. However, even if the parameters of the regeneration niche are known, or are in principle knowable, the outcome does not result from organization in the community except insofar as those plants adjoining the gap, by their differential capacity for interference with the potential colonists, affect the chances of the latter establishing successfully. Similar considerations apply to sedentary animals and to animals with interspecifically defended territories, e.g. fishes on coral reefs (cf. Sale 1980; Ogden & Ebersole 1981). In summary, it appears that it is only on small scales that the effects of chance may in some circumstances override the effects of organization and stochastic processes may be the most important factor determining community composition.

2.6 Assessment of degree of organization

There is no doubt that communities have some degree of organization, at least in all but the early stages of succession in a previously unoccupied habitat. The resources of a habitat are finite and there is inevitably competition between organisms for these resources. Because different species (and often different subspecific taxa) of both plants and animals affect others in characteristic ways, both by having different resource requirements and, indirectly, by modifying the immediate habitat, a community does not consist of a random assemblage of species sifted by the environmental characteristics from those available. So much is scarcely in doubt. It is much more difficult to assess, even qualitatively, how closely organized a particular community is and to compare different communities in terms of degree of organization. There are two principal difficulties, of heterogeneity and of scale.

Consider a completely uniform and sterile habitat which has become available to living organisms. Any departure from random spatial arrangement in the community that develops will be caused by the complex of interactions between the species it contains. In principle, degree of organization could in part be assessed by the amount of association of occurrence between species within the community; appropriate techniques are available (Greig-Smith 1983). Association between species will not, however, take account of the complete absence of species available and capable of establishing in isolation in the habitat but unable to do so in the presence of other species, nor of the occurrence of species not able to establish in isolation because they depend on overall modification of the habitat by other organisms. Potential but absent colonizers are not easy to identify. The converse situation is much more evident; thus, many woodland herbs are dependent on the tree canopy, either directly on the modified microclimate, as is probably true of many species of the ground layer in rainforest, or indirectly by various mechanisms; e.g. the bulbous herb *Hyacinthoides nonscripta* performs best in full daylight and is intolerant of deep shade, but is largely confined to woodland because of the absence there of aggressively competing grasses and of grazing, to both of which it is susceptible (Blackman & Rutter 1954). While animals are necessarily dependent on plants for food, they too may depend on modified microclimate or on structural modification of the soil.

It is not profitable to pursue this hypothetical situation further; no habitat is uniform when examined at the degree of detail relevant to species interactions. In a heterogeneous environment association between species may be due either to a common response, in the same (positive association) or opposite (negative association) sense to environmental difference, or to interaction between species. While techniques of analysis of the association between species at different spatial scales and of describing heterogeneity of environmental measures are available (Greig-Smith 1983), it is rarely possible to identify with certainty those associations due to species interactions.

The discussion above has implicitly been confined to organisms of more or less comparable size. Further problems in assessing the degree of organization in the community arise from the very wide range of size of organisms, from micro-organisms a few microns or less in diameter to trees 100 m tall. Associations between species of the soil microflora have clearly to be examined on a very different scale from that appropriate to the soil mesofauna and these on a different scale from that used for higher plants and vertebrates. Similar difficulties apply where larger plants provide the physical support for smaller organisms; the bryophyte and lichen flora of tree trunks, with its associated invertebrate fauna, forms to some degree a distinctive community within a community, yet interacts as a unit with the higher plants rooted in the soil by, for example, its absorption of mineral nutrients from rain water. Are associations at different scales to be weighted equally in assessing degree of organization and, if not, what relative weightings should be used? How is the community to be divided into appropriate 'subcommunities' of different scales?

Not only are the methodological difficulties severe, but the collection of the necessary information for even one community would be a formidable task. It is not realistic at present to look for more than an approximate and rather subjective comparison of communities in terms of degree of organization. What is important is to be aware of the complexity of interactions within a community and to realize that organization cannot be judged only in terms of the larger plants and animals. Even apparently simple communities may be highly organized; we are not yet in a position to judge.

We may ask finally how useful assessments of degree of organization are, either in contributing to understanding of a community or as an aid to its management. Only after very considerable understanding of the interactions within a community has been attained can organization be assessed; the assessment is a resultant of understanding and one that retains so little of the total information that it is unlikely to be useful even in comparison of communities. Management implies taking action either to maintain the status quo in a community that would otherwise change in composition and/or structure, or to change it in a predictable way. Organization implies that the elimination or reduction in abundance of one species will have more or less complex repercussions on other species. A less highly organized community may be expected to have a more easily predictable response to interference, but prediction will still depend on an understanding of the interactions within that community. Assessment of *degree* of organization appears to have little value.

3 Biotope Structure and Patterning

David W. Goodall

3.1 The definition

The word 'biotope' like so many ecological terms, has been variously used by different authors. To some it is synonymous with 'niche' in a habitat sense (e.g. Hesse *et al* 1937). For Allee *et al* (1949) it could be synonymous with microhabitat: "...a cattle dropping...may be termed a biotope, habitat, habitat niche, a microhabitat...". For others, it is a habitat unit "characterized by a recognizable community of organisms" (Ladd 1977). For Kenneth (1963) it was "an area in which the main environmental conditions and biotypes adapted to them are uniform; a place where organisms can survive; also, microhabitat", while "microhabitat" is "the immediate special environment of an organism...". The Oxford Dictionary definition (Burchfield 1972) is similar: "The smallest subdivision of a habitat, characterized by a high degree of uniformity in its environmental conditions and in its plant and animal life". In all these later definitions, there seems to be very little differentiation between the concepts of 'biotope' and 'microhabitat'; they take no account of the importance of environmental heterogeneity to biota, nor of the enormous differences in environmental scale for different organisms. It seems, on the other hand, that one needs a term for the environment to which a particular species is adapted (cf. 'niche' in its functional sense), and the word 'biotope' seems well suited to this need.

Thus, we define the biotope of a species as *a particular region of space and time within one or more ecosystems, in which an individual of that species may be able to establish itself and complete its life cycle*. The size, duration and internal structure of biotopes differ enormously between different species, and there can be no question of dividing an ecosystem up into discrete biotopes, as if they were cells of space. Even for individuals of a single species, their biotope volumes may overlap to a large extent, limited by such considerations as home range, territory, and direct competition between neighbours for resources.

The previous paragraph is a little vague on the question of whether the word 'biotope' should refer to one particular portion of an ecosystem space, or a class of portions — whether each individual has 'its own' biotope, or whether the species has a conceptual biotope, realized separately for each individual of that species, with a great deal of random variation in characteristics between the individual realizations, and often overlaps between them. The latter seems to me the more useful concept. Thus, the biotope of a species is an abstract concept — a class of concrete entities, each of which is potentially a realization of this concept for a particular individual of the species.

There is a need to distinguish 'habitat' from 'biotope'. As used here, the concept of 'habitat' is independent of species. It is an environment, with its spatial and temporal pattern of varying factors, which may constitute part or the whole of the biotope of various species. One can regard a geographical area, with its extension in time, as divisible into a number of habitat units — which indeed can be subdivided almost indefinitely into microhabitats — while the biotope of a particular species constitutes an overlay, also in both space and time, over this habitat pattern, including a number of habitats and excluding others.

3.2 Abiotic and biotic effects on biotopes

On the global scale, chemical effects of biota in the past can cause large differences in the kinds of biotopes now available. Perhaps the clearest and most important example of this is the change in the earth's atmosphere supposed to have taken place in the early history of life on the planet, when autotrophic carbon fixers evolved from their heterotrophic ancestors. Others are changes in the chemical forms of sulphur, nitrogen and iron caused by many micro-organisms, as well as the incorporation of these and other nutrient elements into organic form by plant life. All these physical and chemical changes in the environment, together with the formation of organic residues, resulting from the past activities of organisms, cause the set of biotopes available for occupation to be greatly modified.

When one considers the effects on biotopes of organisms now living, new dimensions are introduced. The most obvious ones are concerned with trophic relations. Every new species in an ecosystem constitutes a new set of resources for those other species that can make use of them. But there are other effects too — local physical and chemical effects determining the spatial and temporal distribution of all the environmental factors which may affect living organisms. A grassland area with a tree in it has a much wider range of biotopes than an otherwise similar grassland area without any woody vegetation. The microfauna and microflora of a rhizosphere are quite different from those of the soil in the interspaces. The differences may be due in part to the use of root tissue as a resource by some organisms, but are much more ascribable to the physical and chemical effects of the proximity of the roots.

The set of biotopes existing within an ecosystem depends very largely on the organisms which are currently present, or which have been present there during the history of the ecosystem. Except at the stage of primary colonization of a bare mineral surface, or a sterile body of water, biotic effects always modify the underlying physical environment — often to the extent that it is almost unrecognizable in the biotopes available to smaller and relatively immobile species. The influence of climate, geological formations and geomorphological processes on biotopes for soil nematodes, for instance, is negligible. They create the *preconditions* for those biotopes, but the biotopes themselves are the creation of organisms, now and in the past. In fact, only a very small number of species could live successfully in a purely physical environment.

The historical effects of biota on the existing set of biotopes show themselves most clearly in the presence of organic remains of various sorts — either structureless, as in the humus component of soils, or highly structured, as in fallen logs or carrion. These organic residues may either provide food for new members of the biota, or modify its spatial structure and microenvironmental factors. However, apart from these more obvious historical effects, part of the disintegration of rock substrates is due to biotic factors; spatial distribution of water is affected by biota — beaver dams are an outstanding example; and coral reefs can form dry land where ocean had been.

3.3 Biotopes of sessile organisms

The environmental needs of a sessile organism are in general definable in terms of a point of space, though not in time. At any point in time, though the conditions within a given radius of the point of attachment may be relevant, their spatial distribution may not. Thus, for a plant growing in forest shade, it does not matter where within the area covered by the plant the foliage is located as long as the proportion of the foliage in illumination above the light compensation point is sufficient to permit positive production overall. On the other hand, the pattern of variation in water content vertically in the soil may be of the greatest importance in determining whether a particular species can exist there or not. If the water is all beyond the extreme depth of penetration of the root system, it might as well not exist.

For a sessile organism, local changes along the time dimension may be more important than spatial heterogeneity. For algae on soil surfaces in arid areas, the few hours during a day when the surface is moistened by dew may be all-important to survival; and for all desert plants the days during and following the infrequent storm periods determine success in the completion of the life cycle. Another temporal aspect of importance is that the various factors affecting a particular form of activity must be *simultaneously* within suitable ranges if the species is to exist there. If suitable prey are available only at times when the temperatures are too low for predation, they will be immune from attack, and no predator biotope exists. Each of the successive stages of an organism's development must find the requisite conditions in succession in order that the life cycle should be completed. This is very evident, for instance, in the phenophases of development of many plants. Development, flowering and fruit maturation, however, may all have specific requirements in terms of temperature, radiation and photoperiod, and unless they are met in a succession which also has requirements in terms of the duration of stages, development cannot be successfully completed.

3.4 Biotopes of mobile organisms

Mobile organisms have more complex requirements in the patterning of their biotopes. They may require two or more very distinct habitats within a short period of their existence. They may, for instance, need to spend most of their time in a habitat conferring protection from predators, but where their food supplies are not to be found. So their time must be divided between habitats suitable for resting and those suitable for feeding, and they can survive only where these two types of habitat occur in reasonable proximity. The biotope includes both considered together. These requirements for two different habitat types may often be combined with a nocturnal or diurnal pattern of behaviour, so that the different types of habitat or resource are used by day or by night respectively.

There are also larger-scale patterns of resource or habitat requirement by mobile animals, often associated with reproductive phases. Reproduction may call for habitats very different from those used in non-reproductive life. One need only cite anadromy (spawning migration from sea to rivers) and catadromy (spawning migration from rivers to sea) in fishes, and the need of many marine birds and mammals for land on which to nest or give birth. On land, birds often require different habitats during the reproductive season — perhaps for nesting sites, for nesting materials, or for different or additional foods needed by the nestlings. In all these cases, the biotope for the species is a combination of the habitat needed for reproduction and that for non-reproductive life.

The extreme of complexity in time for biotope requirements perhaps is shown by those species which engage in intercontinental migrations, usually following the sun between south and north. At different times of year the species makes use of two or more completely different habitats in different parts of the world, and their spatial relationship would seem to have no importance. The biotope consists of two entirely disjunct portions.

Apart from direct effects on a species, the pattern of habitats in space and time may have important indirect effects through other biota.

The presence nearby of biotopes suitable for an important predator, parasite or competitor may disqualify for a particular species biotopes which — with different neighbours — might be very suitable. On the other hand, the existence nearby of breeding or shelter habitats for prey species may be important to predator species, to which otherwise they would be irrelevant. Birds hawking for midges over a pool may have no other use for the pool.

In such (very numerous) cases where an organism needs two or more different environments to live successfully, the question arises whether, for this species, the different locations in which it spends its life should be regarded as separate biotopes or as parts of one. If one considered only the more extreme examples like migratory birds, the former approach might seem more natural. There seems no essential difference, however, between their behaviour on the large scale in space and time and the small-scale movement of a hunting spider lurking under a stone and darting out to capture prey. Both need two different environments for different aspects of their life. If the bird is occupying two different biotopes, then so is the spider.

It seems best to define biotope in terms of a species, so that it designates the whole living space in which the species operates and completes its life cycle, with perhaps a pattern of intensity of occupancy — the proportion of time spent in different parts of the biotope. Otherwise, if the biotope were simply a portion of space defined without reference to its occupation and use, it would only be by chance that it coincided with the needs of any species, and thus had ecological relevance.

In this interpretation of the concept it is inevitable that biotopes overlap, and indeed are nested. The biotope of a fox may include those of many rabbits, many more grass tussocks, many more beetles, and still more nematodes. It will also overlap with those of other foxes, the intensity of occupation by each in these overlap zones being reduced. An area of landscape could then be described as containing, say ten biotopes for foxes (not necessarily all occupied), 10^3 for rabbits, 10^5 for grass tussocks, and so forth. It is an interesting point that the number of biotopes for one organism may depend on the *occupancy* of biotopes for another. If, for instance, the number of foxes is greatly reduced by disease or hunting, it may be possible for rabbits to live successfully in sites rather unfavourable for burrowing, so that new biotopes become available for rabbit occupation.

3.5 Biotopes versus resources

A distinction between biotopes and resources is rather difficult. Arguments have already been put to the effect that the biotope of a species includes all living and nonliving features of the environment which are relevant to the successful completion of its life cycle. The word 'resources' suggests those features of the biotope that are actually used by the species. It is perhaps possible, in the case of a plant, say, to distinguish between the nitrogen, phosphorus and potassium which are taken up and incorporated into the cellular structure, and the silica, smectite and other minerals among which the root hairs penetrate. But what of radiant energy, some of which is reflected from the leaf, some transmitted through, some used for evaporation of liquid water, and only a small part incorporated in organic material? Is the whole of the radiant energy incident on the plant to be regarded as a 'resource', or only that part which is in fact used in photosynthesis — a proportion which is far from fixed, but depends in part on the light intensity, in part on the history of the plant, and is never 100%, even within the photosynthetic wavelengths?

Space is sometimes called a resource. The only space from which a species actually excludes all others (except parasites) is that actually occupied by its body at a fixed point in time. All space around it — at least in part, if the time dimension is extended — is occupied and thus used as a resource. The space is used as a means of gaining other resources — light, water, nutrients, prey.

'Resources' are regarded as merely particular aspects of the biotope for a particular species. For two species, the biotope may be identical, but a particular feature in it may be a resource for one but not for the other. For a potato plant, soil water supply is a feature of the biotope — a resource — which is of great importance. For an aphid sharing the same biotope, soil water supply (though still a feature of the biotope, and one indirectly of importance to it) is not a resource; but the potato plant itself, another feature of the biotope, is. On the other hand, the aphid population is a feature of the biotope which is of considerable importance to the potato plant, though not a resource for it.

Thus 'biotope' and 'resource' are in my view both species-oriented concepts, 'resources' being a vaguely defined subset of the features of the biotope.

3.6 Patterning of biotopes

Let us now turn to the patterning of biotopes within an ecosystem, using two hypothetical case histories as illustrative examples.

3.6.1 Forest development in primary succession

Let the site be the moraine left behind by a retreating glacier. Here there is a primary differentiation of biotopes at the surface from those at different depths within the moraine. On the surface, at the small scale, there is the differentiation provided by different rock types, with their differing chemistry and surface structure, providing biotopes for occupation by different lichen species. The sunny and shady sides of the rocks provide further differentiation. Surfaces formed of finer particles, with different conditions of insolation, temperature and water supply, may constitute suitable biotopes for algae or mosses, and perhaps even a few higher plants. Biotopes below the surface of the moraine at the early stage of primary succession are probably unsuitable for any species, unless they include organic matter brought down by the glacier from higher altitudes, in which case the interstices between the particles may constitute very small-scale biotopes for bacteria and other microscopic detritivores. The tops of larger rocks in the moraine may perhaps be parts of the biotopes of raptors which choose to nest there and seek their prey in surrounding areas; and hollows under the rocks may similarly be parts of the biotopes of beasts of prey which make their lairs there, while ranging over lower ground for prey.

The development of the first colonists will create biotopes for other species; for other and more complex detritivores, for herbivores, for their predators, and for higher plants more demanding in respect of the water conditions in their substrates. The location and pattern of biotopes for these new species will depend in some cases on the location of the primary colonists. Algae on the surface of finer-particle aggregates may be eaten by snails, whose preferred resting place is on rock surfaces; thus the snail biotope will consist of a combination of rock and fine-particle areas. The whole moraine can be divided into a number of areas where such a combination occurs (each providing biotopes for many individual snails), and other areas which, for one reason or another, do not constitute snail biotopes. For grasshoppers feeding on the higher plants, on the other hand, the whole moraine may constitute a single biotope shared by many individuals. Some of the higher plants themselves may be highly localized in crevices where organic matter has accumulated. Not all crevices will act as accumulation noda; this will depend on the microtopography and on the chances of establishment of primary colonists around. Consequently the biotopes for the more demanding species of higher plants will be localized and sparse. In consequence, again, the species themselves will be localized (perhaps even more so, if their means of dispersion are limited), and the biotopes for any less mobile organisms depending on them (sedentary herbivores like scale insects, for instance) will also be so.

Beneath the surface, biotopes are also patterned — that is, they are not distributed uniformly or randomly in space. Suitable areas for deep root penetration depend on the presence of interspaces or fissures in the rocks filled with finer material, and consequently one may expect that biotopes for plants with a deep-rooting habit will occur in patches. Biotopes for all decomposer organisms will depend on the presence in the soil of organic matter, which will occur in pockets rather than uniformly distributed — the heterogeneity of its distribution being accentuated by the tendency of roots to proliferate in these pockets. But different decomposer organisms are most active at different stages of decomposition, so that within those soil zones where organic matter is available there will be a further partitioning of biotopes according to the age and the stage of decomposition of the organic matter. And the partitioning of biotopes for predators on the primary decomposers will of course depend on their dietary preferences, as well as on such factors as temperature and moisture content of the surrounding soil, so that patterns of biotopes for soil predators will develop based on, and partly dependent on, those for the primary decomposers, but overlapping very extensively for different predator species.

So far, possible differentiation and partitioning of biotopes in time has not been discussed for this hypothetical case history. Owls and eagles may each have part of their biotopes within this developing ecosystem, and spatially the biotopes may even coincide exactly, but temporally they do not (those parts of the biotope used by day and by night being interchanged between the two groups), and in consequence the biotopes also do not coincide in respect of biotic features. An important temporal differentiation of biotopes from quite early in the development of an ecosystem arises from the great constancy of temperature conditions deep below the soil surface, as against the marked diurnal and seasonal fluctuations at lesser depths. As soon as these greater depths become habitable for other reasons (e.g. by the accumulation of organic matter) they become suitable biotopes for a range of organisms that cannot tolerate much variation in temperature.

As succession proceeds, and a shrub community develops, the possibilities for biotope differentiation, and the complexities of patterning in space and time, increase. On the smallest scale, the soil microorganisms find the biotope differentiation due to depth from the surface and particle size distribution, subsequently modified by local accumulation of organic matter, further changed by the activities of root systems of various species of vascular plants. The chemical differentiation of rhizospheres of different species is now well established, so that each vascular plant species provides a new set of biotopes for occupation by particular populations of microorganisms, and the invertebrates feeding on them directly or indirectly.

On the soil surface, the original differentiation between rocks and areas of finer particles has been further complicated by the uneven distribution of litter — leaves and branches — determined in part by the distribution of the plants from which it comes, in part from movements over the soil surface by air or water (including eddies around any projection). Any accumulations of litter provide food for a large number of decomposer organisms and those that feed upon them, as well as shelter for them and for other organisms whose trophic requirements may be met elsewhere. The same needs for shelter may also be met by rocks and the hollows under them. In each case there may be differences between the different sides of a sheltering object, depending on the incidence of radiation, snow accumulation, or prevailing winds — differences in temperature, moisture and light intensity, as well as their biotic consequences. All these differences between biotopes or part-biotopes on the soil surface must further be considered in relation to time. A comparison between temperatures on the north and south face of a rock, for instance, will be quite different by day and by night, in winter and in summer.

Above the soil surface, the development of a canopy of shrub foliage creates additional possibilities of biotope differentiation. A gradient of microclimate, with variations in radiation, temperature and water relations, is established between the open air and the ground surface. The gradients change through the day and the year, and are specific to particular canopy species. In consequence, the ground flora develops patchily beneath the canopy, and local differentiation of biotopes is increased. The great heterogeneity of biotopes within an advanced ecosystem is most marked for the smaller organisms; but even mammals and birds can make use of it. Birds, for instance, are in the first place restricted in their feeding habits, so that they make use of only a limited range of food types, and their feeding activities take place within a part of the biotope defined by the availability of the food. They may be restricted also in the way in which they seek or use the food, which may impose further restraints upon their biotopes. Resting may require particular types of perch or shelter; and at times of reproduction nesting habits will pose additional requirements on the biotope at a particular time of year. The biotope for the species thus needs, for a full definition, the environmental conditions (including biota needed for support, food, etc.) for each of the types of activity in which the species must engage. Unless these requirements are all met in the same general area, no complete biotope for that species exists, and the species will not occur there except as a transient. And, in a rich environment, the biotopes for different species will overlap and intermingle very freely, so that the different species can occupy them while 'getting in one another's way' to only a very limited extent.

Further development of the ecosystem from shrubland to forest entails no great changes in biotope pattern within it. For many types of organism the scale of horizontal pattern will increase, being governed by the size of canopy of individuals of the dominant species. Vertically, the increased height will make possible greater differentiation in habitat among organisms living above the ground surface. And the greater permanency of the bark surfaces of the trees will mean that long-lived organisms, or those whose requirements for establishment are less frequently met, can find a home there.

3.6.2 *A pool ecosystem*

As a second case-history, let us consider an aquatic environment — a pool in a river, say. The bottom consists of a variety of microhabitats depending in the first instance on the depth and clarity of the water and the distribution of stones and finer particles of different dimensions, in the second place on the accumulation of debris as a result of bottom irregularities and eddies, and finally on the biota already established there, providing food, shelter, or other environmental modifications. Local differences in the bottom microhabitats may also arise from shade cast by overhanging banks or trees. There will be special microhabitats within the interstices of the substrate, as well as on its surface, and the distribution of these and their character will depend on the particle size distribution.

In shallow water by the bank, there are biotopes for emergent water plants, whose rooting depends on the substrate structure, while early growth depends on the depth of water and its sediment load; after emergence from the water surface, the plant will depend on the same physical factors as land plants (including, for instance, shade by riparian trees), both for growth and for reproduction. Thus the biotope for these plants (and, *a fortiori*, for any animals dependent on them) will include elements of both the aquatic and the subaerial environment.

Spatial differentiation of biotopes within the water mass is hardly possible, except at its surface (for floating plants and a few insects). Free diffusion and movement within the water mass means that within the limnetic zone it must be considered as a single common biotope for all those organisms that make it their home — phytoplankton, zooplankton and (for at least a part of their time) the nekton. The last may indeed combine in their biotopes special parts of the substrate for shelter or reproduction, together with the water mass for feeding. And there are organisms for which the pool as a whole constitutes only part of their biotope — the insects that spend their larval stages in it, but their adult life elsewhere; the birds that dive into the pool to catch fish; the mammals that live on land but feed in the pool.

Little has been said of the structuring in time of biotopes within the pool ecosystem. Diurnal cycles affect it as on land. The changes of light intensity — here accompanied by little or no change in temperature below water level, and none in humidity or water availability — trigger activity cycles in plants and animals, and result in some biotope differentiation. Seasonal cycles are far more important. The enormous changes through the season in phytoplankton populations which one sees in still waters will be to some extent duplicated in the pool under discussion, and these may lead to some changes in water chemistry, and in herbivore populations. For shorter-lived species, these seasonal changes define the biotope; for one, conditions in April and May may be conducive to activity, while the rest of the year is suitable only for dormancy; for another, the period of activity is the winter. For longer-lived species, the seasonal changes in environmental factors (including the biota) are a characteristic of the biotope. If some of the periods are unfavourable, they must be survived.

A special feature of the seasonal cycle is ice cover, largely cutting the water mass off from the atmosphere above it, and for a while restricting the exchange of gases and preventing the movement of organisms between them. In many such situations, an important feature of the habitat is its susceptibility to changes in water level. These may happen predictably in the spring or unpredictably following heavy rains. In either case, the temporary influence on the habitat is enormous. The depth of water over the bottom and its rate of flow increase, as may turbidity. Some of the organisms may be swept away — indeed it may be only attached species, and those protected in the hollows of the substrate, that are able to remain in place. For all organisms, the biotope includes these periodical and often drastic fluctuations in the environment, superimposed on the average prevailing conditions and the minor changes on a shorter time-scale to which they are subject.

3.7 Conclusion

An ecosystem includes biotopes or part-biotopes for all the organisms which live in it. These biotopes overlap extensively, and vary enormously in scale. They can be described, not only in terms of the average conditions obtaining in them, but specifically in terms of the structure in space and time — the variety of habitats and biota they include, and the spatial relationships of these various elements, and the changes in time to which they are subject. Each organism is adapted to its biotope, and is *ex hypothesi* able to complete its development successfully in it. The biotopes and the organisms occupying them together constitute the ecosystem.

4 Complexity, Diversity and Stability

Jiro Kikkawa

4.1 Introduction

The concept of community embodies the ecological interactions of species populations (Chapter 2). These interactions, consisting broadly of biotope (or habitat), trophic and competitive interactions, produce spatial and temporal patterns of distribution among organisms. The patterns may be recurrent, cyclic, ordered or disordered. We need measures of organization that reflect the sorts of interactions and their magnitude and that allow comparison of communities. Complexity and diversity are two such measures that have been developed by theoretical ecologists (Margalef 1968; May 1973; Maynard Smith 1974). They have been used to predict the stability of communities under various conditions but, as yet, empirical verification is lacking (Paine 1980; King & Pimm 1983).

In this chapter the concepts of complexity and diversity are examined in relation to the stability of communities. I argue that stability is an historical function of diversity, and compare the diversity measures of different communities, relating them to habitat structure. I then present hypothetical relations between (i) the complexity and stability of environment and the complexity, diversity and stability of communities, and (ii) the intracommunity properties of complexity, diversity and stability.

Methods are often integral to the development and understanding of concepts. The understanding of communities in terms of complexity and diversity is constrained by the mathematical definitions of those concepts and by their applicability to real world situations. Comparison of communities is restricted by the parameters chosen, by the accuracy and completeness of information and by available methods of comparison. I discuss some of the methodological problems associated with the development of concepts in community ecology.

4.2 Complexity

Complexity is a function of the number of interconnections between elements in a set. Thus complexity will increase as the number of interacting species in the community increases. All interactions between populations of species that have non-zero outcomes will increase complexity of the community. Such interactions may be classed as competitive, trophic and symbiotic. Competitive interactions are *horizontal*; they involve pairs or sets of species primarily at the same trophic level. Trophic interactions, such as plant–herbivore, prey–predator, host–parasite, are *vertical*, involving species at different trophic levels within the community (Pimm 1982). Symbiotic interactions are often of trophic nature but, rather than creating more vertical connections, tend to compartmentalize food-webs or be nested within particular food-chains.

4.2.1 *Does complexity enhance stability?*

Intuitively, complex communities are considered stable. This is because it is thought that the impact of sudden population change of one species caused by external factors will be cushioned by the large number of interacting species and will not produce drastic effects on other species or the community as a whole. Elton (1958) suggested that such buffer mechanisms existed in tropical rainforests where insect outbreaks were not known. He observed that outbreaks of pests were common in simple communities such as those under cultivation or disturbed directly or indirectly by people. MacArthur (1955) quantified the relationship between food-web structure and community stability, and attempted to demonstrate that a large number of species each with a restricted diet, or a small number of species each with a broader diet, would enhance community stability so long as the abundance of organisms at a lower trophic level was not controlled by organisms at higher trophic levels. In these discussions, stability referred to the ability of the system to absorb perturbations, and it looked as though the notion that complexity enhances stability was substantiated.

4.2.2 *Theoretical considerations*

More recently May and others (see May 1981) have theoretically modelled the behaviour of communities following perturbations. They have paid special attention to the relation between the complexity of a food-web and its stability. 'Food-web connectance' (C) has been used as a measure of complexity. It is expressed as a fraction of all possible links between species,

$$C = \frac{2k}{S(S-1)}$$

where S is the number of species in the food-web and k is the number of trophic links between species. Connectance is a function of the number of vertical links in the food-web and is considered to influence community stability [see May (1973); the concept of connectivity, or connectedness, is similarly based (Maynard Smith 1974; Armstrong 1982)].

The equation for connectance implies that for large numbers of species (S) a given value of C can be maintained only if the number of vertical links (k) increases quadratically in relation to increase in S. This means that with each replication of a food-web structure (doubling the number of species) all possible vertical cross-connections between new and old elements need to occur if the value of C is to remain the same (Fig. 4.1a). Conversely, if the community is divided into compartments with respect to trophic relations as it often is, the value of C decreases with increase in the number of species (Fig. 4.1b). Thus the number of species and connectance may have a hyperbolic relationship,

$$SC = \text{constant.}$$

This seems to hold for many communities (Rejmánek & Starý 1979; Yodzis 1980; Pimm 1982), suggesting that, for specific values of S and C, dynamic stability is generally enhanced by the food-web consisting of loosely coupled subunits.

In these considerations stability is defined in terms of the vector field of a system in n-dimensional space (Lewontin 1969) and as the system's ability to return to a stationary point or an equilibrium after perturbation. Local stability, which is also known as neighbourhood stability or Lyapunov stability, refers to the return of a system to its original point after a minor disturbance. Large disturbances may shift systems away from such locally stable vector fields. If a system that has been subjected to a large disturbance still tends to return towards the original stable point, then that point is said to be globally stable. We would like to know if there are stable points other than the original one, at which systems may arrive after large perturbations. Large perturbations may be induced by addition or elimination of species that have strong or many trophic links. The outcome may be disintegration of the community, return to the original food-web structure after internal adjustment or stabilizing with changed composition of species and new food-web connectance. Resilience

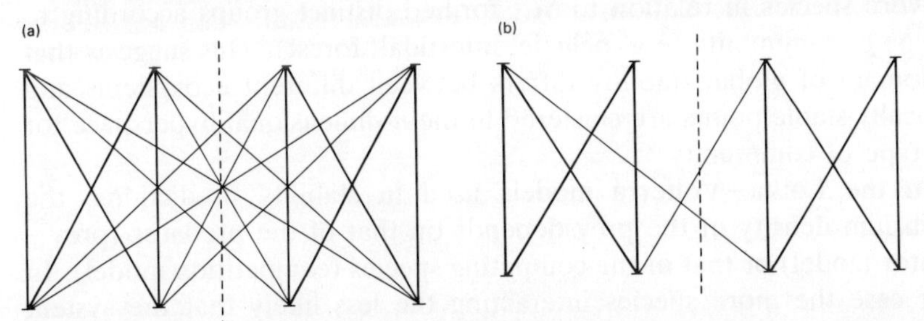

Fig. 4.1. Food-web connectance with two sets of four species, showing (a) all vertical links between species and (b) a tendency for compartmentation.

and persistence are two different measures of stability in units of time (Pimm 1984). Resilience is defined as the time the system (or its components) takes to return to the equilibrium after a perturbation whereas persistence is the time it lasts without change in its components (cf. Holling 1973).

In general, the theoretical work has demonstrated that if food-webs are randomly constructed, (i) the more species in the food-web, (ii) the more links between species and (iii) the greater intensity of trophic interactions, all cause less, not greater, stability (May 1973).

4.2.3 Conditions of stability

Theoretical models based on probabilistic associations of species suggest that additional species populations make complex communities more unstable than they make simple communities. This does not support the holistic view of communities where a dynamic balance is thought to be maintained by the network of species present in the community (cf. Kikkawa 1977). Empirical verification of assumptions and conclusions of theoretical models has been delayed because relevant data are difficult to collect. Perhaps more realistic assumptions will lead to testable hypotheses.

Real food-webs are not likely to show the random connectance that has been assumed in many theoretical studies. Even the most individualistic view of communities would not regard trophic links to be formed at random. If natural communities are the product of evolutionary processes then both the trophic links between members of those communities and the strength of those links are largely products of natural selection. In what way, then, do the links deviate from random expectations? Briand (1983) analysed 40 community food-webs and found great variability in food-web structure. However, he found that in a fluctuating environment the value of C decreased rapidly with increase in S, compared with a 'constant' environment (see also Cohen 1978; Briand & Cohen 1984). Moreover, the food-web configurations, as shown by the percentage of herbivore species in relation to SC, formed distinct groups according to the type of community (e.g. pelagic, intertidal, forest). This suggests that the domain of global stability differs between different ecosystems and the locally stable points are clustered in the n-dimensional hyperspace for each type of community.

In the Lotka–Volterra models used in stability predictions, the equilibrium density of the prey depends on that of the predator (prey–predator model) or that of the competing species (competition model). In either case the more species interacting the less likely that the system remains stable (May 1971, 1972). If, however, other models are used or certain constraints are placed on the models, the prediction may be reversed. Maynard Smith (1974), for example, shows with a model of species abundance distribution that the stability of an ecosystem is increased, in the sense that the chances of its persistence is increased, if there are a large number of weak connections between the component species rather than a small number of strong ones. However, his underlying assumptions that all species have the same equilibrium density and fluctuate with large amplitudes are untenable in the real world. In another model, the predator, instead of controlling the prey population, is allowed to take only those prey that would die from other causes (donor-control model). As in scavengers and detritivores the predators in this model do not affect the prey population, and increased complexity increases the chances that no species will be lost following a species removal (Pimm 1982). Although Pimm considers this model to have only limited applications, the model may reflect food-webs more realistically than the Lotka–Volterra model (see Chapter 7 for the role of detritivores and Chapter 9 for the role of parasites, in reducing the predator control of prey populations).

In earlier work stability was treated in relation to population densities. In the donor-controlled system it is more pertinent to deal with biomass stability. DeAngelis (1975) developed a model representing the dynamics of biomass with realistic assumptions, and demonstrated an increase of biomass stability with increase of connectance in the model. King and Pimm (1983) investigated the relationship between complexity and biomass stability following the removal of a herbivore, using the Lotka–Volterra model. The result confirmed McNaughton's (1977) finding that a herbivore modifies the total plant biomass less in more diverse than in less diverse grasslands. However, this is to be expected, also, if the herbivore specialized more in the former than in the latter. They considered it possible that more connected systems (more competitive interactions) are both more likely to change in composition, yet less likely to change in total biomass, following the removal of a species from the system.

As biomass stability suggests, there may be different kinds of stability for different properties of the community. Armstrong's (1982) theoretical study has shown that prey and predator populations behave differently with increased trophic links and that the decreased stability of the community is due to the predator's, and not the prey's, response to increased trophic links.

Among the non-random interactions, the tendency for species to interact strongly only with those within their compartment in the food-web structure has been studied by Pimm and Lawton (1980). They found that such compartments were not likely to be the result of a requirement for stable food-webs but that the boundaries tended to occur between adjacent habitats.

4.3 Diversity

The analysis of complexity has usually emphasized the trophic structure of communities although, in some studies, competitive interactions have been treated as a component of connectance. The concept of diversity is concerned with the total number of species in the community (species richness) and how individuals are distributed among them (equitability). It is not an easy task to find all the species in a community and most ecologists opt to survey only a section of the community. For example, particular taxonomic groups are treated as though they are communities in their own right. Thus we have studies of plant communities, that usually refer to vascular plants and ferns but occasionally include mosses and lichens. There are studies of ant communities, avian communities and small mammal communities. In other work the divisions are made according to habitat; for example the forest litter community, plankton community, intertidal community and dung community. Where taxonomic problems exist, particularly with small organisms, such as fungi, protozoa, nematodes and soil microarthropods, taxa above the level of species are often used in the description of communities. Higher taxa diversity may be based on the distribution of species among genera or higher taxa and then used to compare the structure of large communities. It is also possible to compare communities using diversity measures of biomass, life forms, life stages or any other ecological grouping.

However the community is defined, and whatever portion of it is surveyed, species richness, the total number of species, is the simplest measure related to diversity. Unfortunately, species richness alone tells us rather little about the organization of communities. Sample size also influences species richness. The number of species increases with the size of area surveyed, even without change of habitat. If individuals are sampled, the rate at which new species are added to the sample may be very different even for communities that have the same number of species. These two relationships will be examined below.

4.3.1 Species–area relationships

Plant sociologists have employed a concept of minimal area to define particular plant associations and to determine the sample size. The minimum area has been taken as the area at which the species–area curve (Fig. 4.2) flattens out or at which the number of species with a frequency occurrence of 90% or greater (i.e. constant species) no longer increases. But the search for a unit of regular species association or of homogeneity of vegetation (Goodall 1952) has met with the methodological problem of choosing quadrat size (contrast Figs 4.2a and b) and the biological problem of understanding spatial pattern of distribution and its scale in species–area relationships (Hopkins 1957; Kilburn 1966; Greig-Smith 1983).

The relationships between size of the area and number of species can be expressed as

$$S = CA^z$$

where S is the number of species, A the area, and C and z constants (Arrhenius 1921). For a known set of values of S and A, the constants C and z may be estimated from the linear regression

$$\log S = z\log A + \log C.$$

The constant C represents the number of species in the unit area of quadrat chosen, while z is the slope of the regression line on the log scale. Within particular geographic regions the biotas of islands fit the equations well though with different z values for different taxonomic groups and regions (MacArthur & Wilson 1967; Diamond & May 1981). Preston (1962) has shown that, under ideal conditions in which the area is occupied by a complete ensemble of lognormally distributed individuals (Section 4.3.2) and the population density does not change substantially over the range of areas being considered, z has a theoretical value of 0.262 (cf. MacArthur & Wilson 1967; Sugihara 1980; Connor *et al* 1983).

While the relationship between species number and area described above has been useful for examining the biotas of real islands and discrete habitat islands, it breaks down for areas that are separated by major geographic barriers and for areas of very different habitat. For these sorts of areas different parameters are needed to compare the flora and fauna.

4.3.2 *Species–abundance relationships*

Instead of using area as a unit for sampling, individuals may be sampled to determine how they are distributed among the species. There is, however, no simple relation between the number of individuals and the number of species to which they belong.

Motomura (1932) found that sample data from the benthos of lakes fitted the geometric series

$$\log n_x = b - ax$$

where n_x was the number of individuals of the species that occupied the xth rank in order of abundance and a and b were constants. In this equation, smaller values of a reflect greater species diversity and codominance by more than one species. Large values of a indicate that rank–log abundance lines have steep slopes and that only a few species contribute significantly to total abundance. This sort of species–abundance

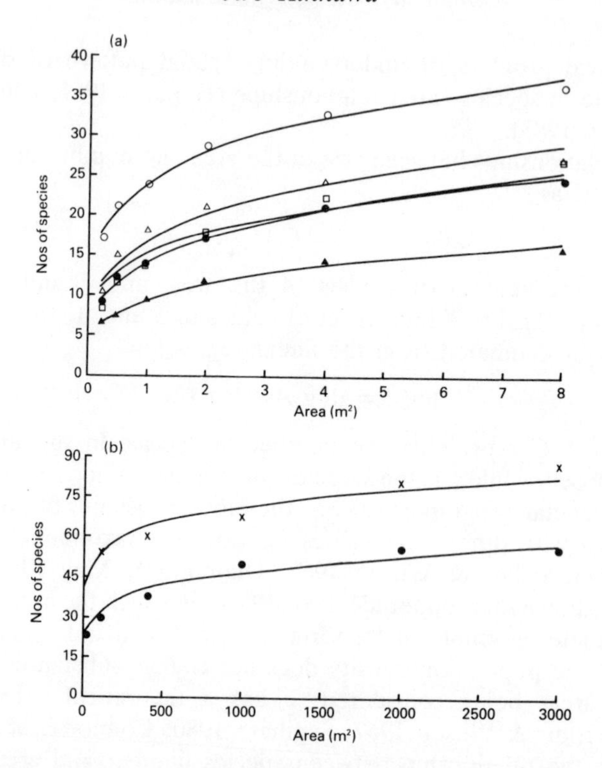

Fig. 4.2. Species–area curves for heathland plants based on (a) small quadrats in eastern Australia (data from Specht 1979) and (b) large quadrats in Western Australia (from George *et al* 1979). The lines are drawn from the fitted regressions. In (a) the open figures indicate dry-heathlands and the closed figures wet-heathlands, with circles, triangles and squares representing data from Wilson's Promontory (Vic.), Stradbroke Island (Qld) and Dark Island (SA), respectively. In (b) the crosses represent samples taken from the laterite soil area and the circles from the deep sand over gneiss, both in open-heathland at Corackerup Nature Reserve.

relationship lends itself to an interpretation that, from the most abundant to the least abundant, species pre-empt the available resources successively with each species utilizing a constant fraction of the resources remaining (niche pre-emption model). Thus the equation can be useful for the analysis of horizontal structure and comparison between communities, and may show how resources are partitioned among the species that belong to one broadly-defined guild. However, the parameters a and b indicate only the combined effects of the availability of resources and the degrees of segregation and competition between species.

In a relatively large assemblage of individuals the frequency distribution of the number of species represented by one individual, two individuals, three individuals, etc, may be fitted to a logarithmic series,

$$\alpha x, \frac{\alpha x^2}{2}, \frac{\alpha x^3}{3}, \ldots, \frac{\alpha x^n}{n},$$

where αx is the number of species with one individual, $\alpha x^n/n$ the number of species with n individuals and x is a constant (Fisher *et al* 1943). The total number of species (S) is the sum of all terms:

$$S = -\alpha \ln(1-x) = \alpha \ln\left(1 - \frac{N}{\alpha}\right),$$

where N, the total number of individuals, is given by $\alpha x/(1-x)$. Thus α, the index of diversity, is given by $N(1-x)/x$, where x generally has a value between 0.9 and 1.0. Unfortunately, random samples cannot represent the entire community because the number of species represented by one individual in the sample is greatest whereas the number of species not represented in the sample is unknown. However, over a wide range of sample sizes in which the discrete logarithmic function matches the frequency distribution of abundances, the diversity (α) with its variance is a powerful index calculable simply from N and S (Williams 1947).

In a very large sample (tens of thousands, rather than hundreds, of individuals), the number of species represented by one individual may not yield the greatest frequency. In fact, if a complete census of a community were possible the species–abundance relationship might reveal a lognormal distribution of the numbers of species in equal increment intervals (in logarithm) from the most abundant to the least abundant (Preston 1948);

$$S_R = S_0 e^{-(aR)^2},$$

where S_R is the number of species in the Rth interval (either direction) from the mode at which the number of species reaches S_0. The abundance intervals containing the number of individuals in successive doublings (i.e. 1–2, 2–4, 4–8, 8–16, . . .) are called the octaves. In reality, the normal distribution of the numbers of species over the octaves is not fully revealed as the left-hand side of the first octave (1–2 individuals) is always truncated (veiled) regardless of sample size. Theoretically, after the mode is unveiled, larger samples reveal more rare species and contain fewer of the rarest species. In other words, the veil line shifts to the left, revealing more of the left-hand side of the normal curve. For most communities the entire curve is never unveiled and we have little knowledge of the lower end of the distribution where the rarest species revealed may consist of new arrivals and near-extinct species, maintaining minimum viable populations.

For the three types of distribution discussed above there has been no proof of any possible biological significance (e.g. competition). The size of the ensemble seems to be the biggest factor determining which model fits the data best (Fig. 4.3). There have been some attempts at developing models based on ecological principles. For example, MacArthur's (1957) broken stick model was based on the assumption that species divided the environment into nonoverlapping niches of randomly allocated size. However, Cohen (1966) showed that this was not the unique condition for the model. Sugihara (1980), on the other hand, put forward an alternative hypothesis to explain the regularities of the canonical lognormal distribution (Preston 1962) in terms of a hierarchical community structure represented by a sequentially divided multidimensional niche space.

4.3.3 *Diversity measures*

Let us now examine the equitability component of species diversity. For a given number of individuals diversity would be highest if all the individuals belonged to different species and lowest if all belonged to one species. In real situations diversity lies between these extremes. Hence, it becomes important to know how common the most dominant species is and the proportion of the assemblage occupied by the rarest species. For comparison of different communities, an index of diversity is needed which is independent of sample size.

Diversity measures, such as α of the logarithmic series, depend on precise mathematical relations between S (the number of species) and N (sample size). Because we wish to compare communities in terms of the distribution of individuals among the species, it would be desirable to have an index that was distribution-free as well as sample-size free.

The most commonly used index of diversity is the Shannon–Weaver function,

$$H' = -\sum^{s} p_i \log p_i ,$$

where s is the number of species and p_i is the fraction of the ith species in

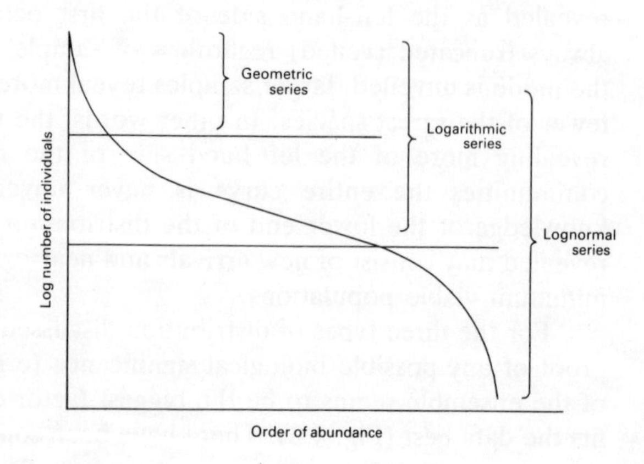

Fig. 4.3. Schematic relations between the number of individuals (log scale) and the species rank in order of abundance, with ranges of different sample size fitting geometric, logarithmic and lognormal series (from Morisita 1961).

the total. The latter is estimated as n_i/N, where n_i is the number of individuals of ith species and N is the sample size. The true diversity of the system, given by Brillouin, is:

$$H = \frac{1}{N}(\log N! - \sum^{s} \log n_i!).$$

But the estimates of H' may be taken as approximations to H, when n_i is sufficiently large for $n_i(\log n_i - \log e)$ to approximate to $\log n_i!$ for all i.

Because H deals with a finite population or a particular assemblage of species with a discrete distribution of individuals among the species, H is considered more realistic than H'. Both H and H' combine the evenness of species distribution with species richness, so that their values increase with increase in the number of species while remaining always $H' > H$. The degree to which the maximum value is realized, therefore, may be a better measure than H or H' when comparing communities. Such a measure seems to be useful when sample sizes are very different or there is a vast difference in the number of species. Unfortunately, proportional measures, such as $J = H/H_{max}$, have mathematical problems inherent in the nature of the data and cannot be used for comparison when the sample size differs (Peet 1975).

If H is consistently large for given S's of particular communities, species richness alone may be sufficient for the comparison of communities. Figure 4.4 plots H against H_{max} ($= \log S$) of bird species for a number of tropical and southern forest plots. H is generally close to H_{max} over a wide range of values (the only exception being an edge habitat where mist-netting in winter produced a large number of one flocking species). For comparison of forest bird communities, then, it would be more profitable to try to obtain as complete a species list as possible for the entire community (it is not very difficult to produce a fairly complete species list) than to make a special effort to count the number of individuals accurately for each species (this is agonizingly difficult) in a forest plot.

For describing the internal structure of a community the simplest index may be a dominance index, which expresses the abundance of the commonest species as a fraction of the total number of individuals. Unlike other indices the dominance index decreases with increase in the species number down to a certain level when it levels off. Another frequently used index is Simpson's (1949) index of diversity (D) based on the sum of the probabilities that two randomly chosen individuals are conspecific with respect to each species in the sample:

$$D = 1 - \frac{\sum^{s} n_i(n_i - 1)}{N(N-1)}$$

where n_i is the number of individuals of the ith species. This index

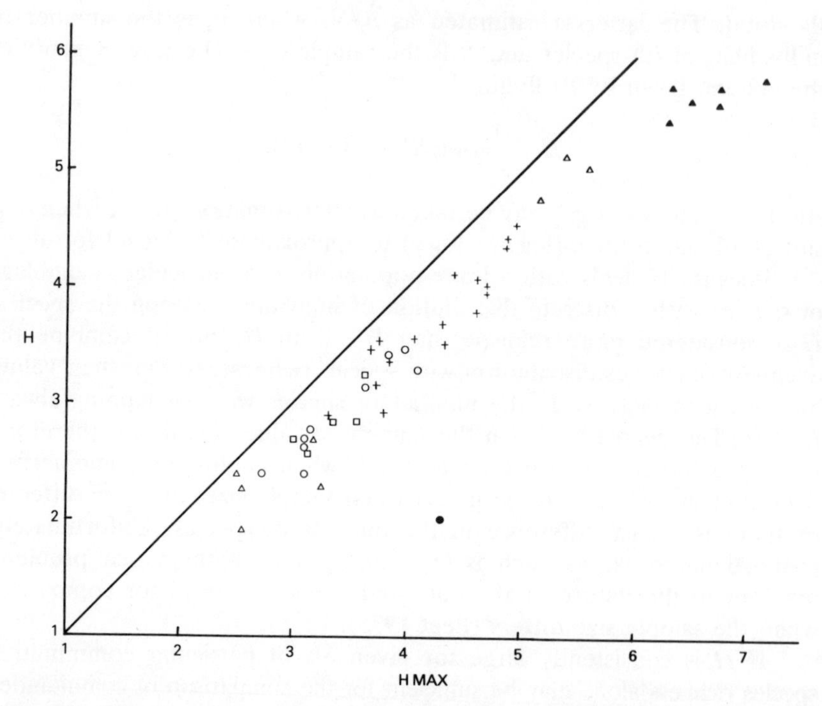

Fig. 4.4. Relations between H and H_{max} calculated from census and mist-netting data for forest bird communities of the tropics and the southern hemisphere. Data source: Australia, census (+) and netting [summer ○, winter (edge habitat) ●], from Kikkawa (1968, and unpublished); New Zealand, census (□), from Kikkawa (1966); Amazon, netting (▲), from Lovejoy (1975); Panama and Bahamas, census (△), from Karr (1971) and Karr and Roth (1971).

diverges enormously when the total number of species is increased, depending on the type of species–abundance relationship used for calculation (May 1975).

For mat-forming sessile organisms that have a patchy distribution or that lack definable individuals, diversity measures cannot be based on estimates of population sizes. Biomass units may be used instead. Wilhm (1968) found substantial differences in the mean diversity of benthic macroinvertebrates between the population-based and biomass-based diversities. This is because biomass and individuals have completely different distributions among the species. Sometimes it makes better ecological sense to use biomass than population estimates. Take herbivore communities of African savannah, for example. Mammals range in size from the mouse (0.02 kg) to the elephant (5000 kg) and birds from the weaverbird (10 g) to the ostrich (144 000 g). Population-based measures are useful if they reflect the role of competition and segregation in community structure as when applied to animals of certain size range within a broad trophic level or a major taxonomic group. Biomass diversity overcomes the size problem and, as in population-based diversity, is most useful when applied to a single trophic level rather than to the entire community.

4.3.4 Diversity and stability

Do diversity measures tell us anything about the stability of the community? Stability, as resilience or as persistence of the system, is affected by the number of interacting elements in the system. Whereas complexity is thought to bring about stability through predator–prey or vertical interactions, diversity is considered to influence stability through competitive or horizontal interactions.

As pointed out in Sections 4.2.2 and 4.2.3, there is, at present, little theoretical support for the intuitive notion that species network brings about stability. However, there is mounting evidence that competition and diversity are inversely correlated (Huston 1979). Intense competition leads to low diversity while weak competition permits coexistence and hence high diversity. Current theories of stability treat vertical and horizontal interactions of species alike. Thus theoretical instability based on random interactions is lessened by loose competitive interactions, or conditions that compartmentalize community structure such as specialization in guild and habitat diversification in ecological succession.

Much discussion of diversity has been centred on its relation to latitudinal gradient (e.g. Pianka 1966a), heterogeneity of habitat (e.g. Levin 1974), gap dynamics (e.g. Grubb 1977) and the frequency of disturbances (e.g. Connell 1978). More recently, the demonstration that competing species coevolve (Roughgarden 1983) provides a strong argument that stability is an historical function of diversity.

Diverse communities do not emerge spontaneously. They have developed and persisted regardless of, or in spite of, their diversity because they have achieved stability (in the dynamic sense) in the course of development. In this view the more diverse the community, the more difficult it is for additional species to establish without destroying the initial stability. What we see in a very diverse community is a result of evolutionary or ecological processes that have selected patterned interactions that do not jeopardize dynamic stability. Formation of guild structure, *K* and *r* strategies, compartmentation of trophic structure, niche differentiation within guilds, competitive mutualism (Pianka 1980), character difference and ecological succession are some of the processes that make 'surviving' interactions highly non-random.

The importance of historical factors in determining species richness of forest birds (hence species diversity, see Fig. 4.4) was demonstrated by

Pearson (1982) who studied the foraging ecology of birds in six lowland tropical forest sites. The total number of species on each 15 ha plot of primary forest was large and ranged from New Guinea with 114 species, through Borneo, Gabon, Bolivia, Peru, to Ecuador with 232 species. This order of species richness was not predictable either from species complementarity, species packing, the presence of non-avian competitors or from forest structure and seasonality. It was best explained in terms of the history of the fauna in Pleistocene forest refuges and the pattern of colonization. In spite of differences in species richness and phylogenetic origin the guild structure of birds in lowland tropical forests was very similar (see Section 5.5.1 for details). Could this mean that stability in species-rich communities had been achieved by coevolved species with highly selected modes of interaction?

Coevolution does not necessarily mean that many species have coexisted unchanged throughout the evolution of a community. Community membership can change with or without evolution. Examine the Australian heathlands, for example. In contrast to the South African heathlands (fynbos) with some 832 species of the Ericaceae, the Australian heathlands contain today about 333 species of the Epacridaceae and a handful of species of the Ericaceae and Vacciniaceae (Specht 1979). In eastern Australia the actual number of plant species developing on an area of 8 m^2 was only 22–36 in dry-heathlands and fewer in wet-heathlands (Fig. 4.2). These numbers were reached in the most productive phase of heathlands following a fire. Competitive interactions reduced the number of species above ground to 20 in a stand 25 years old on Dark Island and 10 in a 50-year old stand on Stradbroke Island (Specht 1979). In spite of such local impoverishments, the Australian heathlands as a whole support a formerly widespread Tertiary flora, of which Fabaceae, Myrtaceae and Proteaceae also have radiated in heathlands and each still maintains more species than the Epacridaceae. Heathland communities as they exist in Australia are a result of both evolutionary and ecological processes involving species of various origin.

The dense ground cover, the generally low status of soil nutrients, the sandy or rocky substrate, the restricted or special nature of food resources and frequent fire all contribute to produce the unique ecosystem of heathlands. Patterns of vertebrate adaptation to this ecosystem have been in flux as the habitat itself has undergone change. Specialization to heathland living must have occurred at various times in the past, as shown by large eggs and direct development in frogs, burrowing in frogs and reptiles and nectar feeding in birds and mammals (Kikkawa *et al* 1979). Besides, the present heathland fauna does not consist entirely of relict species of the Tertiary fauna. Adaptation and specialization should be regarded as continual processes involving taxa from the oldest to the latest colonizers. The heathland community which has developed on recent Quaternary landscapes of coastal areas contains many recent invaders from adjoining non-heath habitats, including introduced species such as the cane toad *Bufo marinus* (see Chapter 10). We may view the development of heathland communities as a continuum of species association over evolutionary (Tertiary to Quaternary) and ecological (pyric succession) time, in which species diversity waxes and wanes as selection of particular interactions among community members 'plays' with dynamic stability.

4.4 Comparison between communities

Communities that share a common biogeographic history have a pool of species associated with the available habitats of the region. Species diversity for the entire region (i.e. gamma diversity) is the sum of all within-habitat diversities (i.e. alpha diversity) estimated for the region. Between-habitat diversity (i.e. beta diversity) measures the extent of change or the degree of difference in species composition along the environmental gradient or among the samples of a set (Whittaker 1975). In the temperate-tropical comparison of species diversity, for example, higher values for gamma diversity in the tropics may be explained in terms of either greater species packing (alpha diversity) or greater habitat specialization (beta diversity) or both.

4.4.1 *Similarity and between-habitat diversity*

Morisita (1959, 1971) developed useful similarity indices (C_λ, C'_λ) from Simpson's measure of diversity (Simpson 1949):

$$\lambda = \frac{\Sigma n_i(n_i - 1)}{N(N-1)},$$

where n_i is the number of individuals belonging to species i in a sample of N individuals. The similarity index, C_λ, between two sites is written as:

$$C_\lambda = \frac{2\Sigma n_{1i} n_{2i}}{(\lambda_1 + \lambda_2)N_1 N_2}$$

where subscripts indicate the sites 1 and 2. If a community data set (sites × species) consists of the row totals (R_i) giving the total number of individuals for each species ($\sum_j n_{ij}$) and the column totals (N_j) giving the total number of individuals for each site ($\sum_i n_{ij}$), with T as the grand total, C'_λ, a corrected form of C_λ, is given by

$$C'_\lambda = \frac{\delta_s}{1 - \delta_s} \times \frac{1 - \bar{\delta}_z}{\bar{\delta}_z}$$

where

$$\delta_s = \frac{\Sigma N_j(N_j - 1)}{T(T-1)}$$

and

$$\bar{\delta}_z = \frac{\Sigma\Sigma n_{ij}(n_{ij} - 1)}{\Sigma R_i(R_i - 1)}$$

The value of C'_λ varies from 0 for no overlap of species between communities to 1 or a fraction greater for communities with identical composition. Thus beta diversity may be expressed as $(1 - C'_\lambda)$.

This index is used below to examine between-habitat diversity of birds in Australia. In northern New South Wales, various vegetation types may be arranged along the mesic–xeric gradient from subtropical rainforest of the coastal ranges to open woodland or savannah of the western plains. Table 4.1 shows changes of species overlap and beta diversity of birds between 8 ha plots examined along this gradient. Species overlap decreases and beta diversity increases more rapidly at the xeric end than at the mesic end of the gradient. These findings confirm that the species association is more predictable in the wet-formations than in the dry-formations (Kikkawa 1968, 1974) and that the high bird species diversity of the semi-arid formations (Brereton & Kikkawa 1963), compared with the wet-formations, is a result of both alpha and beta diversity.

In the above example, geographical proximity did not influence beta diversity between different vegetation types. Can we then place communities of unknown relations along the gradient using the beta diversity index? There were two aberrant sites which did not fit easily into the vegetational gradient. One was a subalpine formation of tall open forest (grassy forest) at Barrington Tops, for which the index was closest to that for tall open forest (wet sclerophyll forest) at Pt Lookout. The other was an exotic pine forest at Armidale which, despite extreme dissimilarity with all other wooded habitats of the region, was placed along the gradient between open forest (dry sclerophyll forest) and tall woodland. This pine forest behaved as though it was part of the natural series of vegetation types in its relation to other habitats.

4.4.2 *Classification and ordination*

For analysis and comparison of complex communities numerical techniques of classification and ordination are the most powerful methods available (Whittaker 1973; Poole 1974; Clifford & Stephenson 1975; Williams 1976; Pielou 1977; Gauch 1982; Greig-Smith 1983). These techniques, broadly referred to as 'pattern analysis' (Williams 1976) are

Table 4.1. Bird species overlap (%, right of diagonal) and beta diversity ($1 - C'_\lambda$, left of diagonal) between plots of 8 ha along the mesic–xeric gradient of vegetation in northern NSW (from Kikkawa 1968 and unpublished).

Locality	vegetation[1]		B1[2]	B2	B7	B8	B14	–	B19	B22	B26	B27	B28	B44	No. of species in 8 ha
Dorrigo	STRF	B1		61.7	52.0	32.6	44.4	15.6	20.0	20.4	9.3	7.8	7.6	3.5	38
Pt Lookout	STRF	B2	−0.06		55.1	29.6	47.2	15.6	20.0	16.1	9.3	10.7	6.0	3.5	38
Pt Lookout	NF	B7	0.37	0.38		42.5	50.0	3.8	29.2	25.0	14.6	18.6	7.6	3.5	38
Barrington Tops	NF	B8	0.47	0.51	0.13		28.3	32.0	16.1	15.0	7.1	6.7	6.1	2.6	19
Pt Lookout	TOF	B14	0.48	0.50	0.35	0.37		14.9	32.3	21.8	15.7	21.4	10.6	5.2	40
Barrington Tops	TOF	–	0.70	0.72	0.55	0.54	0.25		15.4	13.9	7.7	13.5	9.3	6.1	14
Mt Kaputar	OF	B19	0.84	0.84	0.81	0.87	0.42	0.53		28.1	39.7	44.4	21.5	9.8	44
Armidale	PF	B22	0.89	0.90	0.87	0.92	0.52	0.76	0.17		29.7	33.3	20.0	9.1	27
Warialda	TW	B26	0.93	0.93	0.92	0.95	0.78	0.80	0.39	0.25		44.3	29.0	16.7	55
Armidale	TW	B27	0.92	0.92	0.90	0.93	0.79	0.72	0.36	0.42	0.36		41.8	17.9	44
Armidale	W	B28	0.97	0.97	0.92	0.98	0.95	0.92	0.89	0.94	0.85	0.48		25.6	33
Boorooma	OW	B44	0.98	0.99	0.99	0.99	0.97	0.98	0.93	0.97	0.81	0.86	0.87		21

[1] Vegetation: STRF, subtropical rainforest; NF, *Nothofagus* forest; TOF, tall open-forest; OF, open-forest; TW, tall woodland; PF, pineforest; W, woodland; OW, open-woodland.
[2] Site numbers given in Appendix 1, Kikkawa 1968.

subjective in that the revealed pattern is an interaction between properties of the data set and the mind of the user, i.e. a 'pattern-for-an-agent'. In addition, the data set is not regarded as a sample of an undefined parent population or of the universe but, rather, is treated as an entity in its own right. Therefore, a revealed pattern is never correct or incorrect, never true or false — it can only be profitable or unprofitable for the users (Williams 1976). The purpose of extracting patterns using these techniques is to generate, rather than to test, hypotheses.

Comparison of complex communities cannot easily be based on replicate samples, with known statistical properties (see Chapter 11). The more familiar one becomes with complex communities, the more sceptical one grows about the validity of random sampling in estimating habitat features or abundance of organisms. This is particularly true for tropical rainforests where replicate samples (regardless of size) are almost always impossible to obtain. But once random sampling is abandoned as a procedure, selection of study sites and of organisms and collecting methods becomes an important exercise, requiring consideration of all variables which, from available information, might be thought to contribute to the pattern. Thus in pattern analysis one is conscious of the nature of information collected, its representativeness with respect to the purpose of inquiry and its relative accuracy among the sets of data.

For both hierarchical classification and ordination, methods of analysis have developed in relation to characteristics of the data collected rather than the sort of question being asked. Figure 4.5 compares results of ordination and classification for 18 sites in tropical forests of north Queensland in terms of the distribution and abundance of birds (Kikkawa 1982). Abundance was ranked 0 to 8 according to the frequency of capture in mist-nets. The diversity of each community was calculated using the information content,

$$I = N \ln N - \sum_{j=1}^{S} n_j \ln n_j$$

where N is the total number of individuals and n_j the number of individuals of the jth species. Information gain was used as a dissimilarity measure to group sites for hierarchical classification and to ordinate by means of principal coordinate analysis (Williams 1976). The spread of sites in two dimensions (Fig. 4.5a) was interpreted in relation to an altitude-related gradient on axis 1 and a habitat-related gradient on axis 2. In this ordination about half the variation in the data set was explained by the first two vectors. In classification (Fig. 4.5b), clustering of sites indicated separation of lowland and tableland sites and further subdivisions related to vegetation types within each. For the same degree of similarity lowland site groups accumulated more species than tableland site groups; this indicated higher beta diversity in the lowlands than in the tablelands. Thus, a single data set can reveal gradients (ordination) and boundaries (classification) for the comparison of communities.

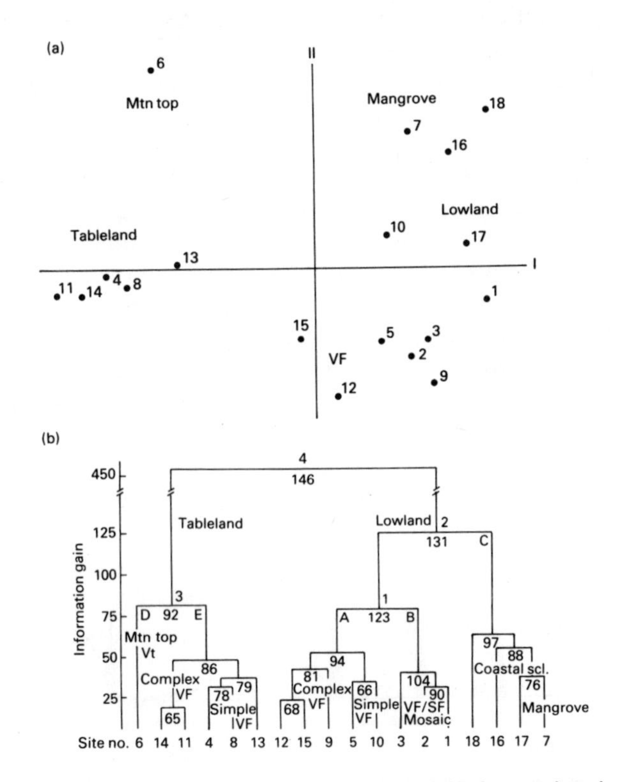

Fig. 4.5. Ordination (a) and classification (b) of 18 forest sites in northeastern Australia in terms of frequency data on birds (from Kikkawa 1982). The number of species is given at each fusion of sites and site groups.

4.4.3 Habitat relations

Measures of between-habitat diversity indicate the degree of change in species associations from habitat to habitat. With detailed knowledge of biology of all the species we could construct or determine niche relations or guild structures for the community and, to a large extent, this should explain the between-habitat diversity. For example, we know enough about the broad requirements of forest birds to categorize their guilds (Chapter 5) and explain their associations with particular habitats, hence local species diversity. Using information on the feeding and nesting requirements of birds, Kikkawa and Webb (1967) proposed the niche occupation types (guild types) for land birds. Within Australia, tree-nesting and tree-feeding herbivores (frugivores) increase in number from subtropical to tropical forests (Kikkawa & Webb 1967) and from semi-arid to wet habitats in the subtropical region (Kikkawa 1974). In New

Guinea, the proportion of tree-nesting insectivores increases whereas the proportions of frugivores, predators and scavengers decrease, with increase in altitude (Kikkawa & Williams 1971). Thus the between-habitat diversity reflects shift in guild types.

If the components of forest habitat could be treated as discrete life forms of plants, and associated bird species could be arranged as in Fig. 4.6, then the diversity, density and similarity of the bird fauna in different forest types which were represented by different combinations of the components could be predicted from the habitat relations of species. This type of model would be helpful in devising pro forma checksheets of habitat components for field use. If such information was obtained from the field every time a species list of animals was prepared, the accumulated information might reveal habitat relations of animals and lead to detailed biological studies of selected species important for the understanding of

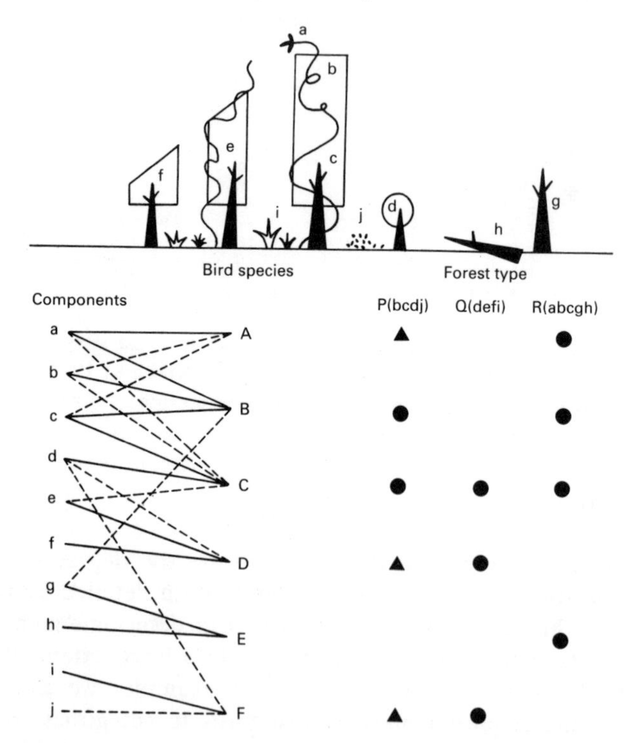

Fig. 4.6. Schematic relations between components (a–j) of forest habitat and bird species (A–F), leading to different diversity and density of birds in three forest types (P–R; no. species: 5, 3, 4, respectively). Solid lines indicate regular association permitting normal density (●) and broken lines indicate marginal association with low density (▲) in forests with relevant components. Diversity, density and similarity when compared between forest types give: Diversity P > R > Q; Density R > P > Q, Similarity P – Q > P – R > Q – R.

community organization or conservation. This pro forma study permits analysis of the components of habitat independently from information on the fauna collected at the same sites. Canonical coordinate analysis (Williams 1976) between the environmental vectors and the animal vectors that have been derived separately by ordination may then suggest important components of the habitat for the distribution or abundance of particular animal species (Kikkawa *et al* 1980; Kikkawa 1982).

4.5 Complexity, diversity and stability

The organization of a community may be described in three main dimensions; namely, biotope (habitat), complexity and diversity. The biotope dimension comprises abiotic and biotic structuring (see Chapter 3), which develops with the stages of ecological succession. Communities are delimited in this dimension by the combination of physical factors and the degree of niche exploitation and differentiation achieved within the constraints of the environmental conditions. The complexity dimension in the restricted sense describes the trophic structure of the community, including intercompartmental links with neighbouring habitats. The diversity dimension describes the information content of the community and, more specifically, the guild structure and competitive patterns. Biomass is an energetic parameter of the community and its breakdown along the complexity and diversity dimensions gives the distribution of energy within the community.

Community organization described by the three broad dimensions should reveal the degree of stability of the community. The following are some hypothetical relations derived from the discussions in this chapter. In the biotope dimension, increase in the heterogeneity (complexity) and stability (short-term and long-term) of the *environment* is considered to influence complexity, diversity and stability of the *community* (Table 4.2a). Complex environments tend to compartmentalize food-web structure and increase species diversity. The internal structure of the community may become vulnerable to perturbation but the community as a whole will persist well in the absence of catastrophes. The short term and long term stability of the environment would have exactly the opposite effects on the community. The short-term stability, with fairly frequent disturbances, will tend to keep community structure simple but permit relatively high diversity. The community would become resilient and even adapted to disturbances (e.g. fire in Australian heathlands). Such a community would lose stability in the sense of persistence. On the other hand, the long-term stability of the environment would enhance complexity, reduce diversity and resilience to perturbations, though it would increase persistence of the community as a whole.

Table 4.2. Hypothetical relations (a) of complexity and stability (short-term, long-term) of the environment to complexity, diversity and stability (resilience, persistence) of the community and (b) among complexity, diversity and stability of the community.

(a) Environment

Community	Complex	Stable	
		Short-term	Long-term
Complexity	Compartment	Decrease	Increase
Diversity	Increase	Increase	Decrease
Stability Resilience	Decrease	Increase	Decrease
Persistence	Increase	Decrease	Increase

(b) Community

Complexity (vertical connectance)			
–	Diversity (within-guild diversity)		
–	+	Stability (resilience)	
+	–	–	Stability (persistence)

Within a community, if the meaning of complexity is restricted to the vertical connectance of food-web and that of diversity to the within-guild species richness, then complexity and diversity are negatively associated with each other (cf. Margalef & Gutiérrez 1983). Thus their relations to stability have opposite signs in Table 4.2b. These relationships are very much simplified and may even prove to be wrong. They are presented here not as conclusions but as propositions.

Ecological interactions and evolutionary processes that produce and modify organization and stability of communities are discussed in the rest of the book.

Part 3
Community Organization

Three chapters in Part 3 take synthetic, rather than analytical, approaches to the understanding of community structure and function, particularly the pattern of resource utilization.

The central theme of Chapter 5 is the guild structure which is examined within and between communities. The convergence of guild structure is seen in the panglobal comparison of tropical rainforest birds. Both historical factors and more immediate ecological factors are considered to influence guild structure and species diversity.

Chapter 6 reviews the vast literature on resource partitioning, both field observations and experiments, and discusses community regulation. Multivariate explanations are offered as an alternative approach to Popperian hypothesis testing.

Chapter 7 examines the role of decomposers in the energy transfer of terrestrial and aquatic communities. An important conclusion reached is that consumers and associated microbes work together to break down the organic matter produced by green plants for their use. The guilds of those consumers are classified according to the strategy used by them to concentrate their microbial or microbe-induced food supply.

5 Guilds and their Utility in Ecology

John Terborgh and Scott Robinson

5.1 Introduction

Analysis of natural communities can be conducted at several levels. At one extreme it is commonplace to phrase questions about whole communities concerning such issues as species/number relationships, species diversity, food-webs and energy flow (Whittaker 1970; May 1973). At the other extreme, one can ask about the effects of individual species on other species, and about the factors controlling the presence and absence of particular species (Diamond 1975; Pulliam 1975; Terborgh & Weske 1975). Falling in between the levels of organization embodied in the community, with its several trophic levels, and the species, is the utilitarian concept of the guild. As applied to ecology, the notion of a guild is a recent one, tracing to the germinal paper of Root (1967), although the essential idea had been formulated in other words by Elton (1927).

The adoption by ecologists of the term guild rests on an analogy between functionally related groups of species and medieval guilds — trade organizations uniting artisans having a common skill and livelihood, such as goldsmiths, wheelrights and candlemakers. The analogy translates into ecology as sets of species that derive their subsistence from common pools of resources. The special value of the concept as applied to ecology is that guilds can be more or less objectively defined independently of the particular species that comprise them. This opens the possibility of making detailed comparisons of the functional organization of different communities, such as those on different continents, even though they may share no species in common.

5.2 How guilds have been characterized

Root's (1967) original definition of a guild as a "group of species that exploit the same class of resources in a similar way" leaves a great deal of room for different interpretations. This probably explains why the guild concept has enjoyed such wide acceptance in ecology. Its flexibility permits applications to many taxa, situations and problems. Examples of taxa and communities to which the guild concept has been applied are: herbivorous insects (Root 1973; Rathcke 1976), parasitic insects (Askew

1980), nectarivorous birds (Feinsinger 1976), forest-dwelling insectivorous birds (Root 1967; Williamson 1971; Landres & MacMahon 1980), amphibians and reptiles (Inger & Colwell 1977; Hairston 1981), grassland plants (Platt & Weis 1977; McNaughton 1978), even plants that share seed dispersal agents (Beatie & Culver 1981). The common denominator in these examples is that the guild members share the same resources, co-occur in the same or overlapping microhabitats and are taxonomically related. Ideally, both guilds and communities should transcend taxonomy, but in practice they almost never do because of limitations of time, methodology and the expertise of the investigators. Only recently have ecologists begun to leap over taxonomic barriers and look at complex multi-taxon communities and guilds (Brown & Davidson 1977; Reichman *et al* 1979).

A schematic diagram illustrating different levels of organization in two well-studied but very different communities (forest dwelling insectivorous birds and herbivorous insects) is shown in Fig. 5.1. In this diagram we have distinguished five major levels of organization: taxon (I), diet or trophic level (II), microhabitat (III), substrate used for foraging (IV) and foraging behaviour (V). Communities can be subdivided into major groupings at each level of organization, but levels IV and V are usually considered guild-level groupings. Any further subdivision usually occurs at the level of the species. For communities such as those of West Indian *Anolis* lizards which consist of very few species, each of which is ecologically distinct (e.g. Moermond 1979), the guild concept loses much of its

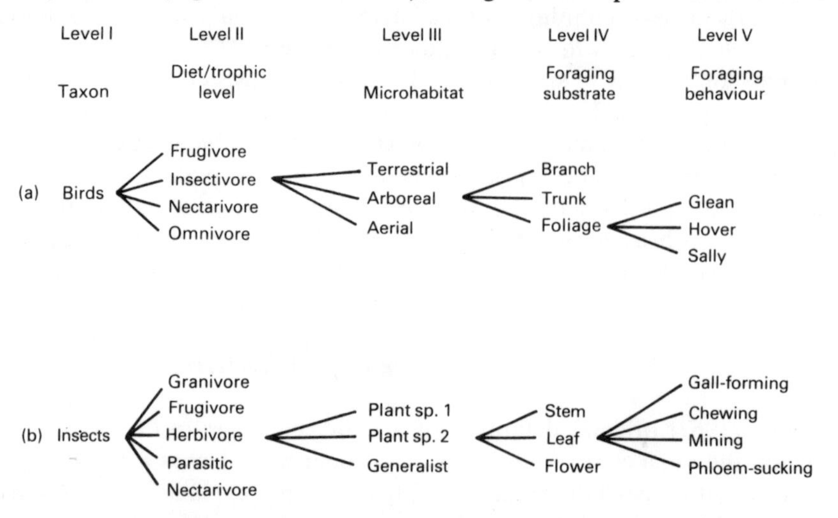

Fig. 5.1. A schematic diagram of five levels of organization of two different kinds of communities; (a) forest birds (e.g. Willson 1974) and (b) old field insects (e.g. Root 1973; Lawton 1982; D. Tonkyn personal communication). Only one of several possible subdivisions of each previous level are shown.

meaning, as each species is essentially in its own guild. Studies of such narrowly defined communities are better carried out at the level of the species.

For the herbivorous insect community illustrated in Fig. 5.1(b), the guilds of insects on a single plant species are often very well defined and intuitively satisfying because of the extreme morphological and physiological specializations required for such activities as leaf mining, gall forming and phloem sucking (Root 1973). However, if the community is defined more broadly to include all herbivorous insects on all of the plants in an old field, for example, community analysis is made vastly more complex by the presence of many generalist herbivores as well as specialists. For this reason, Root (1973) suggested that the best way to study the community organization of herbivorous insects in habitats with many plant species is to break the community into 'component communities', each consisting of the herbivorous insects that feed on a particular plant species.

In an attempt to define guilds objectively a number of investigators have used multivariate analysis to identify sets of species that cluster closely together in a multidimensional space based on morphological, ecological or behavioural variables (Inger & Colwell 1977; Holmes *et al* 1979b; Landres & MacMahon 1980; Sabo 1980). The major advantage of multivariate analysis is that it reduces many intercorrelated variables to a smaller number of independent variables, and identifies the variables that make the greatest relative contribution to the overall community pattern. The main limitation of this procedure, as most of its practitioners expressly acknowledge, is that the guild structure generated by multivariate analysis still depends very much on what resource variables are subjectively chosen for inclusion in the data set.

A distinct approach is to separate guilds via the multivariate analysis of morphological variables, used sometimes in conjunction with ecological and behavioural variables (birds: Ricklefs & Travis 1980; lizards: Ricklefs *et al* 1981). Studies of community structure based on the morphology of the component species have the potential advantage that morphological measurements are independent of the habitat and season, and are not affected by fluctuations in the abundance and distribution of resources that can distort the results of short term studies of foraging behaviour (Wiens 1977; Wiens & Rotenberry 1979). Such studies are burdened, however, by the disadvantage that morphology does not fully predetermine behaviour, and ultimately it is behaviour that defines a species' functional role in its community.

An appreciation of the importance of behaviour has emerged from studies of predatory animals (birds, lizards, spiders, etc.). At the extremes, two distinct modes of hunting can be distinguished: 'passive searchers' that remain motionless for long periods waiting for actively moving prey to come within striking distance, and 'active searchers' that hunt through foliage, leaf litter, etc. for inactive prey (MacArthur & Pianka 1966; Eckhardt 1979). Active searchers tend to forage amidst dense foliage, while passive searchers generally forage in the open (Eckhardt 1979; Moermond 1979; Robinson & Holmes 1982).

Moermond (1979) and Fitzpatrick (1980) have also shown that at least some features of the external morphology of lizards and birds are associated with these and other aspects of their searching behaviour. Root and Chaplin (1976) termed these correlated suits of morphological and behavioural characters 'adaptive syndromes', which are defined as 'coordinated sets of characters'. The identification of such adaptive syndromes, using multivariate techniques that objectively determine which behavioural, morphological and microhabitat variables are intercorrelated, will eventually provide a much more inclusive guild concept that integrates all three levels of organization.

5.3 The utility of the guild concept

There are three major contexts in which the guild concept has been applied: (1) studies of single guilds, (2) studies of single communities, and (3) comparisons of different communities. Our main emphasis will be on the third of these, between-community comparisons. However, we shall begin with a brief synopsis of both single guild and single community studies.

5.3.1 Studies of single guilds

The questions that are generally asked in single guild studies are how species in the same guild partition the resources of a given habitat, and whether there is any evidence that the species in a guild are competing for those resources. Not surprisingly, given the diversity of communities studied, some investigators have found evidence of both resource partitioning and competition within guilds (McClure & Price 1975; Davidson 1977; Noon 1981), while others have found no clear evidence of either (Rathcke 1976a,b; Rotenberry & Wiens 1980; Hairston 1981). Hairston has even suggested that guilds are merely artificial constructs until it has been shown that the members compete for the resources used to define the guild. There is, however, nothing in Root's (1967) original definition which requires that guild members have to compete and that non-guild members do not compete. Studies of resource partitioning within guilds are treated in more detail in Chapter 6.

5.3.2 *Studies of single communities*

The major goal of studies of the guild structure of single communities is to identify the resources that determine the structure of the community. In the most extensive study of this kind to date, Holmes *et al* (1979b) used multivariate techniques to characterize the resource use patterns of the insectivorous birds of a northern hardwoods forest, and to identify the major resource axes that separated clusters of species (Fig. 5.2). The first bifurcation in this community represents the basic microhabitat difference

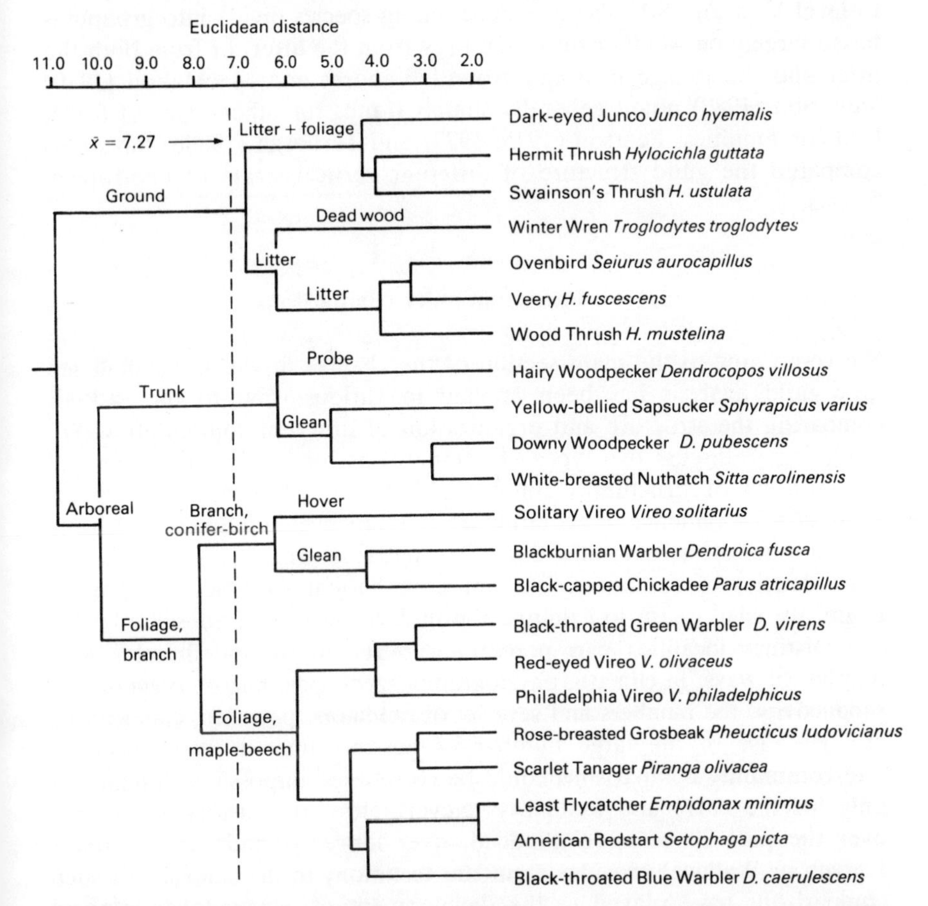

Fig. 5.2. Community dendrogram of Euclidean distances between the major insectivorous breeding birds of a northern hardwoods forest, USA (based on Holmes *et al* 1979b). Dashed line indicates the mean Euclidean distance between all combinations of species pairs. The labels represent the most heavily weighted variables for each bifurcation derived from Varimax rotated factor analysis.

between arboreal and terrestrial species, which corresponds to level III in Fig. 5.1. The second major division corresponds to the differences between foliage-dwelling and trunk-foraging species, which corresponds to level IV in Fig. 5.1. The next division of the community is much less intuitively obvious. The foliage-dwelling species of this particular forest split into two major groupings based largely on their tree species preferences, a rather surprising result given the prevailing wisdom that tree species use is not important in differentiating between bird species, especially at the level of the guild (MacArthur & MacArthur 1961; James 1971; Willson 1974). The final level of organization of the arboreal species is into groupings based largely on foraging behaviour, which corresponds to level V in Fig. 5.1. The ground-dwelling species divide into groupings based largely on whether they take prey from the litter, or from both the litter and the foliage near the ground. Landres and MacMahon (1980) and Sabo (1980) noted generally similar trends for other types of forest bird communities. Pearson (1975, 1977) and Terborgh (1980b) have also compared the guild structure of different vertical strata of neotropical forests.

5.4 Intercommunity comparisons

We come now to the main section of the chapter in which we shall see how guild analysis has been applied in various ways to the task of comparing the structure and organization of different communities. We are going to discuss two types of comparisons, one in which the animal communities of structurally similar habitats on different continents are examined for evidence of evolutionary convergence, and one in which the communities of structurally dissimilar habitats are analysed to reveal dependencies of community variables on habitat variables. Before we begin, we shall pause to examine the underlying assumptions.

Distinct localities can potentially differ in an indefinitely large number of ways: in climate, physiography, geological history, vegetation, productivity, the numbers and severity of predators, parasites, competitors, etc. Because of the large number of possibly uncontrolled variables, intercommunity comparisons could be considered impossibly difficult, or only for the naïve and foolhardy. Nevertheless, the history of ecology over the past 20 years (or, indeed, over longer periods, if one reads Darwin or Wallace) provides abundant testimony to the central role such comparisons have played in the development of our understanding of ecological relationships (MacArthur 1972; Cody & Diamond 1975). How can this fact be reconciled with the extreme complexity inherent in making intercommunity comparisons? The validity and assurance with which one can make such comparisons hinges on what aspects of community structure are being compared and which variables are important to the comparison.

In comparisons of communities occupying similar habitats in separated localities, pertinent variables are controlled as much as possible by careful site selection (Table 5.1). In matching sites half way around the world, certain features such as climate, physiography, and gross morphology of the vegetation (canopy height, etc.) can be specified to a considerable degree of precision, while others, such as the quantity and quality of resources available are only weakly under control, and still others (e.g. plant species composition) are intrinsically uncontrollable. What is most important is that the sites should be representative of their respective biogeographical regions. Given this criterion, the major worry is that local variations in productivity, or plant species composition, could be so pronounced as to obscure the characteristic regional pattern. Such fears, while they should be taken seriously, should not be allowed to inhibit inquiry.

Local (intraregional) differences in community structure are usually less than the contrasts between regions. For example, there are now several localities in the lowlands of western Amazonia that have been intensively surveyed for their resident bird communities. Situated in Ecuador and Peru, each of these localities is hundreds of kilometres from

Table 5.1. Variables implicit in intercommunity comparisons.

Variable	Habitat similar	Habitat different
Climate	Controlled in site selection	May be similar or different depending on distance between sites
Physiography, geology, soil	Potentially controllable in site selection	May be similar or different depending on distance between sites
Structure of habitat (vegetation)	Controlled in site selection	Often independent variable of interest
Resources	Assumed equivalent or sometimes actually measured	Usually uncontrolled, but potentially measurable
Productivity	Assumed equivalent or sometimes actually measured	Usually uncontrolled, but potentially measurable
Competitors, etc.	Usually dependent variable of interest	Usually dependent variable of interest
Evolutionary history	Usually independent variable of interest	Controlled if sites in close proximity, otherwise uncontrolled

any of the others, yet the species lists differ only in minor detail (O'Neill 1974; O'Neill & Pearson 1974; Pearson *et al* 1977; Parker 1980; Terborgh *et al* 1983). Any one of these localities could be taken to represent a 'typical' Amazonian bird community in a comparison with, say, Africa or Malaysia.

Careful site selection is important because it boosts the signal to noise ratio to the highest attainable levels. In intercontinental comparisons the primary variable being examined is the separate evolutionary history of the communities. Secondary variables, such as differences in available resources, must inevitably contribute to the measured contrasts, but it is often difficult to say whether such differences are due to local ecological variability or to historical factors.

In comparisons of different kinds of communities, there is much less precise control over the pertinent variables. If one is examining a series of adjacent or nearby habitats, or a set of islands in an archipelago, the climate and historical background are likely to be similar. But if the sites are far apart, as they would be, for example, in any comparison of tropical and temperate communities, the uncontrolled variables are legion. This does not mean necessarily that such comparisons are frivolous and devoid of interest (as we shall see), but it does caution us not to assign dogmatic or unitary interpretations to the results.

5.5 Panglobal community comparisons

A number of well-travelled investigators have made attempts to study community convergence on a panglobal scale. The basic question is whether, in the presence of nearly identical physical conditions (climate, relief, soil, etc.), evolution will produce biological communities having manifestly similar structure and organization. In other words, starting from different beginnings in the remote past, does natural selection, operating under specified environmental conditions, lead to a repeated, hence predictable, outcome?

While this is a very basic question about evolution, and one that is simply asked, it is not so simply answered. Difficulties arise on at least two levels. First, it is not obvious what features of ecosystems that share few or no species in common should be expected to converge. One can look at gross community-wide statistics on species diversity, food-webs, and plant productivity (e.g. Whittaker 1975; Walter 1973) or at somewhat finer details such as the physiognomy of the vegetation (Richards 1964), the distribution of plant life forms (Raunkier 1934), or the representation of various mammalian orders in the fauna (Keast 1969; Bourlière 1973). Or, pursuing the quest for similarity to the most extreme level, one can ask for each species in one community whether there is a recognizable ecological counterpart in the community to which it is being compared (e.g. Bourlière 1973; Cody 1974; Fig. 5.3).

The second difficulty is one of evaluation. How similar is similar? There is no yardstick or control against which to gauge the results. Whether or not convergence has occurred becomes at least partially a matter of judgement, and this is all too readily coloured by the scientist's own zeal or scepticism. In the absence of any absolute standard, one must resort to a relativistic standard. Are the plant and animal communities that occupy homoclimes (identical climates) at scattered points around the world more similar to each other than they are to the communities of nearby but distinct climates (Walter 1973)? If so, it is safe to conclude that convergence has occurred.

Clearly, at the grossest levels — physiognomy of the vegetation, distribution of plant life forms, energy flow through the food-web — there

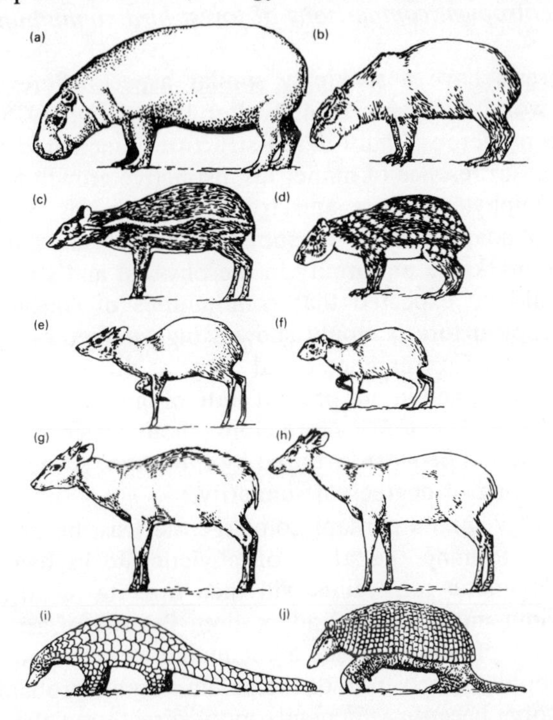

Fig. 5.3. Morphological convergences among African (left) and neotropical (right) rainforest mammals. (a) pigmy hippopotamus (*Choeropsis liberiensis*) and (b) capybara (*Hydrochoerus hydrochaeris*); (c) African chevrotain (*Hyemoschus aquaticus*) and (d) paca (*Cuniculus paca*); (e) royal antelope (*Neotragus pygmaeus*) and (f) agouti (*Dasyprocta aguti*); (g) yellow-back duiker (*Cephalophus silvicultor*) and (h) brocket deer (*Mazama gouazoubira*); (i) terrestrial pangolin (*Manis gigantea*) and (j) giant armadillo (*Priodontes giganteus*). Each pair of animals is drawn to the same scale (from Bourlière 1973).

is abundant evidence of convergence on a global scale, as is emphasized in every elementary ecology textbook. It is when we push the analysis to finer levels that the situation becomes murky. Here the effects of history and geography begin to distort, if not to override, the unifying force of evolution.

We shall now describe two cases in which evidence for convergence was sought in worldwide comparisons: the bird fauna of tropical rainforest and the lizard communities of subtropical desert scrub. These two cases were chosen because the first demonstrates a strong trend toward convergence, but offers as well some important insights into the issue of which features of the communities converge and which do not, while the second provides an outstanding example of non-convergence.

5.5.1 *Pantropical comparisons of forest bird communities*

Tropical forests have a strikingly similar aspect wherever they occur around the world (Richards 1964; Walter 1973; Leigh 1975). Convergent features are numerous: multi-tiered structure punctuated by giant 50 m emergent trees, presence of numerous distinctive growth forms — palms, stranglers, epiphytes, lianes and trunk climbers, as well as arresting morphological adaptations — stilt roots, aerial roots and flaring buttresses. Given this remarkable uniformity in the physical and structural environment, it could be expected that communities of consumer organisms inhabiting tropical forests would show a high degree of convergence in their adaptations. Convergence could occur at the levels of morphology, diet, or behaviour, and in our present state of ignorance, it is not possible to go very far with *a priori* predictions about which of these types of convergence should be most favoured by evolution. At best, we can guess at a mere ranking of degrees of similarity.

Morphology should perhaps converge the least because accidents of history will determine the array of phyletic stocks available in each region. Phylogenetic constraints will then operate in different ways to channel or limit morphological adaptation. Priority of arrival may often affect the ability of taxa to invade a community or to radiate subsequently.

Diet (trophic composition of the community) is probably less dependent on history, because it depends more directly on the structure and productivity of the vegetation. If the energy flow through various tropical forests is more or less similar (Karr 1975; Leigh 1975), then because of averaging over the vast number of plant species that comprise such vegetation, the productivity of the several kinds of resources that are consumed by birds should also be similar. But there is evidence that history can enter at this level too. A prime example is the southeast Asian forests that are dominated by the plant family Dipterocarpaceae. Trees of this family are famous for their synchronous fruiting at intervals of several to many years (Janzen 1974). Because the fruits are available so intermittently, they do not constitute a significant resource for birds. Dipterocarp-dominated forests might thus contain a paucity of frugivores, as indeed Janzen (1974) claims.

The greatest degree of convergence might be expected when communities are analysed on the basis of the behaviour employed in feeding. Because of the striking structural uniformity of rainforests around the world we could expect a high degree of convergence in the spatial distribution of resources, which in turn should be reflected in the details of how the animals present go about obtaining their food. This is because (i) foraging techniques are adaptations more to the spatial distribution of resources than to the quality of the resources, and (ii) behaviour is much less constrained by phylogenetic history than is morphology. Birds with such disparate morphologies as a hummingbird, a nightjar, a woodpecker and a flycatcher may all on occasion sally forth from a perch to hawk a passing insect. Conversely, any species, within limits set by its fixed morphology, may be capable of several kinds of foraging manoeuvres. The common Yellow-rumped Warbler *Dendroica coronata* of North America, for example, may hop on the ground pecking prey from the surface, glean insects from foliage, hover at branch tips to pick berries, or hawk flying insects from the open air. It is this partial independence of behaviour from morphology that assures a good matching of behaviour to resources. Now, let us examine our proposition.

Convergence in morphology

Short of anecdotal examples in which animals have been selected from two communities for their morphological and sometimes ecological resemblances (e.g. Fig. 5.3), there have been surprisingly few efforts to examine the proposition of morphological convergence with objective statistical methods. The most serious attempt to do this is that of Karr and James (1975) who compared the bird communities of forested sites in Panama and Liberia. Although the communities did contain a number of pairs of species that could be recognized as one to one counterparts, it was quickly realized that an attempt to push this matching process very far would lead to failure, if only on the grounds that the two faunas contained disparate numbers of species. Thus, for a given species in Liberia, there might be two in the richer Panamanian community that filled its role. Clearly, it is naïve to expect a complete set of matching counterparts, especially when there is no guarantee that the resource base of the communities is identical.

Karr and James applied multivariate statistical methods to the problem, using standard morphological measurements such as bill dimensions, wing length, tarsus length, etc. Pairs of morphological indices in two-dimensional plots often effected complete segregation of species having distinct ecological roles within each community; e.g. frugivorous vs insectivorous hornbills, or sallying vs hover-gleaning flycatchers. Similar correlations between morphology and ecological roles have also been demonstrated by Keast (1972a) and Fitzpatrick (1980) in their studies of New World flycatchers (Tyrannidae).

Comparing the Panamanian and Liberian communities, Karr and James found that there was considerable congruence in the distribution of species into morphological 'cells', and that the patterns shown by the two tropical communities bore a closer resemblance than either did to a typical temperate community (Illinois, USA: Table 5.2). The tropical forests, for example, contained many more bird species that walk on the ground, searching the leaf litter for insects, fallen fruits, seeds, etc. (species in the left hand column of the table with low wing/tarsus ratios). Nevertheless, there were some major disparities, of which the exclusively New World hummingbirds provided the most outstanding example. Because of their exceptionally small size, tiny tarsi and long bills, hummingbirds had no morphological equivalents in the Liberian community, although they did have obvious ecological counterparts in the nectarivorous but non-hovering sunbirds.

From these very preliminary results, it is difficult to draw any broad conclusions. The tropical communities did pass the test for convergence in that the two sets of species showed more similarities than either did to

Table 5.2. Number of species per cell in morphological space in three forest bird communities (from Karr & James 1975).

		Log_{10} (wing/tarsus)			
Area	Tarsal length (mm)	0.300–0.475	0.475–0.650	0.650–0.825	0.825–1.000
Illinois, USA	0–15	–	1	1	–
	15–30	2	14	9	–
	30–45	–	4	1	–
	45–60	–	–	–	–
Liberia	0–15	–	2	2	1
	15–30	8	21	6	3
	30–45	1	2	2	1
	45–60	–	1	–	–
Panama	0–15	–	2	1	3
	15–30	7	24	10	2
	30–45	2	–	1	–
	45–60	–	1	1	–
	60–75	–	1	–	–
	75–90	1	–	–	–

the temperate community, but that is not saying a lot. A better perspective on the issue awaits the completion of research now in progress comparing the morphological structure of bird communities of many kinds on a global scale (e.g. Ricklefs & Travis 1980).

Convergence in diet

The production of resources that can be utilized by consumer organisms at any site ultimately depends on the vegetation. One might thus suppose, somewhat simplistically, that sites carrying similar vegetation would provide similar arrays of resources, and that the radiation of consumer taxa would reflect this underlying similarity. While these assumptions, really suppositions, have received some empirical support from a limited number of studies (e.g. Leigh 1975), far more data on resource production by natural environments is sorely needed.

If we look at the dietary specializations of understorey bird communities of tropical forests around the world, a number of consistencies are evident (Fig. 5.4; Karr 1980). Insectivores invariably outnumber frugivores, and among insectivores, those that glean foliage predominate over those that practise other capture techniques, such as sallying or pecking prey from bark. Omnivores are less prevalent than specialized insectivores, and are roughly equal in representation to strict frugivores.

When the guild diagrams are quantified in the form of a dendrogram, the relationships of the various communities are seen to be quite close (Fig. 5.5). Replicate samples from the same Panama site are only marginally more similar to each other than to two widely separated Costa Rican samples. The 'guild signatures' of the New World localities form a cluster at about the same level as the two African localities. Not sur-

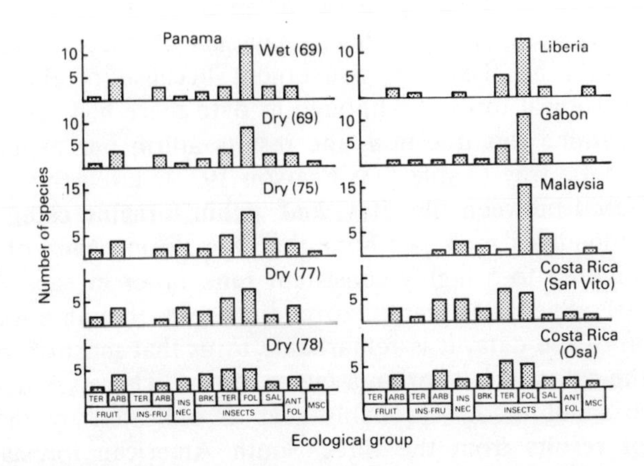

Fig. 5.4. Guild signatures (number of species in each ecological group) represented in 100-bird mist-net samples from the undergrowth of tropical forests (from Karr 1980).

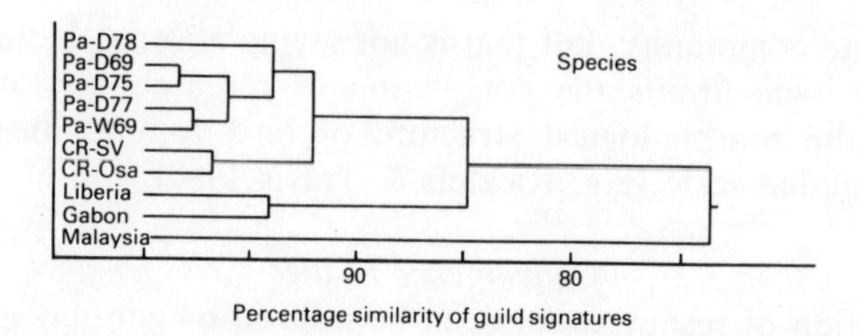

Fig. 5.5. Dendrograms for percentage similarity of guild signatures for 100-bird samples from lowland tropical forests. Pa = Limbo Hunt Club, Panama; CR-SV = San Vito, Costa Rica; CR-OSA = Osa Peninsula, Costa Rica; D78 = Dry Season 1978; W69 = Wet season 1969, etc. (from Karr 1980).

prisingly, larger differences separate the samples taken on different continents, though a minimum degree of similarity of about 73% reflects a strong convergent trend. Nonetheless, there are significant differences. The Old World communities lack the specialized ant-following birds that are found throughout the humid neotropics. Frugivores and bark gleaners are under-represented in the Old World forests relative to Central America, and the Malaysian sample was deficient in understorey frugivores and terrestrial insectivores. Even larger differences have been found in comparing the trophic (dietary) structure of tropical bat faunas around the world (McNab 1971a; Wilson 1973). Such contrasts can be attributed either to distinct patterns of resource production in separate geographical regions, or to historical effects, such as the prominence of army ants in the Neotropics, or the usurpation of trophic roles by distinct taxa in different regions. At present it is not easy to separate these possibilities.

Convergence in behaviour

Earlier we speculated that foraging behaviour, of all community characteristics, would be the most convergent because of the structural uniformity of tropical forests. Although to date there has been only one attempt to examine this question, the results affirm our expectation in quite an emphatic way (Table 5.3; Pearson 1977). Even though species numbers differed between the sites, and within foraging categories, the number of individuals seen per hour that were using each of the nine techniques adhered to a highly consistent rank order in each of the six sites ($P < 0.01$, Spearman rank correlation). In the absence of any normalization of the data, it is remarkable to us that most of the figures in many of the columns fall within a factor of two. There are differences, but they are small though probably real as indicated by the greater uniformity of results from the three South American forests than in intercontinental comparisons. Gleaning, for some reason, was more prevalent on the Old World plots, while trunk foraging was far more common in the New World sites, as Karr (1980) also observed. Lacking data on

Table 5.3. The rank order shown by the mean number of individuals (italics) recorded per hour of observation for each of nine foraging techniques used by birds, and the foraging technique diversity for each of six plots within the tropics ($H' = -\Sigma p_i \ln p_i$, where p_i = proportion of individuals in the *i*th foraging class) (from Pearson 1977).

	Glean	Sally	Snatch	Peck probe	Flower hover	Fruit	Ant follower	Glean/ Sally	Glean/ Snatch	Total no. ind.	H'
Ecuador	1	3	5	4	7	2	6	8	9	4874	1.48
	3.44	*0.71*	*0.32*	*0.63*	*0.06*	*1.31*	*0.29*	*0.04*	*0.01*		
Peru	1	4	5	3	9	2	8	5	7	2539	1.67
	2.15	*0.59*	*0.27*	*0.68*	*0.06*	*1.47*	*0.08*	*0.27*	*0.09*		
Bolivia	1	3	5	3	8	2	5	7	9	1827	1.62
	3.33	*0.70*	*0.36*	*0.70*	*0.15*	*2.79*	*0.36*	*0.35*	*0.04*		
Borneo	1	3	5	6	–	2	–	4	7	2919	1.30
	6.90	*1.27*	*0.48*	*0.35*		*1.73*		*0.75*	*0.20*		
New Guinea	1	4	5	8	7	2	–	3	6	2426	1.48
	4.83	*0.83*	*0.44*	*0.04*	*0.25*	*3.04*		*1.06*	*0.29*		
Gabon	1	2	4	6	9	3	7	5	8	3307	1.42
	5.12	*1.30*	*0.57*	*0.28*	*0.02*	*1.21*	*0.16*	0.38	*0.12*		

resource distributions in each of the forests, it is impossible to ascribe minor differences such as these to specific causes.

Convergence in tropical forest avifaunas: conclusions

We set out to rate the degree of convergence shown by tropical forest bird communities in morphology, diet and behaviour, but ran into obstacles in the course of examining the available data. Because of the large numbers of species and the difficulty of finding them in the recesses of 50 m tall forests, tropical bird communities are difficult and time consuming to study. To do it properly one should spend several months at each site, as Karr (1976) rightly emphasizes. As yet, no one has succeeded in compiling all three types of data for a series of sites on different continents in a way that would allow them to be compared in a uniform fashion, for example, by means of multivariate dendrograms such as that shown in Fig. 5.5. Nevertheless, from the results at hand, it is evident that tropical forest bird communities are highly convergent in all three community characteristics (morphology, diet and behaviour), though it cannot yet be said just how convergent.

There are at least two reasons for thinking that morphological comparisons will generally result in the lowest level of similarity. One is that the major biogeographic divisions of the tropics differ conspicuously in their inherent levels of species richness (Keast 1969, 1972b, 1972c; Amadon 1973), a fact that is reflected in the local (alpha) diversities recorded at single sites as well (Pearson 1977; Karr 1980). All the authors cited above express agreement that the differences are due to historical factors such as the varying amounts of tropical forest habitat extant on the continents, differential effects of Pleistocene wet–dry cycles, presence of more or fewer dispersal barriers that could affect speciation rates, etc. Given significant differences in species numbers in the communities being compared, there will be severe constraints on the evolution of exact ecological counterparts, all other things being equal. Thus, morphological convergence is limited by at least one constraint that does not operate at the levels of diet and behaviour.

Another reason why morphology may fail to converge is that different phyletic stocks may combine roles in distinct ways. A good example is found in the nectar feeding birds of three regions: tropical America, Africa and Australasia. In America there has been a spectacular radiation of hummingbirds (319 species) which are distinctive in their diminutive size (< 2.5 g) and in their ability to extract nectar from flowers while hovering. In Africa, the sunbirds (104 species) perform the same operation, but at a much larger average body size and while perched. In Australasia nectar feeding is dominated by the family Meliphagidae (honeyeaters, 160 species). Unlike hummingbirds and sunbirds, which feed only on nectar and insects, the meliphagids are extremely versatile in their feeding habits, commonly taking fruit as well as insects in addition to nectar (Keast 1968; Terborgh & Diamond 1970). Accordingly, the honeyeaters have evolved a much wider range of sizes, the largest species weighing over 100 g.

Because of the low protein content of nectar, and its poor reliability as a resource over the seasons, nectar feeders must migrate or supplement their diets with other resources (Feinsinger 1976). This fact, plus the 'accidental' discovery of hovering flight by the New World hummingbirds prevents any good morphological correspondence between the world's three major groups of nectar feeding birds. Strong contrasts in mean body size between New and Old World fruit and nectar bats provide another unexplained example of non-convergence (McNab 1971a). The lesson to be derived from these examples is that species will be subject to divergent morphological evolution where they are selecting resources in different combinations. At the dietary and behavioural levels, however, the effects of differing morphologies may be scarcely discernible.

5.5.2 Subtropical lizard communities

Encouraged by our success in finding evidence for convergence in tropical forest bird communities, we might now anticipate that other groups of organisms in other ecosystems would be similarly well behaved. The best tests of this are contained in the work of Cody (1975) on the bird communities of Mediterranean scrub vegetation, which show pronounced indications of convergence, and in the work of Pianka on desert lizards. We shall focus now on lizard communities because these have not converged to nearly the same extent as bird communities.

Pianka studied three desert regions in the subtropical belt: the southwestern USA, the Kalahari of southern Africa and the Great Victoria Desert of Western Australia. In order to include the full range of variation of habitat types, he surveyed 8–10 localities in each region (Pianka 1967, 1969a, 1971). There was no conscious attempt to match sites between continents, but the ranges of climatic variables (annual temperature, rainfall) for the three regions were broadly overlapping, and in some instances there were close resemblances in the physiognomy of the generally shrubby vegetation.

Notwithstanding the similarity of the environments, Pianka found great differences in the size and organization of the lizard communities on the three continents (Table 5.4). In North America the average community contained about eight species, roughly half as many as in Africa (15), and only about a quarter as many as in Australia (28). The desert sandhill habitat in Australia was remarkable in containing 40 species, including 11 in one genus of skinks, *Ctenotus* (Pianka 1969b). This was more than

Table 5.4. Number of lizard species in each of five guilds in three desert regions. The numbers represent averages for 8–10 study sites in each region (from Pianka 1973).

Guild	North America no.	(%)	Kalahari no.	(%)	Australia no.	(%)
Diurnal, terrestrial	5.7	69	6.3	43	14.4	51
Diurnal, arboreal	1.2	14	1.9	13	2.6	9
Nocturnal, terrestrial	1.4	17	3.5	24	7.6	27
Nocturnal, arboreal	0.0	0	1.6	11	2.6	9
Fossorial	0.0	0	1.4	10	1.1	4
Totals	8.3	100	14.8	101	28.3	100

twice the number of species found at any site in Africa or North America.

In addition to these purely numerical differences in species richness, lizard communities in the three regions differed in a number of other respects. Nocturnal arboreal and fossorial (burrowing) species were missing altogether in North America, but 3–4 species per site in these categories was the rule in the other regions. Specialized ant feeders occur in the North American and Australian deserts, while termite specialists are included in the African and Australian faunas. There is an herbivorous species (*Dipsosaurus*) in North America, but not in the other two regions, while North America and Australia but not Africa possess carnivorous forms.

As for ecological equivalents, Pianka (1969a) compared two structurally nearly identical sites in Australia and North America and concluded that each of the five species in the North American site could be closely matched up with one or two Australian counterparts, accounting for a total of seven species on the Australian site. He then continued to say, "if these seven species are then removed from consideration, of the remaining eleven [on the Australian site], one is ecologically a mammal (*Varanus gouldii*), one is an 'insect' (*Ablepharus greyi*), one is a 'worm' (*Ablepharus timidus*), and the other eight species are nocturnal". The conclusion is inescapable that Australian lizards have radiated into ecological roles that either do not exist in the North American deserts, or are filled there by other taxa.

Why are there so many differences between the lizard faunas of the three desert regions? The answer can only be an educated guess, for we had no way of knowing what to expect in the first place. As we might imagine, Pianka attributes most of the discrepancies to historical factors: greater spatial heterogeneity of the Australian desert, differences in the degree of isolation of the land masses, initial presence of distinct phyletic stocks, usurpation of trophic roles by other taxa (birds, mammals, snakes, etc.) to a greater degree in some areas than in others, etc.

Is there any way that these historical factors could have been identified in advance, and their effects anticipated? None that we know of. Nor did Pianka (1966b), who expressly felt after his first experiences in North America that the desert environment was structurally complex enough to provide niches for a maximum of only 10 species. Imagine his astonishment in finding 40 species together in Australia!

Historical factors are perhaps the most elusive element in ecology. We know they are there, but we have not yet been able to get a good hold on them. Clearly, there is a long way to go before the problem of convergence acquires the respectability of a predictive science.

5.5.3 Dissimilar communities: a tropical–temperate comparison

A great deal of effort has been devoted to comparing communities that differ in climate or vegetation, or both. The purpose behind such comparisons is to assess the effects of habitat structure on community characteristics, such as species diversity, biomass, number and size of guilds present, etc. Although the number of variables involved is potentially great, it is usually assumed (or taken on blind faith) that the effects of habitat structure predominate in determining the results (over, say, differences in productivity, types of resources available or microclimate).

Studies have focused on three types of situations involving varying degrees of proximity of the sites being compared. The sequence of vegetational stages in plant succession is one. Examples of each stage (open field, early second growth, etc.) may often occur within a few kilometres, so that the macroclimate is at least held constant. MacArthur's work on the dependence of bird species diversity on foliage height diversity stimulated intense interest in the ways in which habitat structure influences community organization, and is still the classic example of this kind of study (MacArthur & MacArthur 1961).

Environmental gradients offer another opportunity for informative comparisons, but here conditions are not so well controlled because both climatic and habitat variables generally vary over the length of a gradient. Guild analysis has been little used in conjunction with investigations of environmental gradients because the main interest has usually been in the limits of distribution of individual species and in overall changes in diversity (e.g. Whittaker 1967; Terborgh 1971; Brown 1975; Cody 1975).

The third type of comparison is the most adventuresome from the point of view of uncontrolled variables, because the focus is on widely separated localities with conspicuously different characteristics. Most comparisons of this type have been prompted by a continuing fascination with tropical diversity, and the still incompletely resolved question of why the tropics harbour so many more species than the temperate regions (Pianka 1966a). The only practical way this question can be attacked is by comparing selected tropical and temperate communities, even if the risks entailed in interpreting the results are very great. Propelled by the nearly irresistible lure of this issue, a number of investigators have undertaken such comparisons, and the extraordinary wealth of their suggested explanations is in itself eloquent testimony to the hazard of uncontrolled variables. Here are some of the suggestions: more finely divided niches in tropical communities (Klopfer & MacArthur 1960), greater niche overlap (Klopfer & MacArthur 1961), greater structural complexity of the habitat (MacArthur *et al* 1966), greater climatic stability and evolutionary age of habitats (Sanders 1968; Stiles 1978), greater climatic instability leading to accelerated speciation rates (Haffer 1969, 1974), greater species turnover between habitats (MacArthur 1969), presence of guilds that are poorly represented or missing in temperate communities (Orians 1969; Karr 1971), broader resource spectra (Schoener 1971), greater equilibrium species number due to higher speciation rates and/or lower extinction rates (MacArthur 1969; Diamond 1973), greater area per habitat causing higher speciation and lower extinction rates (Terborgh 1973).

Because there are a large number of competing hypotheses, it does not follow automatically that all but one of them are wrong; indeed, it is possible that several or many of them are substantially correct, if only we concede that the tropical–temperate contrast is due to a number of causes acting simultaneously (Terborgh 1977). A major step toward simplifying the problem can be made by realizing that all the above proposals can be classified into two major categories, those concerning differences in ecology (structure of the habitat, availability of resources, etc.) and those concerning differences in evolutionary history (climatic stability, age and area of habitats, speciation and extinction rates, etc.). By focusing on this major dichotomy in the hierarchy of causes, ecology vs evolution, it has proven possible to evaluate, albeit somewhat tentatively, the relative magnitudes of the respective contributions of ecological and evolutionary factors to the diversity of tropical bird communities (Terborgh 1980a).

Taking a temperate forest bird community as the frame of reference, one can ask about the numbers of 'extra' species in various guilds in a corresponding tropical community. The temperate locality selected for this comparison is the most extensive (5000 ha) remaining virgin lowland forest in the eastern United States. It is near Columbia, South Carolina, at a latitude of 33°N, and is now protected as the Congaree National Monument. The forest has a multi-tiered structure and contains more than 35 species of trees, some of which reach heights of nearly 50 m. Tropical plant forms include palmettos, bamboo, numerous climbing vines and hanging lianes, and a modest representation of vascular epiphytes. Indeed, in its great stature and structural complexity, this forest more closely resembles a tropical forest than any other that could be found in temperate North America. Its tropical counterpart in this comparison is an undisturbed Amazonian forest in eastern Peru at 9°S latitude.

How are we to distinguish evolutionary from ecological influences on species diversity? We assume, without avoiding all ambiguities, that resources, and the behaviour that birds use to obtain them, will be reflected in the guild structure of the communities. This structure will be largely, though not entirely, independent of the number of species. The accuracy of the results depends on this independence, even if it is only an approximation. In any case, the guilds define the ecological component of the comparison. The evolutionary component is then represented by the level of species packing within each guild, though to some unknown degree this is an approximation too.

Before proceeding to apply these assumptions to the case at hand, we must consider how to control for the size dimension of each guild niche. A broad spectrum of available resources (very small to very large insects, for example) would provide adaptive opportunities for insectivorous birds of many sizes, while a narrow resource spectrum would not. Indeed, it is well known that there are more large insects in the tropics (Schoener & Janzen 1968; Janzen 1973a). To take into account differences in guild dimensions, we apply one of the most revered, yet least understood, empirical rules of ecology. It states that ecologically related coexisting species, whose functional difference is one of size alone, will differ in length by a factor of 1.2–1.3, or in weight by a factor of roughly 2 (Hutchinson 1959; Schoener 1965; Diamond 1973; Horn & May 1977). Following this rule, this dimension of the guild niche is defined as $\log_2(W_l/W_s)$, where W_l and W_s are the weights of the largest and smallest guild members, respectively. The figure that results is the number of doublings in size that separates the smallest and largest guild members. It will be used to represent the size dimension of the niche space being exploited by the guild. The density of species packing within the guild is now easily obtained as $(S - 1)/\log_2(W_l/W_s)$, where S is the number of species in the guild. More intuitively, the formula gives us the number of guild members per doubling in size within the guild niche.

We are now ready to examine the two bird communities. The temperate forest had 40 breeding species, while the Amazonian site had 207. (See Terborgh 1980a for further details.) These are assigned to a series of trophic–behavioural guilds in Table 5.5.

The tropical community contains some 24 guilds, eight of which (italicized) are not represented in the temperate community. Most of the tropical guilds have five or more species, while most of the temperate guilds have only one or two. The major exception is the foliage gleaning insectivore guild which has 15 members (38%) in the temperate community. Conspicuously lacking or under-represented in the temperate community are frugivores, omnivores, and certain groups of behaviourally

Table 5.5. Guild structure, guild niche dimension and species packing in the bird communities of a tropical and temperate forest (from Terborgh 1980a).

Guild	Tropical forest				Temperate forest			
	No. species	Weight (g) min – max[1]	Guild niche dimension[2]	Packing[3]	No. species	Weight (g) min. – max.	Guild niche dimension	Packing
Carrion	2	1600–3125	1.0	1.0	1	2100	–	–
Raptor								
general (mammals, etc.)	7	380–4250	3.5	1.7	1	1020	–	–
bird	4	100–540	2.4	1.2	1	470	–	–
other (snails, reptiles, insects)	7	160–610	1.9	3.1	1	280	–	–
Owl	5	65–760	3.5	1.1	2	165–780	2.2	0.4
Nightjar[4]	1	40	–	–				
Mast eater								
terrestrial	5	210–1010	2.3	1.8	2	142–6100	5.42	0.2
arboreal	8	170–1150	2.8	2.5				
Frugivore								
terrestrial	3	100–3400	5.1	0.4				
arboreal	18	9–1500	7.4	2.3	1	130	–	–
Nectarivore	8	3–8.5	1.5	4.7	1	4	–	–

Insectivore								
terrestrial	10	11–55	2.3	3.9	2	20–49	1.3	0.8
woodpecker	8	10–220	4.5	1.6	5	25–260	3.4	1.2
bark gleaning	9	13–135	3.4	2.4	1	22	–	–
foliage gleaning	19	9–105	3.5	5.1	15	6–75	3.6	3.8
sallying	27	4.5–108	4.6	5.7	3	12–34	1.5	1.3
aerial	4	10–50	2.3	1.3	1	25	–	–
ant following	6	19–70	1.9	2.7				
dead leaf gleaning	7	10–63	2.7	2.3				
vine gleaning	7	7–85	3.6	1.7				
Frugivore/predator	6	130–540	2.1	2.4	2	87–350	2.0	0.5
Frugivore/insectivore								
arboreal, gleaning	12	14–52	1.9	5.8	1	44	–	–
arboreal/sallying	13	11–85	3.0	4.1				
Frugivore/insectivore/ nectarivore	11	11–360	5.0	2.0				
Total	207				40			

[1]Weight of the smallest and largest guild members.
[2]Computed as $\log_2(W_l/W_s)$, where min. and max. refer, respectively, to the weights of the smallest and largest guild members.
[3]Species packing computed as (the number of species in the guild − 1) ÷ (the included niche). See text for details.
[4]Italicized guilds are found in the tropical forest but are evidently lacking in the temperate forest.

specialized insectivores: ant followers, vine and dead leaf gleaners. These add up to a minimum of 56 species in the tropical community that have no temperate counterparts; they depend on resources that are not produced by the temperate forest (at least during the breeding season), or on a peculiarly tropical foraging opportunity (army ants), or on foraging substrates that are much less prevalent in the temperate forest (vines, dead leaves). From this it is obvious that the tropical forest offers additional ecological roles, which of course must ultimately be filled by evolution through adaptive radiation.

Turning now to the guilds that are represented in both communities, we note that a majority of the temperate guilds (9 out of 16) have only one species, while this is true of but a single tropical guild. It is reassuring to observe that in each of the temperate singleton guilds, the weight of the lone species falls squarely within the range circumscribed by its tropical counterparts. This can be taken as indirect evidence that the model values of the underlying resource spectra are about the same in both communities, though there may still be disparities in the breadths of the resource spectra. It is possible to investigate this point with the seven temperate guilds that include two or more members. In three cases the temperate niches are larger (greater number of doublings in size separating the smallest and largest species), in two cases the tropical niches are larger, and in two cases the niches are similar in dimensions. This suggests that resource spectra may not be systematically broader in the tropical forest.

The greater numbers of species in the tropical guilds could be a result of ecological conditions; e.g. broader resource spectra, or of evolutionary processes (tighter species packing). These two possibilities are separated in Table 5.6. Controlling for the different dimensionality of many tropical and temperate guild niches, we obtain the estimate that roughly half (49%) of the 'extra' tropical species are attributable to increased species packing, about a third (34%) to additional guilds and the remainder (17%) to wider tropical guild niches.

Lest anyone take these numbers too seriously, it is emphasized that their primary value is heuristic in pointing out that tropical diversity has more than one cause, and that both ecological and evolutionary factors are involved. The procedure followed is flawed with ambiguities. If a greater or smaller number of tropical guilds had been recognized, for example, the number of guilds not represented in the temperate community might have been altered. In a few cases non-representation could have been due to sampling (e.g. absence of nightjar in the temperate community). Even the distinction between the ecological and evolutionary contributions to the excess tropical diversity is not precise. Recognizing more guilds would have meant reducing packing levels (evolution) and transferring the species to more new guilds (ecology).

Table 5.6. Apportioning of excess tropical species (from Terborgh 1980a).

Guild	No. additional tropical species	Presence due to: Guild missing in temperate forest	Larger tropical guild niche[1]	Increased species packing[2]
Carrion	1		1	0
Raptor				
general	6		3	3
bird	3		2	1
other	6		2	4
Owl	3		0	3
Mast eater				
terrestrial	3		0	3
Frugivore				
arboreal	17		7	10
Nectarivore	7		1	6
Insectivore				
terrestrial	8		1	7
woodpecker	3		1	2
bark gleaning	8		3	5
foliage gleaning	4		0	4
sallying	24		4	20
aerial	3		2	1
Frugivore/predator	4		0	4
Frugivore/insectivore				
arboreal, gleaning	11		2	9
Guild missing	56	56		
Total	167	56	29	82
Excess (%)		34%	17%	49%

[1](Tropical included niche) × (temperate species packing in that guild) = (no. of species that would be in the tropical guild at the temperate packing value − 1). (This + 1) − (no. of species in the temperate guild) = (extra species attributable to a broader tropical guild niche). Packing in temperate singleton guilds was assumed 1.0.
[2](Total of extra species in the tropical guild) − (no. of extra species due to broader tropical niche) = (extra species due to increased tropical packing).

The larger tropical guild niches in some cases may result from increased species packing (evolution) rather than from broader underlying resource spectra (ecology). The results are not to be taken literally, but neither are they to be regarded as entirely frivolous. There is some truth in this, even if we perceive it only dimly.

5.6 Epilogue

Most of the work discussed in this chapter represents beginnings, often the barest beginnings, in a quest for solutions to some major ecological problems.

On the matter of the convergence of ecosystems, we are groping for the right questions to ask, and for ways to evaluate whether or not convergence has occurred, and if so, whether it is more or less than could be 'expected' under the circumstances. More panglobal comparisons must be made before natural habitats disappear in certain regions of the globe and foreclose the possibility forever. In the future, investigators making these comparisons must look at morphological features, trophic structures and feeding behaviour simultaneously ideally coupled with measurements of available resources over the annual cycle. This will require a large coordinated effort.

Many uncertainties remain in understanding the contrasting diversities of tropical and temperate ecosystems. We have looked here at only one comparison — forest birds. Flemming (1973) has compared forest mammals with quite different results. No doubt comparisons of other taxa — reptiles, trees, insects, etc. — or comparisons of habitats other than rainforest, e.g. grasslands, deciduous forest or alpine tundra, would provide further interesting and perhaps unexpected contrasts. Much of what we know to date comes from studies of birds, yet we have no real assurance that the patterns shown by birds can be generalized to other groups.

A pervasive theme throughout the chapter has been the will-o-the-wisp of 'historical factors.' We know that these are many and diverse in their nature, and that they lurk in the background of every geographical comparison, but as yet we have made only the most tentative start in learning how to separate them from more immediate ecological factors. This remains a formidable challenge. We can anticipate that the challenge will be met in part through further advances in defining and enumerating guilds. For maximum utility, guilds must correspond to natural ecological subdivisions of communities. Future guild definitions, for greater reality, will often have to transcend taxonomic boundaries. Such refinements in the guild concept will lead not only to a better understanding of the internal organization of communities, but also to an appreciation of the environmental constraints that operate within communities to promote evolutionary convergence. Guilds will become the standard currency of ecologists in their efforts to understand community relationships of many kinds.

6 Resource Partitioning

Thomas W. Schoener

6.1 Introduction

Apart from their revolutionary theoretical contributions, the nineteenth century explorer-naturalists — Darwin and Wallace among others — contributed by their diligent collecting much information about the diversity of species. For example, William Bates, best known for his ideas on mimicry, found 700 species of butterflies near Pará (now Belém), Brazil, and in all collected 14 712 species of animals on the Amazon River, of which 8000 were new to science (Usinger 1962). Such numbers of species were almost beyond comprehension to biologists, whose experience then was mostly confined to temperate systems. At the time, and indeed for nearly a hundred years thereafter, the main conceptual issue stimulated by this new information was the origin of such diversity: what process could generate all those species?

In the 1940s, emphasis shifted, at first gradually and then rapidly, away from the question of how species originate to the question of how so many can coexist. This ecological, as opposed to evolutionary, issue is sometimes referred to as the 'maintenance of diversity' problem. Why did coexistence become an issue? Perhaps most responsible were the experiments and ideas of Gause (1934). He found that two very similar species of micro-organisms, when placed together in a variety of simple laboratory settings, did not both persist. Rather, one species outcompeted the other, leading to the loser's extinction. Gause was able in part to interpret his experiments using the mathematical models of Volterra (1926) and Lotka (1932), which predicted competitive exclusion under many circumstances. This work suggested that generally similar species might not be able to coexist in nature without at least one major ecological difference, such as where they obtain most of their food. Indeed, in part stimulated by Gause's work, many biologists began to document what they suspected were the crucial ways in which coexisting species differed in nature.

Such studies, although not so labelled at first, are now called studies of resource partitioning, or of the ways in which species differ in their resource use. The extent to which resource-partitioning patterns in fact result from pressures, evolutionary or otherwise, to avoid interspecific competition is now more of an issue than when most of the studies were carried out. The bulk of this chapter evaluates various aspects of the hypothesis that competition structures resource-partitioning patterns. We shall also consider alternatives such as predator avoidance in Section 6.8, and ask how such explanations can be distinguished from that of competition. Most of the data now at hand are observational, i.e. deal with patterns in undisturbed natural systems, and the bulk of this chapter concerns such data (Sections 6.4–6.6). However, a large number of field experiments on interspecific competition have recently been performed, and we shall see how they elucidate the dynamics of shifting resource use (Section 6.7). We begin with some definitions and methodology (Sections 6.2–6.3).

6.2 The niche as a utilization distribution

Before analysing patterns of resource partitioning, we must briefly review the language currently in vogue, that of the ecological niche. [See Haefner (1981) for rival language.]

As summarized by Pianka (1981), the ecological niche has been defined in various ways. The precursor to the definition used in this chapter is Hutchinson's 'multidimensional hypervolume'. Hutchinson (1957) postulated that a species population can be characterized by the range of some environmental variable (e.g. temperature, humidity) within which it can persist, or more precisely, where its rate of increase r is positive. Several such ranges, taken jointly, form a multidimensional region or so-called hypervolume within which r is positive. For example, assuming no interaction between dimensions, two dimensions form a rectangle and three form a rectangular box. Maguire (1973) has actually measured, in terms of population-growth parameters, the Hutchinsonian niche for several species of lower organisms, though such studies are rare.

Now we most commonly define the niche as a utilization distribution. The utilization can be envisioned as a frequency histogram, as follows. Compute p_{ih}, the fraction of the total use by Species i's population for each category along some resource dimension indexed by h. For example, the p_{ih}'s might represent the fraction of a population's total diet in each of a set of food-size classes; e.g. 0–1 mm, 1–2 mm, and so on. The p_{ih}'s can be plotted as a frequency histogram (Fig. 6.1a); by definition this is the utilization distribution. Often the histogram is represented as smoothed (Fig. 6.1b).

Just as for Hutchinson's hypervolume, the utilization can be multidimensional. Figure 6.1(c) gives an example for two dimensions. We now have $p_{ih_1h_2}$, indicating that each category is jointly defined in two dimensions, say by a prey-size and height interval. An example of a particular $p_{ih_1h_2}$ would be prey between 3 and 4 mm occurring at a height of 5–6 m. Though topology does not allow additional dimensions to be pictured easily, we can generalize symbolically to as many dimensions as

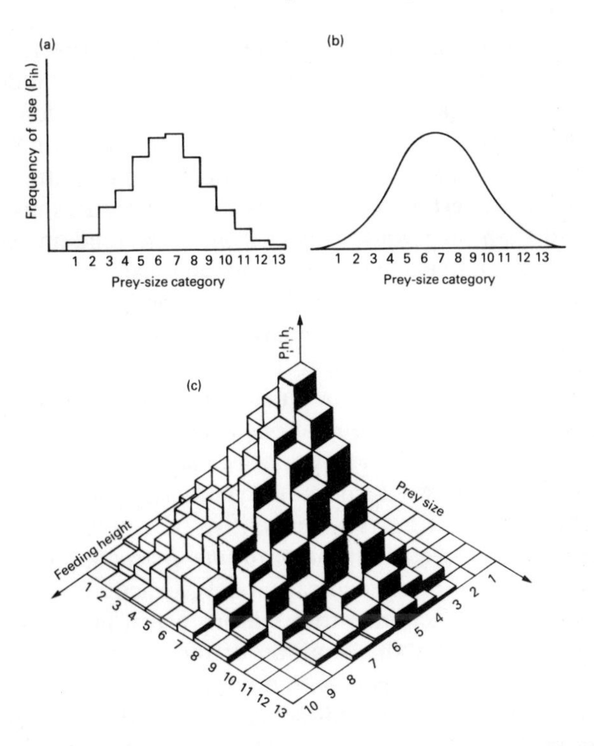

Fig. 6.1. (a) The utilization for Species i as a frequency histogram. This utilization is one-dimensional, where the dimension is prey size. Numbers refer to prey-size categories, indexed by h. (b) The same utilization smoothed. (c) A utilization for two resource dimensions, prey size and feeding height.

we wish, i.e. $p_{ih_1h_2 \ldots h_n}$: each p is jointly defined by as many intervals as there are dimensions.

6.3 Methodology

Historically, we can identify three approaches to the study of resource partitioning.

The *qualitative* approach, which characterized most earlier work, attempts to specify the outcome of competitive processes from biological intuition and common sense. Reasoning directly from Gause's results, it predicts that co-occurring species should differ 'substantially' from one another in at least one major ecological characteristic. In terms of evolution, species whose geographic ranges come to overlap should diverge 'substantially'; this may be reflected in some morphological trait, called an indicator, as when species diverge in prey size and concomitantly in the size of their trophic apparatus.

A difficulty with the foregoing approach is that species having no interactions (e.g. not in competition) or evolving entirely independently of one another are expected to be different to some extent (Schoener 1974a). In other words, randomly contrived communities will have species that show differences. The second, *null-model* approach, attempts to deal with this problem by testing whether observed differences are greater or more patterned than one would expect from some random model. Such tests can involve relatively complex constructs called neutral (Caswell 1976) or null models. Null models make specific assumptions about where the randomness occurs (there are many possibilities) and produce null hypotheses against which data are to be tested.

A second difficulty with the qualitative approach is the use of imprecise qualifiers (placed in quotations above); for example, just how much is 'substantially'? Conceivably species may differ from one another more than enough to falsify some null hypothesis, yet still not differ as much as might be expected given competitive pressures and the width of the available resource spectrum. The third, *theoretical* approach, attempts to specify, by means of mathematical models, exactly how great species differences should be. In addition to producing actual numbers, theoretical models attempt to confirm the intuition of the qualitative approach; as we shall see, confirmation is usually, but not always, achieved. Curiously, the null-model and theoretical approaches have operated largely independently of one another; the assumptions of null models often are guided by what statistical procedures are available rather than theoretical models of biological processes, and this has caused some recent criticism (Roughgarden 1983). Moreover, the 'logical primacy' (*sensu* Strong *et al* 1979) of a random or null hypothesis has been challenged on several grounds (Roughgarden 1983; Quinn & Dunham 1983); in epidemiology and pest control, the alternative hypothesis that something nonrandom is going on typically has economic primacy (Toft & Shea 1983). On the other hand, the null-model practitioners (e.g. Simberloff & Boecklen 1981) argue that biologists have often asserted that differences between species were significant without any kind of statistical test.

Presently, because the three approaches are largely incarnated as different investigators, the field seems tumultuous, at times almost chaotic. Things will doubtless stabilize as the best elements of each approach — biological intuition, statistical rigor and theoretical legitimacy — are fused into one. But in considering what follows, the reader should keep in mind that in some cases the approaches are presently at odds.

6.4 The natural history of resource partitioning

The kinds of resource-partitioning differences between species, or kinds of dimensions, can be grouped under the three general headings of habitat, food type and time (Pianka 1969b). Habitat differences are differences in space. These include relatively coarse-scaled or *macrohabitat* differences (e.g. vegetation zone) and relatively fine-scaled or *microhabitat* differences (e.g. perch diameter); of course, the two terms define the ends of a continuum, and the boundary must vary between kinds of organisms depending upon how wide ranging individuals typically are. Food-type differences involve food size, food hardness, and food taxon, among others. Temporal differences can be either diel (i.e. activity time) or seasonal (i.e. phenology).

Resource partitioning is usually studied between sympatric members of a group of species having some general similarity in their trophic roles; such groups are called 'guilds' (Root 1967). A host of examples of resource partitioning exist, mostly from animals (but see below). Schoener (1974a) found 81 studies through 1973 involving three or more animal species ranging from protozoans to lions. This list must be much longer by now: though in 1981 Toft (1985) found 76 such studies involving reptiles and amphibians alone. Hence only a few examples can be given here.

Perhaps the most historically important cases involve Darwin's finches of the Galápagos islands; for example, three species of *Geospiza* differing greatly in bill size eat foods of differing hardness and size (Lack 1947; Bowman 1961). A similar case is reported for Christmas Island terns (Ashmole 1968), where food size is most closely correlated with bill thickness (Fig. 6.2). Free-living flatworms (triclads) partition resources by prey type and, to a lesser extent, depth (Reynoldson & Davies 1970). Similarly, the colourful cone-shelled *Conus* molluscs, which poison and eat large prey (including fish!), differ in subtidal areas mainly in prey type and to a lesser extent microhabitat (Kohn & Nybakken 1975); prey size may also be important (Leviten 1978). Poison-arrow frogs (dendrobatids) of Peruvian rainforests separate into an ant-eating and non-ant-eating guild, within each of which prey size is partitioned (Toft 1980). Wandering spiders separate mainly by seasonal activity time, though size and habitat are also important (Uetz 1977). Ontario lake fishes separate in many ways: prey taxon, prey size, water depth, distance from shore, and both daily and seasonal activity times (Keast 1970a). Certain bumblebees have tongues of different lengths which are adapted for corollas of different lengths (Heinrich 1976; Inouye 1978); flowers may also compete and 'resource-partition' pollinators, and the two groups coevolve specializations (e.g. Pleasants 1980). In Caribbean reef corals, the degree of heterotrophy vs autotrophy is different for each species (Porter 1976). In the most diverse communities of *Anolis* lizards, food size, structural habitat (perch height and diameter) and climatic habitat are of equal importance (review in Schoener 1977).

In the above-mentioned survey of resource-partitioning in animals

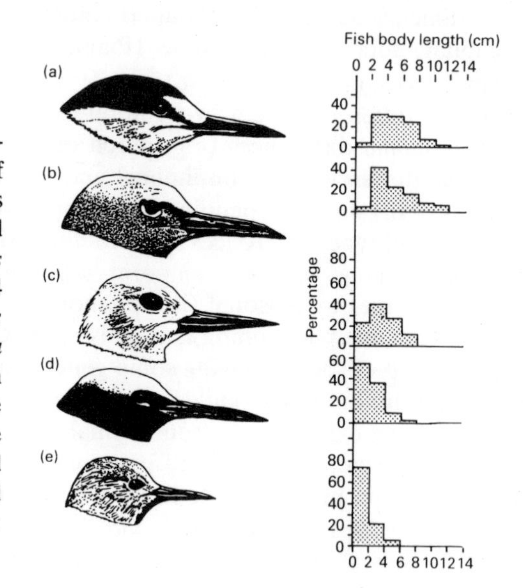

Fig. 6.2. Length–frequency distributions for fish in the diets of five species of terns on Christmas Island. (a) *Sterna fuscata* (based on 803 items). (b) *Anous stolidus* (139 items). (c) *Gygis alba* (224 items). (d) *Anous tenuirostris* (1911 items). (e) *Procelsterna cerulea* (702 items). The diagram includes both items that could be measured accurately, and those that were partly digested and whose lengths were estimated (from Ashmole 1968 & Lack 1971).

(Schoener 1974a), habitat was the most common mode of partitioning, followed by food type. Time was a distant third.

The predominance of habitat over food type can be explained in several ways. The first argument, enshrined in the 'compression hypothesis' (MacArthur & Wilson 1967; Schoener 1974b), is from optimal foraging theory. When species first begin to overlap spatially, the range of food types eaten by each should expand or remain the same and the species thereby should become more similar in diet; once found, a food item worth eating in the absence of competition is worth eating in its presence. However, species should specialize in the kinds of habitat patches visited, as each would deplete the food in its preferred patch type to the point at which other species would not find it profitable to forage there. The compression hypothesis does not apply to evolutionary time, however, as then species can evolve morphological specializations enabling them to feed optimally on different food types. Secondly, macrohabitat (e.g. vegetation type) is almost infinitely partitionable: an individual's home range can easily fit within a particular macrohabitat, but it cannot specialize too much on prey types without losing prohibitively much time and energy getting from one acceptable prey to another (e.g. Schoener 1974a). For example, *Tamiasciurus* squirrels cannot afford to specialize on food type because they transport food long distances to their caches, and such transport is expensive in energy and time (Smith 1981). Thirdly, larger animals should often find it efficient to eat smaller as well as larger food items; this will produce nested prey-size utilizations, rather than separate and adjacent ones (Schoener 1969a; Wilson 1975).

Temporal partitioning is, in theory, even less likely. No energetic gain can be derived from not feeding during most time periods, and temporal specialization should occur only if the risk of predation is large relative to the need for food energy. Even then, all species may specialize on the same time period.

Curiously, in a survey of three desert lizard faunas (North American, Kalahari and Australian), Pianka (1973) found that whether habitat, food type and time are important dimensions varies with the continent. Average microhabitat overlap is low everywhere, whereas food type overlap is only low in North America and Australia, and temporal overlap is only low in Australia. Interestingly, these overall relative importances are about the same as found for animals as a whole. Moreover, overall niche overlap is lower, the more diverse the fauna (see Section 6.5.2).

Other trends in resource partitioning have been found (Schoener 1974a).

First, predators separate more by daily activity time than do other trophic groups. Possibly this is because prey species are active at different times; therefore, temporally separated predator species can deplete different resource populations. The same is not true of the food of herbivores; a leaf available during the day is also available at night. Recently, Huey and Pianka (1983) examined several guilds to see if a tendency to be active at the same diel time implied a tendency to concentrate on the same prey taxa. They were able to show this for desert lizards, though the trend was not overwhelming. However, no significant relation held for raptorial birds or for snakes. Schoener (1983a) has given some reasons why in theory the relation might not be expected to be strong.

Secondly, terrestrial ectotherms (e.g. lizards) relatively more often separate by daily activity time. Such animals are particularly sensitive to changes in climatic factors, and such changes are particularly marked in terrestrial, as opposed to aquatic, environments.

Thirdly, lower animals separate more by seasonal activity than do vertebrates. Species whose generation times are long, unless they practise dormancy, cannot separate by the time of year at which they are active. A conspicuous exception to this trend is amphibian larvae, which separate most commonly by phenology (Toft 1985). But here is the exception that proves the rule: the juvenile stage of amphibians is usually short compared with the adult stage.

Fourthly, animals feeding on prey items large in relation to their own size tend to segregate by prey type; snakes are an excellent example (reviews in Schoener 1977; Toft 1985). One possible explanation is that a necessary condition for specialization is that a relatively small number of prey items is required per unit time.

Resource partitioning in autotrophic plants is on average less clearcut than in most animals. Such species often require the same resources — light, carbon dioxide, nutrients and water — and extreme specialization is generally impossible (Harper 1969; Werner 1979). However, plant species do differ in some of the same dimensions that separate animal species. In a group of old-field annuals, for example, Parrish and Bazzaz (1976) found both seasonal differences and partitioning by rooting depth in soil. The former type of difference is especially common among plants, and other types of habitat differences, especially macrohabitat, also occur (review in Werner 1979). Moreover, some plants (e.g. certain algae), and perhaps many plants, vary in their population-growth response to different *combinations* of the same nutrients, and this can provide a basis for coexistence (e.g. Titman 1976).

6.5 Between-species patterns: observations and tests

6.5.1 *Regular spacing along a single dimension*

In 1959 Hutchinson listed seven groups of coexisting species, ranked according to size of trophic apparatus, in which the ratio of the larger to smaller member of adjacent pairs was *c.* 1.3. Although Hutchinson's original list was small, evidence from certain other guilds (e.g. Diamond 1975; studies cited in May 1973) was soon gathered that seemed to support his generalization. Because the size of the trophic apparatus is generally strongly correlated with mean food size [e.g. in birds (Hespenheide 1973, but see Rotenberry 1980), lizards (Pianka 1969b), frogs (Toft 1981), tiger beetles (Pearson & Mury 1979)], Hutchinson's generalization suggests that species have their utilizations regularly spaced along the prey-size dimension. If so, then the data are qualitatively consistent with the hypothesis that competition is the cause, though other selective pressures producing regular differences are not ruled out (Section 6.8).

This simple conclusion has generated two kinds of criticisms, one theoretical, the other statistical.

First, in theory, ratios (or differences on a log scale) need not all be equal. The nature of the pattern should depend on the shape of the resource spectrum (Roughgarden 1976; Slatkin 1980; Case 1982) and whether or not utilizations systematically change shape as a function of position. Were the resource spectrum strongly peaked, utilizations should be more widely spaced at the tails than in the middle. Were niche width to increase strongly enough with increasing absolute size, as where larger species are more generalized in prey size or contain more size classes in their populations (Schoener 1974a; Wilson 1975), ratios should increase with increasing absolute size. Indeed, some evidence for the latter prediction exists in lizards and certain birds (Schoener 1965, 1970, but see Schoener 1984).

Secondly, several investigators have argued that a careful statistical investigation of the size data accumulated since Hutchinson's 1959 paper very often does not support constant ratios (e.g. Roth 1981; Simberloff & Boecklen 1981). The Simberloff and Boecklen critique, which is especially extensive, employs two statistical tests to show that most real size data cannot be distinguished from sizes randomly sampled from a uniform distribution. This distribution is defined on the (log) size interval bracketed by the minimum and maximum (log) sizes observed in the guild being tested. It specifies that over that interval all (log) sizes are equally likely to occur. Two criticisms of this approach are as follows. First, a better distribution would be one that is peaked — real size distributions are often lognormal (e.g. Schoener & Janzen 1968; Case *et al* 1983; Schoener 1984). Random sampling from a uniform, rather than peaked, distribution generates fewer small ratios and for certain tests may favour acceptance of the 'no-difference' hypothesis. Moreover, evolution viewed theoretically as a branching process produces a peaked distribution (Colwell & Winkler 1984). Secondly, rather than test each guild separately, some aggregative test should be done — in fact, such tests do change conclusions (Schoener 1984; Biehl & Cody, in manuscript).

Hutchinson's original data included the archetypal example of resource partitioning, the Galápagos finches (Section 5.4). Not surprisingly, recent criticism has centred on this example. Using a null-model approach, Strong *et al* (1979) concluded that "little if any evidence" for greater-than-random size differences exist. Strong *et al*'s procedure was to form artificial or 'null' guilds by drawing taxa randomly from a pool comprising the taxa actually found on the islands, then to compare mean ratios for the null guilds to those for the real guilds. Their method of generating null guilds has been criticized on a variety of grounds, ranging from computational through statistical to biological (Grant & Abbott 1980; Hendrickson 1981; Case 1983; Schoener 1984; Colwell & Winkler 1984; Grant & Schluter 1984). Both Hendrickson (1981) and Grant and Schluter (1984) were able to find more cases where sizes were significantly different from random expectation than did Strong *et al*, though the number is far from overwhelming. Those criticisms that are most general, that is which also apply to studies similar to Strong *et al*, are as follows:

(1) The species forming the pool are ecologically dissimilar in ways other than size and therefore would not be expected always to show large size differences (Grant & Abbott 1980; Schoener 1984; Biehl & Cody in manuscript).

(2) The null procedure of Strong *et al* assumes equal dispersal abilities; in fact, if dispersal ability and size are correlated, properly constructed

null communities will have species more similar to one another than those of Strong *et al* (Grant & Abbott 1980; Colwell & Winkler 1984).

(3) If past competition among the species of an archipelago has exterminated certain species too similar to other species to coexist with them, then random drawing from the remaining, relatively different, species underestimates the role of competition (Case 1983; Colwell & Winkler 1984).

Trying to avoid the first problem while retaining a sample size giving reasonable statistical power, Schoener (1984) constructed all guilds of 2–5 species that could be formed from the world's species of accipiter-like hawks and computed size ratios. He compared the resulting null distributions of ratios with those of actual guilds using a Kolmogorov–Smirnov test. In most cases real ratios were significantly larger.

In a second study using among other methods the Strong *et al* procedure, Case *et al* (1983) found greater-than-null size differences between West Indian birds. Conclusions from this and the hawk study are conservative, in that the above-mentioned problems (2) and (3) are not averted, yet size differences are still significantly larger than expected.

As a final example, one which illustrates all three approaches discussed in Section 6.3, we consider Lesser Antillean *Anolis* lizards. Each major island in this archipelago has either one or two species (Schoener 1969b, 1970; Williams 1972; Roughgarden *et al* 1983). One-species islands, with one exception, contain *Anolis* of a particular size (*c*. 70 mm length for males) called the 'solitary' size. Two-species islands contain lizards which, with one exception (St Maarten), bracket the solitary size and, moreover, are very different from one another (ratios of 1.4 to 2.2).

This situation appears for the most part to agree with the common-sense interpretation of Gause's results (Section 6.3), and it has been so viewed (Schoener 1969b, 1970; Williams 1972). Exceptions are mostly not explainable by simple qualitative arguments and can be considered noise.

A null-model approach, however, might ask whether the size differences on two-species islands are greater than one would expect by chance. Simberloff and Boecklen (1981), making the assumptions given above, concluded that they are not. (Biologists may gasp at this conclusion, given the large ratios!) A second test, in which they ask whether ratios are greater than a preconceived number, finds significance in one case of two. However, the Simberloff–Boecklen test, because the assumed uniform distribution is rescaled for each guild, is actually more suitable for testing whether regular spacing occurs than for testing whether ratios are large in some universal sense (Schoener 1984).

The same data can be analysed using the method of Strong *et al* (1979) for generating null guilds (Schoener in manuscript). All possible pairwise guilds are computed from the species pool consisting of all the island species. Now the actual ratios are significantly larger than the randomly contrived ones. Thus depending on the choice of null model, one or another conclusion is possible.

The preceding analysis has the two problems (2 and 3) discussed above for similar such analyses. How much difference does this make? Fortunately, for northern Lesser Antillean *Anolis* we know with reasonable certainty what the ancestral source island was: Puerto Rico (Williams 1972). Therefore, we can construct our null communities from all Puerto Rican species taken two-at-a-time, and again compare these to actual Lesser Antillean communities. Interestingly, the median ratio of the null distribution so produced is about the same as that from the previous, Strong *et al* procedure, and statistical significance is about the same. But the variance of the new null distribution is much greater: very large and very small *Anolis*, which together generate very large size ratios, do not occur on the small islands at all, perhaps not so much because of their poorer dispersal abilities as because such specialists are not favoured there. In short, problem (3), that null ratios from randomized island faunas might be too small because similarly sized competitors eliminate one another everywhere, is counterbalanced by a tendency for extreme-sized species not to occur on the islands at all.

Finally, Lesser Antillean *Anolis* can be examined from the viewpoint of a detailed theoretical model (Roughgarden *et al* 1983). In this model, solitary species evolve toward a single size, and their islands can be successfully invaded by much larger species. Because the larger species are competitively superior (Section 6.6.2), they cause the original species to displace in size downward and eventually go extinct. Thus Roughgarden *et al*'s model is one in which two-species islands are not in evolutionary equilibrium and one-species islands are invadable. This model accounts for more than just the solitary-size rule and the large size differences between species on two-species islands. It predicts that exceptions to the solitary size should be too large (the one known exception is), that islands should exist where the smaller species is on the verge of extinction and close to the larger species size (this is true of St Maarten), and that no three-species islands should exist (they do not). In short, a detailed theoretical model can predict phenomena that must be considered noise by the other approaches and that would actually diminish the significance of statistical tests. Statistical verification of such a model, of course, is at the most general level impossible, though details of its assumptions and predictions can be tested.

6.5.2 *Limiting similarity*

General considerations

The idea here is that species can be no more similar in their utilizations than a certain amount, this being the 'limiting similarity,' if they are to coexist. Hutchinson's (1959) data, *inter alia*, are consistent with this notion if the constant ratio is large enough. So are guilds whose species are not equally spaced along a size dimension, provided all differences are large. Simberloff and Boecklen (1981) attempt to test for a minimum difference in their compiled size data and find little evidence for it. However, the smallest ratio passing their test, and thereby supporting the existence of a limiting similarity, depends on the total size range spanned by the particular group of species examined (Schoener 1984; Section 6.5.1). A second way to look for limiting similarity using a null model is given by Schoener (1984) in a study on hawks: what fraction of ratios from all guilds combined is above a certain number compared to the fraction expected from random association? Except for species pairs, hawks give little evidence for a single value of limiting similarity.

As we now discuss, failure to find much evidence for a single value of limiting similarity holding for all species in all localities is not surprising in view of very recent theoretical models, though it was not always so. Models that generate limiting-similarity predictions fall into three classes: deterministic–ecological, stochastic–ecological, and deterministic–evolutionary.

Deterministic–ecological models

Ecological models ask, given an established or equilibrial community, what species, if any, can (i) invade and (ii) persist? No evolution is allowed: utilizations remain constant throughout the competitive process and such immutable species either fit together or go extinct. Deterministic models are those not incorporating any kind of environmental fluctuation or 'sampling-error' variability due to small numbers of individuals.

The first limiting-similarity model, that of MacArthur and Levins (1967), is of this type. It asks the simplest sort of question: suppose two species have their utilizations located along a single dimension (Fig. 6.3) can a third species be sandwiched in between? The model assumes bell-shaped (normal) utilizations, equally spaced and equally shaped. Most importantly, it assumes that a formula for niche overlap can be used to compute α, the competition coefficient. In symbols

$$\alpha_{12} = \sum_h p_{1h} p_{2h} / \sum_h p_{1h}^2 \qquad (1)$$

where α_{12} is the effect of Species 2 on 1, relative to the effect of Species 1

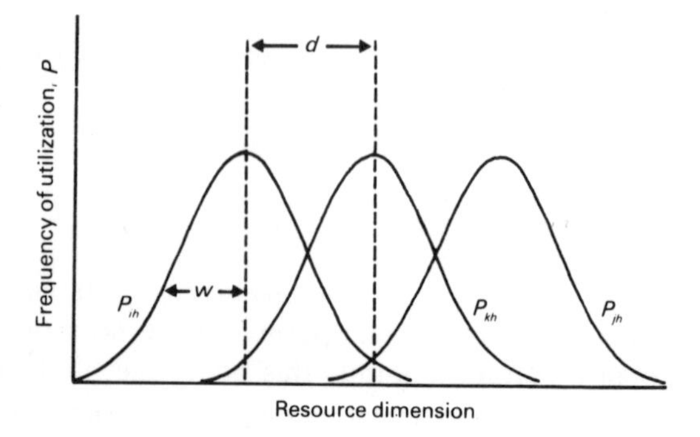

Fig. 6.3. Utilizations for Species *i*, *j*, and *k*. The utilizations are spaced *d* units apart and have standard deviation or width *w*. The MacArthur–Levins (1967) treatment asks whether Species *k* can be inserted between Species *i* and *j*.

on itself; the formula for α_{21} has all numbers reversed. The *p*'s are defined in Section 6.2. In the Lotka–Volterra model of competition, two kinds of quantities, α's and *K*'s (carrying capacities), determine the outcome of competition between two species (see review in Pianka 1981). Fixing the *K*'s, we can calculate the α's that are just small enough to allow coexistence. Rather than express limiting similarity in terms of α, however, we can back-calculate using equation (1) and express it in terms of the utilizations. In particular, if *d* is the distance between utilization peaks and *w* is utilization width (specifically its standard deviation), then limiting similarity can be expressed in units of *d*/*w*. May (1973, 1974) has expanded the MacArthur–Levins analysis to include both persistence and invasive success and has calculated limiting *d*/*w*'s for more than two species. Results for three species are given in Fig. 6.4; note that limiting *d*/*w* centres about, but is somewhat larger than, 1.0, the exact value depending on the *K*'s. Calculations for more than three species give a somewhat larger limiting *d*/*w*. This is because the more species there are, the greater the summed competition from all the species. This summed competition is called 'diffuse competition' (MacArthur 1972); some evidence for it exists in desert lizard communities, where species are more loosely 'packed' (*d*/*w*'s larger), the more total species in the community (Pianka 1974).

The limiting *d*/*w*, although relatively robust under certain kinds of variation in assumptions (May 1974), is highly sensitive to four such kinds of variation. First, as niche shape varies from the normal distribution toward a thick-tailed, peaked distribution (leptokurtic), limiting *d*/*w* diminishes (Roughgarden 1974a). Secondly, if α is calculated in sensible ways other than given by equation (1), limiting *d*/*w* diminishes (Abrams 1975). Thirdly, limiting *d*/*w* diminishes under frequency-dependent predation, defined as predation in which the more numerous the prey population, the more likely any given prey individual is to be eaten (Roughgarden & Feldman 1975). Fourthly, mathematical models of competition other than the Lotka–Volterra one (i.e. those producing non-linear zero-isoclines) generally allow a smaller limiting similarity (Schoener 1974d, 1976, 1978).

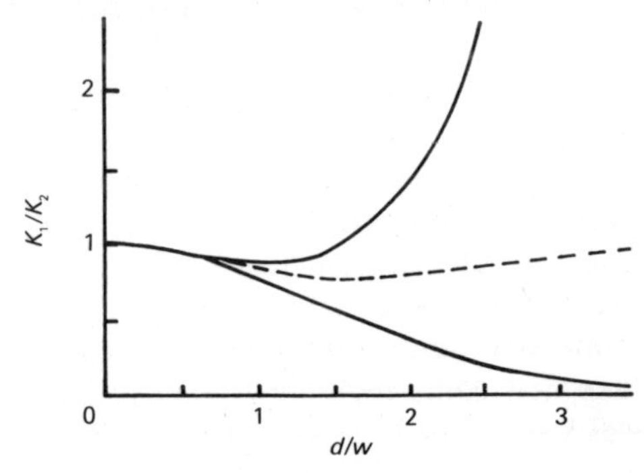

Fig. 6.4. The range of resource spectrum shapes (*K*'s) which permits a three-species equilibrium community, as a function of niche overlap, *d*/*w*. Here we assume a symmetric spectrum, $K_1 = K_3$, so that a one-parameter shape characterization (K_1/K_2) is possible. Again the dashed line indicates the condition where all three equilibrium populations are equal (from May 1973).

Stochastic–ecological models

In these models environmental fluctuations are incorporated into population-dynamics equations such as that of Lotka and Volterra. In the first such model, May and MacArthur (1972) concluded that: (i) any degree of fluctuation, no matter how small, predisposes a less stable (no coexistence) outcome, and (ii) small-to-moderate amounts of fluctuation have about the same destabilizing effect. Recently, Turelli (1981) performed a wide-ranging, detailed analysis of the effect of environmental stochasticity, bolstering his analytical models with extensive computer simulation. His analysis contradicts May and MacArthur's conclusions: slightly fluctuating environments are very similar to deterministic (no fluctuation) ones; models for the latter are therefore valid for the former. Only for large fluctuations do stochastic models give predictions different from deterministic ones, and then the effect may be either destabilizing or stabilizing, depending on the mathematical structure of the model used.

Few data are available to test such models. Inger and Colwell (1977) found that reptiles and amphibians were most tightly packed (d/w smallest) in evergreen forest, the most climatically predictable of the three sites they studied in Thailand. This is consistent with certain of the Turelli models, though the nature of the environmental variation in real sites was somewhat different from the models' idealization.

Deterministic–evolutionary models

Communities of species at their limiting similarity are saturated — further invasion is impossible. But in the absence of a sufficient number of invading species, resident species may coevolve their utilizations such that d/w exceeds the ecological limiting similarity. Roughgarden (1976) constructed the first model of this process; he allowed niche position, but not niche width, to evolve. For a given such width, values of d/w at evolutionary equilibrium are larger, the greater the width of the available resource spectrum. The fewer the number of species in a guild, the farther apart are their utilizations, and d/w's can be as much as twice the limiting d/w from ecological models. Moreover, if both the number of species and resource variance increases from one community to another, d/w may actually increases with increasing number of species; this provides an alternative explanation for the above-mentioned data of Pianka on ecological overlap.

Case (1979) used the Roughgarden formalism as a basis for predicting sizes of *Cnemidophorus* lizards on small islands in the Gulf of California. Sizes on one-species islands were predicted quite closely. The model for two species gives two (alternative) stable equilibria; which equilibrium is attained depends on historical factors, i.e. the sizes of the species when they first come together. One of Case's predicted two-species equilibria was very close to sizes found on the existing two-species islands.

In a more recent treatment, Case (1982) allowed both niche position and width to evolve, the latter via within-phenotype variation (definition in Section 6.6.3). In this analysis, variation in the width of the resource spectrum had less effect than in Roughgarden's, but it was still quite marked. Case concluded that only certain island guilds have *d*/*w*'s large enough to be consistent with coevolutionary models. Large *d*/*w*'s are expected for islands, in particular, since invading species are relatively scarce.

A plethora of theory now exists for limiting similarity. [For a recent review see Abrams (1983).] Its upshot is that no single value of limiting similarity is likely to hold for all species and all situations. On the other hand, optimistically we can say that, given a knowledge of the relevant characteristics of the species and the environment, the appropriate limiting similarity can be specified by the theory. Whether enough such knowledge can be realistically gathered is a question that mostly remains for the future, though Case (1979) has made a promising start.

6.5.3 Patterns in several dimensions

Up to now, we have dealt with differences between species along a single dimension. However, in virtually all guilds studied, several dimensions appear to be important in separating species; the modal number is 2–3 (Schoener 1974a). Two qualitative propositions can be made about species differences in a multidimensional situation, given that niches are overdispersed because of competitive pressures (Schoener 1974a).

The first is that, as the number of species increases, the number of important dimensions should increase. This proposition is weakly supported by the above-mentioned survey (Section 6.4) of resource partitioning data in all animals; it is more strongly supported in an analysis confined to reptiles (Schoener 1977). Few tests, however, are available, in part because of the difficulty of identifying 'important dimensions'. The simplest theoretical argument for this result is ecological: the more dimensions, the smaller the possible overlap in utilizations. Pacala and Roughgarden (1982a) have recently given a coevolutionary argument for the same phenomenon: under some circumstances, multidimensional separation occurs at evolutionary equilibrium for an $(n+1)$-species but not an n-species community.

The second proposition, not entirely unrelated to the first, is that species similar in one or more dimensions should be dissimilar in others; niche separation should be complementary. Much evidence supports this proposition. Cody (1968) found that in grassland-bird communities, species tended to separate along three dimensions such that the total separation for a given community was roughly constant (Fig. 6.5a). Schoener (1968) and Kohn and Nybakken (1975) found for *Anolis* lizards and *Conus* gastropods, respectively, that pairs of species occupying similar habitats have different diets and vice versa. In the lizard example, pairs of sex-age classes (e.g. adult males and adult females) within the same species showed the opposite relation (Fig. 6.5b). The *Anolis* lizards on two-species islands of the Lesser Antilles, discussed in Section 5.5.1, also show complementarity: species less separated by size are more separated by habitat and vice versa (review in Roughgarden *et al* 1983).

A way to implicate complementarity of separation indirectly was devised by Gatz (1979) in his analysis of three stream-fish communities in eastern North America. Similarly to Cody's result, Gatz showed that the average multidimensional niche overlap, as well as a multivariate measure of morphological distance, were approximately constant for the three communities. Null models indicated that both more large and small differences occurred than expected by chance; this is consistent with small differences along one dimension and large differences along another.

6.5.4 *Distributional complementarities*

Patterns in species occurrences

Sometimes species very similar in their ecological requirements occupy disjunct geographic ranges. On continents, this often manifests itself as parapatric (abutting or slightly overlapping) distributions. For example, the giant *Anolis* lizards of both Hispaniola and Cuba, comprising three and five species respectively, are parapatric (e.g. Schwartz & Garrido 1972), so much so that they were formerly considered subspecies. Distributional separation can act as a niche dimension so far as complementarity is concerned. This was demonstrated quantitatively for Greater Antillean *Anolis* (Schoener 1970). Species with overlapping distributions are more dissimilar in size than those with allopatric or parapatric distributions. Species more dissimilar in structural habitat (perch height and diameter) are also more dissimilar in size, but for the same degree of structural-habitat similarity, size differences are still greater in sympatry.

While parapatric distributions can thus be interpreted as evidence of competition, alternatively they may represent incipient speciation. Complementary distributions among a set of isolates, such as islands, are more convincing, particularly where isolates each having only one of a set of species are interspersed. Diamond (1975) calls such distributions 'checkerboards'. He has documented numerous examples, both from water-surrounded islands (Fig. 6.6) and montane 'habitat islands'. Diamond (1975) also noted that certain trios of species never occurred on

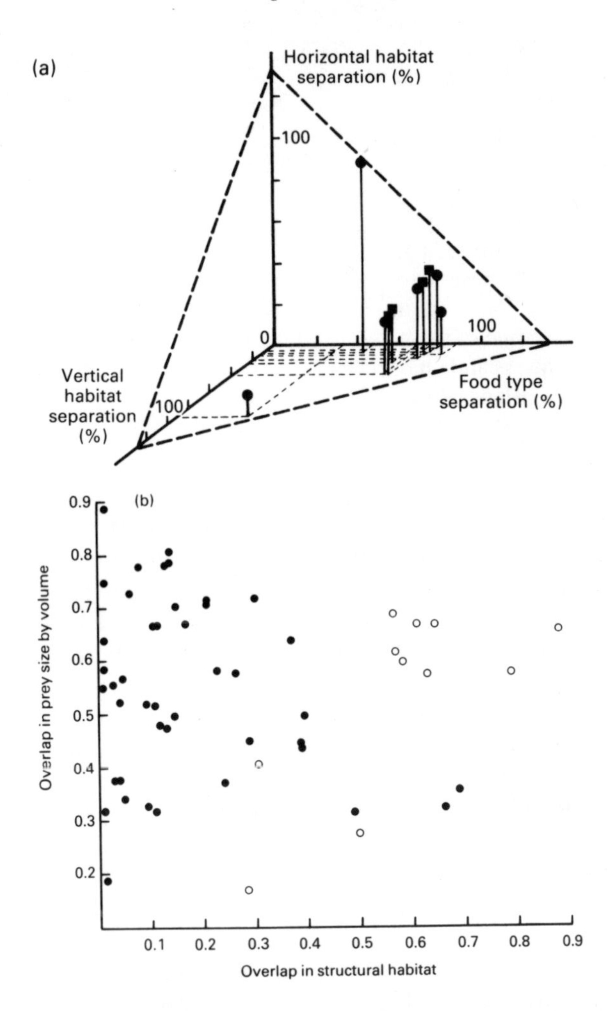

Fig. 6.5. (a) Cody's grassland bird communities, showing how total separation in horizontal habitat, vertical habitat, and food type is constant, but the relative proportion of the three types of separation varies. Squares are South American communities, circles are North American (from Schoener 1974a after Cody 1968). (b) Similarity in prey size (weighted by prey volume) plotted against similarity in structural habitat for all pairwise combinations of classes of Bimini *Anolis* lizards. For interspecific pairs (●), similarity in habitat implies dissimilarity in prey size and vice versa, whereas the reverse is true for intraspecific pairs (○) (from Schoener 1968, 1974a).

the same island, and he calculated the probability that this could be due to chance for particular such 'forbidden combinations'.

These and others of Diamond's 'assembly rules' were vigorously challenged by Connor and Simberloff (1979). Among other things they

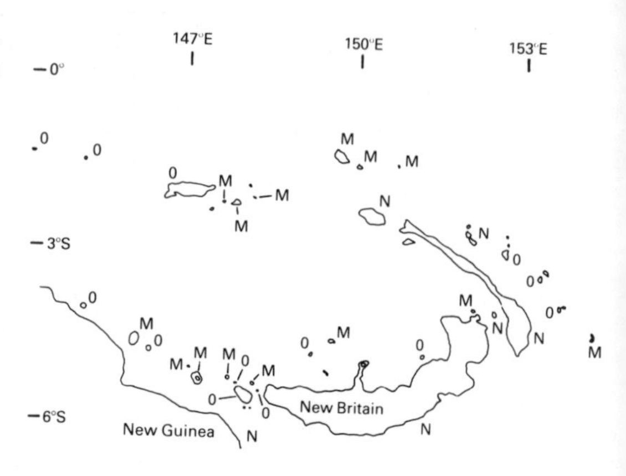

Fig. 6.6. Checkerboard distribution of small *Macropygia* cuckoo-dove species in the Bismarck region. Islands whose pigeon faunas are known are designated by M (*Macropygia mackinlayi* resident), N (*Macropygia nigrirostris* resident), or O (neither of these two species resident). Note that most islands have one of these two species, no island has both, and some islands have neither (from Diamond 1975).

argued that, by chance alone, certain species combinations should be missing from all islands, this number depending on the number of islands, on the number of species in the whole system and the number inhabiting each island, and on other conditions. Connor and Simberloff concluded, using data on birds and bats, that there were usually no more pairs or trios missing from the entire system than a null model would predict.

Gilpin and Diamond (1982) mounted a fierce counter-attack to the criticisms of Connor and Simberloff. Among other things, they argued that (i) except for birds of the New Hebrides, Connor and Simberloff's tests refute their own null hypothesis of random occurrence, usually at a level of $p < 10^{-8}$, and (ii) for technical reasons, the randomization procedure of Connor and Simberloff is deficient, and in particular that it does not detect checkerboards as nonrandom. Using a different procedure based on loglinear models, Gilpin and Diamond showed that both the New Hebrides and Bismarck avifaunas were nonrandomly distributed. In the Bismarcks, more species pairs had both negative (occurred together less often than expected by chance) *and* positive associations. However, in the New Hebrides, only positive associations were more frequent than chance expectation. A negative association is consistent with a competition hypothesis. The most likely explanation of a positive association is that the species have a habitat requirement in common, and that this habitat is restricted to certain islands. For example, certain aquatic birds occur only on islands with large bodies of freshwater.

Participants in the debate over species co-occurrences on islands are still very far from agreement, as the lengthy exchange in Strong *et al* (1984b) illustrates.

Patterns in species abundances

While competition may sometimes be expected to produce complementary occurrences, it should much more commonly act simply to reduce abundances of the competing species (Gilpin & Diamond 1982). Almost no studies have examined complementary abundances, a recent exception being a study by Toft *et al* (1982) on the five species of ducks breeding on subarctic ponds. They found that abundances of certain pairs of species showed strong negative associations from one pond to another (Fig. 6.7). Species most similar in their breeding times and habitat preferences were most complementary in their abundances, despite a tendency for habitat type to be related to pond identity. Toft *et al*, employing the contingency-table method of Cohen (1970), also found complementarities in occurrences. In this method, ponds are classified into four categories: (i) having both species A and B; (ii) having neither species; (iii) having A but not B; and (iv) having B but not A. A simple chi-square test will show whether occurrences are significantly complementary.

6.5.5 *Temporal variation in the degree of resource partitioning*

We have already seen how species can differ diurnally or seasonally in their resource use (Section 6.4). Here I discuss another kind of temporal pattern, that is, variation in the degree to which the species of a community, taken collectively, concentrate on the same or different resources.

Smith *et al* (1978) found that all 12 of the guilds they surveyed had greater niche overlap during the 'fat' (resources abundant) than during the 'lean' season. Generally the fat season corresponds to spring–summer in temperate areas or the rainy season in the tropics. For example, in their own field study Smith *et al* found that during the rainy season, Galápagos finches tend to concentrate on the then abundant easily handled seeds and fruits. Recently, Schoener (1982) increased Smith *et al*'s list and included year-to-year, as well as within-year, variation. Only two of 30 cases did not support the trend established by Smith *et al.*

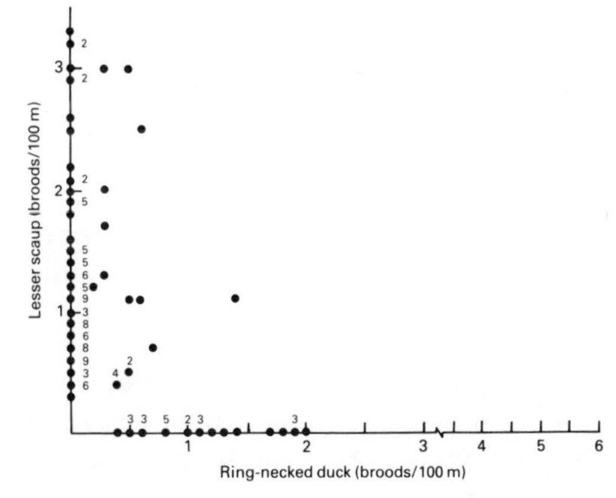

Fig. 6.7. Density (broods per 100 m) of lesser scaup (*Aythya affinis*) and ring-necked duck (*A. collaris*) along transects on ponds of all sizes, 1962–65 (from Toft *et al* 1982).

The following scenario, as yet mostly hypothetical, shows how this pattern in seasonal variation in ecological overlap can support the notion of interspecific competition (Schoener 1982, and references therein). During lean times, species are most likely to be in competition, and at that time selection should act especially strongly to produce those sorts of specializations that result in niche separation. During fat times, certain types of food become more profitable to use than those for which the species' trophic phenotypes were specifically selected. Moreover, the now profitable food types are the same for a number of species, and the species all converge upon them. According to optimal foraging theory (Schoener 1974b; Pyke *et al* 1977), food items that are spatially concentrated or easily handled are especially profitable, and such foods are known often to be differentially present during the fat season. As lean times return, each species retreats to the food-resource set to which its phenotype was selected.

The result on seasonal overlap may be criticized on the basis that if one or a few types of resources differentially increase in abundance during the fat season, species that randomly select resources would still automatically overlap more at that time. It is impossible to know how important this factor is in general, but measures of resource overlap that divide use (the p's of Fig. 6.1) by percentage availability (Schoener 1974c, as in Toft 1980) minimize this problem and so are greatly preferable for the above sort of comparison.

6.6 Patterns within species

So far we have examined only community-wide patterns in resource partitioning. Now we examine how particular species change their niche characteristics with a changing context of potential competitors.

6.6.1 Character displacement

Character displacement occurs when two species, similar in some trait where allopatric, diverge in that trait where sympatric (Brown & Wilson 1956; see Grant 1972, for a broader definition). The ecological explanation for character displacement is that divergence in a morphological trait, for example, may indicate divergence in the species utilizations, as a result of selection minimizing interspecific competition.

Some of the more famous cases of character displacement occur in Galápagos finches (Lack 1947; Grant & Schluter 1984), already discussed in relation to aggregate size differences within communities (Section 6.5.1). Character displacement is simply the evolutionary process whereby such differences can come to exist (see below).

Another classical example of character displacement occurs in two species of nuthatches (*Sitta*) of Asia Minor. Bill lengths and plumage traits are similar in allopatry and dissimilar where the species overlap. An ecological basis for this character displacement has never been found (Grant 1975). Moreover, Grant (1972) has argued that the pattern is explainable as parallel clinal variation rather than displacement in sympatry (Fig. 6.8). It can be counter-argued that the clinal variation itself results from gene flow away from sympatric populations selected to diverge because of competition. However, given the great distances involved, this is perhaps unlikely.

In a somewhat similar case, two species of warblers (*Dendroica*) have been studied in eastern North America (Ficken *et al* 1968). Where the species overlap, the bill length of the behaviourally subordinate species diverges away from the other's, giving a bill-length ratio of about 1.3 (Hutchinson's ratio; Section 6.5.1). The diverging species becomes a specialist in probing into loblolly pine cones. Again, there is some controversy over this example (Grant 1972).

An especially well documented case of character displacement occurs in the fossorial lizard genus *Typhlosaurus* of the Kalahari semi-desert (Huey & Pianka 1974). Head size and proportion in one species shifts away from the other in sympatry. The shift is paralleled by a shift in prey size (Fig. 6.9).

Another well-studied case involves the particle-feeding molluscs *Hydrobia*, occurring in Danish mudflats (Fenchel 1975b). Larger species ingest larger particles. Two species occur in sympatry in 15 locations and in 17 locations occur alone. Wherever the two occur together, they displace in size by about the same amount, such that the ratio of shell lengths is about 1.5 and d/w is about 1.0 (Fig. 6.10). All displacements must have taken place after a fjord collapsed 150 years ago, as this allowed one species to invade the other's range. Fenchel and Kofoed (1976) have shown in laboratory experiments that identically sized individuals of different species are equivalent competitors. The basis for character displacement in *Hydrobia* has recently been challenged by Levinton (1982) who grew variously sized *Hydrobia* equally well on particles of

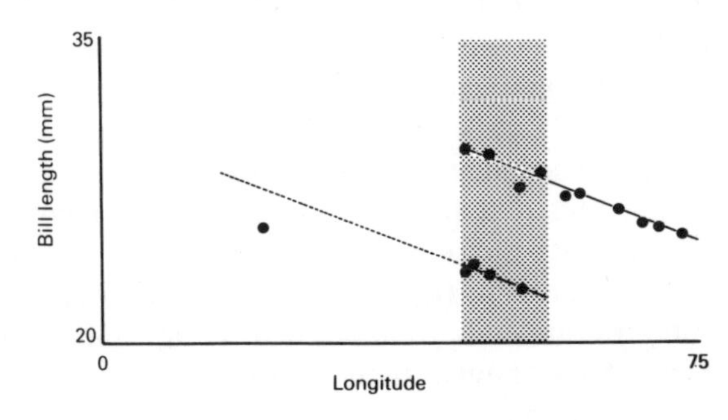

Fig. 6.8. Geographical variation in mean bill length of *Sitta tephronota* (upper) and *S. neumayer* (lower). The broken lines are projections of calculated (solid) regression lines fitted by the least squares method. Stippling indicates sympatry (from Grant 1972).

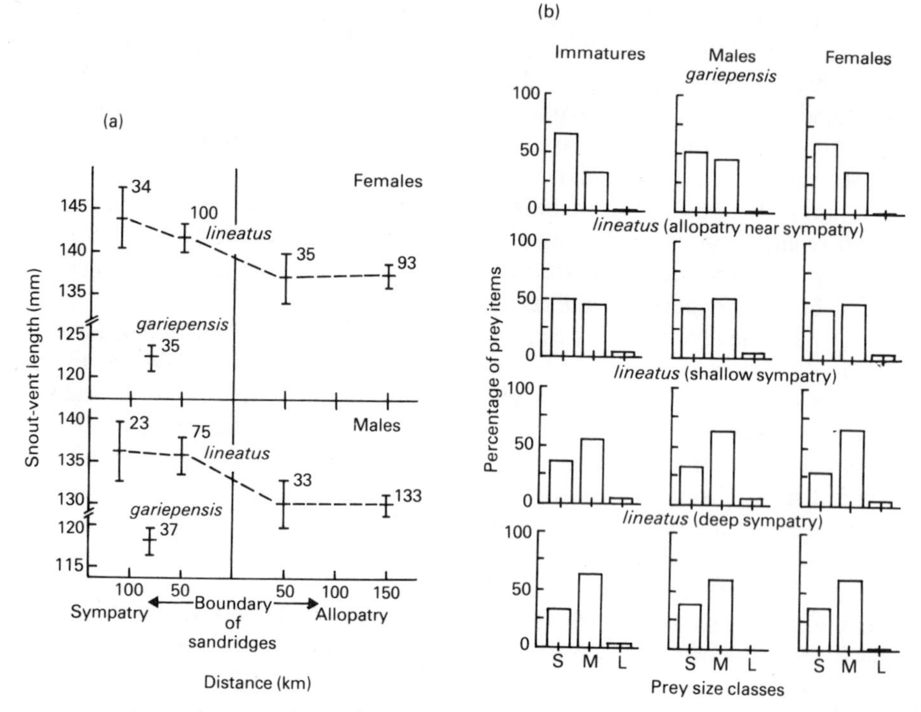

Fig. 6.9. Character displacement in *Typhlosaurus* skinks. (a) Mean ± 2 s.e. of males (below) and females (above) vs approximate average distance of populations from sandridge boundary. Numbers are sample sizes. (b) Percentage prey items from each of three size classes: S (small), M (medium) and L (large). Sympatric *lineatus* eat larger prey than allopatric *lineatus* and than the smaller *gariepensis*. Sample sizes: *gariepensis*, immatures 493, males 855 ($\bar{X}_{SVL} = 116.1$), females 1279 ($\bar{X}_{SVL} = 122.7$); *lineatus* (allopatry near sympatry), immatures 637, males 655 (130.1), females 544 (136.9); *lineatus* (shallow sympatry), immatures 866, males 1442 (135.6), females 2241 (141.6); *lineatus* (deep sympatry), immatures 488, males 638 (136.1), females 1130 (144.0) (from Huey & Pianka 1974).

different sizes. Levinton suggests that mode of feeding as correlated with shell size, rather than particle-size differences, may be its basis.

As a final example, McNab (1971b) has argued that certain cases of Bergmann's Rule in mammals are more likely examples of character displacement than of physiological adaptation to climate. Bergmann's Rule states that individuals of a species increase in size with increasing latitude. Its physiological explanation is that larger animals can retain heat longer. But, (i) the effect is physiologically miniscule; (ii) ectotherms show Bergmann's Rule and (iii) larger animals still lose absolutely more heat, even though their per-gram loss is less. Mustelids (the weasel family) provide the most convincing case of character displacement: the

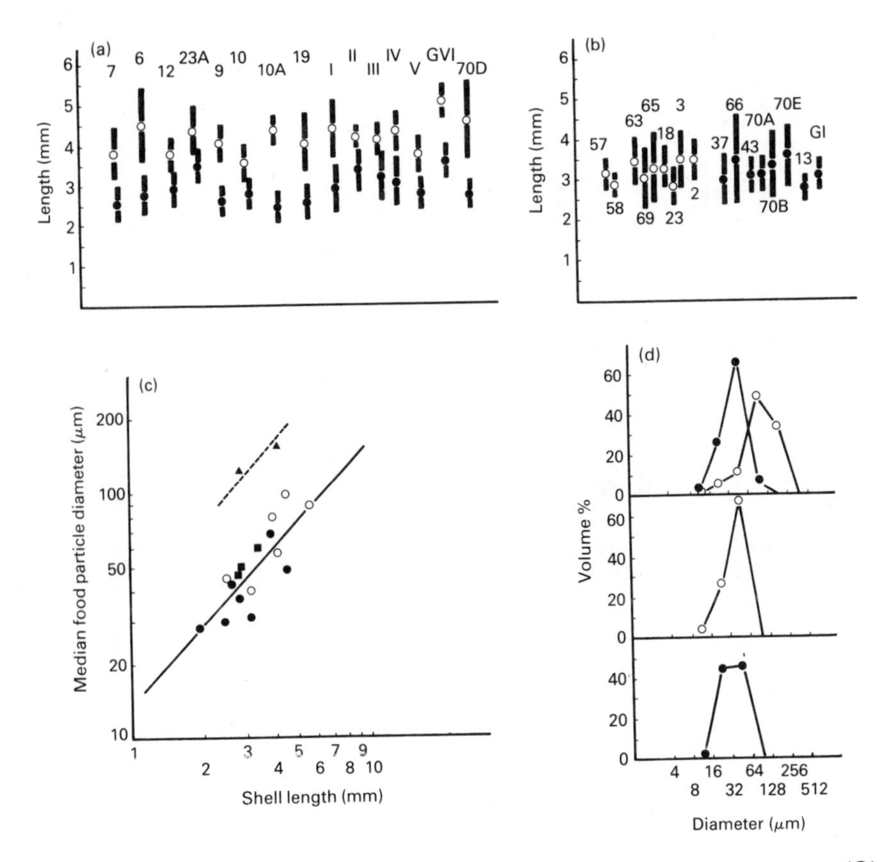

Fig. 6.10. (a) Average lengths of *Hydrobia ulvae* (○) and of *H. ventrosa* (●) from 15 localities in the Limfjord where the two species coexist. The vertical bars indicate one standard deviation. All samples are from the summer of 1974. (b) Average lengths of *H. ulvae* and of *H. ventrosa* from 17 localities in the Limfjord where one of the two species occurs alone. All samples are from the summer of 1974. Legends as for (a). (c) Median diameter of ingested food particles of the four species of hydrobiids plotted against shell length. ▲ *P. jenkins*, ● *H. ventrosa*, ■ *H. neglecta*, ○ *H. ulvae*. (d) Food particle size distributions (volume %) for the populations of *H. ulvae* (○) and of *H. ventrosa* (●) from a locality where the species coexist (top) and from two localities where they live alone (from Fenchel 1975b).

largest species drops out at high latitudes and the smaller species increase, somewhat quantumly, in size there (Fig. 6.11). In short, here is a case where a situation formerly interpreted as a cline is interpreted now as character displacement, the opposite of the *Sitta* case discussed above.

The most rigorous attempt to separate the effect of potentially competing species on geographic character variation from the effects of clinal variation and habitat change is that of Dunham *et al* (1979) on

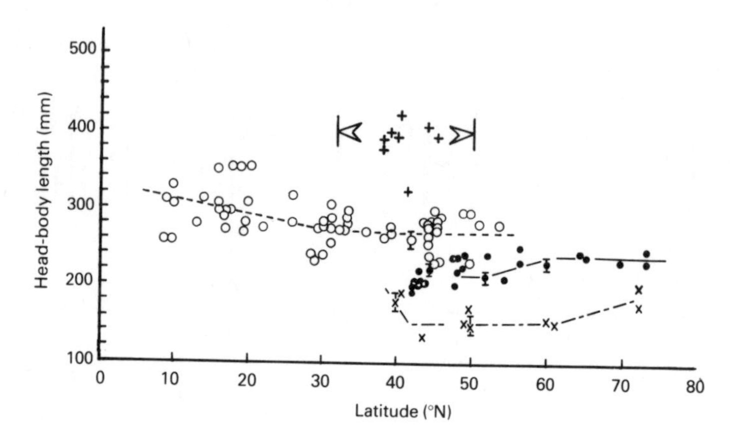

Fig. 6.11. Head-body lengths of male weasels as a function of latitude + *Mustela nigripes*, ○ *M. frenata*, ● *M. erminea*, × *M. nivalis* (from McNab 1971b).

mountain suckers (catostomid fishes). Using multiple regression, they analysed variation in number of gill rakers and number of vertebrae. Apparently, number of gill rakers indicates food-particle size, and number of vertebrae, by being related to swimming hydrodynamics, indicates habitat. Competitors were more important as correlates of character variation than factors such as latitude, longitude, elevation and stream properties.

Theoretical models of coevolution, discussed in Section 6.5.3, are automatically models of character displacement. Additionally, some of our intuitive notions about character displacement have been contradicted in another theoretical treatment, that of Slatkin (1980). In Slatkin's approach, both the mean and variance of the phenotype distribution (and therefore the utilization) are subject to evolutionary change; only between-phenotype variation (definition in Section 6.6.3) is considered. The simplest model assumes that the available resource spectrum is bell-shaped and that identical phenotypes from different species have identical efficiencies of resource use. This model is unable to produce strong divergence and sometimes produces convergence. The problem is that selection for an increased variance can often halt or turn about an initial divergence in the means. Slatkin finds that two non-obvious conditions are necessary to produce extreme examples of displacement such as are found in certain real organisms. First, if the resource spectrum is skewed, one species at equilibrium ends up under the spectrum's peak, its variance constrained from expanding further; the other species occupies the spectrum's long tail (Fig. 6.12). A second condition, even more effective, is if identical phenotypes from the two competitor species do not have identical efficiencies of resource use. One way in which the second condition may be produced is if the species differ along resource

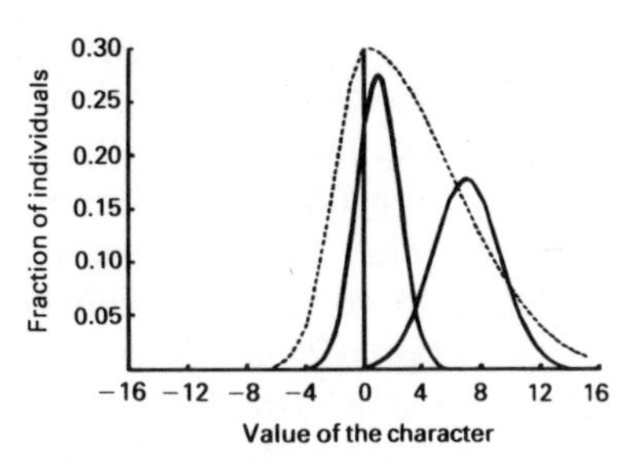

Fig. 6.12. Evolutionary equilibrium for case where the resource spectrum (dotted distribution) is skewed. This situation gives considerable character displacement (from Slatkin 1980).

dimensions other than those correlated with body size, say certain habitat dimensions. Then the actual prey species composing a given prey-size class are not the same for the two competitor species. In short, Slatkin's work shows that some of the qualitative arguments concerning the effect of competition are difficult to justify with specific mathematics.

Species may also become more similar in size in areas of sympatry. Such convergence is theoretically expected under special conditions (review in Slatkin 1980). In *Anolis* lizards, almost all cases of convergence involve a decrease in size; for species with broadly overlapping prey sizes, a decrease in resource abundance may favour a decrease in body size. This is because the gain from reduced metabolic requirements is greater than the loss from becoming more similar to a competing species (Schoener 1969a).

6.6.2 *Habitat shift*

Niche shift refers to a change in central tendency; e.g. in the mean or mode, of the utilization. The morphological changes discussed in the previous section were generally, though not always, hypothesized or shown to be associated with food-size shifts. Species may also shift their habitats in apparent response to competition; not only is this in some ways theoretically more likely than food-size shifts (Section 6.4), but it is potentially more observable as well.

Habitat shifts have been observed in a variety of organisms. In Tasmania, where taxa of birds specialized to feed on trunks are absent, species from other avian taxa shift into this zone (Keast 1970b). In New Guinea birds, the commonest type of shift is altitudinal; species also shift in vegetation type and foraging height (Diamond 1970). Shifts in diet are much rarer. A pattern similar to that in New Guinea occurs in birds of the New Hebrides (Diamond & Marshall 1977). In certain of these bird species, as well as in ants (Wilson 1961), shifts from low to high altitude or from secondary to primary forest are especially common, as would be expected for colonizers of remote islands.

While response to interspecific competition is one hypothesis for niche shifts, changes in resource availability is a second (e.g. Bowman 1961). Schoener (1974d, 1975) measured vegetational changes from one locality to another in a study of structural-habitat (perch height and diameter) shift in *Anolis* lizards. He found that habitat shifts in apparent response to competition can be both enhanced *and* obscured by interlocality vegetational differences. Note in Fig. 6.13 that *A. sagrei* does not appear to shift downward on Bimini in the presence of a higher species, but that, because Bimini has relatively tall trees, *A. sagrei* in fact would be expected to occur at greater heights than it actually does. This study also showed that: (i) the more similar species are in other niche dimensions (food size, climatic habitat), the more likely they are to shift structural habitat in apparent response to one another, and (ii) smaller forms tend to shift more in response to larger forms than vice versa. The second result is explainable both on the basis of exploitative competition (larger forms eat more) and interference competition (larger forms win in aggressive encounters).

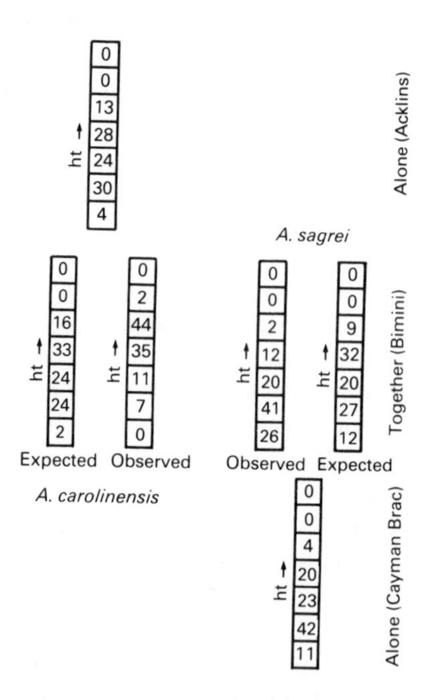

Fig. 6.13. Habitat shift in *Anolis sagrei* and *A. carolinensis* on Bimini. Notice how *A. sagrei* changes little in the actual distribution of heights occupied, but that when adjustment is made for the greater occurrence of high vegetation on Bimini, shift is more marked (from Schoener 1975).

6.6.3 *Niche expansion*

Niche expansion can be defined as an increase in the variance, or width (w), of the utilization. Intuitively, a reduction in numbers of competing species should cause niche expansion in a given species because intraspecific competition becomes relatively more predominant. Expansion is especially favoured when competing species on all sides of a given species' utilization drop out.

In contrast to niche shift, unambiguous examples of niche expansion, as plausibly affected by interspecific competition, are rare (Beever 1979). Crowell (1962) studied three species of birds that were much denser on species-poor Bermuda than on the species-rich mainland. The Bermudan populations showed no new foraging behaviour, and their niche widths, if anything, were slightly smaller than on the mainland. In Rabenold's (1978) study of spruce-fir forest birds, niche width for a given species was larger in places having fewer species. Such places had less abundant food, and Rabenold argues that foraging-strategy considerations (Section 6.4) caused niche expansion there as an individual response. The gastropod *Conus miliaris*, occurring virtually alone on Easter Island, has an expanded diet and possibly depth range but not microhabitat (Kohn 1978). Certain *Anolis* lizards show a relatively small climatic breadth in species-poor communities (review in Schoener 1977), but striking exceptions exist (Hertz 1983). The evidence for prey-size expansion in *Anolis* is equally mixed (Roughgarden 1974b, summary in Schoener 1977). The most striking example in support of niche expansion is perch height in *Anolis sagrei* (Lister 1976a; Fig. 6.14).

The niche variance of a species population can be decomposed into two components, that due to variance *between phenotypes*, and that due to variance *within phenotypes* (Van Valen 1965). Roughgarden (1972, 1974b) has shown that, as in the analysis of variance, total population variance equals the between-phenotype variance plus the mean within-phenotype variance. Roughgarden's (1972) analysis was the first mathematical treatment of niche expansion. It models change in the between-phenotype variance. [The theory of within-phenotype variance is to a large extent the theory of optimal foraging (see also Case 1982), and this has been reviewed above (Section 6.4).] Niche width (Var N) in an asexual population equals Var K – Var α, where K reflects the resource spectrum and α is the competition function — the larger Var α, the more quickly α decreases with increasing phenotypic distance. In contrast to an asexual population, a sexual population is severely constrained in the ability to expand its niche: sexual reproduction has an averaging effect, causing regression toward the mean. For sexual species, Var N approximately equals twice the phenotypic variance of the offspring distribution.

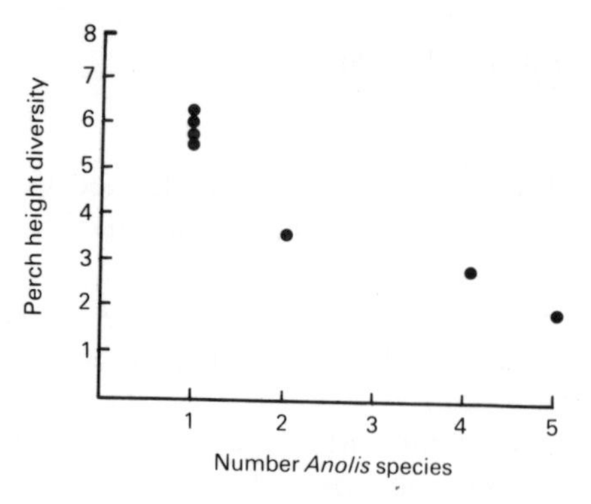

Fig. 6.14. Perch height diversity ($1/\Sigma p_i^2$, where p_i is the fraction of observations in height category i) vs number of *Anolis* species for populations of female *A. sagrei* ($r = -0.94$, $P < 0.01$). The more sympatric species, the lower the perch height breadth (from Lister 1976a).

The 'niche-variation hypothesis' formalizes the relationship between between-phenotype variation and interspecific competition (Van Valen 1965). It states that, as interspecific competition declines, utilizations should expand and associated morphological indicators should become more variable — an increase in between-phenotype variance. Van Valen supported his hypothesis with five avian examples: bill dimensions were more variable in populations showing habitat expansion. Later attempts to confirm this hypothesis in birds were either unsuccessful or only partly successful (review in Beever 1979). Perhaps the most impressive support is found in Rothstein's (1973) analysis, in which morphological variability declined with increasing species diversity.

Possibly the most intensive test of the niche-variation hypothesis is that of Lister (1976a,b) in *Anolis* lizards. His study considered three niche dimensions and associated morphological traits: (i) perch height and number of toe lamellae, (ii) temperature and scale size, and (iii) prey length and head length. In all three, utilization variance increased, the fewer the coexisting species (e.g. Fig. 6.14). Yet only for prey size did morphology become more variable as well. Even the latter does not support the niche-variation hypothesis, as maximum head length increased in the absence of competition, and this automatically implies a greater head-length variation (and even coefficient of variation), given the way lizards grow (Lister & McMurtrie 1976).

Though he had no direct evidence, Lister concluded from his analysis that the observed niche expansion must be almost entirely within-phenotype. Other studies (summary in Case 1982) also show that within-phenotype expansion is by far the most important component of niche expansion. However, while relatively rare, dietary differences between individuals in the species of prey they prefer are known (Curio 1976). An intriguing example is the oystercatcher (*Haematopus ostralegus*) in which either 'stabbers' or 'hammerers', distinguished by how they kill prey, exist

(Norton-Griffiths 1968). Moreover, Davidson (1978) showed for the polymorphic ant *Veromessor pergandei* that variability in worker size is inversely correlated with the number of potentially competing species (Fig. 6.15). In neither of these examples is the phenotypic variation reflected in a corresponding genetic variation. But variation in dietary preferences within the same population having a strong genetic component is extensively documented in garter snakes *Thamnophis* (Arnold 1982).

For the most part, the important between-phenotype variation in populations occurs between sex and age classes. In particular, several studies strongly support an increased sexual size dimorphism in areas of reduced interspecific competition (Selander 1966; Rothstein 1973; Schoener 1977 — but see Willson 1969; Wiens & Rotenberry 1980). Just as for different species, one can argue that the sexes within a species should differ in one or more niche dimensions to avoid intraspecific competition, and that this will be favoured when interspecific competition is reduced.

The same kind of argument can be made for age classes. For example, Enders (1976) points out that successive arthropod instars often have size ratios of about 1.3, identical to Hutchinson's between-species ratio (Section 6.5.1). Van Horne (1982) found that juvenile deer mice (*Peromyscus maniculatus*) consumed more low-quality food when adults were dense than otherwise — a within-species niche shift similar to the between-species shifts discussed in Section 6.6.2. While numerous other examples of niche differences between age classes exist, many must simply be by-products of necessary features of the life cycle, e.g. growth in size (Schoener 1977). Moreover, for both sex and age differences, behavioural strategies (mating, juvenile avoidance of cannibalism by adults) play a large, perhaps dominant role compared with intraspecific competition (e.g. Stamps 1983).

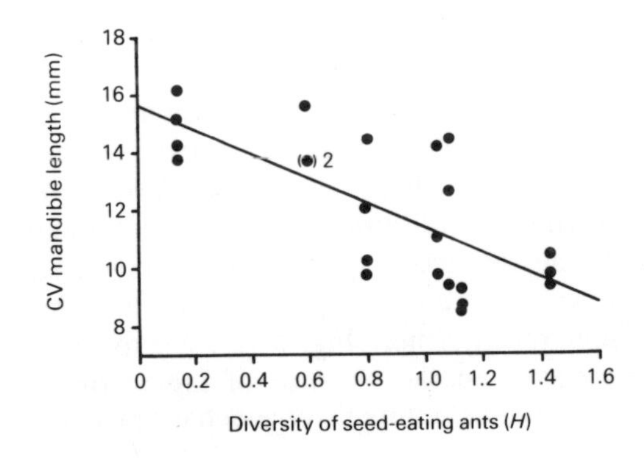

Fig. 6.15. The relationship between the within-colony coefficient of variation in mandible length and the species diversity of seed-harvesting ants (*Veromessor pergandei*) in the community ($r = -0.71$; $P < 0.01$). Each point includes the measurements of 30 workers from one colony (from Davidson 1978).

6.7 Experimental studies of resource partitioning

6.7.1 Experimentally produced changes in the niche

Field manipulations of populations are now relatively routine. At least 164 such studies of interspecific competition alone existed by the end of 1982 (Schoener 1983b). These manipulations have sometimes resulted in changes in species utilizations.

By far the most common niche change is in habitat: 22 such field experiments have been done, some quite detailed. The majority involve rodents, a relatively elaborate example of which are studies by Grant (1969, 1971). *Microtus pennsylvanicus* and *Clethrionomys gapperi* co-occur over much of the North American continent, on which the former is restricted to grassland and the latter to woodland. On Newfoundland, where *Microtus* is absent, *C. gapperi* occurs in both grassland and woodland. In experimental enclosures on the continent, *C. gapperi* expanded into grassland when *M. pennsylvanicus* was removed but not when maintained (Grant 1969). A similar set of experiments and results were later obtained with *Peromyscus* and *Microtus* (Grant 1971). The competitive mechanism was apparently territorial aggression.

Another set of experiments producing habitat shift, but by an entirely different mechanism, was performed by Werner and Hall (1979) on sunfish (*Lepomis*). Each of three species coexisting over much of North America was shown experimentally to prefer when alone the vegetated zone of ponds rather than the sediment or open-water zone. However, when placed together, only the green sunfish (*L. cyanellus*) remained in the vegetation; the pumpkinseed (*L. gibbosus*) shifted to sediments, and the bluegill (*L. macrochirus*) shifted to open water. Werner and Hall showed that the green sunfish was most efficient at feeding in the vegetation, and that, when it was present, vegetation was no longer as profitable for the other two species as was some other habitat. Thus here the principal mechanism is exploitative competition. The habitat shift closely corresponds to that envisioned by the compression hypothesis (Section 6.4).

Though demonstrations are much rarer, habitat shifts have also been experimentally produced in lower animals, including parasites. Holmes (1961) introduced tapeworms (*Hymenolepis diminuta*) and spiny-headed worms (*Moniliformis dubius*) together and separately into laboratory rats. In the absence of spiny-headed worms, tapeworms expanded their habitat toward the anterior part of the gut, where their carbohydrate food was apparently richer (Fig. 6.16). No reciprocal shift by spiny-headed worms occurred in the absence of tapeworms. In this relatively unobservable system, the competitive mechanism is unknown.

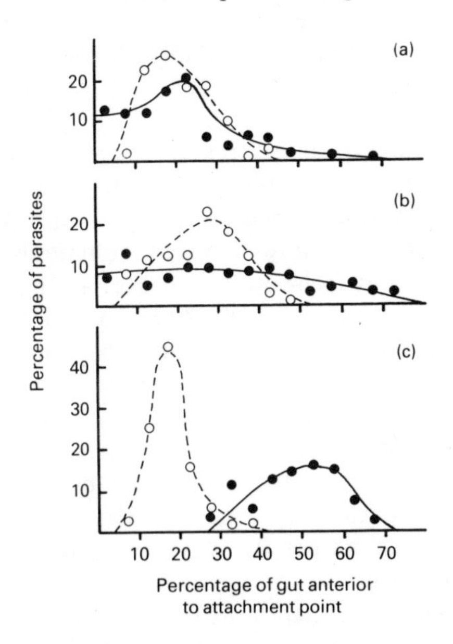

Fig. 6.16. Effects of single-species [(a) low density, (b) high density] and mixed infections (c) on the intra-intestinal distribution of the tapeworm *Hymenolepis diminuta* (———, ●) and the acanthocephalan *Moniliformis dubius* (○, – – – –). The data points are from Holmes (1961). The curves were drawn by eye (from Colwell & Fuentes 1975).

Experimentally produced shifts along niche dimensions other than habitat are much rarer. I know of only two such cases for activity time. The intertidal starfish *Leptasterias* and *Pisaster* compete for food (Menge 1972). *Pisaster*, the larger species, also interferes with *Leptasterias* by attacking it with tiny pinchers called pedicellariae (Menge & Menge 1974). During days when experimentally added *Pisaster* were inactive, *Leptasterias* were more active. The second case in which activity time was experimentally curtailed by the introduction of another species involves ants; here the mechanism is territorial aggression (Stebaev & Reznikova 1972). I know of only one direct experimental study of a competitively caused change in prey size, that by Pacala and Roughgarden (1985) on *Anolis* lizards. When two species relatively unalike in size were placed together in field enclosures, a variety of competitive effects occurred, one of which was that the larger lizard took *smaller* prey in the presence of the smaller species than when alone. It also ate fewer prey per unit time. These results appear to be consistent with the compression hypothesis, especially as the experiment was very short-term, lasting only a few months. Indeed, the relative rarity of experimentally produced shifts along time and diet dimensions is in general consistent with foraging theory (Section 6.4).

6.7.2 *When does resource overlap imply competition? Experimental investigations*

Although competition has often been produced in the laboratory between species overlapping in their resources (e.g. Gause 1934; Titman 1976), similar experimental demonstrations are less easily produced in the field. Moreover, equation (1), the expression most commonly used to compute the competition coefficient from ecological overlap, has been questioned as appropriate for the field (Colwell & Futuyama 1971; Vandermeer 1972). Certain field experiments have directly or indirectly addressed this question (Schoener 1983b). In six experiments, high ecological overlap was associated with high competition (as estimated from the experimental dynamics) and vice versa; all involved microhabitat or food-type dimensions, five exclusively so. In contrast, low overlap in macrohabitat has in almost all experiments performed been associated with high competition or vice versa. Thus experimentally, overlap in macrohabitat seems a relatively poor estimate of the competition coefficient. Though reasons for this may vary from one situation to another, an important one is that macrohabitat separation may often be the result of competition: by territorially excluding one another, or by wiping one another out over large areas in other competitive ways, species may come to differ in macrohabitat.

Even though resource overlap may be basically related to the competition coefficient, equation (1) may be too simple to estimate it adequately. An actual derivation of this equation results in more kinds of terms than just the *p*'s: in particular, terms for how much prey an individual of each competing species consumes, caloric values of the prey types, and relative abundances of prey types in the environment are included. [See Schoener (1974c) for this expression.] Thus researchers need to ensure that an adequate overlap formula is being tested. Recent papers evaluating various overlap measures are, among many others, Abrams (1980), Linton *et al* (1981), and Slobodchikoff and Schultz (1980).

Finally, a theory of limiting similarity based on equation (1) and the Lotka–Volterra equations (Section 6.5.2) may not be adequate for describing how resource overlap affects competition. Alternative models (e.g. Schoener 1974d, 1976, 1978; Titman 1976), more mechanistical in their representation of resource partitioning, now exist, and such models can be successfully applied to field situations (e.g. Belovsky 1984, see review in Schoener 1983b).

6.8 Factors other than competition affecting resource partitioning

Ecological communities are structured by many processes, only one of which is competition. These include: (i) other biotic interactions, such as predation (Connell 1975) and mutualism (Boucher 1982), (ii) abiotic factors, such as wave disturbance (e.g. Paine & Levin 1981) and climatic effects (e.g. Andrewartha & Birch 1954; Wiens 1977), and (iii) migration (e.g. Kareiva 1982). Moreover, competition itself is often not exploitative, that is for resources, but may involve the collection of effects labelled interference: territoriality, poisoning, injury or death from encounter, and overgrowth.

Though interaction and coregulation must often occur (see below), considerable controversy exists as to which process is most important (see review in Schoener 1982).

For example, the community composition of coral-reef fishes has been explained in two ways. Sale (1977) and Sale and Williams (1982) believe it is primarily structured by migration. In this scheme, individuals of different species are relatively interchangeable so far as particular units of space are concerned; when an individual disappears, the species replacing it is a function of the larval frequency in the vicinity at the time, itself relatively unrelated to local conditions. On the other hand, Anderson *et al* (1981) find coral-reef fish communities to follow classical resource-partitioning patterns, suggesting the predominance of exploitative competition. Both views allot considerable proximate importance to interference competition (interspecific territoriality) as well. Theoretical treatments of the coral-reef system are available that bolster both views (Chesson & Warner 1981; Abrams 1984). For both explanations, the greater the temporal variability in population parameters, the more likely is species coexistence.

A second controversy involves communities of birds occupying arid vegetation. Cody (1968, 1974), studying grassland communities, was able to interpret most of his data in terms of competition (Section 6.5.4). Wiens and Rotenberry, on the other hand, studying shrub-steppe communities, were unable to do so. Species co-occurring at local sites did not show year-to-year variation in population size that was explainable as a tight tracking of a variable resource base, nor did they show much regularity in resource-partitioning patterns, especially with respect to food size (e.g. Wiens & Rotenberry 1980; Rotenberry & Wiens 1981). The communities studied by Cody, and by Wiens and Rotenberry, respectively, were largely in different habitats, but one suspects that this is not the only reason for the discrepancy.

A final example involves the salamander genus *Desmognathus*. Hairston (1980a), studying this genus in the Appalachian mountains of North Carolina, found competition inadequate as a sole explanation for its niche structure. Certain habitat and size shifts expected on the basis of competition did not take place. Moreover, the group shows an evolutionary trend away from habitats of most intense predation, i.e.

those occupied by predatory aquatic salamanders. Though the ecological evidence was mostly negative, Hairston concluded that predation was the most important factor structuring the *Desmognathus* guild. On the other hand, Krzysik (1979) was able to interpret a similar guild of salamanders located slightly farther north, in the Pennsylvanian Appalachian mountains, largely according to classical competition theory. Size relations between sympatric species were very close to Hutchinson's ratio (Section 6.5.1), although size indicated microhabitat rather than food size in this group. Values of d/w for prey-size utilizations were substantially smaller than 1.0, probably because species utilizations were multidimensional. Habitat shifts and changes in niche width were interpretable in terms of the presence or absence of competing species. Incidentally, competition has experimentally been shown important in other salamander guilds (Hairston 1981; Jaeger 1974; and references), though the experiments have not yet ascertained its mechanism.

Interspecific differences in resource use, though often resulting from exploitative competition, may also result from other processes. We have already seen how territoriality can produce macrohabitat separation (Section 6.7). Selection for reproductive isolation was for a while the principal hypothesis alternative to that of competition for explaining size differences between species (review in Schoener 1982). However, other modalities — colour and pattern — would seem to be more appropriate for reproductive isolation than size, as the latter affects so many vital functions. Interestingly, an interpretation of species differences in terms of reproductive isolation can have much the same structure as one involving competition, as Greenfield and Karandinos (1979) showed in their study of moth pheromones. In that study, 'dimensions' were complementary, but less so than in the strongest ecological examples (Section 6.5.4).

Predation also can produce differences between species. A predator that concentrates more on prey of a certain appearance, the more encounters it has with that prey, can favour the evolution of 'aspect diversity', in which prey species evolve to appear unalike (e.g. Clarke 1962; Rand 1967). Indeed, such aspect diversity can also be treated in ways analogous to competitively structured niche differences. For example, Ricklefs and O'Rourke (1975) found that tropical Lepidoptera were no more similar in appearance on average than temperate ones, even though they occurred in communities with more species. Instead, the total 'aspect space' occupied by tropical species was greater. Exactly the same seems true of resource use and associated morphological characters of tropical birds in comparison to temperate ones (Karr 1971; Orians 1969; Schoener 1971). Tropical bird species are less apt to feed exclusively on animals, and they feed on a greater range of food sizes than do temperate species. In addition to their evolutionary effect, predators can directly or indirectly cause differences in habitat use over ecological time: Paine (1981) gives examples for barnacles (but see Newman & Stanley 1981), and the necessity to find suitable cover for certain rodent species can result in habitat specialization (Brown 1975).

Whether predation or competition is most important in structuring a particular guild may depend upon that guild's trophic position. In the seminal paper on this subject, Hairston *et al* (1960) argue that carnivores and autotrophs are more likely to be controlled by competition, whereas phytophagous herbivores, because of their intermediate trophic position, are more likely to be controlled by predation. A rather substantial amount of support, especially from field experiments, exists for their view, particularly if granivores and nectivores are included with carnivores (Schoener 1983b). Some striking cases of resource overlap occur in phytophagous insects; e.g. certain leafhoppers (Ross 1957) or a guild of tropical hispine beetles that all feed on *Heliconia* (Strong 1982). Lawton and Strong (1981), recently reviewing the literature on phytophagous insects, conclude that predation, especially by parasitoids, is likely to be the major controlling factor. However, strong competition among a few phytophagous insects has been experimentally shown (leafhoppers: McClure & Price 1975) or inferred by comparisons with null models (grasshoppers: Joern & Lawlor 1980).

The situation for large phytophagous herbivores, however, is likely to be ambiguous or even reversed: such species may not have much predation upon themselves because of their size. For example, Belovsky (1984) found strong competition between moose (*Alces alces*) and snowshoe hare (*Lepus americanus*) in forests on Isle Royale. He also (personal communication) has shown that large herbivorous mammal species of western North American grasslands (e.g. bison) can compete with one another and even with grasshoppers. Belovsky hypothesizes that resource partitioning in this group is accomplished by size: small species require more digestible food but are able to subsist on smaller packets of food than large herbivores.

No doubt in many guilds several factors, e.g. competition and predation, are acting in concert to produce niche differences (e.g. Toft 1985). Resource partitioning among hermit crabs is one such example: here, the resource is shells, and these appear in most communities to be in short supply and hence the object of interspecific competition (e.g. Vance 1972; Abrams 1981b). Yet the major function of the shell is as protection against predators and physical factors, and indeed, different shell morphologies are differentially useful in this regard (Bertness 1981b). Thus the threat of predation can motivate competition.

In theory, competition and predation can act together to regulate a community; Vance (1978) has shown this for species exhibiting a trade-off between competitive ability and ability to avoid predators. Moreover, the importance of competitors, as opposed to predators or physical factors, can oscillate seasonally (Koch 1974) or vary from one ecosystem to another (Menge & Sutherland 1976). Indeed, our sometimes explicitly but more often covertly Popperian attitude toward hypothesis testing has suppressed the development of multivariate explanations in ecology (Quinn & Dunham 1983). In the future we shall doubtless see a less unifactorial explanatory approach for ecological phenomena such as resource partitioning.

7 Trophic Relations of Decomposers

Yasushi Kurihara and Jiro Kikkawa

7.1 Introduction

We have seen that resources are partitioned by organisms belonging to different guilds (Chapters 5 and 6). In terms of trophic functions the resources represent utilizable energy and chemicals that bind the energy in special forms to support vital life processes. Apart from the chemosynthetic bacteria, which utilize inorganic matter as their source of energy, the autotrophs are photosynthesizers which produce organic matter using radiation energy from the sun. They are the producers of almost all ecosystems. All other organisms are heterotrophs, which obtain energy from organic matter; those utilizing living organisms are grouped as consumers (herbivores, carnivores and parasites) and those utilizing dead organic matter are grouped as decomposers (scavengers and detritivores). Although specific food-chains (such as predator chains, parasite chains and decomposer chains) or a complex food-web of a community may be constructed realistically with species studied, the trophic levels of a community are considered abstractions when they are grouped as Producers → Consumers → Decomposers to indicate the general flow of energy. While energy dissipates, matter is circulated in the biosphere. Thus Decomposers → Producers indicates the recycling of elements. Such idealized relations of trophic levels have been challenged in recent years as the role of microbes has been studied in a variety of ecosystems.

In this chapter we show that the trophic organization of a community (Fig. 7.1) is commonly one in which the primary consumers or so-called herbivores obtain much of their required energy from decomposers, and the secondary consumers (carnivores) from decomposer-based food-chains. The proportion of plant biomass used directly by the herbivores is small compared with the proportion that dies and enters the decomposer-based chain. True herbivores, which digest plants with their own enzymes, are not numerous or comparable in biomass to the producers in stable ecosystems. From the plant's point of view the harvesters, grazers and browsers are predators whereas those causing damage from inside the living plants are parasites. The nectar feeders and fruit eaters are often, but not always, pollinators and seed dispersers, respectively. Seed eaters are predators though they may also be dispersers. Thus the arms race or coevolution between plants and animals produces intricate energy relations within the community.

In this chapter we shall examine the role of decomposers in the terrestrial and aquatic communities along the proposed pathways of energy (Figs 7.1 and 7.7).

7.2 Terrestrial communities

7.2.1 *Primary production as a source of food*

The terrestrial ecosystem is characterized by the presence of enormous plant biomass. Primary production is dependent on solar radiation, but only about 3.7% of available light energy is used by the temperate forest and 1.4% by the tropical rainforest, where available solar energy is ten times as great as in the temperate region (Petrusewicz & Macfadyen 1970). Both the biomass and gross productivity of tropical forests are much greater than those of comparable temperate forests but, as respiration by tropical plants is increased disproportionately, the net production is similar to that of temperate forests. This means that the amount of energy becoming available to consumers as a result of production in mature forests is not necessarily greater in the tropics than in the temperate region.

In most terrestrial ecosystems much of the net primary production enters the decomposer chain without being utilized by the herbivores. Petrusewicz and Macfadyen (1970) estimated from various IBP studies that 50% of net production in temperate forests and 63% of net production in tropical forests entered the decomposer systems directly. Another IBP study revealed that, of the 9.8 t/ha annual net primary production above ground in a mixed deciduous forest, only 1% was consumed by herbivores while 66% entered the soil-litter system as dead wood and leaf litter (Satchell 1974).

Pferiffer and Wiegert (1981) found a significant relationship between the annual net primary production and the total consumption by herbivores

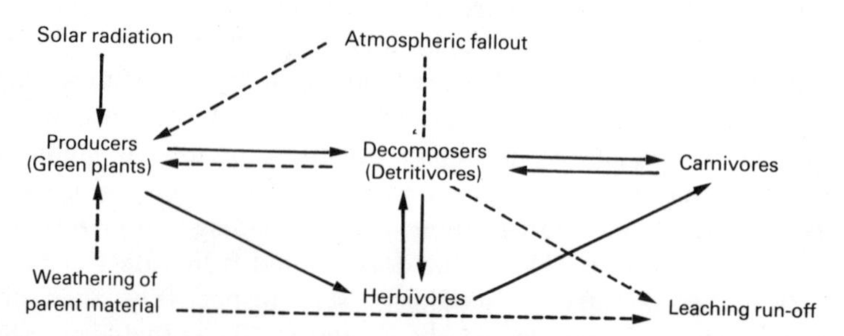

Fig. 7.1. Energy flow (———→) and independent nutrient movements (– – –→) through the trophic levels of a terrestrial community.

in grassland devoid of ungulate populations. The energy consumption by herbivores averaged about 8% of the net primary production. Similar proportions of grass production in Serengeti were consumed by grasshoppers but the total consumption by herbivores increased up to 38% (short grass area) when the ungulate populations were included. Grazing by cattle or sheep in the northern temperate grasslands produced similar results. Thus the proportion of grassland production utilized by herbivores is small compared with direct utilization by detritivores. Also, of the part utilized by herbivores a greater proportion is decomposed by gut microbes of ungulates than assimilated by leaf chewing and sap sucking insects. Perhaps the most significant direct pathway from primary production would be to the seed predators (Janzen 1971; Smith 1975). Many species of insects, birds (Kendeigh & Pinowski 1973) and small mammals (Hayward & Phillipson 1979) in the semi-arid regions of the world depend on plant seeds, particularly nutrient-rich and toxin-free grass seeds, as their main sources of energy. Two species of harvester ants, *Chelaner whitei* and *C. rothstein* in Australia, for example, consumed as much as 25.2 kcal/m^2 of seeds in a good year, which is an substantial portion of the total energy consumption by herbivores in any arid zone ecosystem (Davison 1982).

Other major sources of plant food in terrestrial ecosystems are nectar, fruits and nuts, which have all developed through coevolution between plants and animals. Many insects have evolved mechanisms to cope with or even to utilize plant allelochemicals for non-nutritional purposes (Reese 1979).

In the following sections we examine the roles of two major decomposer groups — detritivores of forest floor and rumen microbes.

7.2.2 *Detritivores of forest floor*

By far the greatest detritivore activity in the terrestrial ecosystems occurs on the forest floor. Dead organic matter accumulates in peaty swamps and sphagnum bogs where the rate of decomposition is extremely slow, but in most forest environments a steady state of turnover is reached after the formation of steppe matting or stratification of soil with an organic layer (A horizon). The amount of accumulated organic matter in the forest floor is greatest in shrub tundra (83.5 t/ha) and decreases with decreasing latitude through taiga (30–45 t/ha), deciduous broad-leaf forests (15–30 t/ha), subtropical forests (10 t/ha), meadow steppe (6–12 t/ha) to tropical rain forests (2 t/ha) (Rodin & Bazilevich 1967). In spite of large amounts of litter fall in tropical forests there is generally little accumulation of organic matter on the forest floor. The ratio of biomass between the forest floor and the litter fall (Fig. 7.2) reflects a high turnover rate of biomass in the tropics.

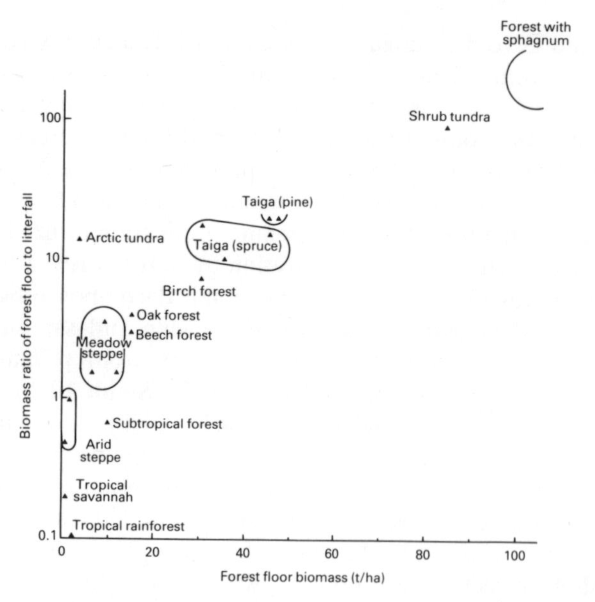

Fig. 7.2. Ratio of forest floor biomass to litter fall in relation to the amount of forest floor biomass in different forest types (data from Rodin & Bazilevich 1967).

Relative role of fauna in litter breakdown

The breakdown of leaf litter is due to abiotic (mechanical and chemical) and biotic (microbial and faunal) processes. Bocock *et al*'s (1960) field experiments suggested that the activity of soil macrofauna was significant in the disappearance of leaf litter. The soil macrofauna consists of earthworms, snails, millipedes, isopods, dipteran larvae and other arthropods feeding mainly on plant litter and converting it into fragments. Although their respiration accounts for only a small percentage of the total forest floor metabolism, their indirect control of decomposition processes through fragmentation of litter by ingestion and through their interactions with soil microbes is considered important in regulating the energy flow (Crossley 1977). Also, because of their relatively large size their biomass energy is transferred from the decomposer food-chain to the predator chain through the insectivorous birds and mammals of the forest floor, thus contributing to the higher trophic levels of the forest and slowing down further the return of nutrients to the soil.

It is possible to demonstrate the effect of soil fauna on the decomposition of leaf litter by using litter bags with varying mesh size to exclude certain size ranges of the fauna or by treating the forest floor with naphthalene. Zlotin and Khodashova (1980) dried the litter to a standard weight and placed it in wooden frames covered with Capron nets. A mesh size of 10 mm permitted all macrofauna, mesofauna and micro-organisms access to the litter, but the 1.1 mm mesh excluded the macrofauna and the 0.05 mm mesh excluded the macrofauna and mesofauna. The mesofauna consisted of mites, most enchytraeid worms, nematodes and Collembola. In other experiments they placed the litter in the frame on a small mesh sieve surrounded by a barrier of naphthalene and moistened the sample with clean toluene every 7–10 days. This reagent eliminated microbial activities in the sample, thus only the abiotic degradation of the litter was possible. The results (Table 7.1) showed that abiotic factors accounted for 8–23% and biogenic decomposition 77–92% of the total decomposition of the forest litter. This was in contrast to the decomposition of grass and herb litter for which the abiotic factors were largely responsible. In biogenic decomposition the most significant part was played by the complex of macrofauna — earthworms, millipedes and woodlice. The significance of mesofauna was somewhat less; the most influential being enchytraeid worms. The role of micro-organisms alone in decomposition of leaf litter was minimal.

However, soil macrofauna ingests a small portion of the available litter and can assimilate only a fraction of the ingested food. For instance, woodlice consume leaf litter corresponding to 20–82% of their body weight in a day but assimilate 4–53% of the food intake (Gere 1956; Hartenstein 1964; Watanabe 1967; Saito 1969; Dallinger & Wieser 1977; Soma & Saitô 1983). In the floor of an evergreen broadleaf forest three species of woodlice were estimated to consume 99.6 kcal/m^2, about 2% of the annual litter fall (Saito 1969). According to Edwards (1974) assimilation efficiency of millipedes varies similarly (3.8–50% of the ingested food), and ingestion by *Glomeris* (millipede) represented 1.7–10% of annual litter fall in oak forests (Bocock 1963).

Although most soil animals can digest proteins, fats, sugars, starch and hemicellulose, the main components of leaf litter are cellulose and lignin which require special enzymes for decomposition. Absence of these enzymes would account for their low assimilation rate in animals. Beck and Friebe (1981) examined the enzyme activity of gut extracts from millipedes and woodlice and found that the millipede *Polydesmus angustus* was able to hydrolyze cellulose, hemicellulose and pectin, whereas the

Table 7.1. Percentage contributions of abiotic factors and soil organisms to the decomposition of forest and grassland litter (from Zlotin & Khodashova 1980).

	Oak	Aspen	Elm	Grass/herb
Abiotic decomposition	12–23	14	8	66–73
Biotic decomposition				
microorganisms	11–15	5	7	10–18
mesofauna	24–37	21	37	6–12
macrofauna	29–49	60	48	3–12

woodlouse *Oniscus asellus* hydrolyzed only starch. However, Hartenstein (1964) had earlier found the activity of cellulase in the digestive tract and hepatopancreas of *O. asellus*, and Terada and Oshima (1970) demonstrated assimilation of cellulose powder in another species of woodlouse, *Armadillidium vulgare*. The problem is the digestion of ligno-cellulose *in situ*. Reyes and Tiedje (1976b) found that the woodlouse *Tracheoniscus rathkei* was unable to assimilate the ligno-cellulose fraction of C-labelled plant material.

Interactions between microbes and soil arthropod fauna

Enzymes capable of breaking down cellulose or lignin have not been found in phytophagous insects. Leaf cutting ants (Attini) harvest leaves in tropical forests of America for their underground fungus gardens. They keep pure cultures of fungi for food by weeding and tending. Because no aerial hyphae or fruit body develops the identification of the fungi is a problem, but they are thought to belong to the ascomycetes and the basidiomycetes which break down cellulose and lignin, respectively. Thus, leaf cutting by *Atta colombica* in Costa Rica was estimated to reduce gross primary production of the forest by 1.76 kcal/m^2/day (Lugo *et al* 1973) and the fungus garden of this species in Panama was considered to concentrate 1.7% of the total annual energy flow through leaf litter fall (Haines 1978).

Cellulose comprises 40–62% and lignin 18–38% of wood in dry weight. The rest of wood is made up of hemicelluloses, a mixture of polysaccharides including hexosans and pentosans, and a small amount of protein. Many species of termites (Mastotermitidae, Hodotermitidae, Kalotermitidae and Rhinotermitidae) harbour symbiotic flagellates (Protozoa) in their hindgut for cellulose fermentation and digestion (Honigberg 1970). In a eucalypt forest in South Australia, Lee and Wood (1971) found approximately 600 individuals of the termite *Nasutitermes exitiosus* with a biomass of 3 g/m^2, which were estimated to consume 16.6% of the total annual fall of sticks and logs or 4.9% of the total annual litter fall.

Termites of the subfamily Macrotermitinae build sponge-like combs with their primary faeces or macerated plant material to grow symbiotic fungi *Termitomyces* (basidiomycetes), which break down lignin (Sands 1969; Lee & Wood 1971). There is strong evidence that the first stage of digestion involves the breakdown of cellulose in the initial passage through the gut and the faeces containing lignin are built into the fungus combs which are consumed when lignin is broken down by fungi (Sands 1969).

These symbiotic associations between arthropods and microbes suggest the importance of their combined effects, even outside the symbiotic relations, on the decomposition of litter. Fragmentation of egested litter material must influence its subsequent decomposition by microbes and microbes in turn change the palatability of leaf litter and themselves become a food source for the meso- and macro-fauna.

Fresh faeces of millipedes and woodlice show higher bacterial density or biomass than the food litter (Reyes & Tiedje 1976a; Anderson & Bignell 1980; Hanlon 1981b). Anderson and Bignell (1980) suggested that increases of bacterial populations after passage through the gut were primarily the result of rapid growth of the indigenous litter bacteria rather than a population of specific gut symbionts. Reyes and Tiedje (1976a) suggested that the gut of soil animals favoured the growth of bacteria and influenced the subsequent microbial colonization and hence the rate of litter decay in faeces. Nicholson *et al* (1966) examined faecal pellets of *Glomeris* and found that bacterial numbers and oxygen uptake increased during the first 2 weeks of decomposition and then decreased to a relatively constant level. Van der Drift and Witkamp (1960) reported that evolution of carbon dioxide from faecal pellets of caddis flies reached a maximum about 5 days after egestion and then fell during the following 3 weeks to the level observed in the whole leaves. They considered that mechanical breakdown of leaves by the animals was important for stimulating microbial activity in the faeces. Hanlon (1981a) measured oxygen consumption of bacteria isolated from the faeces of *Oniscus* and growing separately on leaf litter of five different particle sizes. Bacterial respiration increased rapidly to a maximum at 6 days after inoculation, then declined rapidly. The maximum levels were correlated inversely with particle size. Nicholson *et al* (1966) found a marked succession amongst the 45 species of superficial fungi recorded from the decomposing pellets of *Glomeris*. Mucoraceous fungi, especially *Mucor hiemalis* and *M. ramannianus*, were common up to the fourteenth day of decomposition, after which they were replaced by a wide range of Fungi Imperfecti and sterile forms.

Although bacteria have a wide range of decomposing capabilities, fungi are considered to be the major decomposers of cellulose and lignin (Gyllenberg & Ekland 1974). Factors influencing the proportions of bacteria and fungi colonizing a resource may affect the subsequent microbial succession and the rate of litter decay. For example, Plowman (1979, 1981) found a singificant difference in pH leading to different modes of litter breakdown between a subtropical rainforest and an adjacent sclerophyll forest with a tall canopy of *Eucalyptus* and *Tristania*, in southeast Queensland. She suggested that tannins formed by polymerization in the low pH protect the leaf litter with a high polyphenolic content from bacterial decomposition. Thus the decomposition of sclerophyll litter is slower and favours fungi and termites, whereas mineralization of rainforest litter is aided by bacterial action. The species configuration of mites (Cryptostigmata and Mesostigmata) was different between the litter and soil faunas of the two forests.

Do forest soil animals then have definite preferences for different species of leaf litter? The earthworm *Lumbricus terrestris*, with a habit of pulling surface litter into its burrows, was used in selection experiments by Satchell and Lowe (1967). They cut discs from the lamina of leaves of broadleaf species and pieces of conifer needles and put them in containers placed on soil, so that the number of litter items removed could be recorded. The results showed that the leaf materials derived from elder, elm, alder, black poplar, Myrobalan plum, ash and sycamore were highly palatable, those from birch, spindle, aspen, field maple, lime, guelder rose, Japanese cherry, hazel, horse chestnut and walnut were moderately palatable, and those from Virginia creeper, beech, oak, spruce, pine and larch were unpalatable to the earthworm.

In the unweathered litter, the order of palatability is strongly associated with the tannin content of the leaves. Thus the microbial degradation of tannins is important in increasing the palatability of litter during weathering. Cameron and LaPoint (1978), in a laboratory feeding experiment, demonstrated that woodlice consumed ground, leached leaves of tallow at a much faster rate than they did untreated leaves, and suggested that the tallow leaves were not utilized by reducers until tannins were leached out and the physical structure was altered by abiotic and/or microbial actions. Satchell and Lowe (1967) also examined water soluble extractives in the leaves to see if these, unlike insoluble high molecular weight compounds, possessed certain chemotropic or nutritive characteristics. They found a broad relationship between palatability and carbohydrate content, but individual litter materials did not always conform to the general pattern; alder litter increased in palatability with weathering but decreased in soluble carbohydrate content, whereas weathered larch, which was less palatable than weathered beech, contained a considerably higher percentage of soluble carbohydrates.

Plant toxins have inhibitory effects on decomposers generally and slow the rate of litter decomposition (Middleton 1984). The type and amount of polyphenols were also suggested as determinants of palatability (King & Heath 1967; Edwards 1974), but the extensive preference experiments conducted by Neuhauser and Hartenstein (1978) failed to establish any correlation between the phenolic content of leaves and the feeding preferences of millipedes and woodlice.

The decomposition of litter by fungi brings about changes in its chemical composition and physical property. All saprophytic fungi can metabolize simple carbohydrates, but only the ascomycetes and the deuteromycetes are considered to have cellulolytic enzymes while the basidiomycetes break down lignin. Hudson (1968) proposed a generalized schema for the fungus succession on plant tissues. The first group to appear is the parasites invading living tissues. Some of them persist until the host becomes senescent or dies. Next come the primary saprophytes, including sugar fungi which metabolize only easily available sugars. The secondary saprophytes are mostly ascomycetes and deuteromycetes but basidiomycetes may also be involved.

Saitô (1965, 1966) found two sequential patterns in the decomposition of beech litter. In sequence one the brown leaves decayed into yellowish, then mouldy and finally fibrous leaves by the actions of basidiomycetes, *Collybia* spp. and *Mycena* spp. In this process, vigorous degradation of lignin and an increase of water soluble extractives were observed. In sequence two, slow degradation of cellulose occurred through the actions of other fungi. Table 7.2 shows a succession of fungi which produced similar decaying processes in the pine needles (Soma & Saitô 1979). In general, the darkening of leaves (sequence two) occurred progressively through brown needles in the litter layer to dark brown needles and blackish brown needles in the fermentation layer. On the other hand, the decolouring process (sequence one) was characterized by the overwhelming growth of basidiomycetes, producing yellowish needles and compactly compressed, decomposed needles in the fermentation layer. These two sequences, however, were so closely linked that it was virtually impossible to separate them in the field.

How do these processes affect the palatability of litter for soil animals? Saitô (1966) found masses of faeces of soil animals under the hymenomycete-infested leaves in the litter layer (sequence one). On the other hand, the brown and greyish brown leaves undergoing the brown rot process (sequence two) were not favoured by soil animals. When the larvae of the scatopsid fly were bred on the two types of decaying litter in petri dishes, the mouldy leaves were eaten and transformed into faeces rapidly while the brown and greyish brown leaves remained almost intact. Latter and Howson (1978) reported the *Calluna* litter inoculated with basidiomycetous mycelium gave the best growth of enchytraeid worms though the mycelium alone caused the highest death rate of the worms.

Woodlice and millipedes are known to utilize decomposed wood. Neuhauser and Hartenstein (1978) used fungi, such as *Polyporus versicolor*, *Lenzites trabea* and *Fomes pini*, to decompose wood blocks in the laboratory and then fed them to woodlice and millipedes. None of these 'artificially' decomposed materials was ingested, though the naturally decomposed wood was readily consumed. The laboratory decomposed wood was apparently sturdier than the naturally decomposed wood and the animals were not able to macerate the wood fibres.

Some of the macrofauna are known to consume fungal materials. Lewis (1971) found fungal hyphae and spores in the gut of millipedes. The gut of woodlice inhabiting coastal pine plantations consisted mainly of pine needles and partly of fungal hyphae (Soma & Saitô 1983). The proportion of fungal hyphae varied with season, age of animals and fungal density in the litter layer. The woodlice, inhabiting the mouldy

Table 7.2. Major species of fungi (excluding mycorrhizal fungi) associated with decaying of pine needles (from Soma & Saitô 1979).

	Living needles	Brown needles	Dark brown needles	Blackish brown needles	Yellowish needles	Yellowish needles (decayed)	H layer
Aureobasidium pullulans (de Bary) Arn	c	f					
Cladosporium spp.	c	f					
Pestalotia sp.	c	f					
Phoma sp.	f						
Lophodermium pinastri (Schrad.) Chev	a	a					
Cenangium sp.	a	a	a	f			
Desmazierella acicola Lib		f	a	a			
Endophragmia alternata Tubaki & Saitô			a	a			
Unidentified-F1051			a	a			
Kriegeriella mirabilis Höhn			c	a			
Marasmius androsaceus (L. ex Fr.) Fr			a	a			
Collybia confluens (Fr.) Quél					a	a	
Collybia dryophila (Fr.) Quél					f	f	
Mortierella ramannianus (Möller) Linnem				f			c
Penicillium spp.				c			a
Trichoderma spp.				c			a

a: abundant, c: common, f: frequent

needle layer contained much more fungal hyphae in their gut than those inhabiting the fresh needle layer. Moreover, the proportion of hyphae in the gut content rose to 2.5 times the volume of mouldy needles. When the pieces of mouldy needles and those stripped of mycelial wefts were presented, the woodlice consumed all mycelial wefts on the mouldy needles and a substantial amount of those needles themselves, while leaving the bare needles intact. Soma and Saitô (1979) compared the population density of the woodlouse *Porcellio scaber* between the mycelial zone produced by basidiomycetes and the zone of outer scanty mycelia. As shown in Table 7.3 the population density in the concentrated mycelial zone was significantly higher than that of the outer zone in every plot. These high densities were caused by aggregations on the fleshy basidiocarps and the mouldy needles. A considerable number of woodlice aggregated on the basidiocarps of *Lactarius hatsudake* and *Suillus* spp., particularly in the hollow stem of *Lactarius* after the central part of its depressed cap had been eaten out.

Fungi accumulate substantial concentrations of nutrients in forest ecosystems. Fungal rhizomorph tissues contain significantly greater concentrations of calcium, potassium and sodium and fungal sporocarps contain significantly greater concentrations of copper, potassium, sodium and phosphorus than forest floor leaf litter, either hardwood or conifers (Cromack *et al* 1975). Stark (1972, 1973) found that calcium concentrations in fungal rhizomorphs and sodium concentrations in both rhizomorphs and sporocarps were several-fold higher than in tree foliage. Both fungal hyphae and rhizomorphs generally contain calcium concentrations which are much greater than their ambient substrates. Fungi typically contain some 3–5% dry weight of nitrogen and some 5–10% of other inorganic nutrients; by contrast, leaf litter contains on average less than 1% nitrogen and 3.5% other nutrients, and woody tissues much less nitrogen (0.1–0.2%) (Harley 1971). Concentration of nutrients also occurs underground by the actions of nitrogen fixing bacteria and root fungi. Further concentration may occur as a result of translocation and defaecation by the soil macrofauna. Thus the woody substrate on the forest floor may trap

Table 7.3. Population density (number of individuals/25 × 25 cm) of *Porcellio scaber* around basidiocarps (mycelial zone) and 50 cm outside (non-mycelial zone) (from Soma & Satiô 1979).

	Lactarius hatsudake plots						*Suillus* spp. plots			
	A	B	C	D	E	F	A	B	C	D
Mycelial zone	33	25	18	16	9	7	68	33	22	12
Outer zone	5.3	1.3	0.3	2.3	1.0	0.0	10.3	2.7	1.7	2.0

upward movement of minerals and become a nutrient sink (Ausmus 1977).

Some of the mesofauna feed exclusively on fungal hyphae (Harding & Stuttards 1974). Hartenstein (1962) reported that oribatid mites preferred *Trichoderma*, *Cladosporium*, *Phialophora* and *Stemphylium* to *Aspergillus* and *Penicillium*. Mitchell and Parkinson (1976) also concluded that *Trichoderma* were the most suitable food sources for oribatid mites. Booth and Anderson (1976) examined the growth and fecundity of *Folsomia candida* (Collembola) feeding on cultures of basidiomycetes incubated with different N levels. The rate of moulting and egg-laying increased as a result of increased availability of nitrogen to the fungi, indicating the influence of fungal food quality. They were also affected by the species of fungus presented as food to the Collembola. Anderson (1975), who studied a succession of oribatid mites in decomposing leaf litter, found that a number of mites in the litter bags fed on fungal materials during the initial stage of decomposition and later fed on decaying materials.

In summary, the decomposition processes of forest floor litter convert largely unavailable energy and nutrients of primary production into new forms and biomass, which are then exploited by animals of higher trophic levels. The amount of nutrients trapped in the living tissues of decomposers and predators is small compared with the enormous plant biomass, but evidence is accumulating that detritivores play a vital role in regulating the mobility of essential nutrients (Fahey 1983; Vogt *et al* 1983). In a mature ecosystem a steady state will be maintained so long as leaching losses are not great. The nutrient balance may be achieved by weathering of parental material or atmospheric inputs (Jordan 1982).

7.2.3 Microbes of the rumen

Cellulose is the chief structural polysaccharide of primary producers, which cannot be utilized by herbivores without cellulolytic enzymes of symbiotic microbes. Pregastric fermentation may have originally developed to detoxify plant secondary compounds, but the anaerobic gastro-intestinal fermentation enabled production of volatile fatty acids (VFA) for use by the host herbivores (Van Soest 1982). The cellulose decomposers in the digestive tract of herbivores are bacteria and protozoa which occupy a trophic level between plants and 'herbivores'. In ruminants they are found in the rumen and the reticulum. Briefly, the plant material after ingestion and some mastication enters this region where it is acted upon by bacteria and protozoa. The food is then brought back to the mouth for further mastication (chewing the cud). When fully masticated the food material is returned to the rumen with large quantities of saliva.

The food particles now present a larger surface area for the fermentation activity of microbes. This active fermentation yields VFA, such as acetic, butyric and propionic acids, and lactic and succinic acids, which are neutralized by sodium bicarbonate from the saliva; if this were not so, the acids would kill the microbes. Lactic and succinic acids are further metabolized by bacteria to give VFA. Foodstuff protein is degraded to peptides and amino acids, and the latter can be degraded to VFA and ammonia. VFA — an important source of energy for the ruminant — passes directly to the bloodstream through the rumen walls. Ammonia formed from urea in saliva or protein in foodstuff is synthesized by bacteria into their own protein, or is absorbed by the animal from the rumen, and finally excreted as urea in the urine. The other products of the fermentation are the gases, methane and carbon dioxide, most of which are regurgitated into the air. At the same time a continuous stream of food residues, bacteria and protozoa, passes on to the omasum. Entering the abomasum, the protozoa and probably most of the bacteria are immediately destroyed by the acid environment of this chamber secreting hydrochloric acid and digestive enzymes. Food materials then follow a normal course through the intestines.

Protozoa

Various specialized bacteria and ciliate protozoa are found in the rumens of cattle, sheep and goats, and in camel, reindeer and wild ungulates. Other types have been known from the caecum and colon of the horse, African elephant, rhinoceros, American tapir, guinea-pig, rabbit, vole, chimpanzee, gorilla, and from the stomach of hippopotamus and kangaroo (Moir 1968).

The total number of protozoa in the rumen of a sheep on a concentrate and hay diet was estimated to be of the order of 10^6/ml rumen content (Kurihara *et al* 1968). They fall into two guilds; the oligotrichs and the holotrichs. The holotrich ciliates, *Isotricha* and *Dasytricha*, are mainly concerned with the fermentation of soluble sugars, though the larger *Isotricha* also digest small starch granules (Oxford 1951; Heald & Oxford 1953; Mould & Thomas 1958). Of the oligotrich ciliates, the larger species, (e.g. *Polyplastron multivesiculatum*, *Diplodinium denticulatum*, *Eudiplodinium neglectum*, *E. maggii*) are known to ingest plant particles including starchy and cellulosic materials while the smaller *Entodinium* species ingest mainly starch granules and contain an α-amylase (Abbou Akkada & Howard 1960). The principal products of fermentation of carbohydrates are acetic, butyric and lactic acids, carbon dioxide, hydrogen and intracellular polysaccharides of amylopectin-type (Heald & Oxford 1953; Gutierrez 1955).

In spite of intensive search no protective form or cysts of the ciliates have been found, which would have supported a mode of transfer from one animal to another through the ration (Eadie & Howard 1963). Although many species are not host-specific, direct mouth contact seems to be the main mode of transfer between animals.

For some time, doubt has been cast on the role of the protozoa in the normal rumen as bacteria can apparently perform all metabolic processes carried out by the protozoa. The absence of ciliate protozoa appeared to have no significant effect on the general maintenance of calves and lambs (Eadie & Howard 1963). Pounden and Hibbs (1950) reported, however, that a rough coat and more potbellied appearance was associated with protozoa-free calves.

Bacteria

The number of bacteria in the rumen has been estimated by counts of wet preparations, in a counting chamber. Kurihara *et al* (1968) found the number of free bacteria to vary, depending on time after feeding by the host, from 10 to 60×10^9/ml rumen content in sheep fed on hay and concentrate. The development of media for the isolation of rumen bacteria, and the use of washed cell suspension have led to a clearer conception of the role of many rumen bacteria. The method identified a number of morphological types which can hydrolyse cellulose, starch, hemicellulose and sugar. The important cellulolytic guild bacteria so far isolated and identified are: *Bacteroides succinogenes*, *Ruminococcus flavifaciens, R. albus, Clostridium longisporum, Butyrivibrio fibrisolvens*. The enzyme systems of these bacteria were studied in pure cultures and VFA, carbon dioxide, hydrogen and lactic and succinic acids were obtained as the ultimate products of their cellulolytic fermentation (Hobson 1963; Hungate 1966).

A number of cellulolytic bacteria are also amylolytic, e.g. *Clostridium lochheadii*, some strains of *Bacteroides succinogenes* and most strains of *Butyrivibrio fibrisolvens*. The purely amylolytic species are *Streptococcus bovis*, *Bacteroides ruminicola*, *B. amylophilus*, *Succinomonas amylolytica* and *Selenomonas ruminantium*. The ultimate products of the fermentation of starch include VFA, succinic and lactic acids, carbon dioxide and hydrogen (Hobson 1963; Hungate 1966).

Utilization of microbes and their products by the host animal

A large proportion of the energy required by the host animal is derived from the ruminal fermentation. Baldwin *et al* (1969) computed a yield of 3.92 moles of VFA/kg of alfalfa hay fed on their model and this value corresponded to 56% of the digestible carbohydrates. In Carroll and Hungate's (1954) calculations the amounts of VFA absorbed from the rumens of cattle on hay, or hay plus grain rations, represented about 70% of the total energy requirements of the animals.

On the other hand, nitrogen may become available to the host animal either by absorption from the rumen, or by passage to and absorption from the intestines either with, or without, further enzymatic actions. Rumen microbial protein absorbed from the intestine is made of good quality amino acids and of high digestibility. McNaught *et al* (1954) have found that dried cattle protozoans when fed to rats are a better source of protein than dried rumen bacteria or brewer's yeast. The protozoal protein is richer in essential amino acids, notably lysine, than the bacterial protein (Weller 1957).

The rumen bacteria and protozoans visibly disintegrate in the abomasum and proximal part of the small intestine. Tracer studies using the stable isotope ^{15}N enabled quantitative estimation of nitrogen transactions through an animal under normal nutritional conditions. Nolan (1975) presented a quantitative model, based on the tracer studies, of nitrogen transfer through the stomach of a sheep (Fig. 7.3) as part of a whole-animal model. In this model, about 61% of the non-ammonia nitrogen entering the duodenum is derived from the microbial nitrogen compounds.

The favourable conditions provided by the rumen permit growth of many species of bacteria and protozoa which in turn decompose otherwise indigestible materials to VFA, a main source of energy for the host, and also synthesize microbial protein, providing the host with amino acids not otherwise obtainable or available in sufficient quantity. This symbiotic relationship, mutualism, is between the *total microbial community* and

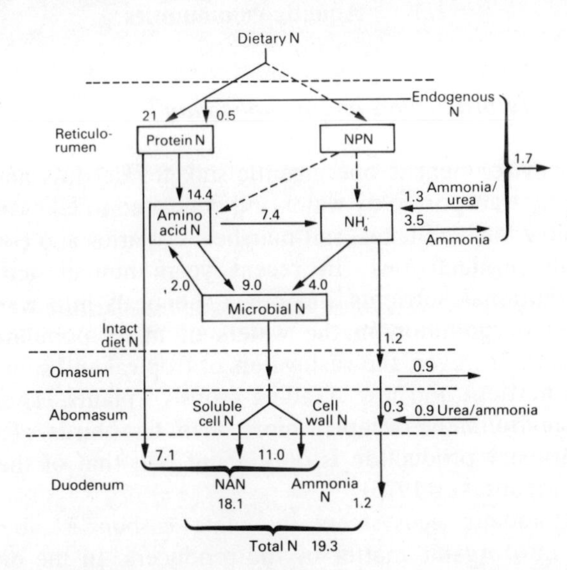

Fig. 7.3. A nitrogen flow chart through the stomach of a sheep, showing examples of main transfer rates in g N/day (NPN: non-protein N, NAN: non-ammonia N) (from Nolan 1975).

the ruminant host. Individual microbial species may perform functions that are not essential to the host. However, the role of each is not simply with the host but also with the other members of the total community. The evolution of such microbe–herbivore symbioses, with complex and obligatory relationships among microbe species as well as between microbes and their hosts (Hungate 1966; Van Soest 1982), may have been dependent on the evolution of social systems in the herbivores to permit transmission of microbial populations from generation to generation (Troyer 1984).

The role of microbes in the rumen is much more complex than that of the symbiotic bacteria or flagellates of termites acting as agents of cellulose fermentation. Termites may digest some bacterial cells and protozoa (especially at the time of moulting), but not to any significant degree. Also, they do not appear to have methane bacteria which in the rumen produce methane from hydrogen and carbon dioxide.

Hungate (1966) likened the utilization of micro-organisms by the ruminant to plankton feeding by filter-feeders in the ocean. The ruminant as a plankton feeder cultures its own 'plankton' under highly productive conditions, thus concentrating the food supply in a small volume and harvesting it continuously. Besides consuming the microbes this 'plankton feeder' utilizes some of the waste products formed during their growth.

7.3 Aquatic communities

7.3.1 *Primary production and its utilization*

The aquatic environment is one gigantic sink for detritus and nutrients, which are easily transported by water and deposited in lakes and seas. As a result, shallow lowland lakes, salt marshes, estuaries and coastal waters maintain high productivities. In recent years human activities have introduced additional nutrients and other chemicals into water, causing eutrophication or pollution in the waters of many populated regions. Away from land, however, the vast waters of tropical and subtropical seas contain little nutrient and low standing crops of plants. Thus, although the aquatic environment occupies more than two-thirds of the earth's surface, its primary production is only about half that of the terrestrial environment (Whittaker 1975).

In most aquatic ecosystems inorganic carbon is abundant and incorporated into organic matter by the producers. In the organic form carbon is mostly found as the dissolved organic carbon (DOC) but is also partly suspended as the particulate organic carbon (POC). The carbon in the tissues of living organisms forms only a small fraction of the total organic carbon in the environment. The amount of carbon fixed by the producers is often used as a measure of net primary production (1 g C is worth about 10 kcal) in aquatic environments. It varies from less than 1 g C/m^2/yr in some polar and alpine lakes and temperate rivers to 2500 g C/m^2/yr in tropical eutrophic lakes and coral reefs (Likens 1975; Lewis 1981).

The producers of the aquatic ecosystems may be divided into five major groups: the phytoplankton, the macrophytes, the periphyton, the symbiotic algae and the chemosynthetic microbes.

The phytoplankton is the major primary producer of open waters. In lakes its net productivity varies from 50 to 450 g C/m^2/yr (Likens 1975), whereas in the euphotic zone (<50 m) of continental shelves it is in the order of 200–300 g C/m^2/yr (Walsh 1981). In the sea the lowest phytoplankton production of less than 30 g C/m^2/yr has been recorded from the subtropical gyral region (Koblentz-Mishke *et al* 1970) where nitrogen and possibly phosphorus are considered to be limiting, and the highest of more than 1000 g C/m^2/yr from the Peruvian upwelling (Walsh 1981). In temperate regions the seasonal fluctuation in productivity is great and maximum phytoplankton growth (bloom) occurs in spring or early summer depending on the temperature cycle. The zooplankton usually becomes abundant about 1 month after the phytoplankton bloom, but as we shall see later much of its energy requirement is met from the detritus-based food-web. On the northeast continental shelf of North America, about one-third of the annual primary production in organic carbon is transferred to zooplankton and two-thirds to the detritus pool of the bottom (Walsh 1981). In the tropics the rich nutrients of the upwellings produce phytoplankton blooms. Off the coast of Peru an enormous amount of this primary production once supported the world's biggest biomass of anchovies. Today the utilization of phytoplankton by anchovies is much reduced (anchovies have been overfished). Excess phytoplankton is exported from the bloom centre and its energy transferred to detritivores. Outside the bloom patch it is largely consumed by zooplankton (Table 7.4).

The macrophytes are the macroscopic vascular plants and algae, either rooted or floating. Macrophyte production is seldom used by grazing animals, and in temperate springs and rivers freshwater macrophytes often die at the end of the growing season and enter the detritus food-web (Tilly 1968; Mann 1975). The dugong *Dugong dugon* and the manatees (*Trichechus*) are the only mammals that depend solely on aquatic vascular plants for food. The dugong in northern Australia consumes seagrasses of all available genera (*Halodule*, *Halophila*, *Cymodocea*, *Thalassia*) (Marsh *et al* 1982), and digests the cellulose with help of symbiotic microbes in the midgut caecum and the 30 m long colon where high concentrations of VFA appear (Murray *et al* 1977) as in

Table 7.4. Fate of primary production in pelagic ecosystems (from Sorokin 1981).

	Phytoplankton production (Kcal/m^2/day)	Energy transfer (%)			
		Microzooplankton	Mesozooplankton	Anchovies	Detritivores
Japan Sea (mid-June)	4750	33.1	12.2	0	54.7
Peruvian upwelling (Chimbote)					
Inside bloom patch	87 000	0.7	1.3	14.7	83.3
Outside bloom patch	17 900	11.2	82.1	6.7	0

horses and elephants. The intertidal and subtidal fleshy algal macrophytes are grazed by benthic and epiphytic herbivores (Underwood 1980), but as with salt marsh plants and mangroves they contribute mostly to the detritus food-web. Encrusting coralline algae also belong to the benthic primary producers which are little utilized by herbivores.

The periphyton consists of filamentous algae (green algae, blue-green algae, diatoms) growing on or in submerged substrates, sometimes forming a dense algal mat. This is the main source of food for aquatic herbivores, such as chitons, limpets, snails, sea-urchins and herbivorous fishes. In many detritus-based stream communities, the only primary producers may be periphytonic mosses not utilized by herbivores (Fisher & Likens 1973). In autotrophic lotic communities, however, diatoms may produce as much as 40 kcal/m^2/day, which can support the herbivorous fish *Plecoglossus altivelis* at a density of 5.4 individuals/m^2 (Kawanabe *et al* 1959). Otherwise, the grazer and scraper guilds of macroinvertebrates (gastropods and case-bearing insect larvae) ingest about 10% of primary production in the river ecosystems (Cummins 1975). In the temperate seas very few littoral fishes utilize periphyton (Wheeler 1980), but in the tropics many parrotfishes (Scaridae), surgeonfishes (Acanthuridae) and damselfishes (Pomacentridae) keep the filamentous algae very closely grazed (Ogden & Lobel 1978). A remarkable growth of periphyton has been observed on the beach rock of a coral cay by excluding the reef fishes in an exclosure experiment (Stephenson & Searles 1960).

The symbiotic algae are characteristic of coral reef ecosystems. These are endosymbiotic dinoflagellate algae, known as zooxanthellae, and are found in the polyps of all hermatypic corals and in the mantle of giant clams *Tridacna*. Together with the periphyton and other phytobenthos they are responsible for very high productivities of 300–5000 g C/m^2/yr (highest recorded, 11 680 g C/m^2/yr in Hawaii) in the otherwise nutrient-poor environment. Nitrogen is fixed by several organisms from both the atmosphere and water, enriching the reef community (Lewis 1981). Production is generally greater than the total respiration in the reef ecosystem. This means that organic matter accumulates in the lagoon and is also exported to open waters where the standing crop of phytoplankton is small. The zooxanthellae inside the corals are protected from herbivorous reef fishes while providing soluble products of photosynthesis to the corals and apparently controlling the recycling of nutrients.

Chemosynthetic microbes are not significant producers except in the recently discovered abyssal community around hydrothermal vents near Galápagos (Karl *et al* 1980). Unlike the low biomass communities of entirely heterotrophic organisms in the deep sea, the dense animal populations of this sea-floor spreading centre, 2550 m deep, are supported by chemosynthetic bacteria which use geothermically reduced sulphur compounds emitted from the vents to synthesize organic matter from carbon dioxide.

Although the fate of aquatic primary production is varied, much of it enters the detritus food-web directly. In the following we shall examine the trophic relations of detritivores.

7.3.2 *Detritus in aquatic environments*

Because organic materials are close to neutral buoyancy in sea water, the aquatic environments, especially the coastal waters, contain a large quantity of suspended organic matter. The nonliving portion of it is significantly greater than the living portion and includes fragments of plant and animal tissues, faecal material and particles formed from dissolved organic matter in the water. Apart from the fragments of plankton and faecal pellets that can be identified morphologically under a microscope, there are two kinds of amorphous particles known as flakes and flocs (flocculent aggregates) (Riley 1963, 1970). Both flocs and flakes are thought to be produced in the water by the condensation of dissolved organic matter on surface films. Macroscopic particulate matter is commonly seen in the water by divers and is often called marine snow. There is a continuum of particulate size down to microscopic aggregates. They sometimes contain living microbes (Pomeroy & Johannes 1968), but are not readily utilized by planktonic organisms. In feeding experiments, for example, a common copepod, *Calanus heligolandicus*, could not be made to utilize any such structureless particles with detrital carbon (Paffenhöfer & Strickland 1970).

We shall now examine the role of faecal material and plant detritus in the trophic relations of aquatic organisms.

Faeces as a food source

Coprophagy in aquatic invertebrates has been noted by many authors and has recently received specific attention. The observed coprophagy rates vary widely, depending on the type of animal ingesting faecal pellets and the type of pellets being ingested (Table 7.5) (Frankenberg & Smith 1967). The potential value of coprophagy in meeting the metabolic needs of the ingesting organisms can be obtained by estimating the required metabolic rates and the metabolic rates supportable by ingested faecal organic matter. The values in Table 7.5 are only gross estimates but suggest that organisms could obtain a substantial portion of the energy required for maintenance by ingesting faecal pellets.

According to Frankenberg and Smith (1967) the faeces containing large quantities of organic matter are consumed more rapidly than those having small quantities. Johannes and Satomi (1966) reported that the faecal pellets produced by the shrimp *Palaemonetes pugio* fed on the diatom *Nitzschia closterium* were rich in assimilable protein compared with the faecal pellets derived from ingested faeces (Table 7.6). As the food residues were converted into living cells of assimilable bacteria in the gut, the faecal pellets were largely in the form of intestinal bacteria and not undigested diatoms. Most of the particulate organic residues arising from digestion were converted to food by the time the intestinal contents were egested (about 30 min after the ingestion). When living diatoms were not available, the shrimps reingested their own faeces. Thus faecal pellet production coupled with coprophagy is an important energy transfer mechanism in the marine ecosystem.

Table 7.5. Ingestion rates and their contributions to metabolic rates of coprophagous marine invertebrates and fishes (from Frankenberg & Smith 1967).

Species ingesting faecal pellets	Species producing faecal pellets	Ingestion rate (% body weight/48 hr) Mean ± SE	Metabolic rate provided by metabolism of ingested faeces (%)
Palemonetes puqio (shrimp)	*Penaeus setiferus* (shrimp)	83 ± 27	154
Pagurus longicarpus (hermit crab)	*Penaeus setiferus*	70 ± 5	213
Pagurus annulipes (hermit crab)	*Crassostrea virginica* (oyster)	33 ± 1	166
Pagurus longicarpus	*Onuphis microcephala* (polychaete)	23 ± 2	25
Hydrobiid spp. (snails)	hydrobiid spp.	21 ± 1	43
Palaemonetes pugio	hydrobiid spp.	14 ± 2	85
Fundulus majalis (Killifish)	*Penaeus setiferus*	11 ± 0.3	19
Pagurus longicarpus	*Mugil cephalus* (mullet)	11 ± 0.5	31
Fundulus majalis	*Crassostrea virginica*	9.5 ± 2	31
Glycera dibranchiata (polychaete)	*Mugil cephalus*	9.2 ± 5	255
Palaemonetes pugio	*Onuphis microcephala*	8.2 ± 3	13
Pagurus longicarpus	*Callianassa major* (burrowing shrimp)	6.7	9
Callinectes sapidus (crab)	*Mugil cephalus*	5.5	26
Nereis (Neanthes) succinea	*Crassostrea virginica*	5.3 ± 4.3	16
Glycera dibranchiata	*Penaeus setiferus*	3.6 ± 3.0	86

Table 7.6. Chemical composition (% dry weight) of faecal pellets of *Palaemonetes pugio* fed on diatom and fed on faeces (from Johannes & Satomi 1966).

Constituent	Faecal pellets derived from ingested diatoms	Faecal pellets derived from ingested faeces
Organic carbon	20	12
Protein (N × 6.25)	28	14
Carbohydrate	13	–
Lipid	2.5	–
Phosphorus	1.7	0.9
Ash	26	–

Honjo and Roman (1978) found, however, that the microbial enrichment in faeces was not derived from the gut bacteria, but from the external infestation of microbes and their colonization of the interior of faecal pellets. In the faecal pellets obtained from laboratory cultures of the copepods *Calanus finmarchicus*, *Acartia tonsa* and *A. clausi* which were fed with bacteria-free species of coccolithophores (approximately 10 μm in diameter) and diatoms (approximately 6 μm and 50 μm in diameter), individual bacteria cells of gut origin were not found in the interior of faecal pellets when examined under the scanning electron microscope. The surface of faecal pellets was covered by a protective surface membrane, which was colonized rapidly by bacteria upon exposure to sea water. The surface membranes were thus degraded within 3 hr of exposure at 20°C. A typical pattern of bacterial development was that at first small bacterial cells of less than 0.5 μm appeared on the membrane surface and then these grew to or were replaced by larger bacteria with time. As the surface membrane was broken down by bacteria, the contents of the pellet were exposed. Ciliates would be seen swimming about the decomposing faecal pellets and eventually bacteria would colonize and cause disintegration, dispersing the undigested contents.

Pomeroy and Deibel (1980) reported on the succession of microorganisms on the faecal pellets of pelagic tunicates fed with concentrates of natural seston and an axenic diatom culture. Fresh faeces consisted of partially digested phytoplankton and other inclusions in an amorphous gelatinous mass. After 18–36 hr, a population of large bacteria developed in the mass and in some of the remains of phytoplankton contained in the faeces. From 48–96 hr, protozoan populations arose which consumed bacteria and some remains of the phytoplankton in the faeces.

Newell (1965) observed that faecal pellets of the intertidal prosobranch *Hydrobia ulvae* initially contained very low protein concentrations but growth of non-photosynthetic micro-organisms in the faeces resulted in a marked increase in protein content over a period of 3 days. These micro-organisms apparently obtained their nitrogen from the overlying waters and their energy from the nitrogen-poor organic compounds in the faeces. The newly incorporated protein was largely assimilated when *Hydrobia* ingested these pellets (Fig. 7.4). This indicates that the faecal pellets are reingested usefully only after their nitrogen content has been increased. Johannes and Satomi (1966) reported that a prawn, *Palaemonetes pugio*, reingests its primary faecal pellets soon after defecation, while 'doubly-ingested' faeces which have smaller quantities of carbon and nitrogen than original faeces are not re-eaten for several days. In any case, because large quantities of carbon and nitrogen are contained in these pellets when they are consumed, their role in the trophic relationships is great. This role may be especially important in benthic habitats where large quantities of faecal pellets are concentrated. Here the interactions between microbes and larger invertebrates, such as polychaetes, molluscs and crustaceans are akin to those between microbes and soil macrofauna on forest floor where they contribute energy and nutrients to higher trophic levels.

Coprophagy among fishes associated with coral reefs is common and takes place in water columns (Robertson 1982). Faeces of carnivores with high calorific contents (protein and lipid) are consumed in greater proportion than those of herbivores with low calorific contents (carbohydrate) (Bailey & Robertson 1982).

Plant detritus

Macrophytic algae and vascular plants growing in the intertidal zone

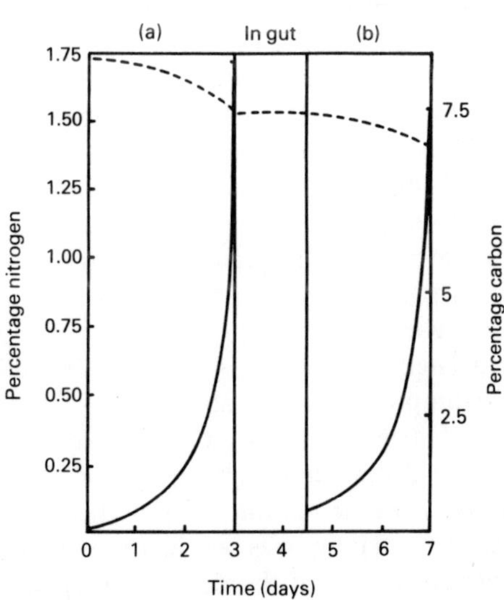

Fig. 7.4. Percentage organic carbon (– – – –) and nitrogen (——) in faeces cultured at 18°C in sea water under a noon light, showing dramatic increases of nitrogen in the primary (a) and secondary (b) faeces of the intertidal prosobranch *Hydrobia ulvae* (from Newell 1965).

contribute large amounts of detritus to estuarine and coastal ecosystems. Many early workers speculated on their nutritive value to the benthos. For example, Ekman (1947) hypothesized that the basic source of food of the benthos in the shallow water was the organic matter derived chiefly from the decay of rooted vegetation. Dexter (1944) stated that the detritus derived from *Zostera* was the basis for the food-chain associated with eel grass beds and documented the decline in secondary productivity when eel grass was killed by the 'wasting' disease at Cape Ann, Massachusetts.

The plant detritus is composed of complex carbohydrates and intractable lignins which most macrobenthic detritivores cannot assimilate. Hylleberg Kristensen (1972) incubated extracts from 22 species of marine detritus feeders and omnivores (molluscs, crustaceans, annelids and echinoderms) with 29 different carbohydrates, and estimated enzyme spectra from chromatograms. The results showed that hydrolysis of structural polysaccharides in most cases was weak or absent, indicating little utilization of structural polysaccharides by invertebrates. Hargrave (1970) examined digestion of cellulose and lignin-like compounds by an amphipod, *Hyalella azteca*, with radioisotope methods and concluded that *Hyalella* was unable to digest either of these compounds. Therefore it is possible that 'ageing' involves depolymerization of complex organic materials by bacteria and fungi.

Microbial enrichment of detritus

Nitrogen is probably the major limiting factor in most detritus-based systems. Most detritus entering coastal regions is low in nitrogen. Even in coastal areas that receive detritus from algae, which are generally higher in nitrogen content than that derived from vascular plants, the nitrogen content of detritus in the sediment becomes a nutritionally limiting factor. As we have seen, microbes are stripped from faecal pellets in the gut of benthic feeders but otherwise the new faecal pellets egested are essentially unchanged. These pellets are recolonized by microbes that use the pellets as a surface and a carbon substrate to fix inorganic nitrogen from the environment. This protein enrichment is considered necessary for detrital sources with low nitrogen content to become utilizable. In such cases, 'ageing', i.e. breakdown and transformation into available substances over periods, by microbes, is necessary to improve food quality. Tsuchiya and Kurihara (1979) observed that a polychaete, *Neanthes japonica*, did not ingest the fragments of marsh reed until they had been attacked and decomposed by microbes.

Using ^{14}C Tenore (1975) investigated the effect of initial particle size (0.18–2 mm) and 'ageing' (0–54 days) of detritus derived from the eel grass *Zostera marina* on net incorporation by the polychaete *Capitella capitata*. There was an increase in the rate of net incorporation with increasing age of the detritus of all the different particle sizes (Fig. 7.5).

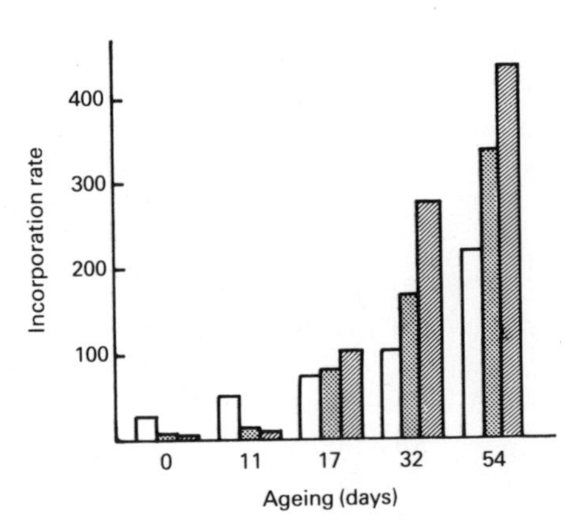

Fig. 7.5. The effect of 'ageing' of different initial particle-sizes (0.180 mm open bar, 0.405 mm stippled bar, 2.000 mm shaded bar) eel grass detritus on the net incorporation rate (μg dry wt/mg dry wt) by the polychaete *Capitella capitata* (the average for 40 worms) (from Tenore 1975).

This decay experiment also showed an initial rapid decrease then a slow recovery of nitrogen content and corresponding changes in C : N ratios. The increase in incorporation with ageing might be related to the low initial nutritional value of detritus derived from eel grass and the later development of associated microbes richer in nitrogenous compounds.

The detritus, depending on its source, has different calorific and biochemical characteristics that affect its decomposition by microbes and detritivores (de la Cruz & Poe 1975; Gunnison & Alexander 1975; Hanson 1982). Tenore (1977) examined the nutritional value of detritus derived from the marsh grass *Spartina alterniflora*, the eel grass *Zostera marina*, the rockweed *Fucus*, the red alga *Gracilaria* and Gerber's mixed cereal, to the growth of the polychaete *Capitella capitata*. The standing crop of polychaetes increased with increase in the level of detritus of all types (Fig. 7.6a). The best index of their nutritional value was the amount of nitrogen supplied to the polychaetes (Fig. 7.6b).

Tenore (1981) also estimated the nutritional value of different ratios of 32 sources of detritus to macroconsumers by measuring the growth of *Capitella capitata* at the maximum population density sustained by a given set of environmental factors. Multiple regression was used to estimate the relative contribution of organic nitrogen and calorific content (both total and available) to *Capitella* biomass. In general, nitrogen and 'available' calorific content, but not total calories, were significant contributors to *Capitella* biomass. Here the 'available' calorific content is defined arbitrarily as that portion of the total calorific content hydrolized by 1 N HCl for 6 hr at 20°C that approximates the weak hydrolytic digestive capabilities of marine polychaetes. Below a nitrogen ration of *c*. 100 mg $N/m^2/day$, nitrogen has a greater influence than the available calorific content on the growth of the polychaetes. For high nitrogen rations the available calorific content was the primary variable predicting biomass, but the nitrogen content was always a significant secondary contributor.

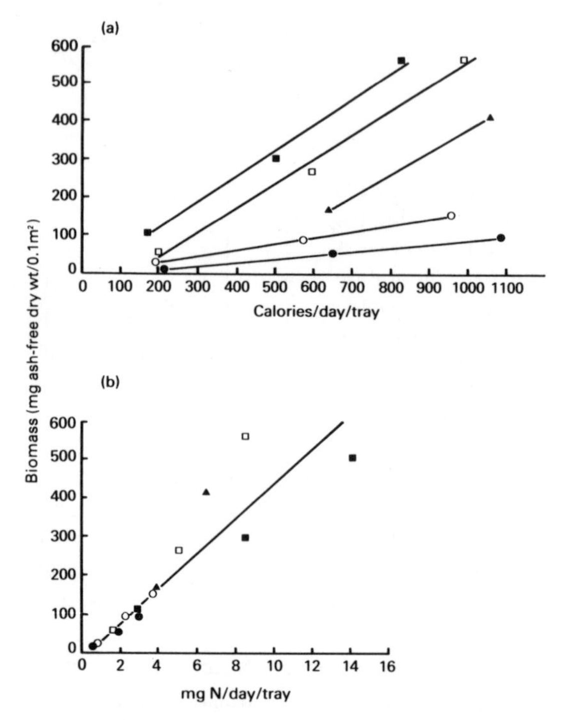

Fig. 7.6. Biomass of *Capitella capitata* at three food levels of five different types of detritus: (a) based on calories; (b) based on amount of nitrogen. ▲ Mixed cereal, ○ eelgrass, ■ *Gracilaria*, ● marsh grass, □ rockweed. Each value is based on eight replicates (four each at 10°C and 20°C) (from Tenore 1977).

Haines and Hanson (1979) estimated the potential food value of detritus derived from several marsh plants by measuring the protein increase with ageing of detritus and found that after 28 days the maximum yield of microbial biomass in proportion to the total ash-free dry weight (AFDW) was 29% for *Spartina*, 22% for *Salicornia* and 15.5% for *Juncus*. The conversion efficiency of plant detritus to microbial biomass ([g AFDW microbes/g AFDW plant material consumed] × 100) in aerobic, nitrogen-enriched cultures was 64.3% for *Spartina*, 19.4% for *Salicornia* and 55.6% for *Juncus*. They also showed that the extractable (protein) nitrogen increased during decomposition while there was little or no change in the non-extractable nitrogen. The proportion of the nitrogen increase ascribed to protein was 87% for *Spartina* detritus and over 100% for *Juncus* and *Salicornia* detritus. Thus the nitrogen content and the nutritional value of the decaying plant material increase with ageing as a result of microbial activity.

Fenchel (1970) studied the composition of the microbial communities living on detrital particles derived from the turtle grass *Thalassia testudinum*.

The abundance of organisms was approximately proportional to the total surface area of the detritus in field samples. There were about 3×10^9 bacteria, 5×10^7 flagellates, 5×10^4 ciliates and 2×10^7 diatoms, consuming $0.7 - 1.4$ mg O_2/hr per gram dry weight of detritus. The succession of microorganisms on the *Thalassia* detritus starts with a fast growth of bacterial populations which, after a few days, decline as the protozoans (first the flagellates and later the ciliates) increase. In these experiments populations of diatoms did not attain densities comparable to those found in the field samples of detritus.

The amphipod *Parhyalella whelpleyi* feeds on the detrital particles from *Thalassia testudinum*. The amphipods held detritus particles with their appendages while they tore bits of particles with their mouthparts. These bits were either swallowed or spread around. Microscopic observation of faecal pellets up to 0.4 mm long showed that the bits of detritus were completely undigested; often bits of intact *Thalassia* tissue several hundred microns long were seen (Fenchel 1970). Freshly formed faecal pellets are almost devoid of microorganisms. *Parhyalella whelpleyi* feeds on detrital particles and on its own faecal pellets but it only extracts microorganisms; the dead plant residue passes undigested through the intestine.

Heinle *et al* (1977) studied the nutritional value of marsh plant detritus to survival and egg production of the copepods *Eurytemora affinis* and *Scottolana canadensis*. The copepods fed with detritus that was not autoclaved sometimes survived well and produced offspring in greater numbers than unfed controls, while those fed with autoclaved detritus never did, i.e. the copepods did not survive well or produce eggs.

7.3.3 Bacteria-based marine food-web

Bacteria as food of benthic feeders

From the above results it is clear that bacteria in mud serves as an important source of food for benthic animals. The mud-dwelling animals probably derive an appreciable amount of their nourishment directly from ingested bacteria.

Moriarty (1976) estimated the biomass of bacteria ingested by two benthic feeders, the mullet *Mugil cephalus* and the prawn *Metapenaeus bennettae*, by measuring the muramic acid level in their gut contents. Mullet -vere found to have a large amount of muramic acid, which is a cell wall component characteristic of procaryotic organisms. The ratio of muramic acid to ash was greater in the stomach than in the rectum in the mullet and greater in the proventriculus than in the mid-gut in the prawn. It was calculated that about 30% of the muramic acid in the bacterial cell walls is digested and that a higher proportion of bacterial cell contents may be digested. In laboratory experiments, using radioisotope techniques, five species of bacteria and one blue-green alga were digested and assimilated by the prawns. The highest percentage assimilation was over 90% and this was interpreted to indicate that micro-organisms passing into the digestive tract were nearly completely digested (Table 7.7).

Digestion of bacteria by other detritivores has been demonstrated by Wavre and Brinkhurst (1971) who found a marked reduction (60–90%) in the number of bacteria in mud after passage through the gut of the tubificid worms *Tubifex tubifex* and *Limnodrilus hoffmeisteri.* Kikuchi and Kurihara (1977) also observed that the presence of tubificids, *Limnodrilus socialis* and *Branchiura sowerbyi*, suppressed the number of heterotrophic aerobes in the paddy field soil.

Kurihara (1983) has demonstrated that the polychaete *Neanthes japonica* can consume a large quantity of sludge, containing bacteria, extracted from domestic sewage treatment plants. The sludge ingestion rate varied with the body weight of the worm. Worms of 1 g wet weight consumed sludge at the greatest rate of 500 mg/day/100 cm^2 when kept at a density of 30 individuals (30 g)/100 cm^2. Those weighing 0.5 g each ingested sludge at the maximum rate of 700 mg/day/100 cm^2 when the density was 30 individuals (15 g)/100 cm^2 and those weighing 0.1 g each had the maximum rate of 900 mg/day/100 cm^2 at the density of 90 individuals (9 g)/100 cm^2. Based on these values, Kurihara (1983) calculated that the daily waste production by an urban inhabitant (12 g of sewage sludge) could be consumed by 720 worms weighing 1 g each and living in a 0.24 m^2 sand bed, by 510 worms of 0.5 g in 0.17 m^2, or by 1200 worms of 0.1 g in 0.13 m^2. He operated a pilot plant with an artificial tidal flat simulating the conditions of a natural estuarine tidal flat and maintaining the polychaete habitat. He proposes the mass culture of *N. japonica* with sewage sludge as a new biological method of treating sludge to help solve a serious disposal problem in modern urban management.

Table 7.7. Assimilation of bacteria and a blue-green alga by the prawn *Metapenaeus bennettae* (from Moriarty 1976).

Bacteria	No. of trials	Assimilation (% ^{14}C) Mean	Range
Isolate no. 1	5	91	87–97
Escherichia coli	4	85	84–86
Pseudomonas fluorescens	4	93	90–95
Enterobacter aerogenes	5	96	95–98
Bacillus subtilis	2	84	80–88
Blue-green alga	6	63	48–87

Nutritional significance of non-living materials

It now seems that at least in shallow water ecosystems the detritivores are unable to use the nonliving plant material of detrital particles directly and must rely on bacteria, fungi or other microbes to convert the plant tissue into microbial biomass which then becomes a major food source for the benthic feeders. When detritus is ingested by an animal, the microbes are stripped off and the bulk of the detritus passes through the gut of animals unscathed. Microbes are also largely responsible for the increase in nitrogen in ageing detritus, turning it into a good quality food for animals. Microbial densities alone, however, cannot account for observed increases in detrital organic nitrogen content (Christian & Wetzel 1978) or satisfy the energy requirements of macroconsumers (Cammen 1980).

Cammen (1980) calculated a partial carbon budget for a population of the polychaete *Nereis succinea* living in a salt marsh of North Carolina and found that the estimated intake of microbial carbon was not sufficient to fulfil the carbon requirement of the polychaete worms at any time during the year. If assimilation efficiency for total microbial carbon was the same as the digestion efficiency for detrital heterotrophs (57%), then only 26% of the annual carbon requirement could be accounted for; even if 100% of the ingested microbial carbon was assimilated, only 45% of the total carbon requirement would be met (Table 7.8). One possibility is that extracellular matter is produced by microbes and this makes up the rest of food for detritivores.

Hobbie and Lee (1980) postulated that a significant fraction of sediment carbon may be derived from microbes in the form of cellular debris, mucopolysaccharides etc., and this fraction may be readily avail-

Table 7.8. Annual carbon budget for a population of the polychaete *Nereis succinea* occupying $1 m^2$ (from Cammen 1980). Carbon input from microbial biomass has been estimated assuming an assimilation efficiency (AE) of 57% and 100%. Values are in grams C per m^2.

	57% AE		100% AE
Carbon required:			
production		2.1	
respiration		9.4	
Total		11.5	
Carbon assimilated:			
bacteria	1.0		1.7
microalgae	0.9		1.5
total microbial biomass	3.0		5.2
percentage of requirements met	26%		45%

able to consumers. This hypothesis would explain discrepancies between the energy requirements of detritivores and the microbial food available to them. It would also explain differences between the amount of nitrogen accumulating in decomposing detritus and the nitrogen actually present inside the microbes. Microbial production adds to the nutritional quality of detritus by increasing the organic nitrogen pool. Of course, the decompositional role of microbes (bacteria and fungi) that depolymerise complex organic materials into calorific substrates utilizable by macroconsumers cannot be underestimated (Tenore *et al* 1979). Many types of detritus contain organic substances that are themselves readily available to macroconsumers (Tenore 1980). 'Available' algal detritus from *Gracilaria*, including plant protein, is incorporated by *Capitella capitata* to a greater extent than structurally complex 'unavailable' *Spartina* detritus (Tenore *et al* 1979). Thus energy and organic nitrogen from detritus may be available to the detritivores directly, or from microbial biomass or microbially processed substrate before utilization by macroconsumers.

Bacteria in open water

What are the sources of energy for heterotrophic bacteria in the sea? It has long been recognized that marine phytoplankton releases extracellular material into seawater during photosynthesis. Some earlier studies reported high exudation rates of up to 70% of the total primary production (Watt 1966; Nalewajko & Marin 1969; Choi 1972; Al-Hansen *et al* 1975). Positive correlations between the photosynthetic rate and bacterial activity suggested that the extracellular material might exert major stimulatory effects on bacterial populations (Derenbach & Williams 1974; Sorokin 1981; Griffiths *et al* 1982).

Larsson and Hagstrom (1979) studied the release of dissolved organic carbon (DOC) from phytoplankton during photosynthesis and the utilization of this carbon by planktonic bacteria, using $^{14}CO_2$ and selective filtration. Natural sea water was incubated in the laboratory for detailed studies, and followed *in situ* for the estimation of carbon uptake. They found that labelled organic carbon in the phytoplankton and bacterial fractions continued to increase almost linearly and that the uptake of organic carbon by bacteria was negligible in the absence of phytoplankton cells. The labelled substrate taken up by the bacteria, therefore, was considered to be extracellular material from the phytoplankton. About 65% of the labelled organic carbon was found in the phytoplankton fraction (>3 μm), about 27% in the bacterial fraction (0.2–3.0 μm) and the remaining 8% as DOC (<0.2 μm). The measured annual primary production was 93 g C/m^2, of which phytoplankton accounted for 61 g C/m^2, bacteria 25 g C/m^2 and exudates 7 g C/m^2. About 45% of annual primary production released as DOC was utilized by bacteria. Thus, healthy, actively growing phytoplankton species release a considerable portion of their photo-assimilated carbon into the aquatic environment and DOC can be taken up by heterotrophic bacteria and either degraded or incorporated into bacterial cells, which in turn become available to higher trophic level organisms.

Jensen (1983) also quantified the phytoplankton release of DOC and its subsequent assimilation by planktonic bacteria using the procaryotic inhibitor, streptomycin, in the Danish estuary of Randers Fjord. From 34 to 90% of the released carbon was transferred to the bacteria and the bacterial metabolism of extracellular material ranged from 3 to 30% of total primary production.

Macrophytes also release substantial amounts of DOC into the surrounding water under a variety of conditions; living seagrasses release 2–10% of their daily photosynthetic products as DOC (Brylinsky 1977; Penhale & Smith 1977; Wetzel & Penhale 1979). In a community of the temperate seagrass *Posidonia australis*, DOC accounts for 48% of the annual loss of carbon (Kirkman & Reid 1979). Macrophyte detritus also releases large quantities of DOC; detached leaves of the freshwater macrophyte *Scirpus subterminalis* can rapidly lose 30–40% of their original carbon by autolysis and leaching (Otsuki & Wetzel 1974) and the rapid weight loss of decomposing *Thalassia testudinum* leaves is a result of leaching of DOC (Zieman 1975).

Robertson *et al* (1982) used laboratory simulations to compare the microbial processing of DOC and macroparticulate organic carbon derived from the tropical seagrasses *Thalassia testudinum* and *Syringodium filiforme*. The results showed that dried leaves of *Thalassia* and *Syringodium* released 12.6% and 19.4%, respectively, of their organic carbon as DOC during three days of axenic leaching. When inoculated with microbes the DOC was rapidly converted to bacterial aggregates ranging in size from a few micrometres to a few millimetres. Large populations of ciliates and flagellates also developed, presumably feeding on the unaggregated bacteria.

Bacteria as food of zooplankton

Bacteria play an important role in the productivity of aquatic ecosystems because of their ability to convert DOC into cell substances and to synthesize protoplasm. Bacteria are the primary link between the detrital carbon and consumers. Although protozoa and certain higher-order consumers, such as *Daphnia*, can feed on single bacteria (Peterson *et al* 1978), many animals are not capable of trapping individual cells (Fenchel 1975a). Therefore, the phenomena of protozoan grazing and bacterial aggregation may increase the contribution of DOC to secondary productivity through the formation of food particles of more suitable size for a variety of marine consumers (Seki 1972; Alldredge 1976; Berk *et al* 1977).

Sorokin and his colleagues (see Sorokin 1978) measured the ingestion of ^{14}C-labelled bacteria (both free-living and attached), in terms of daily ration by various zooplankton species. The rations during feeding of fine filterers, such as the cladoceran *Penilia avirostris*, gastropod veliger larvae or hydroids, correspond to 50–100% of the body carbon per day. In these animals, metabolic requirements amount to 10–20% of body carbon. The bacterial food assimilated can easily meet the metabolic losses, even at concentrations 3–5 times below the optimum level. For example, in *P. avirostris*, the threshold concentration is 5 mg C/m^3; the amount of assimilated food at the optimum concentration is about three times higher than the metabolic losses incurred.

King *et al* (1980) presented the clearance rates for any zooplankton eating free-living bacterioplankton. The larvacean (planktonic tunicate) *Oikopleura dioica* was fed with ^3H-labelled natural assemblages of marine bacterioplankton. Grazing rates ranged from 1 to 100 ml/day/individual and were highly dependent on body size. In large floating enclosures, seminatural populations of *O. dioica* were used to determine the impact of the larvacean on the bacterial populations and to estimate the amount of bacteria ingested by the larvaceans. The results showed that bacteria were a major source of food for the larvaceans. Field populations may, in some instances, ingest 100% of their weight of bacterioplankton, representing 25–50% of their daily ration.

Bacteria as food of benthic filter feeders

ZoBell and Feltham (1938) experimented with the mussel *Mytilus californianus* to find out the value of bacteria as food. The mussels in eight jars were fed exclusively with *Rhodococcus agilis*, eight others received *Bacillus marinus*, and the mussels in the eight control jars were not fed. In nine months the mussels fed on *R. agilis* gained an average of 12.4%, those fed on *B. marinus* gained 9.7% and the fasting controls lost an average of 16.3% in weight. It was estimated that between 5 and 10% of the solid matter of bacteria was assimilated by the mussels in this period. They also dissected the digestive diverticula, stomachs, styles and intestines of several large well-nourished mussels and triturated them with sand. The resulting extract was filtered through four thicknesses of cheese-cloth and a drop of the extract was added to a drop of bacterial suspension in the small vials. After 2–6 hours' incubation at 25°C stained smears of the mixture were examined microscopically. The cells of all except three of the 31 different species of marine bacteria tested were dissolved by the extract. It was thus concluded that *Mytilus californianus* ingested and digested bacteria.

Reiswig (1975) demonstrated the potential importance of free bacteria in the diets of sponges inhabiting coastal waters. Samples of ambient aquarium water (inhalant) and oscular stream (exhalant) of the sponges *Haliclona permollis* and *Suberites ficus* were analysed for bacteria.

Bacterial removal efficiencies were 77% from direct counting. The free bacterial biomass retained by *H. permollis* and *S. ficus* would appear to represent a major portion of their diets. In fact, using a conversion rate of 10^7 bacterial cells to 1.1 μg organic matter (ZoBell 1963; Jørgensen 1966) it was estimated that a retention rate of 5.7×10^8 cells/l of water pumped by *H. permollis* represented 63 μg organic matter per litre. This was more than the entire food requirements of the sponge.

Many marine protozoa, including ciliates and non-pigmented flagellates, are bacteriovores which, in turn, are preyed upon by larger forms. Thus a food-web involving a general pathway of energy in the order DOC → bacteria → phagotrophic protozoans → zooplankton → plankton feeders → swimming predators may be proposed for pelagic communities (Fig. 7.7). As in the rumen the protozoans are the primary predators of bacteria in the sea. These phagotrophic protozoans, in turn, will be consumed by microphagous zooplankton. This creates another trophic level with its concomitant loss of energy between the recyclers of DOC, the bacteria and protozoa. Zooplankton then transfers bacterioplanktonic carbon via its biomass to larger filter-feeding animals or juvenile fish.

7.4 Conclusions

The foregoing discussion points to one very important aspect of community organization; that is, microbes and animals (herbivores, scavengers, detritivores and coprophages) work together to break down the organic carbon fixed by the producers.

From the point of view of animals, the microbes break down indigestible plant materials, such as cellulose and lignin, or toxic allelochemicals and make them available for consumption. Microbes also improve the nutrient status of detritus and themselves become sources of food. Even if microbes and their products are available for consumption, their quantities are often too small to be utilized directly by the metazoans. Their energy, therefore, must first be concentrated.

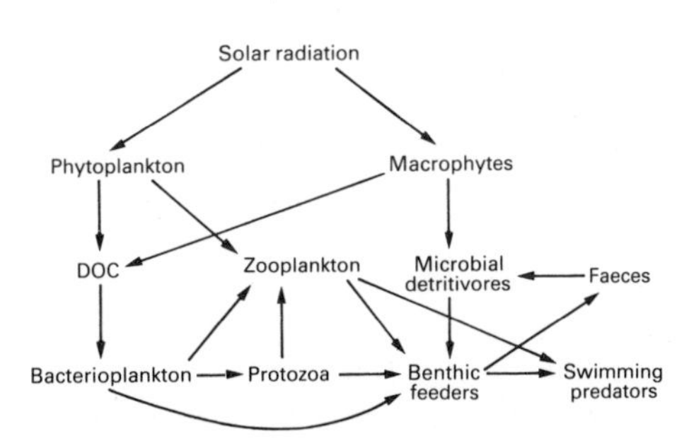

Fig. 7.7. Energy flow through a simplified marine food-web.

We have seen that phagotrophic protozoans have this role and form an important trophic level in the marine ecosystem. The consumers' foraging strategy is to obtain such food supply in sufficient concentration to support the animal biomass. We may classify the guilds of those consumers that utilize microbes and their products as a major source of energy, according to the strategy used to concentrate this food supply.

(1) Filter feeding. This mode of feeding is not restricted to microbe feeders. Microscopic food is gathered and strained by tentacles or other ciliated or meshed structures from water. Many sessile and burrowing animals concentrate their food supply in the aquatic environment by filter feeding.

(2) Deposit feeding. Detritivores inhabit the environments in which a large amount of particulate organic matter (e.g. plant detritus, animal faeces) accumulates and permits colonization and rapid propagation of microbes. Forest floor and estuarine tidal flats are such environments, where the consumers require no special food gathering device for ingestion as concentrated microbial food supplies are readily available.

(3) External culturing (gardening). This includes microbial enrichment through coprophagy in macrotermitine termites and other soil arthropods. Fungus gardening by leaf-cutting ants is another example, where resource microbes are cultured in the media provided externally by the animals.

(4) Internal culturing (symbiotic). The so-called 'herbivores' that utilize cellulose culture symbiotic microbes in their internal media, such as the rumen fluid, to break down the plant tissue. Also, many species of termite culture symbiotic flagellates in their gut for fermentation and digestion of cellulose from wood. As in external culturing the internal media are supplied continually with food materials for the microbes. This is the most sophisticated guild which can control its food supply externally by behavioural mechanisms and internally by physiological mechanisms.

The hermatypic corals and zooxanthellae represent a unique symbiotic relation between the consumer and the producer. The corals culture algae internally for food supply and have dispensed with heterotrophic microbes as an intermediary.

For most ecosystems, the concept of secondary production must incorporate the combined role of decomposers in making the energy of primary production available to animals.

Part 4
Ecological Processes

The first three chapters in this section address the principal biotic interactions recognized by Elton as key processes in community organization. Chapter 8 adopts a 'compare and contrast' approach in seeking constructive generalizations from studies of insect and mammal communities, and *inter alia* points up the difficulty of establishing meaningful correspondence between taxonomic and ecologically functional taxa. Chapter 9 asks two major questions of animal parasites: what is the nature of parasite communities and what influences do these parasites exert on host communities? Chapter 10 seeks to elucidate the significance of predatory interactions in a range of terrestrial and aquatic communities, and emphasizes predator impact, both within their own and within adjacent trophic levels.

Chapter 11 provides a very thorough exploration of current experimental work on competitive interactions in the field and attempts to provide a rigorous rationale for future analysis from the bewildering array of available data and evidence, some of which is not amenable to unequivocal interpretation. In a similar vein but using a broader approach, Chapter 12 examines and finds wanting some of the models of ecological succession that have either persisted in, or have more recently colonized, the literature on dynamic community change. A number of definitional questions are traversed before an attempt is made to cobble together some of the essential components of a modern theory of ecological succession, components which themselves are scale-dependent and which not infrequently depend for their elucidation on the compatibility and coherence of frequently disparate lines of evidence.

Part 4
Ecological Processes

8 The Organization of Herbivore Communities

John H. Lawton and Malcolm MacGarvin

8.1 Introduction

Examiners often ask hapless candidates to 'compare and contrast'. It is a useful technique, and one that we have adopted in this chapter to examine communities of organisms that feed on green plants.

Herbivores come in a bewildering variety of sizes and shapes; cows, caterpillars and copepods are all herbivores, making it neither sensible nor possible to describe the organization of all communities of herbivores within the confines of one chapter. Instead, what we have chosen to do is to pick two extremes — insects and mammals — and to describe and contrast aspects of their community organization.

There are good reasons for restricting the scope of our attention in this way. Mammalian herbivores are often large, beautiful to look at, and are therefore presumed to be important. But the number of species in the world is really rather small. For example, living ungulates (Proboscidea, Perrissodactyla and Artiodactyla) muster little over 200 species. Even rodents have less than 2000 species. Insect herbivores, in contrast, number many tens of thousands, perhaps millions of species (Strong *et al* 1984a), although a comparison of the two groups by biomass would reveal less dramatic differences.

These contrasts in size and total species richness — vast numbers of small insects, many fewer but larger mammals — present a challenge for community ecologists. Why are there so many more species of insects than mammals? How are local communities of insects organized compared with local communities of mammals? How many community and species diversity patterns documented for the one group generalize to the other? If they do not, why not? Is it because entomologists ask different questions to mammalogists, or is it because the underpinning ecological processes are different? How important is the obvious size difference between mammals and insects in generating ecological differences between them?

As shrewd examiners are aware, comparing and contrasting tests understanding. More than anything else, writing this chapter has made us aware how many aspects of herbivore community organization are not well understood. Fortunately, however, we can discern some general patterns and make some useful generalizations. The chapter is organized as follows: Sections 8.2 and 8.3 are devoted to herbivorous insects, and examine processes that determine how many species co-occur, and how they interact. Section 8.4 then looks at mammalian herbivores, comparing and contrasting the organization of mammalian assemblages with insect assemblages. Section 8.5 concludes with some speculations and unsolved problems.

We have chosen to use the word 'community' in a loose and liberal way, to mean an assemblage of co-occurring species with the potential to interact. It is convenient, but clearly subjective and arbitrary, to distinguish 'communities of herbivorous insects' and 'communities of herbivorous mammals'. We make some token remarks in Section 8.5 to the effect that 'real' communities — whole ecosystems — contain both mammals and insects and many other taxa as well. It is, however, better to study manageable bits than to get lost in unmanageable diversity.

8.2 The determinants of species richness for insects on plants

Community ecologists must often explain why a particular site has more or fewer species than another. For insects that feed on the living tissues of higher plants [phytophagous or herbivorous insects *sensu* Strong *et al* (1984a)] an important determinant of species richness is the abundance of their host plant on a continuum of scales from geographic down to local habitat patches. A useful measure of host plant abundance is the area over which it grows; in other words, phytophagous insect species richness can often be predicted by a species (of insects)–area (of host plants) relationship (Lawton 1978; Strong 1979; Strong *et al* 1984a). Correlations between number of species and size of habitat are widespread in ecology (Connor & McCoy 1979; May 1981). Often, but not always, the logarithm of the number of species (S) increases linearly with the logarithm of habitat area (A) so that:

$$S = cA^b \quad (1)$$

where c and b are constants. Data for phytophagous insects are no exception.

8.2.1 *Local effects: the size of plant patches*

Wherever discrete patches of a plant have been studied the number of phytophagous insect species has been found to decline as patches get smaller. Three comparable examples are shown in Fig. 8.1a for the herbivores on patches of juniper shrubs, *Juniperus communis* (Ward & Lakhani 1977), bracken fern, *Pteridium aquilinum* (Rigby & Lawton 1981) and rosebay willowherb (fireweed), *Chamaenerion angustifolium*

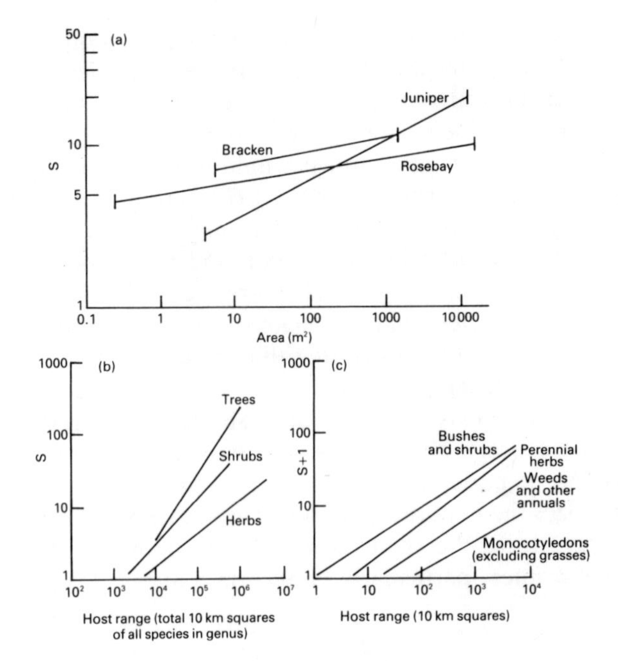

Fig. 8.1. (a) The number of herbivorous insect species (S) present in local plant patches of different area, plotted on a log–log scale. The methods used for bracken and rosebay willowherb were directly comparable. That for juniper measured area as the number of bushes at each site. Here, we deliberately make a minimum estimate of one juniper bush = 4 m^2. If they are larger than this it will shift the line down and to the right relative to the others. Adapted from Ward and Lakhani (1977) (juniper), Rigby and Lawton (1981) (bracken) and MacGarvin (1982) (rosebay). (b) and (c): The total number of insect species (S) on British plants; for genera of plants (b) and for individual species (c). The pattern is one of the number of insects species decreasing with the decline of plant architectural complexity. [From Strong and Levin (1979) and Lawton and Schröder (1977).]

(MacGarvin 1982). These studies deal with entire insect–herbivore communities over a wide range of patch sizes. The bracken and rosebay slopes (b in equation 1) at 0.09 and 0.07, are remarkably shallow compared to the gamut of species–area relationships (reviewed in Connor & McCoy 1979). The slope for juniper (approx. 0.24) is much closer to the average value found in most other species–area studies (approx. 0.3; Preston 1962; MacArthur & Wilson 1967; Connor & McCoy 1979). It remains to be seen if other studies of local host plant patches of different sizes also generate very shallow slopes, because different rates of species accumulation as areas change in size may be due to different mechanisms operating in different systems.

Whilst local species–area relationships for insects on host plants are now reasonably well documented, they are not sufficiently well understood to account unequivocally for subtle or even quite substantial differences in b. However, they can be generated where many of the individual species of herbivores become rarer per unit of habitat on smaller patches (i.e. rarer per plant or per m^2) (Lawton 1978; Strong *et al* 1984a; Kareiva 1983), either because insect death and/or emigration rates are higher on small patches (which offer less protection and fewer total resources) or because birth and/or immigration rates are reduced for similar reasons. Hence individual species tend to decline to extinction at some characteristic average patch-size, generating the species–area relationship with fewer species on smaller patches.

Explaining the existence of local species–area relationships by changes in the average abundances of component species, which are in turn a function of population birth, death, immigration and emigration rates, emphasizes the importance of population dynamics in understanding community structure. We shall return to this important point at several places in the chapter, particularly in Section 8.3.

8.2.2 *Local effects other than area*

In many ways phytophagous insect species–area relationships are a special case of the *resource concentration hypothesis* (Root 1973) which states that "herbivores are more likely to find and remain on hosts that are growing in dense or nearly pure stands (and) that the most specialized species frequently attain higher relative densities in simple environments". Two aspects can be distinguished; the density of a host plant relative to other species, and its absolute density. A growing number of studies have tested the influence of both, particularly on specialist insect species [i.e. species feeding on only one or a small number of host species, reviewed by Bach (1980a), Kareiva (1983), and Stanton (1983)]. Mixing other plants with the host typically depresses the number of specialist phytophagous insects per host plant, perhaps again, because individual population sizes are depressed by the increased difficulties of finding and staying on scattered hosts (e.g. Tahvanainen & Root 1972; Bach 1980a,b; Rausher 1981). Of course, the *total* number of species should be higher in the mixed vegetation, because there are more species of plants each with their own associated insects (Murdoch *et al* 1972; Southwood *et al* 1979). Generalist species numbers per plant, by contrast, may be enhanced in a polyculture (Risch 1980).

The results of differences in simple host plant density are, in contrast, multifarious. Some species are certainly more abundant on high density plots, but many Lepidoptera oviposit preferentially on low density patches or isolated plants, while other species are apparently indifferent to changes in host density (Kareiva 1983; Strong *et al* 1984a). In consequence, there are apparently no simple generalizations that one can make about the way in which insect species richness changes with host plant density.

In sum, local diversity of phytophagous insects is determined in part by size of host plant patches, modified to a greater or lesser extent by host density and the presence of other plant species. Basically, local patches of host plants 'sample' or 'attract' different numbers of insects from a regional pool of species, influencing total population sizes of each species within each patch, and in the limit determining their presence and absence.

Presumably, local community diversity should then be influenced by the species richness of the regional pool of insects from which local communities are assembled. What determines the total number of insect species found on a host plant within any particular geographic region?

8.2.3 Large scale effects

The determinants of phytophagous insect diversity on different species of host plants over geographic scales has stimulated active and sometimes acrimonious debate. Entry into the literature can be made via Kuris *et al* (1980), Lawton *et al* (1981), Rey *et al* (1981), Claridge and Wilson (1982a,b), and Blaustein *et al* (1983).

Four determinants of phytophagous insect diversity on a regional scale are, in approximate order of importance: (i) the areal extent of the host (its regional abundance); (ii) its size and growth form ('architecture'); (iii) the number of closely related host plants (in the same genus or family) growing in the same geographic area; and (iv) the length of time the plant has been present in that area, either as a native or an artificially introduced species (Fig. 8.1b,c). Total species richness is enhanced by a large geographic range (Strong 1979), on plants of large size and high architectural complexity (e.g. trees) (Lawton & Schröder 1977; Strong & Levin 1979; Lawton 1983), on plants with many relatives in the same area (e.g. Connor *et al* 1980; Claridge & Wilson 1982a), and on plants that have been growing in an area for reasonable lengths of time (e.g. Blaustein *et al* 1983). Total species richness is reduced on plants of small stature, with small ranges, etc. [For a review of these problems see Strong *et al* (1984).]

Of course, arguing whether these regional patterns of diversity determine local patterns or *vice versa* is like trying to decide whether eggs or chickens come first. There must be continual interplay between regional and local diversity (e.g. McCoy & Rey 1983). Nevertheless it will probably generally be true that the richest local communities of plant-feeding insects will be drawn from rich regional communities (Cornell 1985). Many more species of phytophagous insects coexist on an individual British oak tree (*Quercus robur*), for example (this oak is widespread, architecturally complex, and a long-established member of the British flora), than coexist on a patch of bluebells or bracken of comparable biomass. Much the same point is true for bracken in different geographic regions. Local communities of bracken in Britain (where bracken is geographically widespread) have many more insect species than local communities in New Mexico (where bracken is rare), even though local patches of similar size and structure are being compared (Fig. 8.2a) (Lawton 1982, 1984a,b). The result is a large number of apparently 'vacant niches' on New Mexico bracken (Fig. 8.2b) with important consequences for our understanding of the mechanisms determining community structure. We return to these points in Section 8.3.4.

8.2.4 Introduced plants and colonizing insects

One of the ways to learn about processes determining the regional and local diversity of phytophagous insect communities is to observe the colonization by insects of plants accidentally or deliberately introduced by people. Crops have been particularly well studied in this context, and reveal, in the case of sugarcane (*Saccharum officinarum*) and cacao (*Theobroma cacao*) that colonization by insects is remarkably quick (Strong 1974; Strong *et al* 1977; Kuris *et al* 1980; Rey *et al* 1981).

The number of species colonizing sugarcane or cacao in any particular geographic region is strongly dependent on the area planted, with the great majority of colonists being recruited locally rather than insects moving from country to country or continent to continent with their host. In other words, introduced plants appear to acquire fauna mainly by insects 'shifting hosts'. But colonization does not continue at a high rate indefinitely. For sugarcane, once area planted has been allowed for, it is impossible to distinguish insect species richness between regions where cane has been grown for as long as 2000 years or as little as 150 years.

Turnipseed and Kogan (1976) described the source of the colonists of the leguminous soybean *Glycine max* as mostly a mixture of polyphagous species and those whose previous diet included other legumes. A much smaller number came from different hosts, even from different plant families. These patterns are typical of most introduced plants (e.g. Goeden & Ricker 1968; Root & Tahvanainen 1969; Goeden 1971, 1974; Wheeler 1974; Bournier 1977; Chiang 1978; Berenbaum 1981). However, there are exceptions, because some introduced plants are extremely slow at acquiring any insect herbivores. These plants are interesting because they are often very distinct from the native flora; examples include cactus

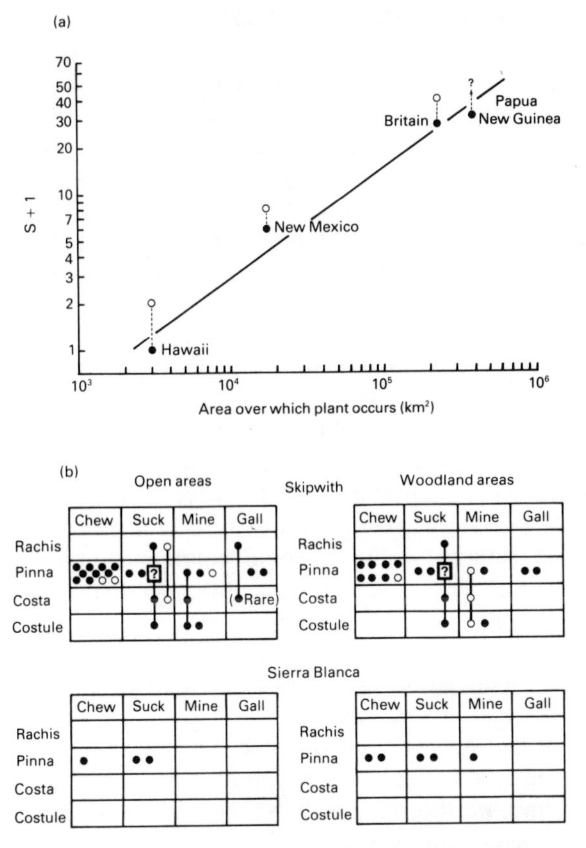

Fig. 8.2. (a) Number of species of insect (S) certainly (●) or possibly (○) feeding on bracken in four parts of the world where bracken grows naturally. The species–area relationship calculated on the solid dots is statistically significant ($r = 0.98$; $0.05 > P > 0.01$) (from Lawton 1984b). (b) Feeding niches of insects at sites within two of these geographic regions (Skipwith in Britain and Sierra Blanca in New Mexico). Feeding niches are characterized by how and where the insect attacks the plant. Each dot, or set of joined dots, represents one species. Open circles at Skipwith are rare species, present in less than 50% of several years of study. Note the much smaller number of species in the local community at Sierra Blanca, where the total species pool in the geographic area is also smaller (a), generating many vacant niches (from Lawton 1982).

Opuntia, introduced from South America to South Africa and Australia, *Eucalyptus* species introduced into North America and Europe from Australia, and oak *Quercus* introduced into the southern hemisphere (Connor *et al* 1980; Moran 1980; Moran & Southwood 1982; Southwood *et al* 1982). In other words, rates of colonization of introduced plants seem to be influenced by two factors: (i) the area planted, or colonized by a plant and (ii) the taxonomic, phenological, biochemical and morphological 'match' between the introduced plant and the native flora. These are the same variables that were emphasized in studies of the species richness of herbivorous insects associated with native species of plants (Section 8.2.3).

Of course 'host switching' by insects onto new host plants is only the first step in the development of a phytophagous insect fauna. Over much longer periods closer associations occur and often appear to lead to the formation of new species of insects, confined to just one species of host plant. The actual rates of speciation, the time taken for speciation (two different things) and the mechanisms are far from clear (Mayr 1963; Bush 1975; White 1978; Hammond 1980; Futuyma & Mayer 1980; Key 1981; Williamson 1981; Futuyma 1983), and lie outside the scope of this chapter. Nonetheless, the ultimate structure of phytophagous insect communities on native plants must to a greater or lesser extent be determined not only by the colonization of suitable hosts by insects, but also by subsequent evolution and speciation. We remain extremely ignorant about the latter processes. We return briefly to some comments about speciation in Section 8.5.

8.3 Population dynamics and community structure

As Section 8.2.1 suggests, a detailed understanding of community structure demands knowledge of the population dynamics of component species. Determining the number of species in a system should only be the beginning of community ecology study, not the end. Birth, death, immigration and emigration rates combine to determine characteristic levels of abundance, patterns of population fluctuations, and in the limit the presence and absence of species; and the nature of species interactions, or lack of them, play a fundamental role in determining how particular communities work. Hence we now need to consider population processes in more detail, focusing first on some theoretical possibilities.

8.3.1 Theoretical possibilities

Two important questions in population dynamics are why and how much population fluctuates, and how populations are regulated. As a broad generalization, populations are regulated and controlled by *density-dependent* processes, and perturbed and disturbed by *density-independent* processes (Begon & Mortimer 1981; May 1981). The form and relative magnitudes of the density-dependent and density-independent events in each species' life history ultimately determines community predictability, i.e. the constancy of species' relative and absolute abundances from one year (or generation) to the next. Moreover, the nature of the density-dependent control determines which species interact, and how they do so, in particular whether interspecific competition is important in structuring the community.

Now, depending upon details of their population dynamics, groups of species may behave in one of several ways.

(1) The most extreme possibility is for density dependence to have no effect, or only very feeble sporadic effects at any density on most of the component populations of a community. The result is a shambles of randomly fluctuating populations. Local extinctions, large changes in abundance, and very high densities should be commonplace in this uncontrolled world; that is, communities should have very little, if any, predictable structure.

(2) A second possibility is for populations to be controlled by density-dependent processes (but not those operating with significant time lags, causing large-amplitude fluctuations), with component populations operating more or less independently of all other populations in the trophic level. This might happen for several reasons. Each population could be resource limited but the community nowhere near saturated with species, for example because of lack of time for species to evolve to fill available niches, or because isolation limits the invasion of potential colonists, or because host plants and individual species' populations are patchily distributed. Patchy distributions markedly reduce the likelihood of effective interspecific competition (Shorrocks *et al* 1979; Atkinson & Shorrocks 1981). Alternatively, natural enemies may keep each population well below the point where resource limitation becomes important. In each case, the result is a community that is reasonably constant in composition and species abundance, but is not influenced to any significant degree by competitive interactions between species on the same trophic level.

(3) A final class of ideal communities is that in which interspecific competition for limiting resources is a persistent and important feature of community organization. The communities have a predictable structure [and hence resemble those in (2)], but are more or less saturated with species; component populations are not kept rare relative to available resources by natural enemies and invasion of new species is prevented by competitive exclusion.

These three possibilities, and yet others that we have not discussed, represent no more than high points on a multidimensional continuum. Most ecologists now accept that different sorts of organisms and habitat templets (Southwood 1977) combine to generate real communities throughout this continuum, and populations may be drawn from more than one of the ideal sets.

It is also important to realize that interspecific competitive interactions [in (3)] may be contemporary and amenable to study by experimental manipulation; or they may be 'ghosts' (Connell 1980), detectable because species' ecologies and morphologies have evolved to minimize competitive effects; or they may lie somewhere between these two extremes.

These are the possibilities; what actually happens in phytophagous insect communities?

8.3.2 *Population dynamic mechanisms and community predictability in assemblages of phytophagous insects*

Published life tables for individual species of phytophagous insects have been collated by Caughley and Lawton (1981) and Strong *et al* (1984a). Dempster (1983) gathered together data for one important group, the Lepidoptera. All these life tables use *k*-factor analysis (Varley *et al* 1973) to identify density-dependence and density-independence operating across generations, and because there are problems with this technique it would be wise to draw only broad conclusions (see Hassell 1985).

Where density-dependence was detected or suspected in these life-table studies natural enemies exceeded intraspecific competition for food as agents of density-dependence by a ratio of 2:1, implying that regulatory processes in herbivorous insect communities more often operate *vertically* through the food-chain rather than *horizontally* between herbivores on the same trophic level (i.e. competitors). However, of the 31 studies gathered together by Strong *et al* (1984a), density-dependence between generations was either too feeble, too sporadic, or both, to be detected in between one-half and one-third of them. So a high frequency of feeble density-dependence may result in communities that are rather unpredictable in space and time.

Are these results and predictions from life-table studies reflected in studies of the structure of entire assemblages of phytophagous insects? The answer is a qualified yes; qualified because data on concurrent changes in the abundance of species at the same site over reasonable periods of time are few. [See Connell and Sousa (1983) for a general discussion on this point, embracing organisms of all types.] Data on bracken-feeding insects and grassland leafhoppers are examples of the limited information available for phytophagous insects. The community of bracken-feeding insects at Skipwith in northern England has moderately constant structure (Fig. 8.3a). Despite fluctuations in numbers, rare species tend to be rare in most years and common species are almost always common. The result is a significant correlation between species' rank abundances comparing each year with all other years for which there are data. But there are exceptions; 1977 was an unusual year and the

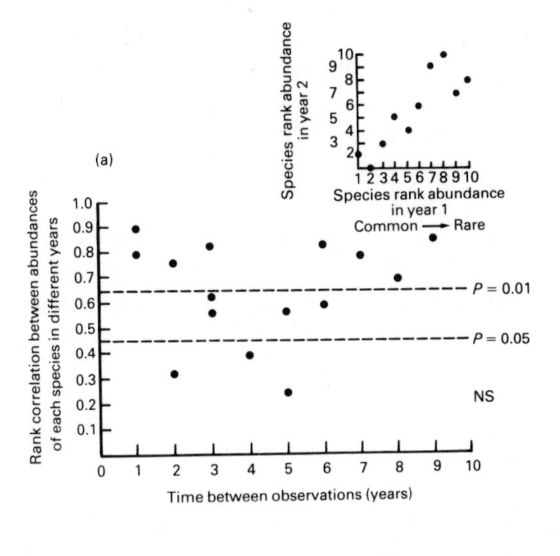

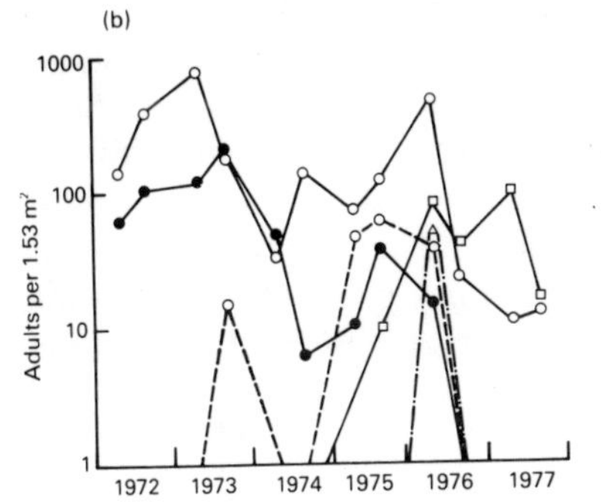

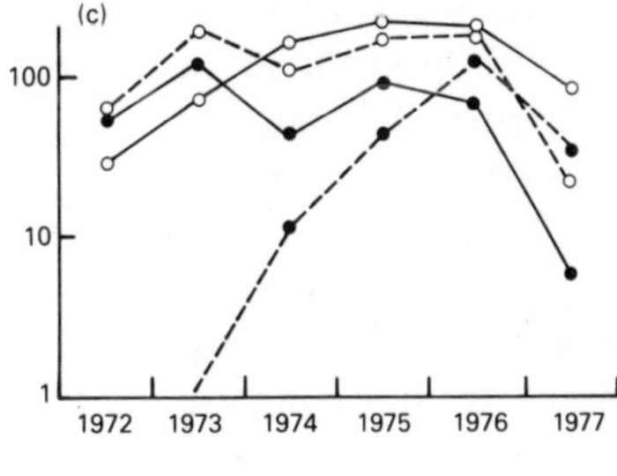

Fig. 8.3. (a) Spearman rank correlation coefficients between species abundances for the community of phytophagous insects feeding on bracken at Skipwith Common, Yorkshire. Each point on the graph is a correlation between the two years, as shown in the inset for a hypothetical 10 species community. All possible pairs of years for which data are available between 1972 and 1981 have been compared. Significant rank correlations between most pairs of years indicate that rare species have remained rare, and common species common; the community has a reasonably predictable and constant structure (from Lawton 1984a). (b) and (c): Adult population densities of nine species of grassland leafhoppers. Each different combination of dots and lines represents one species. (b) Bivoltine species; (c) univoltine species (from Waloff & Thompson 1980). Species presence, absence and abundance in this grassland leafhopper system is much less predictable than (a).

three non-significant correlations in Fig. 8.3a all involve comparisons with this year. Nine species of leafhoppers, part of a grassland community in southern England, showed a less stable relationship (Fig. 8.3b). Only four of these were found throughout the survey, with three otherwise rare or absent species occasionally becoming very common (Waloff & Thompson 1980; Lawton 1984a; Strong *et al* 1984a).

These limited data accord with the results of *k*-factor analysis. Phytophagous insect communities are not a shambles of randomly fluctuating populations; but nor are they highly deterministic and predictable in structure. It would be useful to have more data like that of Fig. 8.3 from different parts of the world.

8.3.3 Vertical and horizontal interactions in food-webs

An important implication of the life-table studies already alluded to in the last section is that intraspecific density-dependent competition for food is less important (by a ratio of approximately 2 : 1) than natural enemies as a process regulating the population of phytophagous insects. Since *interspecific* competition is unlikely without *intraspecific* competition, a corollary of insect life-table studies is that communities of phytophagous insects will not often be significantly structured by strong interspecific competition. This is not to say that occasional pairs of plant-feeding insects do not compete for food, only that most species, most of the time, do not.

On the face of it, such logic seems to accord well with common sense, given that "green plants are abundant and largely intact" (Hairston *et al* 1960). Others argue that resource limitation could be far more subtle with large parts of plants quite unsuitable as food, either because of low nutritional value, because they are chemically or physically defended in some way, or simply too large, small, or tough to exploit (van Emden & Way 1973; Southwood 1973; Crawley 1983; Strong *et al* 1984a).

Evidence for the role of interspecific competition in phytophagous insect communities is reviewed by Lawton and Strong (1981) and Strong *et al* (1984a). They conclude that despite the possibility of subtle and poorly understood interactions via the chemistry of the host plant, or poor nutrition, the balance of evidence at the moment points to a secondary role for interspecific competition as a force structuring communities of phytophagous insects.

Of course, amongst several well-studied groups of phytophagous insects are pairs of species that do compete. One example is due to Rathcke (1976). This study revealed that although the larvae of nine of 13 co-occurring species of stem-boring insects (eight Coleoptera, two Diptera and three Lepidoptera) showed greater than 70% overlap in resource exploitation with at least one other member of the guild, only two of these species had any perceptible negative influence on each other. They were a mordellid beetle larva and a microlepidopteran caterpillar (*Epiblema* sp.) found together in the stems less often than expected by chance. Laboratory observations indicated that the significant difference between observed and expected co-occurrences resulted from interference competition. Whenever larvae of these two species encountered one another within stems, the mordellid would attack, injure and eventually kill *Epiblema*. Rathcke was even able to provide an estimate of the competition coefficient ($\alpha = 0.24$) for the impact of the mordellid larvae on *Epiblema* caterpillar populations. The reciprocal influence was negligible. The effect of the mordellid on *Epiblema* was therefore the only significant competitive interaction within the guild.

Broadly similar results have been obtained by others. Gibson (1980) and Gibson and Visser (1982) found that only two from seven species of grassland mirids (Hemiptera) competed significantly; with seven species, this system has potential for 21 competitive interactions between pairs of species. In similar vein, Seifert and Seifert (1976, 1979) and Seifert (1982) analysed in detail the insects in *Heliconia* bracts in two different areas. Species included not only herbivores, but also detritivores, nectivores and predators. The majority (28 out of 40) of pairwise species interactions within these communities were not significantly different from zero. Several of the significant interactions were mutualistic rather than competitive. There were only five significant competitive interactions involving herbivores, and these were not particularly intense.

That species may compete occasionally, although most of the time they do not, is revealed in Kareiva's (1982) study on three species feeding on collards in cultivation — two flea beetles (*Phyllotreta cruciferae* and *Ph. striolata*) and a white butterfly *Pieris rapae*. In 21 experiments, Kareiva found only two significant interspecific competitive effects, at the highest densities of planted collards. *Pieris* populations depressed the abundance of *Ph. cruciferae*, which in turn depressed the abundance of *Ph. striolata*. Reciprocal effects were negligible.

Undoubtedly one important reason for this state of affairs is that natural enemies keep many species rare, relative to the available food supply. Parasitoids (mainly small parasitic wasps) are probably of crucial importance (Hassell 1978), but a host of other enemies may also have a significant impact, including predatory insects and spiders (e.g. Leston 1973; Kiritani & Kakiya 1975; Varley *et al* 1975; Skinner & Whittaker 1981) true parasites and disease (Anderson & May 1980; Anderson 1981), birds (Holmes *et al* 1979a; Loyn *et al* 1983) and even bats (Rees 1983).

Some indication of the impact of natural enemies can be obtained from cases of successful biological control of former insect pests. Biological control has been most effective against insects accidentally introduced by people from one continent to another where, relieved of natural enemies, exotic insects can become major pests. Releasing insect parasitoids from the pests' country of origin has often proved to be a very effective way of controlling them (DeBach 1964, 1974; Huffaker 1971; Varley *et al* 1973; Clausen 1978; Papavizas 1981), and for our purposes constitute an ecological experiment on a grand scale. As Beddington *et al* (1978) show, the release of one species of parasitoid may be sufficient to depress population numbers of pests to one-hundredth, or even less of their former abundance. There are no grounds for believing that parasitoids do not have a similar impact on native insects, although the reverse experiment of removing parasitoids and seeing if host-species outbreak in their natural environment has not been attempted under controlled conditions. This caveat aside, strong depression of the numbers of herbivorous insects by natural enemies — particularly parasitoids but also probably predators and diseases — is more than sufficient to explain why interspecific competition for food is not a process of major importance structuring communities of herbivorous insects.

8.3.4 Vacant niches and asymptotic colonization curves

The generally feeble and sporadic nature of interspecific competition between plant-feeding insects poses an interesting theoretical dilemma. Introduced plants are often quickly colonized by insects (Section 8.2.4), with most 'colonization curves' seemingly asymptotic — rapid at first but subsequently slowing to a trickle. That this must be so is obvious from the study of native plants; insects very rarely switch hosts between members of a native flora, although examples are known (Strong *et al* 1984a). The easiest way to explain why colonization slows to a trickle is to invoke interspecific competition. As niches fill up with species, it becomes harder and harder for new species to invade (May 1981).

The paradox is that many long established native plants, on which colonization by new species of insects appears to have effectively stopped, display conspicuous vacant niches. Bracken in New Mexico is a good example (Section 8.2.3; Fig. 8.2b). Other examples are not difficult to find. Within groups of closely related, well established plants, parts that are used on one species may be ignored on a close relative. For instance, a few Umbelliferae, unlike the majority, do not support any species of swallowtail butterfly (Slansky 1973) while others for no apparent reason have no leaf-mining and stem-mining agromizid flies (Lawton & Price 1979). Introduced plants are often attacked in parts left unexploited in other areas (Strong *et al* 1984a), and so on. In other words, it is extremely difficult to believe that classical 'niche pre-emption' and interspecific competition for food limit colonization of most plant species by insects.

An alternative model to account for asymptotic colonization curves might be called the 'pool exhaustion hypothesis' (Lawton & Strong 1981). Here it is the pool of potential colonists that is exhausted, not resources; or to extend an analogy borrowed from Hutchinson (1965), it is suitable actors that are in short supply, not good parts. A number of species have a high predilection for colonizing a plant spreading or introduced into a new area, either because they are highly polyphagous or because the plant is functionally quite close to their normal host(s). Such insects find and exploit the plant quite quickly. The vast majority of insect species in the regional pool have a vanishingly small probability of ever being able to make the necessary biochemical, physiological and behavioural jump on to the new host. Of course, some species are intermediate between these two extremes. Hence, the regional pool of potential colonists is quickly exhausted, and colonization is asymptotic.

8.4 Mammalian herbivores: comparisons and contrasts with insects

Bowing to the power of the insect's vastly superior numbers, and displaying our own prejudices as entomologists, we intend to say much less about mammals than we have said about insects. Nevertheless, we hope to say enough to convince readers that comparisons and contrasts between insect and mammalian herbivores are worth making.

8.4.1 Population density and body size of mammals

One of the most intriguing, but poorly understood community patterns documented for mammalian herbivores is Damuth's (1981) discovery of an inverse relationship between the logarithm of each species' mean population density plotted against the logarithm of mean adult body mass (Fig. 8.4a). For 307 species of mammals from a wide range of habitats throughout the world (deserts, temperate and tropical grasslands, forests, etc.), the overall slope of the relationship is −0.75. Groups of mammals from individual communities apparently display similar patterns. In other words large mammals on average live at lower population densities than small mammals. However, the relationship is more interesting than this because metabolic rates and food energy requirements of individual mammals scale approximately as the 0.75 power of their body weights. In consequence, the amount of energy used by a population of herbivorous mammals is approximately constant and independent of body size. Communities of mammalian herbivores, Damuth suggests, are assembled from species, each one of which uses about an equal quantity of food energy. He suggests that "interspecific competition act(s) over evolution-

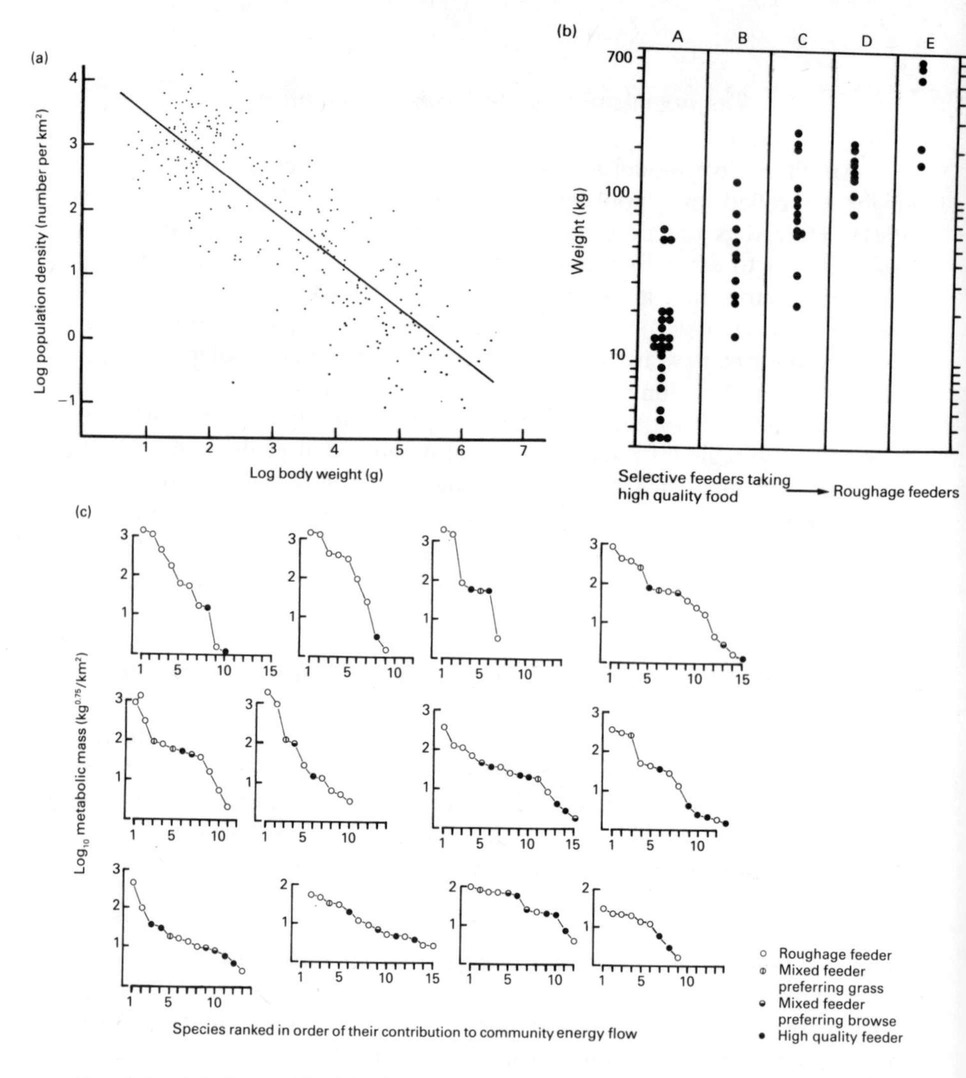

Fig. 8.4. (a) Damuth's (1981) relationship between the weight of mammals and their abundance. This, in combination with the opposed trend of a 0.75 power relationship between body size and metabolic rate suggests that the energy consumption per unit area for a species is constant and unrelated to body size. Damuth's data are apparently contradicted by information on the big mammal communities of Africa, dominated by large roughage feeders; e.g. (b) and (c). (b) illustrates the relationship between the size of large herbivorous mammals and the quality of their diet (smaller mammals eat higher quality diets). Bearing this in mind, (c) shows that large bulk feeders account for a greater part of the energy flow than the smaller high quality feeders at a wide range of sites of differing character. Each small graph in (c) is one community of mammals coexisting in one habitat. Damuth's hypothesis predicts approximately equal values for all species, rather than large contributions to community energy flow by some species, and progressively smaller contributions by others [from Damuth (1981), Jarman's (1974) study of African Bovidae, excluding the Caprinae, and Cumming (1982)].

ary time to keep energy control of all species within similar bounds", and goes on to postulate that similar relationships probably hold for other taxa.

These intriguing data are a classic example of workers in different fields asking different questions, because nothing like Damuth's relationship has even been postulated by entomologists, let alone tested. More recent detailed studies with mammals (Fig. 8.4b, c; Cumming 1982) do not conform well to Damuth's predictions; however, similar trends *are* maintained across all metazoan taxa, vertebrate and invertebrate, spanning 13 orders of magnitude in body size, but with considerable scatter round the fitted regression (Peters 1983; Peters & Wassenberg 1983). In very general terms, therefore, populations of phytophagous insects presumably conform to something like Damuth's relationship. However, this does not necessarily mean that the patterns are generated by interspecific competition, either between mammals or between insects, as Damuth postulated, although other explanations have not been put forward. As is often the case in science, the most intriguing data are the most difficult to explain, and generate more questions than answers!

8.4.2 Species–area relationships

A considerable part of Section 8.2 was devoted to explaining differences in insect species richness on individual species of plants using species–area relationships. Mammals in general are much larger and more polyphagous than insects, so that it is neither sensible nor practicable to relate mammalian species richness to individual species of plants. Rather the nature and scope of the question must be adjusted to encompass fundamentally different biologies. Often this adjustment is made implicitly rather than explicitly, which is a pity, because hidden shifts of emphasis hide much interesting biology. For example, why are herbivorous mammals in general more polyphagous than herbivorous insects?

In general mammalogists seem to have been much less concerned than entomologists to explain why some communities contain many more species than others. Indeed, with fewer communities anyway, the determinants of mammalian species richness could appear less deterministic than is the case for insects, resulting primarily from accidents of evolution, geology and human intervention. Unlike insects, for example, area of habitat plays an uncertain role: of five studies (excluding bats) listed by Connor and McCoy (1979), three show a significant increase in mammalian species richness with habitat area, and two do not. The very shortage of such studies in Connor and McCoy's extensive compilation of data suggests that species–area relationships have not been a question of much concern to mammalogists though the studies of Miller and Harris

(1977) and Soulé *et al* (1979) are interesting exceptions. Is this lack of interest in species–area studies due to research tradition, or does it reflect a genuine difference in biology?

8.4.3 Organizing forces in communities of mammalian herbivores

The literature on communities of mammalian herbivores has a focus and emphasis quite different from that encountered in entomological work. In particular, it emphasizes the detailed interaction of each species of mammal with its food plants, and often assumes a major role for interspecific competition and a secondary role for predation, disease and parasitism. Again we might usefully ask whether these differences reflect real differences in biology. We briefly explore such problems in this section, but first digress and ask whether communities of mammalian herbivores are likely to be more predictable in their structure than insect communities.

A reasonable guess would be that mammalian populations fluctuate less than insect populations, and hence give rise to more predictable stable communities (see Sections 8.3.1–8.3.2) because mammals are more '*K*-selected' than insects (Pianka 1970; Southwood 1981). However, Connell and Sousa's (1983) data provide no clear support for this prediction, with little or no intrinsic differences in the variability of insect and mammal populations, once differences in generation times are allowed for. Hence, if our interpretation of Connell and Sousa's data is correct (their data are entirely for small mammals up to rabbit size; large mammals may be different) there are no strong grounds for believing that communities of herbivorous insects are necessarily less structured and predictable than those of mammals.

Plant–herbivore interactions

Herds of large African mammals are amongst the richest mammalian assemblages in the world, and among the best studied. Here the large amounts of plant material removed by ungulates and proboscideans and the apparent ineffectiveness of predators has led to an emphasis on the role of plant–herbivore interactions as determinants of community structure (Vesey-Fitzgerald 1960; Bell 1970, 1971, 1982; Schaller 1972; Sinclair & Norton-Griffiths 1979; Walters *et al* 1981). Most of the herbivore biomass, and the greatest impact on vegetation, comes from the bulk feeders (Fig. 8.4b; Laws 1970; Laws *et al* 1975; Sinclair & Norton-Griffiths 1979; Bell 1982; Cumming 1982). The amount of material removed can be remarkable: wildebeest *Connochaetes taurinus* passing through one area of the Serengeti over 2–3 weeks removed 85% of green foliage. In areas protected from their grazing for a number of years species composition of the vegetation changed, favouring those plant species that could overgrow their compatriots (McNaughton 1976, 1979). This was dramatically illustrated in the 1890s when rinderpest, accidentally introduced by the Europeans, arrived in the Serengeti and destroyed 95% of cattle, buffalo *Syncerus caffer*, and wildebeest as well as affecting giraffe *Giraffa camelopardalis* and other species (Ford 1971; Sinclair 1979a). In combination with the abandonment of cultivation, these mortalities resulted in the spread of dense thickets [although as Cumming (1982) points out, there are difficulties in separating the impact of wildlife and cattle in Africa, so effects may not mainly be due to a reduction in wild ungulate populations].

Where grazing profoundly alters vegetation structure and composition, the balance and species composition of ungulate communities may be affected. For example, bulk feeders, by clearing the coarse vegetation, allow smaller species to follow and exploit areas from which they are otherwise excluded (Vesey-Fitzgerald 1960; Bell 1970). In theory, zebra remove the coarse top stem, allowing wildebeest access to higher quality forage followed by Thomson's gazelle which benefits from the herbs in the areas grazed by the wildebeest (Bell 1970, 1971; McNaughton 1976). In practice, relationships between species may be more fuzzy (Fig. 8.5), and in a later paper Sinclair and Norton-Griffiths (1982) considered that such facilitation is relatively unimportant in this system.

Minor difficulties of interpretation aside, the impact of large mammals on vegetation, and hence ultimately on mammalian community structure itself, has been much more thoroughly explored than the corresponding effects of insect herbivores. But again we suspect this reflects research traditions rather than fundamental differences in biology. Biological control of weeds by insects dramatically highlights the fact that certain insect species can have major impact on the distribution and abundance of host plants (Harper 1969; Huffaker 1971; Caughley & Lawton 1981), whilst recent experiments suggest that herbivorous insects retard and alter patterns of plant succession (Brown 1982; McBrien *et al* 1983). Effects on associated insect herbivores in such systems must be profound, as host plants change densities and patterns of distribution (Section 8.2). But compared with mammalogists, entomologists have been very slow to study these effects.

Interspecific competition

Studies of herbivorous mammals frequently conclude that interspecific competitive interactions play an important role in limiting species distributions and abundances. Small rodents (mice, voles, gerbils and chipmunks) appear frequently to compete, with good experimental evidence from a wide range of habitats and vegetation zones (Schoener 1983b). Here appears to be a real, major difference between insects and

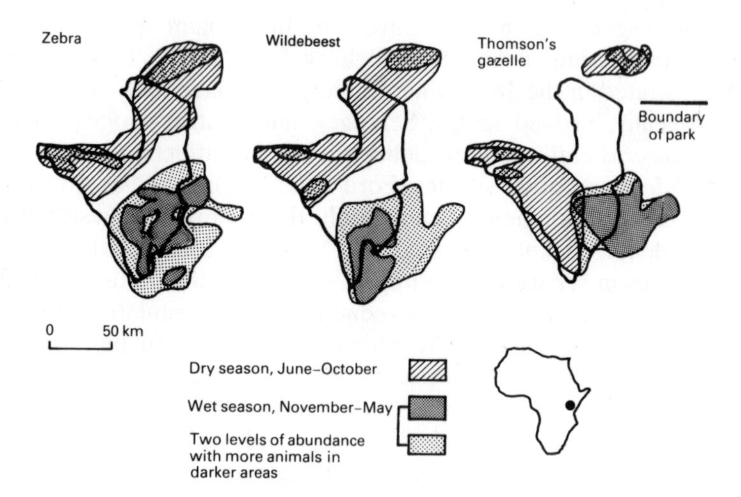

Fig. 8.5. A test of the 'grazing succession' hypothesis. In general, migratory species spend the wet season on the Serengeti plains and the dry season in the woodlands, but there are differences in the areas the three species occupy. Summarized from a detailed 10 km square survey between 1969 and 1972 by Maddock (1979). The grazing succession of zebra followed by wildebeest and then by Thomson's gazelle was not shown clearly by correlations calculated in this study, except that at the end of the wet season when Thomson's gazelle occupied parts of the plain previously used by the other two species.

mammals. It implies that regulation of small mammal populations by predation, parasitism and disease is generally rather ineffective. We have no idea why.

Data for large mammals are more equivocal. Again, some of the most interesting information comes from the Serengeti, where rinderpest was finally eliminated in 1963. Since then the zebra population, unaffected by rinderpest, has remained more or less stable (Fig. 8.6; Sinclair 1979b). However, wildebeest and buffalo, released from this mortality expanded rapidly (Fig. 8.6). Sinclair (1979b) writes:

> "By the late sixties, both populations were levelling off, as a result of mortality induced by the lack of food in the dry season. Food appeared to be limited both by low dry season rainfall and by grazing. However, in 1971, there began a series of years of high dry-season rainfall, which produced a superabundance of food for the animals and allowed the populations to increase a second time. The buffalo...levelled off for four years in the early seventies, a longer period of time than the levelling out exhibited by wildebeest, which was already increasing again by 1972. *This supports the hypothesis that interspecific competition was taking place* [our italics], with buffalo suffering at the expense of wildebeest.

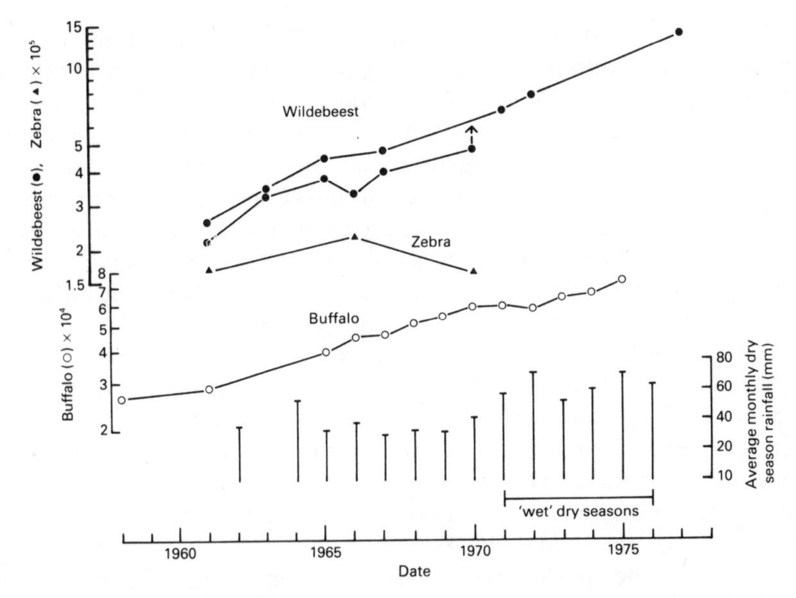

Fig. 8.6. Population dynamics of mammals in the Serengeti since the elimination of rinderpest. Two sets of data from the same wildebeest (●) population are shown (top from Sinclair 1979b, bottom from Schaller 1972), together with those of zebra (▲) and buffalo (○). The arrow indicates a minimum estimate. Rainfall in the dry season is also shown together with the run of abnormally wet 'dry' seasons. In this type of plot an unaltered birth rate and exponential population growth is indicated by a straight line [data from Sinclair (1979b) and Schaller (1972)].

During this same period other, residential, species such as topi *Damaliscus lunatus*, kongoni *Alcelaphus buselaphus*, and impala *Aepyceros melampus* did not increase; again, Sinclair (1979a) suggests, because of interspecific competition with the wildebeest and buffalo.

Whilst Sinclair's interpretation of these data might well be correct, others are possible. Thus, when looking at Fig. 8.6, rather than seeing the pause in the growth of the buffalo population, and a static population of other species as evidence for competition, we are impressed that other species were so little affected by a ten-fold increase in wildebeest! If this is a strongly competitive system, it is a curious one. Indeed, Sinclair (1985) no longer considers competition tenable as the dominating evolutionary process for this ungulate community. However, Hilborn and Sinclair (1979) had predicted that the wildebeest population would stabilize between 1.0 and 1.5 million if the typical dry seasons of the late 1960s returned. This happened between 1977 and 1982 (Sinclair & Norton-Griffith 1982; Sinclair *et al* 1985), and resource limitation was suggested to be the simplest explanation of its cause though others were possible.

Until recently, studies of large mammals have tended to dismiss predation and disease (with one notable exception of introduced rinderpest) as effective controlling agents. Hence food limitation and interspecific competition are seen as the inevitable, dominant processes in population regulation and community organization. One problem is finding communities where man has not had a serious effect on mammalian predators. However, predators *can* play a large part in regulating populations of large mammals. In reviewing the studies of wolves, in relatively undisturbed communities in east North America, Pimlott (1975) concluded that they were a major mortality factor in the prey population (Edwards 1983); similarly Hornocker (1970) identifies the puma *Felis concolor* as a major influence on their prey. Perhaps the most interesting studies are by Schaller (1967), who describes the tiger *Panthera tigris* as a major factor limiting populations of chital *Axis axis*, gaur *Bos gaurunus*, sambur *Cervus unicolor*, and barasingha *Cervus duxanceli*, in Kanho National Park, India, which at the time of the study probably had the most diverse large mammal fauna, for its size, in Asia.

In Africa, the spotted hyena *Crocuta crocuta* controlled the population of wildebeest and zebra in Ngorongoro crater, taking, respectively, 11% and 9% of the population per annum, compared with approximately 1% of each in the Serengeti (Kruuk 1972), while Schaller (1972) estimated that the five main predators accounted for 20 000–30 000 Thomson's gazelle per annum in the Serengeti, and probably played a major part in regulating the populations of this and other residential species. Indeed, in the light of Schaller's data, Sinclair and Norton-Griffiths (1982) now think it likely that lion and hyena predation regulates the Serengeti zebra population. In Nairobi Park lions were responsible for a decline in wildebeest numbers during the 1960s until, with wildebeest numbers low, they turned to other species (Foster & Kearney 1967; Foster & McLaughlin 1968; Schaller 1972).

In sum, predators can be effective controlling agents in communities of large mammals, in Africa and elsewhere. Hence it follows that such communities are not necessarily strongly structured by interspecific competition for food. Moreover, virtually nothing is known about the role of disease in the population dynamics of these beasts, but there are tantalizing references, such as to 'quite frequent' diseased zebra and Thomson's gazelle (Kruuk 1972) in the Serengeti. Anthrax, sarcoptic mange, hoof gangrene (in wildebeest); anthrax (in zebra); and anthrax, sarcoptic mange, haemonchus, lungworm, pleuritis, nephritis and orchitis (in Thomson's gazelle) are among the many diseases that make life unpleasant for ungulates, but whose consequences for population dynamics appear totally unknown.

Obviously these are difficult problems to solve. It is easier to raise alternative explanations than it is to do experiments and make observations that distinguish between them. But this much is clear: it cannot unequivocally be claimed that large mammal communities are structured mainly by interspecific competition for food, whilst phytophagous insect communities are not. The ecological processes at work in assemblages of species drawn from two very different corners of the ark, may not after all, be as different as many people have thought.

8.5 Concluding remarks

Globally, and within any local community, the number of species of insect herbivores is vastly greater than the number of species of mammals. Why? There are several (not mutually exclusive) explanations, all of which ultimately rest on the fact that insects are small and mammals are bigger. Small size means that insects can specialize by feeding on plant parts that are impossible for mammals to use, inside stems or leaves for example. That is, there are many more ways for insects to make a living on and inside plants, and hence more insects (Southwood 1978; May 1978). (As an aside, note that even within the mammals, small species are much more selective feeders than large species, e.g. Fig. 8.4b; Jarman 1974.) Moreover body size and generation times are positively correlated; big organisms have long generation times (Southwood 1981). Hence we might expect rates of evolution to be faster in *absolute* time in those groups of organisms with short generation times — insects. Higher rates of evolution in insects than mammals may generate more species of insects at any moment in time (Fowler & MacMahon 1982).

Finally, small body size means that effective isolation of gene pools can take place over very small physical distances, particularly in species confined to one species of host plant for most of their lives. Isolation of subpopulations again promotes speciation (Bust & Diehl 1982; Wood & Guttman 1983; Strong *et al* 1984a).

Whether these size-related differences in biology are sufficient on their own to explain global and local differences in species richness of mammalian and insect herbivores is open to debate. But at the moment, there are no other explanations.

Ecosystems jumble together the contents of taxonomists' carefully sorted collections. Most terrestrial communities contain a few species of herbivorous mammals and many species of insects. Are we therefore wrong to consider insect assemblages ('communities') separately from mammalian assemblages? In the first instance probably not; considerable progress can be made by specialists in each field ignoring other groups — as we hope this chapter makes plain. But in the end artificial, taxonomically imposed boundaries will have to be broken down if our understanding of pattern and process in community structure is to advance (e.g. Brown *et al* 1979).

For example, what effect do mammalian herbivores have on insects in the same habitat? By changing the composition of vegetation (Section 8.2) altering its 'architecture' by grazing (Lawton 1983) and by trampling, perhaps accidentally eating (Hayes 1981) and generally disturbing insects, mammalian herbivores may significantly affect insect herbivores. But virtually nothing is known about such problems, still less about if and how insects influence mammals. A few pioneer studies consider both taxa [e.g. Sinclair (1975); Crawley 1983; and compare Butcher (1982) with Bell (1982)]. More work that ignores taxonomic boundaries would greatly enhance our understanding of herbivore community organization.

9 Communities of Parasites

John C. Holmes and Peter W. Price

9.1 Introduction

Parasites may be defined broadly as organisms living in or on other living organisms, obtaining from them part or all of their organic nutriment, commonly showing some degree of adaptive modification, and at least potentially causing some damage to their hosts (Price 1980). By this definition, several of the chapters in this book relate to parasite–host systems and Chapter 13 is devoted entirely to the evolution of parasite communities. We restrict our focus in this chapter to the kind of animals typically studied by parasitologists. We will consider communities of animals parasitic on other animals, addressing two main questions: what is the nature of parasite communities, and what influences do these parasites exert on host communities?

In this chapter we use the term 'community' to mean a group of organisms in a particular place, without any preconceptions on whether they interact or not. In Section 9.2 we also concentrate on the organizing factors on one trophic level, particularly niche occupation and the number of coexisting species, and not on other community characteristics such as food-webs, energy flow, or abiotic influences. In Section 9.3 we broaden our coverage to consider the influences of parasites on some of these characteristics.

9.2 The nature of parasite communities

Views on whether, or by what mechanisms, communities are organized are undergoing fundamental re-evaluation at present (e.g. Hanski 1982; Järvinen 1982; Schoener 1982; Simberloff 1982; Strong *et al* 1984b; Price *et al* 1984). Clear empirical studies are necessary to resolve the conflicting views. Parasite communities have several features (examined in Section 9.2.1) which can enable them to contribute significantly to these emerging concepts. It is therefore interesting to examine the concepts that have been applied to communities of parasites of animals. This is undertaken in Sections 9.2.2–4. In each of these sections, we first present the concepts as developed by the original authors, then examine some of the assumptions and predictions of each, and finally suggest how they may be distinguished in real communities. We regard these concepts as a viable set of alternative hypotheses (not mutually exclusive), each of which may be correct at least part of the time, and several of which may apply to the same community part of the time [the multiple causation of Hilborn and Stearns (1982)]. Given the wide range of parasite species, types of host–parasite interactions, life cycles, species richness per community, and habitats used, it is likely that the relative importance of the various processes will differ in various kinds of communities. The challenge is to identify the conditions under which one process or the other becomes important and others unimportant. The field will advance most rapidly if rigorous simultaneous testing of alternative hypotheses is undertaken by many investigators using, as far as possible, an experimental approach.

9.2.1 *Features of parasite communities*

The first three features discussed below combine to make patterns readily recognizable and demonstrable in parasite communities. The third and fourth features incorporate pitfalls to be avoided in interpreting empirical data on parasite communities.

Definition of resources

Accurate identification of resources at a level relevant to the organisms under study greatly helps the elucidation of community processes, but has been very difficult in the many communities of generalists so frequently studied by ecologists. For parasites, it is even more difficult to identify exactly how much of what specific resources they are using; however, a large proportion of those resources are strongly correlated with the host species, or the particular part of the host, that is exploited. Resources at these last two levels, at least, can be defined accurately.

Replicated habitats

A host species represents a habitat for its parasites. Host species within the same taxon are structurally similar with equivalent sets of organ systems available for colonization by parasites. Although much genetic and physiological diversity may exist within and between host populations, hosts provide basically homologous habitats. This feature allows two powerful approaches to community ecology. First, the most similar habitats are provided by individuals of the same species; examination of infracommunities (defined below) in different individuals can provide sufficient replicate measures on independent communities to allow a statistical approach to community analysis. Patterns therefore can be differentiated from random events; for example, Bush and Holmes (1983) developed a statistical test to determine whether species of intestinal helminths are randomly distributed along the intestine, and show that those in lesser scaup *Aythya affinis* are not.

Secondly, a strong comparative approach to community ecology (e.g. Price 1984a) can be developed. If a host species harbours a parasite community of ten parasite species in one area and only two in another, or if one host species supports 20 parasites and another species in the same taxon supports only five, then we can ask the question: If 10 species (or 20 species) could be supported by certain hosts, why is this number of species not present in other hosts in the same taxon? The parasites of one host species are used to indicate what are the available resources or microhabitats in or on that host. We can therefore estimate the number of ecological niches available in or on other host species in the same taxon, observe the number of these that are filled, and ask why the remainder are apparently vacant. This requires adopting Hutchinson's (1957) view, which regards a niche as the complex of environmental characteristics permitting a species to persist and reproduce indefinitely. It must be noted, however, that the resources or microhabitats provided by the host constitute only part of those characteristics. All parasites require suitable intermediate hosts, suitable conditions for free-living stages of the life cycle, or conditions suitable for the transmission of the parasite from host to host; these may be more limiting than the characteristics provided by the host.

Specialization

Most parasite taxa contain species that are relatively specialized compared with free-living generalists like birds and mammals (Price 1980). For some, the degree of specialization is great; for example, 85% of monogenean species on British fishes are known from only one host species (Kennedy 1974), and 87% of bird lice are known to use only one host species in Israel (Theodor & Costa 1967). Other parasites are not so specific (Holmes & Price 1980). Overall, the following generalizations have been made for parasites in communities of similar host species (Freeland 1983): (i) most parasite species are successful in relatively few hosts, (ii) parasite species common in one species of host are not usually common in others, (iii) host species from the same community do not usually have the same parasites, and (iv) parasites that are shared by different host species are usually found at statistically different frequencies.

One feature determining the degree of specificity of a taxon is the extent to which individual parasites are able to select their host. Active colonizers, taxa which actively seek out and invade their host, such as the monogenean or arthropod ectoparasites mentioned above, can choose between and within host species, and can have their decision-making guided by genetically programmed information (see review by MacInnis 1976). Passive colonizers, taxa which colonize their hosts through the host's ingestion of eggs, cysts or infected intermediate hosts, such as most intestinal helminths of vertebrates, are unlikely to be able to reject one host and try again. For the latter, the constraints on specialization are more severe, and such parasites should face strong selection pressure to increase the probability of reaching hosts in which they can develop and to develop in those they regularly reach (see review by Holmes 1976). As a result, some of these parasites are true generalists, with their breeding populations spread evenly across a wide variety of host species (e.g. Butterworth 1982). Most, however, are found in a narrower range of host species, and in very different numbers, as indicated by Freeland (1983). Some even have the bulk of their reproducing individuals in only one species (e.g. Leong & Holmes 1981; Kontrimavichus & Atrashkevich 1982).

The preceding refers to specialization at the level of the host species. Parasites may also be specialized at the level of the particular site in the host that is occupied (termed the microhabitat level in the rest of this chapter) (Crompton 1973; Holmes 1973; Hair & Holmes 1975; Rohde 1979). Holmes (1973) has suggested that much of the differentiation of niches of parasites is at the microhabitat level. Most of the concepts at the infracommunity level (Section 9.2.2) deal with this aspect of specialization. It should be noted, however, that not all parasites are microhabitat specialists. Some can inhabit a variety of organs (e.g. Holliman *et al* 1971), others can occupy virtually the entire small intestine (e.g. Mettrick & Dunkley 1969; Hobbs 1980), and still others even undergo extensive daily (e.g. Arai 1980; Hobbs 1980) or seasonal (e.g. MacKenzie & Gibson 1970) migrations within the host.

In communities where specialists predominate, predictions are more readily generated on such topics as niche relationships, overlap, and expected species number. As parasites become less specialized, their communities become more complex and harder to predict. Communities of parasites composed of different kinds of species would be likely to display different kinds of interactions. By providing communities with different kinds of species, parasite communities can be instrumental in deriving concepts on the significance of specialization for community structure. Some of the differences in the concepts outlined in Section 9.2.2 may relate to the investigators' experience with such different systems.

Hierarchical communities

With host individuals, host populations, and communities of hosts providing resources for parasites, there is an obvious hierarchy of organizational levels to be dealt with in parasite communities. Esch *et al* (1975) recognized this hierarchy in distinguishing between *infrapopulations* and *suprapopulations* of parasites. The *infrapopulation* of a parasite is the population found in an individual host. Similarly, populations of all parasite species within a single individual host constitute an *infracommunity* of parasites. It is at the infracommunity level that all data are collected, and any direct interaction of parasites must take place. All too often, such interactions have been searched for in collapsed data sets, such as distributions summed across infracommunities, in which much of the relevant biological detail has been lost. Studies of potential interaction should be directed to the infracommunity level, as, for example, in Hair and Holmes (1975). Studies at this level, especially using experimental manipulations, are needed.

The significance of any interaction of parasites depends in part on the frequency with which they occur. This will be determined at the next community level, the *component community*. Root (1973) defined a component community as an assemblage of species associated with some microenvironment such as those in tree holes, rotting logs or on a particular host taxon. The study of component communities has received much attention from ecologists because they are relatively simple and moderately discrete. For parasites, all the infracommunities in a population of hosts would make up a component community. It is to this level that most parasitologists have directed their attention; surveys of parasites of particular hosts or host groups are legion. At this level Hanski (1982) has distinguished between *core* species (frequently occurring across infracommunities and numerous within them) and *satellite* species (infrequent and few), suggesting that the number and distribution of the former, but not the latter, may respond to biotic interactions. Bush & Holmes (1986) demonstrate that in at least one case these ideas also apply to component communities of parasites, and that the core species are mainly specialists.

Just as the *suprapopulation* (Esch *et al* 1975) of a parasite comprises all individuals of a species in an ecosystem, including members in the definitive hosts, intermediate hosts and free-living stages, the *compound community* comprises all the suprapopulations in an ecosystem. Root (1973) defined the compound community as a complex mixture of component communities that interact to varying degrees, such as those in a meadow, a wood lot, or a pond or lake. At this level exchange of parasites between host species may be investigated (e.g. Neraasen & Holmes 1975; Leong & Holmes 1981). It is also at this level that characteristics of, or changes in, the host community can have their most profound effects (e.g. Esch 1971; Holmes 1979).

In this chapter, we will keep these distinctions clear, and deal with concepts appropriate to each level separately. Since data are collected at the infracommunity level, and since it is at this level that many of the predictions are the clearest, we will focus on predictions made at this level.

9.2.2 Concepts at the infracommunity level

Concepts at this level focus on what determines abundance and distribution of parasites within a host individual, and especially on the mechanisms producing restricted niches within those communities.

The concepts

Holmes (1973) reviewed the large literature on coexistence of helminths, concluding that *interactive site segregation*, or competition between species within a host that resulted in occupation of discrete microhabitats, was uncommon, *competitive exclusion*, or competition that prevented co-occurrence in the same host individual, was moderately common, while *selective site segregation*, or evolved differences between microhabitats of coexisting species, was very common. The evidence for much evolved niche diversification led Holmes (1973, p. 344) to hypothesize that "parasite faunas are not 'young' or 'pioneering' ones, but mature communities whose diversity has been established to an important extent through biotic interactions". These ideas are embodied in the *competition hypothesis*, which states that competition has been an important organizing force in parasite communities through evolutionary time and still acts in contemporary or ecological time.

Several authors have suggested that evolved niche differences between related species have been selected for by factors associated with their reproductive biology. Sogandares-Bernal (1959) and Martin (1969) suggested that occupation of discrete niches was important as a reproductive isolating mechanism preventing hybridization. Rohde (1979, 1982), having worked extensively with the highly specific monogenean gill flukes of marine fishes, argued that narrow niche occupation was essential in the maintenance of intraspecific contact for mating purposes. This would account for the high host specificity and site specificity on a host in parasite species that typically have sparse populations of small individuals. These views may be classed as the *population concentration hypothesis*.

Price (1980) viewed parasites as highly specialized relative to free-living organisms. Their relatively small size and limited mobility indicate they must exploit a coarse-grained environment, which should lead to specialization (Levins 1962). Coevolutionary processes should also result in specialization. Such processes should lead to species of parasites being individually and independently adapted to narrow niches. The view that individualistic responses of species to available resources is the major process resulting in accumulation of species in communities and their maintenance was expressed early in the conceptual development of community ecology (e.g. Gleason 1926; Ramensky 1926). We refer to this as the *individualistic response hypothesis* (Price 1984b).

The assumptions and predictions

Before presenting the assumptions and predictions of these three models, we must reiterate several caveats. Several authors (Rohde 1979; Brooks

1980a; Connell 1980; Simberloff & Boecklen 1981) have made the points that because species are taxonomically distinct it is likely that their ecology will be different, including occupation of different niches, and that competition is not a necessary prerequisite to distinctness. Therefore, it is not enough to demonstrate differences and then infer competition as their cause. These authors emphasized that competition must be demonstrated as a process and any inference about past evolutionary change through competition must be approached with extreme caution and considerable rigour.

However, all other processes potentially acting on the community must be equally subject to the same rigorous scrutiny. For example, it cannot be assumed that parasites must be specialists, either to a narrow range of host species or to specific microhabitats in or on a host species (evidence above). In addition, any organism that regularly occupies a specific microhabitat (for whatever reason) should develop some adaptations to that microhabitat; the mere presence of these adaptations cannot be taken as evidence for the individualistic response hypothesis.

Given these caveats, the assumptions and some of the predictions, as we perceive them, contained within each hypothesis are shown in Table 9.1. It is apparent from this distillation that the hypotheses predict one of two types of community: the assumptions and predictions of the competition hypothesis differ markedly from those of the other two hypotheses, and most of these differences stem from the assumptions regarding the relative colonizing abilities of parasites. The competition hypothesis is appropriate to communities composed of species with a high probability of colonizing the host, and consequently with relatively high populations, thus a high probability of interacting. It predicts saturated, equilibrial communities structured largely by biotic interactions, with species evenly dispersed in resource space and responding to the presence of other guild members. We refer to these as interactive communities.

The conditions under which such communities could be developed, especially the relatively high population densities, would be expected to exert conflicting selection pressures on the parasites. Intraspecific competition should lead to selection for individuals able to broaden the parasite's niche. Interspecific competition would tend to restrict or narrow the niche. Stochastic differences in population or infracommunity structure would be expected to favour first one, then the other. As a result, parasites should evolve to exploit effectively even temporary opportunities.

The other two hypotheses are appropriate to communities of species with a low probability of colonization, thus with low populations and a reduced probability of interaction between species. They predict unsaturated, non-equilibrial communities, with species individualistically dispersed in resource space and insensitive to the presence of other guild members. We refer to these as isolationist communities.

Table 9.1. Assumptions and predictions on mechanisms involved in the organization of parasite infracommunities, with particular reference to mechanisms producing restricted niches.

	Competition hypothesis	Population concentration hypothesis	Individualistic response hypothesis
Assumptions			
(1) Colonizing ability	High	Low	Low
(2) Important evolutionary processes in the past	Interspecific competition	(a) Reproductive isolation (b) Finding mates	Adaptation to host and microhabitats within host
(3) Important ecological processes acting currently	Interspecific competition	(a) Reduce hybridization (b) Small population sizes	Small population size
(4) Equilibrium status	Equilibrium	Non-equilibrium	Non-equilibrium
(5) Relative roles of:			
(a) past competition	Very important	Unimportant	Unimportant
(b) current competition	Important	Unimportant	Unimportant
(c) individualistic responses	Weak	Strong	Strong
Predictions			
(1) Vacant niches	No	Yes	Yes
(2) Distribution of species in resource space	Regular (core species)	Random (closely related species separated)	Independent of other species*
(3) Realized niche (cf. fundamental)			
(a) increased when population increased	Yes	Yes	Yes
(b) reduced in presence of other guild members	Yes	No	No
Type of community	Interactive	Isolationist	Isolationist

* Significantly different from regular; random or clumped, depending on distribution of resources or evolutionary history.

9.2.3 Concepts at the component community level

Concepts at this level focus on what determines species richness of the parasites of a host population, or more frequently, of a host species.

The concepts

Brooks (1980a) emphasized the need for a strongly historical, evolutionary approach. Phylogeny of hosts and parasites constrain the development and organization of communities. Brooks noted that many studies have shown that parasite phylogenies tend to mirror host phylogenies, suggesting that speciation in a host lineage results in cospeciation in the parasite lineage: "The phylogenetic relationships of. . .parasites reflect a history of allopatric speciation and coevolution. If such represents a general pattern, most parasite faunas (or communities) can be considered co-evolving units" (Brooks 1980a, p. 198). Brooks further argued that parasite communities developed early in the phylogeny of host groups and that members of these communities have tracked the cladogenic development of host groups. Cospeciation with the host group has produced parasite communities with a long phylogenetic history of association with the host's clade. Brooks's view may be called the *cospeciation hypothesis* (Brooks 1980b).

MacArthur and Wilson (1963, 1967) developed a theory of island biogeography: the species richness on an island is determined by an equilibrium between the rate at which species colonize the island and the rate at which existing species become extinct. (Note that this species richness equilibrium is different from and does not necessarily presuppose the equilibrium population concepts at the infracommunity level used in the tables.) The theory was based on two empirical observations; the first was that larger islands support more species than smaller islands. They hypothesized that area alone has a direct and positive effect on the number of species in a community because a larger area has a higher probability of receiving colonists, and such colonists will survive longer because the larger populations would have a lower probability of extinction resulting from competition, enemies, or stochastic events. Other processes producing the same effect are an increase in habitat diversity with area (Williams 1964), or simply that larger areas sample a wider variety of potential colonists, therefore accumulating more species per unit time (Connor & McCoy 1979). This hypothesis has been applied to infracommunities of parasites by defining 'island size' as the size of the host individual (e.g. Dogiel *et al* 1964), but most applications have been to the component community level, defining island size as the size of the host population (Freeland 1979a) or the host range (Dritschilo *et al* 1975; Price & Clancy 1983). According to the *island size hypothesis*, larger 'islands', however defined, should have more species of parasites.

The other attribute MacArthur and Wilson (1967) used to develop their theory of island biogeography was the distance of the island from a source of colonists — the further the source, the fewer the colonists. 'Distance' was in effect a measure of the difficulty of reaching the island. For parasites, the difficulty encountered in reaching a host is frequently a function of its food habits [which influences both exposure to parasites and ability of the parasites to develop (Freeland 1983)] or other aspects of its way of life (Dogiel *et al* 1964); the 'ecological distances' facing parasites vary markedly between host taxa, and sometimes even between individuals. (See discussion on influences of differences in host behaviour on exposure to parasites in Section 9.3.) The *island distance hypothesis* regards the difficulty of invasion to be a prime determinant of parasite communities.

Wilson (1969) suggested that, through time, island (and other) communities go through four phases: the non-interactive, interactive, assortative, and evolutionary phases. The non-interactive phase occurs in young communities and involves individualistic colonization by species uninfluenced by competition or enemies because populations are low and few enemies have colonized; species number has not reached an equilibrium. In the interactive phase, an equilibrium is reached as competition and enemies drive some species to extinction, competitive exclusion is important, and species number declines. The assortative phase leads to a higher equilibrium number because sorting of species through interaction results in tighter species packing of those species most capable of co-existence. The final evolutionary phase results in a yet higher equilibrium number of species because coevolution among competitors leads to refined adaptation to other environmental factors and therefore to narrower niche occupation per species; more species are able to coexist on the available resources. Applications of these ideas to component communities of parasites focus at the evolutionary time scale, generally assuming that host capture, that is, an evolutionary adaptation of a parasite to a new host, is required. Rohde (1978a, 1979) noted that monogenean gill flukes on fishes in north temperate seas have ecological niches just as narrow as those in tropical waters. In a study which illustrates the powerful comparative approach available to students of parasite communities, Rohde (1978a,b, 1979) showed that in tropical waters individual fishes may commonly harbour parasite communities numbering 5–7 species, whereas in cold seas fish species are known to have no monogeneans or only a single species. Rohde (1978b) concluded that many vacant niches are available for future colonization by parasites in north and south temperate fishes and that given enough time these niches will be utilized. Price (1980) suggested that many parasite communities exist in this non-interactive phase, where an equilibrium number of parasite species maintained by equal rates of colonization (by host capture) and extinction has not been reached. These investigators subscribe to the *time hypothesis*: younger communities have fewer species than older communities because evolutionary time is required for the increase in species number.

The assumptions and predictions

The assumptions and predictions of these hypotheses, for the same set of characteristics used for the infracommunity level hypotheses, are shown in Table 9.2. Note that there are two sets of assumptions and predictions for the island size hypothesis, one (a) based on stochastic extinction events due largely to differential population sizes, the other (b) at least partly based on biotic interactions. Three features should be noted. Firstly, the cospeciation hypothesis assumes only a strong coevolutionary specialization of the parasites; since the hypothesis is independent of assumptions on ecological features (Brooks 1980a), it makes no predictions for the characteristics we are examining here. Secondly, the island biogeographic hypotheses are concerned largely with the number of species present, not their ecological relationships; they make no predictions for some of the characteristics. They do make some ecological assumptions, however, which allow prediction of the type of community to be expected. They predict isolationist communities where systems are young, where stochastic events determine parasite extinctions, or where the host is difficult to colonize; interactive communities are predicted only where such constraints are unimportant. Thirdly, the basic tenet of the cospeciation hypothesis, that evolving host species carry with them parasite communities derived from their ancestors, is not only reasonable, but also casts serious doubt on the application of the time hypothesis to host species-level component communities, since a new species, unlike a new island, does not start out devoid of parasites. [For other arguments against the time hypothesis, see Kuris *et al* (1980), Rey *et al* (1981), and especially Boucot (1983); for similar doubts on the applicability of other hypotheses based on the island biogeographic model to such communities, see Kuris *et al* (1980).] The time hypothesis would obviously still apply to host species recently introduced into a new habitat.

9.2.4 *Concepts at the compound community level*

Concepts at this level focus on what mechanisms determine the distribution and abundance of parasites within and between lakes or other habitat units. Parasitologists have directed relatively little attention to compound communities. This is understandable, since considerable time and effort are required to gain sufficient evidence to elucidate pattern in a single component community, let alone the many that make up even a relatively

Table 9.2. Assumptions and predictions on the mechanisms involved in the organization of component communities of parasites, with particular reference to mechanisms determining the number of parasite species harboured by a host species.

	Cospeciation hypothesis	Island or patch size hypothesis	Island or patch distance hypothesis	Time hypothesis
Assumptions				
(1) Colonizing ability of parasites	No assumption	(a) Low (b) High	Low	(a) Existing parasites: no assumption (b) Others: low
(2) Important evolutionary processes in the past	Phylogenetic tracking of host clades by parasite clades; specialization	(a) None (b) Interspecific competition	None	Host capture, specialization
(3) Important ecological processes acting currently	None	(a) Stochastic events (b) Interspecific competition, enemies, and stochastic events	Difficulty of colonization	None
(4) Equilibrium	No assumption	(a) Non-equilibrium (b) Equilibrium	Non-equilibrium	Non-equilibrium

(5) Role of:				
(a) past competition	No assumption	(a) Unimportant (b) Important	Unimportant	Unimportant
(b) current competition	No assumption	(a) Unimportant (b) Important	Unimportant	Unimportant
(c) individualistic responses	Strong	No assumption	No assumption	Strong
Predictions				
(1) Vacant niches	No prediction	(a) No prediction (b) No	Yes	Yes
(2) Distribution of species in resource space	No prediction	(a) Independent (b) Regular	Independent	Independent
(3) Realized niche (cf. fundamental)				
(a) increased when population increased	No prediction	No prediction	No prediction	No prediction
(b) reduced in presence of other guild members	No prediction	(a) No (b) Yes	No prediction	No prediction
Type of community	No prediction	(a) Isolationist (b) Interactive	Isolationist	Isolationist

simple compound community. The few studies that have appeared deal largely with lake habitats, and mostly with the component communities in fish. [See Freeland (1983) for terrestrial examples.]

The concepts

Wisniewski (1958) and Sulgostowska (1963) suggested that the overall parasite community within an ecosystem such as a lake is characterized by the parasites of the numerically dominant host species. Populations of parasites adapted to the abundant hosts will be considerably greater than those of parasites of the less abundant hosts. Other things being equal, both abundant and less abundant hosts will therefore encounter more of the parasites of the abundant hosts than those of the less abundant ones. Since the probability of establishing in the other host should vary directly with exposure, the less abundant hosts would be expected to harbour more of the species adapted to the abundant hosts than vice versa. Since this hypothesis describes an uneven exchange of parasites between abundant and less abundant hosts, we refer to it as the *exchange hypothesis* (Leong & Holmes 1981).

Halvorsen (1971) and Wootton (1973) emphasized that given species of fish have very similar parasites in very different lakes; the relationships between fish and parasites are essentially constant. This hypothesis is similar to the cospeciation hypothesis, but does not preclude the presence of generalist parasites or of parasites acquired through host capture. We refer to it as the *constant fauna hypothesis*.

Kennedy (1978) has emphasized differences between lakes associated with historical factors, especially those related to colonization events. He stressed the influences of the age and size of the lake and its distance from other lakes. This is the *island biogeographic hypothesis*; it could be divided into the same three hypotheses discussed under component communities.

The assumptions and predictions

As far as the infracommunity characteristics considered earlier are concerned, the assumptions and predictions of the island biogeographic hypothesis have been covered above, and those of the constant fauna hypotheses are essentially the same as those of the cospeciation hypothesis. Most predict isolationist infracommunities. Those of the exchange hypothesis appear to be similar to those of the interactive island size hypothesis or the competition hypothesis, and predict interactive communities.

More significantly, however, the exchange hypothesis suggests a re-evaluation of interactive communities, in that it predicts two quite separate sets of member species. The first is the set of species adapted to the host and to each other. These should be widespread and abundant — the *core species* of Hanski (1982). The predictions of the interactive communities apply to this set. The second is the set of species acquired by exchange from ecological associates. They should be sporadic in occurrence and less abundant — Hanski's *satellite species*. These species may be completely random elements, or may fit into temporarily (or stochastically) 'empty' niches. Through exchange of parasites, one host's core parasites become another host's satellite parasites (Butterworth 1982).

9.2.5 Synthesis

Tables 9.1 and 9.2 may well oversimplify and overgeneralize the properties of each hypothesis and the resultant parasite communities. In addition, the classification of communities into only two types, interactive and isolationist, is probably too crude to be of lasting utility. However, we think these are useful first steps towards identifying characteristics that can be studied in nature in order to distinguish types of communities and the hypotheses appropriate to them. An important start would be to determine whether vacant niches occur in the community, whether species (especially core species) are independently or regularly (i.e. dependently) distributed in resource space [e.g. along the length of the small intestine (Bush & Holmes 1983)], and whether realized niches are independent of or more restricted in the presence of other guild members; all would help to distinguish between interactive and isolationist communities.

More fundamentally, the differences between interactive and isolationist infracommunities appear to be consequences of different assumptions on colonization probabilities. A considerable body of literature in parasitology (summarized in Price 1980) assumes that colonization probabilities (i.e. transmission rates) limit parasite populations. Bradley (1974) disagrees, suggesting that limitation through low transmission rates applies only to parasites at the edge of their range or in unfavourable habitats. In more favourable areas nearer the centre of their range, parasite populations would be controlled by other mechanisms, primarily by reducing the number of parasites able to develop. A voluminous literature on the significance of immune responses for the population dynamics of parasites of humans or domestic animals (e.g. Anderson & Michel 1977) support Bradley (although concentrations of individuals are frequently higher than in more natural systems). More information on rates of exposure to parasites in natural communities is needed. The 'tracer' or 'sentinal' animal technique used in studies on domestic animals (e.g. William *et al* 1983; Esch *et al* 1976) provides one good way of obtaining such information. In this technique, known uninfected animals are released (or caged) in the environment for short periods to determine the periods during which, and the rates at which, transmission takes place.

Many features could limit transmission rates, thereby producing isolationist communities. The hypotheses outlined above suggest that time (for introduced species or for compound communities in habitat islands), physical distance (for habitat islands), a small host population size, food habits or other aspects of the way of life of the host which make colonization difficult, or, by implication, severe physical or chemical stresses in a particular microhabitat could do so. Only detailed studies on the most significant aspects of adaptation will distinguish between these mechanisms producing isolationist communities.

9.3 The role of parasites in natural communities

The role parasites play in natural communities depends to a large extent on their effects on individual host animals. The most obvious, but not the most common effect, except in insect-parasitoid systems, is to kill the host. The literature is replete with examples. Many of them suggest extensive annual mortality, such as the many studies on insect parasitoids, a nematode (*Crassicauda* sp.) in the brain of spotted dolphins *Stenella attenuata* (Perrin & Powers 1980) or an ectoparasitic mite (*Hydryphantes tenuabilis*) on an aquatic hemipteran *Hydrometra myrae* (Lanciani 1975). Others suggest extensive, but irregular mortality, such as lungworms (*Protostrongylus* spp.) in the bighorn sheep *Ovis canadensis* (Forrester 1971), winter ticks (*Dermacentor albipictus*) on the moose *Alces alces* (Samuel & Barker 1979) or mange (*Sarcoptes scabei*) on coyotes *Canis latran* or wolves *Canis lupus* (Todd *et al* 1981).

Sublethal effects, where the host is not killed directly, but its fitness is reduced, are far more common. Losses in production constitute a major problem caused by parasites in domestic livestock (Coop 1982). That the same losses occur in wild populations is suggested by the frequently reported correlation of poor condition with higher parasite loads [e.g. nematode larvae (*Contracaecum aduncum*) in the liver of the cod *Gadus morhua* (Petruschevski & Shulman 1961)]. It has been shown experimentally that the reduced production is frequently due to reduced food intake, reduced digestive efficiency, or reduced absorptive capacity (see review by Symons & Steel 1978). Reduced condition can result in decreased stamina, leading to decreased competitive ability (Freeland 1983; Rau 1983), increased predation (Holmes & Bethel 1972), or reduced numbers and postnatal survival of young (Dunsmore 1981). Parasites can also reduce recruitment by castrating their host (Kuris 1974) or by other direct attacks on the reproductive system [e.g. destruction of the mammary gland tissue by a nematode (*Mammanidula asperocutis*) in shrews (*Sorex* spp.) (Okhotina & Nadtochy 1970). Parasites, especially those in intermediate hosts, can also increase susceptibility to predation by altering the behaviour of the infected host (Holmes & Bethel 1972; Bethel & Holmes 1973, 1977).

Whether the outcome of a particular host–parasite combination is death, disease, or has no discernible effect depends on a number of factors, at least in vertebrates (Hanson 1969). These factors can be categorized under the virulence of the parasite, the dosage (or degree of exposure to the parasite), and the resistance of the host (Fig. 9.1). The virulence of the parasite depends largely on the genetic interactions of hosts and parasites (Holmes 1983; Anderson & May 1982; Bremermann & Pickering 1983). In this section, we will focus on the ecological features acting on the exposure to parasites and on host resistance.

It should be obvious that exposure to the infective stages of a parasite is likely to increase with higher densities of host or parasite. However, those infective stages are often clumped (Keymer & Anderson 1979), and may be limited to patchily distributed foci of infection [Carey *et al* (1980); for an extensive development of this idea see Pavlovsky (1966)]. The implications for exposure are explored by Bradley (1972). Individual differences in host behaviour can also result in major differences in exposure (e.g. Combes 1968). As a result of these features, exposure to parasites is not usually even across a host population, but instead is concentrated on a small proportion of the individuals. Host resistance to those parasites is also uneven: resistance is generally reduced by malnutrition (Gordon 1960; Anderson 1979b), stress (Ould & Welch 1980; Anderson & May 1981), and low social status (Barrow 1955). All of these

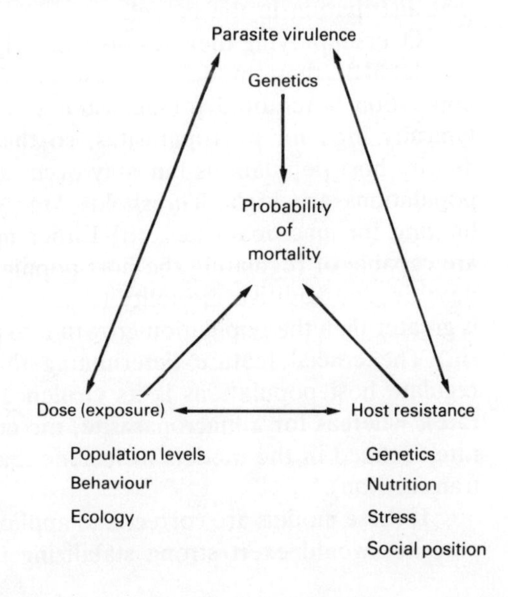

Fig. 9.1. Characteristics which interact to determine the probability of pathogenesis or mortality due to parasites. The three major characteristics are shown at the apices of the triangle; under each are factors which significantly influence that characteristic. Each characteristic affects mortality directly (single-ended arrows) and in combination with the other features (double-ended arrows).

factors interact to produce the usual clumped pattern of distribution of individuals, with most of the parasites concentrated in a few hosts. In addition, for a potentially lethal parasite it produces the characteristic pattern of many infected individuals, some sick, and few dead.

The consequences of these effects on individual hosts for the structure and function of host communities will be covered in the next four sections.

9.3.1 *Influences on host populations*

In an extensive series of recent papers, Anderson and May (1978, 1979; Anderson 1979b; May & Anderson 1978, 1979; May 1983) have investigated the potential of parasites to regulate host populations, or to maintain host populations at reduced levels. In their models, Anderson and May distinguish between microparasites (those having direct reproduction within the host, so that a single infection event can produce a large infrapopulation) and macroparasites (those without such reproduction, so that repeated infections are necessary to produce large infrapopulations). For microparasites (such as viruses or most protozoans), the models distinguish only three classes of hosts: susceptible, infected and immune individuals. For macroparasites (such as ticks or most helminths) the models must incorporate parasite numbers, since almost all host–parasite interactions, including the probability of death, depend on the number of parasites present. The models assume a clumped distribution, normally a negative binomial, and incorporate parameters of that distribution in the model.

Oversimplifying their results (the original papers should be consulted for details), their models suggest the following. (i) A threshold host population is required for persistence of the parasite. This threshold is typically high for microparasites, so that they can persist only in high density host populations but may occur as episodic epizootics when host populations are high. Thresholds are typically low and generally not limiting for macroparasites. (ii) Either microparasites or macroparasites are capable of regulating the host population, provided that at high host densities the additive mortality (that over and above 'normal' mortality) is greater than the population growth rate (births minus normal mortality). (iii) The crucial feature determining the ability of a microparasite to regulate host populations is its virulence (the disease-induced mortality rate), whereas for a macroparasite, the crucial feature is its reproductive rate (defined in the models to include features of both reproduction and transmission).

If these models are correct and applicable to real-world systems, then parasites would exert strong stabilizing influences on host populations.

Yodzis (1981) noted that intraspecific stability was one of the main factors promoting stability (return to equilibrium) of natural food-webs. The stabilizing influence at the host population level should therefore promote community stability.

However, Bradley (1982), in a critical review of the application of epidemiological models to real world situations, has pointed out that "all models have proved less robust than their authors hoped" (p. 333). Recently, some of the assumptions of the models have been questioned. Holmes (1982) concluded that there is little evidence that parasite-induced mortality, at least in vertebrate populations, is actually additive. The influences of malnutrition, stress, and low social status on host resistance combine to concentrate the effects of parasites on those classes of animals which would be expected to bear the heaviest mortality rates anyway (Errington 1946; Jenkins *et al* 1963, 1964). Parasite-induced mortality in these classes should be compensatory (i.e. take the place of 'normal' mortality). The influences of parasites on the nutrition and competitive abilities of their hosts can drive those hosts into the same classes. Such parasites may well act as selective agents, determining *which* individuals are allocated to the high risk classes, but it is only if the parasites significantly increase the *number* of individuals in these classes that parasite-induced mortality would be additive. Negating the assumption of additive mortality negates the ability of the parasite to regulate the host population.

Getz and Pickering (1983) questioned the general applicability of the assumption that transmission is dependent on the density of susceptible hosts; they suggested that for venereally transmitted parasites (and some others) transmission may be dependent on the proportion of susceptible hosts in the population. In their model, negating the assumption eliminated the ability of the parasite to control host populations.

In neither of these cases, however, does negating the assumption exclude the ability of the parasite to help stabilize the host population. Getz and Pickering's model showed that in conjunction with some other regulating factor, the parasite could still bring the host population to an equilibrium below that set by the other factor. Parasites acting as relatively efficient executors of excess animals (where mortality is compensatory) could speed up regulation of population sizes by reducing lag times (due to the short generation times in the parasites) or even by changing scramble-type competition to contest situations. Either should act as a stabilizing influence.

9.3.2 Influences on host species coexistence

Holmes (1982; see also Freeland 1983) pointed out that there is one major class of situations in which parasite-induced mortality is clearly additive and significant — parasites spread from hosts in which they are well adapted to hosts in which they are poorly adapted. The most obvious examples involve parasites brought in with introduced animals. The best documented cases are of parasites introduced with fish. The protozoan agent of whirling disease (*Myxoma cerebralis*) was introduced into North America with brown trout *Salmo trutta*; it has become a serious disease of rainbow trout *Salmo gairdnerii* primarily in hatcheries, but occasionally also in the wild (Hoffman 1970). A monogenean (*Nitzschia sturionis*) introduced with its sturgeon (*Acipenser stellatus*) host into the Aral Sea depressed native sturgeon (*Acipenser nudiventris*) populations enough to halt commercial fishing for two decades (Bauer & Hoffman 1976).

Where the introduced parasite produces extensive mortality in a foundation species [a dominant competitor, key predator or other species which plays a large role in defining the structure of a community (Dayton 1972)], the consequences for the community may be extensive. For example, the cestode *Ligula intestinalis* was introduced into Slapton Ley (a lake in southwestern England) when crested grebes *Podiceps cristatus* began nesting there in 1975. Larval stages of the tapeworm heavily infected juvenile roach *Rutilus rutilus*, the dominant cyprinid, causing heavy mortality and reduced recruitment of roach. A major increase in the population of rudd *Scardinius erythrophthalmus*, a competing cyprinid, resulted. This was followed by a decrease in *Ligula*, an increase in roach and a decrease in rudd, then a second cycle of increased *Ligula*, decreased roach and increased rudd (Burrough *et al* 1979). This pattern suggests a damped oscillation toward a new equilibrium in the cyprinid community. It also indicates that an introduced parasite need not encounter a new species of host to have a significant effect (*Ligula* is found in roach elsewhere in Britain); a new population of hosts is sufficient, presumably due to genetic influences (Anderson & May 1982).

Some invasions have had much more far-reaching effects. Warner (1968) suggested that introductions of avian pox virus, bird malaria (*Plasmodium* spp.), and the mosquito (*Culex pipiens fatigans*) vector of both into the Hawaiian Islands was a major factor in the extinction of significant numbers of endemic land birds and the limitation of the rest to areas above the range of the mosquito. Rinderpest, probably introduced into Eritrea with military oxen (Ford 1970), swept across Africa in the late nineteenth century, producing massive mortality in cattle and native ungulates, and leaving distributional anomalies that still persist (Spinage 1962).

Animals introduced into new areas can be severely affected by parasites of native animals. Trypanosomes (*Trypanosoma* spp.) maintained in wild ungulates and transmitted by tsetse flies (*Glossina* spp.) make cattle-raising difficult (or impossible) in the tsetse belts of Africa (Ford

1971). Holmes (1982) and Freeland (1983) give other examples. Freeland assessed parasites as barriers to invasion and concluded that successful invaders either have parasite susceptibilities quite different from those of native animals, or carry parasites that severely depress the fitness of individuals of one or more resident species.

Barbehenn (1969) suggested that parasites could be used as 'weapons of competition' on a more local scale. He referred to Park's (1948) classical study on competition between grain beetles (*Tribolium* spp.), and the ability of a sporozoan (*Adelina* sp.) parasite to reverse the usual outcome by a greater inhibitory effect on the reproduction and survival of the dominant competitor, and suggested that such effects could either preclude species overlap in nature or, if reciprocal, could reduce populations of both potential competitors sufficiently to allow them to coexist. Cornell (1974) pointed out that in the former class of situations, the parasite and the host to which it gives an advantage would be selected for as a unit. Broekhuizen and Kemmers (1976) provide an excellent example: European rabbits *Oryctolagus cuniculus* maintain a stomach nematode (*Graphidium strigosum*) which is highly pathogenic to hares *Lepus europaeus*. With the aid of their worms, the rabbits are able to displace the hares, but only in wooded areas, to which the worms are restricted by the requirements of their free-living larval stages. Holmes (1979, 1982) gives other examples.

Holmes (1983) has suggested an additional feature of potential importance — an existing complex of well adapted parasites may prevent the establishment of parasites that might depress the fitness of the host more significantly. That protection is well known for bacterial assemblages (e.g. Weinack *et al* 1981), and has been demonstrated for protozoan assemblages (Hein 1976) and helminths (LeJambre 1982; Holmes 1983). Protection of this nature would be most likely with interactive parasite communities. This prediction should be tested.

In the preceding examples, parasites served to reduce the number of coexisting host species. As pointed out above, Barbehenn (1969) suggested that parasites may also act to increase that number by reducing populations, thus preventing competitive exclusion. In addition, parasites may act as a factor selecting strongly for the development of mutualisms. For example, Pierce and Mead (1981) describe a system in which a cryptic lycaenid butterfly (*Glaucopsyche lygadamus*) caterpillar provides honeydew to ants *Formica fusca*, which in turn protect the caterpillar from parasitoids and secondarily from predators. When ants were excluded with tanglefoot, the rate of attack (and mortality) of the caterpillars by braconid and tachinid parasitoids nearly doubled.

In a similar example, Sankurathri and Holmes (1976) have shown that in field situations, the obligate 'commensal' oligochaete, *Chaetogaster l. limnaei*, can significantly reduce invasion of snails by trematode miracidia or cercaria. Since the germinal sacs resulting from invasion by miracidia frequently castrate (or less frequently kill) the snail host (Kuris 1974) and penetration by extensive numbers of cercaria can kill snails (Morris 1970), this protection could be a significant benefit to the snail. Smith (1968, 1979) has shown that some colonies of icterid birds (*Zarhynchus wagleri and Cacicus cela*) tolerated cowbird (*Scaphidura oryzivora*) brood parasites, the nestlings of which fed on ectoparasitic fly (*Philornis* spp.) larvae, the major mortality factor for nestlings of the icterids. Colonies of the icterids which nested in association with stingless bees (*Trigona* sp.), which provide some protection against the flies, did not tolerate the cowbirds.

Other examples include the 'tickbirds' *Buphagus erythrorhynchus* of Africa (Moreau 1933) and the wide variety of 'cleaners' [mostly shrimps or fishes (Feder 1966)] which remove ectoparasites from large ungulates or fishes, respectively. By far the best studied system is that of the marine cleaners. Rohde (1982) reviewed the evidence supporting the following generalizations: (i) cleaners are widespread, (ii) they do remove parasites, (iii) some cleaners feed little (or not at all) on free-living organisms, (iv) cleaning stations seem attractive to fishes, (v) host fishes cooperate with the cleaners, and (vi) some cleaner mimics parasitize the cleaning mutualism. He concluded that, although present evidence on whether the mutualism is ecologically important for host populations is ambiguous, many of the features of the mutualism suggest that "some must have great ecological importance and a long evolutionary history" (p. 86).

Some of the influences discussed in this section, especially the use of parasites as weapons of competition or as agents selecting for the development of mutualisms, provide tight linkages among small groups of species in the community. Such groups of tightly linked species, with little or no linkages with other species in the community, have been postulated to be a stabilizing feature (May 1972).

9.3.3 *Influences on predator–prey interactions*

Parasites can influence rates of predation through their general debilitating effects, especially those reducing the stamina or responsiveness of the infected individuals. As indicated above, these effects are most severe in the heavily infected individuals, and often interact with the host's social behaviour, food limitations, or other stresses to produce a class of individuals which experiences high predation rates (and other mortality). The effects are generalized, and may involve either adult or larval parasites.

In other cases, parasites can increase predation rates by increasing the conspicuousness, disorienting, or altering the normal behaviour patterns of infected individuals. These effects are commonly associated with larval or juvenile parasites which cannot complete their development until the infected host is ingested by a suitable final host, and can be considered adaptations of the parasite which facilitate transmission (Holmes & Bethel 1972). The behavioural changes may be fairly specific (Bethel & Holmes 1973) and can result in predation concentrated on a limited set of predators, including the definitive host.

These effects can influence community processes in three ways. First, they reduce the cost to the predator of catching prey, thereby increasing the efficiency of energy transfer along the food chain. Debilitated or disoriented prey are easily caught (Cram 1931; Locke *et al* 1964); conspicuous prey are more efficiently caught (Carter 1968; Camp & Huizinga 1979); and altered behavioural patterns of infected intermediate hosts, such as the 'skimming' of gammarids (*Gammarus lacustris*) infected with *Polymorphus paradoxus* (Bethel & Holmes 1973, 1977) or the circling of ants (*Camponotus* spp.) infected with *Brachylecithum mosquensis* (Carney 1969), make them highly susceptible to predation by dabbling ducks or insectivorous birds, respectively. Where the prey is affected by adult parasites, which generally cannot transfer to the predator, the energy saved is a clear gain to the predator. Where larval parasites which are able to mature in the predator are concerned, part of this gain may be offset by an increase in the cost of maintaining the parasites acquired, or of maintaining the resistance mechanisms preventing their establishment. The cost would be expected to be highest when feeding on the most heavily infected prey; in at least one case, oystercatchers *Haematopus ostralegus* feeding on clams *Macoma balthica* infected with the metacercariae of *Parvatrema affinis* (Hulscher 1973), there is evidence that the predator can detect and discard individuals with very heavy infections.

Secondly, some of the behavioural changes may re-route energy transfers along more efficient lines. Pearre (1979) found that chaetognaths (*Sagitta* spp.) infected with hemiurid metacercariae remained in the upper, lighted layers of the ocean, grew faster (due to more and different foods available there and to their castration by the trematode), and were less likely to be eaten by invertebrates and more likely to be eaten by planktivorous fishes (the definitive hosts of the parasite). The net effect appears to be greater efficiency of energy transfer to fish.

Thirdly, predation is concentrated on those individuals of lowest reproductive value in the prey population. Slobodkin (1968) pointed out that a 'prudent' predator should do just that; Stearns (1977) suggested that one of the major selection pressures on the life history characteristics of a prey species is to minimize the value of the stage or stages on which predation is concentrated. Parasites may be involved in accomplishing the task through three mechanisms. Where parasites castrate their hosts (as in the chaetognath example above), the same individuals are rendered both expendable and more susceptible to predation. Where parasites debilitate their hosts, that debilitation, and especially its effects on the competitive ability and social status of the host, can markedly reduce the reproductive value of affected individuals. Alternatively, the influences of low social status, stress, or malnutrition on resistance to parasites can increase parasite loads in the low value individuals. Whether the effect of the parasites is potentially additive (as in the first two) or definitely compensatory (as in the last), the effects of parasitism and predation would tend to be concentrated on the same set of individuals, those that have a relatively low reproductive value.

However, not all behavioural changes increase predation rates. Some lepidopteran larvae infected with bacteria or fungi crawl to exposed areas on the top of their host plants; their behaviour is clearly related to the transmission of these parasites to other larvae (Shapiro 1976). Others invaded by parasitoids, which do not mature in predators, show similar behaviour patterns (Shapiro 1976; Smith Trail 1980); predation on these insects would kill the parasites, not aid their transmission. Smith Trail suggested that the 'suicidal' behaviour of the parasitoid-infected larvae could increase predation on them, thus preventing the spread of the parasite to surrounding conspecific insects, and thus could be considered a kin-selected host adaptation. However, Stamp (1981) argued that the aposematic nature of the larvae in the examples used by Smith Trail was inconsistent with increased predation, and demonstrated that the 'suicidal' behaviour of the larvae in one of her examples actually reduced hyper-parasitism, and thereby increased the fitness of the parasite (*Apanteles euphydryidis*), not that of the host (*Euphydryas phaeton*). Stamp considered the behavioural changes to be adaptations of the parasite (Fritz 1982). Effects like these would not have the community-level consequences discussed above.

9.3.4 *Influences on patchiness*

Environmental patchiness plays an integral role in current concepts of community ecology (Dayton 1971; Osman 1977; Connell 1978; Hanski 1982; Tilman 1982). Parasites not only respond to environmental patchiness, but also can intensify or even create that patchiness.

Pavlovsky (1966) emphasized that parasites generally require specific sets of conditions for optimum transmission from host to host. For hookworms of people, for example, transmission requires sufficient faecal deposition to seed the area with viable eggs, soil conditions with sufficient water to prevent drying of the eggs or larvae but not enough to drown them or to wash them out of the surface layers, temperatures high enough for development of the larvae but not above (or below) their tolerance levels, and bare feet (or other exposed skin) for the larvae to penetrate. Even in areas where hookworm populations are high, those conditions are all met only in scattered, localized settings — the 'nidi' or 'hot spots' of infection (Schad *et al* 1983.)

Other parasites may not be so severely limited by the defacation habits of their hosts (or the habit of wearing shoes), but all are limited by the requirements of any free-living stages, by those of any intermediate hosts, and the necessity for the right host to be in the right place at the right time. An example of a rather complicated system is that of the virus of Colorado tick fever (Carey *et al* 1980). The virus cycles between mammals (a wide variety are satisfactory) and the wood tick *Dermacentor andersoni*. The distribution of the tick is determined by a complex of the availability of suitable hosts (for each of three feeding stages) and soil temperatures and moisture conditions suitable for the survival and development of the free-living non-feeding stages of the tick. The distributions of the mammals are determined by a complex of features of soil depth, vegetation patterns, and behavioural interactions between the mammals. Carey *et al* (1980) analysed these variables and showed how the distribution of the virus is dependent upon the "overlap in the distinct ecotope hypervolumes" of the ticks, deer mice *Peromyscus maniculatus*, least chipmunks *Eutamias minimus*, golden-mantled ground squirrels *Spermophilus lateralis* and porcupines *Erethizon dorsatum*.

In this example, the prime concern was to identify those sets of conditions conducive to the transmission of the virus, so that areas with those conditions could be avoided in establishing camp-grounds (or such conditions already existing in an established camp-ground could be modified) to prevent human infections. The significance of the virus to community processes was not investigated. In other cases, such as that of the stomach worm of rabbits discussed earlier (or any other example of a parasite used as a 'weapon of competition'), at least some aspect of species diversity in the patch has been affected, intensifying the distinction between the patch and the surrounding areas.

Parasites can also create patches through epizootics that markedly affect the entire organization of the community through their effects on foundation species. The example of rinderpest is a good one; the epizootic referred to earlier killed vast numbers of cattle and native ungulates, allowing areas that had been maintained as pastoral land through grazing pressure to revert to tsetse scrub (Ford 1971). A more localized example concerns a bacterial or fungal parasite that caused widespread mass mortality in dense populations of slow growing sponges (*Ancorina alata* and *Polymastia fusca*), opening up patches of unoccupied space for other invertebrate colonists (Ayling 1981).

The preceding sections have outlined those features of parasites which should allow them to influence community structure or function.

Evidence that parasites do have such influences, at least occasionally, has also been provided. However, the existing data are totally insufficient to estimate the extent to which parasites do influence host communities. Community-level studies are complex; including parasites in those studies can increase complexity considerably, as evidenced by the study of Carey *et al* (1980). Further development in this field will require close cooperation between ecologists and parasitologists; such cooperation would be highly desirable.

9.4 Conclusions

Despite the variety of hypotheses applied to communities of parasites, the assumptions and predictions of these hypotheses lead to the recognition of only two types of communities, which we have designated interactive and isolationist. We have also stressed the need to distinguish core from satellite species of parasites, since their responses to the host, and to other parasites, may differ greatly. We believe that progress in understanding communities of parasites will be made most rapidly by testing among alternative hypotheses simultaneously, ideally using the same hosts and environments with both types of communities present. These tests will necessarily involve detailed understanding of parasite natural history so that the selective pressures on such things as niche occupation can be perceived. As more host–parasite systems are studied, preferably at several geographic locations per system, concepts and hypotheses will need to be refined and modified. Refinement on the dichotomy of community types will also probably be necessary as understanding increases. Such refinement may well involve some classification of how host resources change through time and how patchy they are (e.g. Price 1984b), or of characteristics of the host community in which the parasite community is imbedded (e.g. Holmes 1979).

The effects of parasites on host communities have received little attention in the ecological literature. Advances in theoretical and conceptual development can be rapid, and more challenging than for predator impact, since parasites have a more diverse array of effects on host populations, as we have discussed above. In particular, parasites common to two or more hosts with differential virulence in those hosts are likely to have profound consequences for the distribution and abundance in host communities.

The complexities of the life cycles of many parasites, or those of the interactions between the parasites and their hosts, are fascinating but daunting for the non-specialist. The reluctance to tackle these complexities can be reduced by early exposure to host–parasite systems. Teachers in general ecology should make an effort to introduce some examples of parasite communities, or of the effects of parasites on host communities, early in a student's training. We feel that ecology in general would benefit from this increased awareness of parasite systems, and more contributors to understanding parasite communities, and their influences on host communities, would emerge.

10 Prey–Predator Interactions

Roger L. Kitching

10.1 Introduction

There is an oft-quoted axiom of systems analysis which states that the dynamics observed at one level of resolution in a hierarchy of such levels will be produced by mechanisms acting at the level beneath. In ecological terms this suggests that we should find the explanations for community structure and dynamics in processes operating at the level of the population. Prey–predator interactions are one class of such processes. This chapter seeks firstly, to establish the importance of predatory interactions as determinants of community parameters and, secondly, to suggest in general terms the mechanisms concerned.

G. Evelyn Hutchinson (1957) defined the ecological niche of an organism as the *n*-dimensional hypervolume defined by the intersection of the exploitation or tolerance curves which describe the many two-dimensional interactions between the species and the biotic and abiotic factors which influence its well-being. This concept has been the foundation, often implicit, in much subsequent development of community theory. Transposing the notion from one species and *n* environmental dimensions to *n* species and one environmental dimension has produced current notions of species 'packing' based on the idea that different species, as a result of competitive interactions, partition the 'resource' represented by that particular environmental dimension. Chapter 6 discusses the development and application of this approach. The theoretical ideas of species packing and the underlying processes of resource partitioning and its morphological manifestation, character displacement, are based almost entirely on a consideration of competition as the basic population process involved at the population level. Undoubtedly, there are good pragmatic reasons for this narrower viewpoint than the general one of Hutchinson. I believe, however, that ecologists investigating processes of competition in the field almost from the first have found that other processes, most notably predation, have had very substantial modifying effects on potential competitive interactions and the resulting community structure.

Connell (1961a), for instance, in his now classical studies of the competition for space occurring between the intertidal barnacles, *Balanus balanoides* and *Chthamalus stellatus*, identified the very important role of the predatory gastropod, *Thais lapillus*, in moderating the intensity of competition as well as acting as a limiting factor for *Balanus* at the lower end of its distribution. Although seldom demonstrated so clearly, this moderating effect seems a near-universal property of generalist predators. The role of predatory beetles such as *Ptomaphila lachrymosa* and *Creophilus erythrocephalus* preying upon carrion-feeding blowfly larvae belonging to the genera, *Calliphora, Chrysomyia* and *Lucilia*, must reduce the intensity of interspecific competition which has been demonstrated to occur among them (Fuller 1934; Ullyett 1950). Similarly, the insectivorous *Dendroica* group of warblers described by MacArthur (1958), regarded as a classical example of resource partitioning following interspecific competition, must have a substantial effect, singly or severally, on modifying competition among the phytophagous insects that are associated with the spruces in and around which the warblers live.

Paine (1966) extended the ideas and observations of Connell showing very clearly that certain predators or top predators play a crucial role in maintaining the structure of food-webs within natural communities. Like Connell, Paine used intertidal communities to support his thesis, comparing selected webs across temperate, subtropical and tropical zones in North America. He identified what he later called 'keystone' predators (Paine 1969), the removal of which led to intense competition within lower trophic levels and, ultimately, substantial simplification of the food-web concerned.

Turning to the behaviour of the predators themselves, Murdoch (1969) identified the process of switching, turning attention away from the one-prey–one-predator situation of previous theorists, to focus on multiple prey situations. Basically Murdoch and others (e.g. Royama 1970) perceive predators as 'switching' their prey-preferences from one species to another as the abundance of the prey species changes. Murdoch *et al* (1975), for instance, demonstrated that guppies selected either *Drosophila* or tubificid worms depending on which was the most abundant. Similar phenomena have been described for a range of predators from *Stentor* to hawks, snails to ladybirds. Murdoch and Oaten (1975) provide an extensive review of switching and its population consequences.

So, predators modifying competitive interactions, keystone species determining community structure and predatory pressures changing with the abundance of species in the trophic level below, all mitigate against the notion that competitive interactions are necessarily paramount in determining community structure. In this chapter, this contention will be examined further in a series of case histories and attempts will be made to draw generalizations from them. Each case history has been selected to bring out particular points of importance while covering a representative range of terrestrial and aquatic communities. I have divided the selected examples into 'natural' and 'manipulated' situations recognizing, however, that this may impose discreteness on what is essentially a continuous situation.

10.2 Natural communities

Most natural communities that have been examined present a web of interactions among trophic levels which appear fixed and immutable (Cohen 1978). These are not the case histories that shed most light on the role of predators in communities. What I have sought to present here are special cases, natural experiments in a sense. In one instance (water-filled tree holes), an intercontinental comparison provides a two-treatment global experiment from which can be inferred the role of predators on the diversity and abundance of species at lower trophic levels. In another (sea otters in North Pacific kelp beds), the ramifications of the presence or absence of a particular predator has repercussions up and down its own food-web and in adjacent systems. In the third 'natural' situation I shall describe (ants and their associates on Australian *Acacia* foliage), the dereliction of isolating two-species interactions is demonstrated and the immutability of the oft-used labels, competitor, predator, mutualist, is shown to be misleading in the extreme. Examples from manipulated communities (the deliberate or accidental introduction of exotic species into well-established communities) show some of the more dramatic effects of the presence of generalist predators.

10.2.1 Water-filled tree holes — predators and the structure of the primary consumer layer

Water-filled tree holes are widespread habitats in a variety of moist forest or woodland situations and, invariably, have a small but very specialized community of animals living in them (Kitching 1971, 1983a). These communities are of particular significance for our present purposes for three reasons: firstly, they are sufficiently simple to allow complete food-webs to be worked out; secondly, their members show substantial taxonomic similarities from site to site even on an intercontinental scale; and, thirdly, one well-worked example can reliably be said to contain *no* metazoan predators.

Taking up the third point first, studies carried out on tree hole communities in northern Europe (Röhnert 1950; Kitching 1971, 1983a) identify the food-web shown in Fig. 10.1. Tree hole communities in the northern regions of North America are very similar (Snow 1958; Fish & Carpenter 1982). Like other tree hole communities, those of the northern Holarctic are allochthonous, basal energy entering the habitats as litter fall and run-off (Carpenter 1982). Unlike other such communities, however, these are without predators and the endemic species appear to partition the basic food resource on the basis of particle size. This partitioning at once suggests competitive interactions and it seems likely that

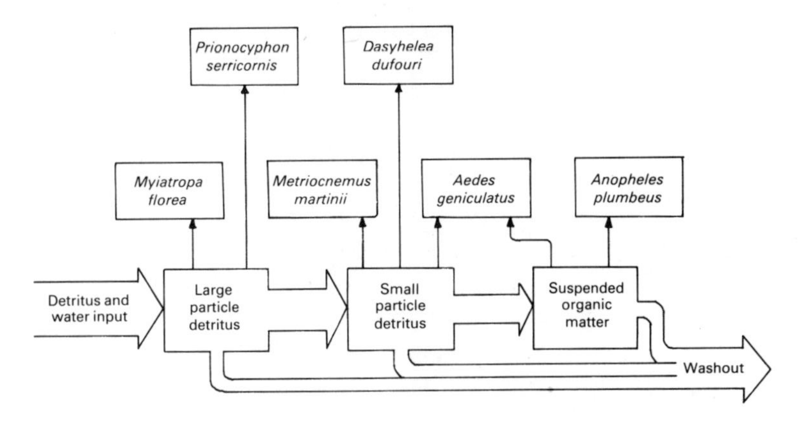

Fig. 10.1. Schematic representation of the food-web present in English water-filled tree holes (from Kitching 1983a, after Kitching 1971).

the origin and maintenance of the temporal separation between the two species of midge, *Metriocnemus martinii* and *Dasyhelea dufouri*, is competitively based. The separation between other 'adjacent' pairs of species, however, probably reflects their taxonomic origin with *in situ* competition involved in the maintenance of separation where minor overlap occurs.

Water-filled tree holes in the subtropics of eastern Australia present an illuminating contrast with the British and American ones. The food-web shown in Fig. 10.2 is based on work of Kitching and Callaghan (1982). Although faunistically and trophically more complex than those studied in detail in Europe, the two are remarkably similar in physico-chemical characteristics. They offer a very similar surface area, average pH, conductivity, and annual litter input to their inmates. They differ thermally of course and in the seasonality of litter input (Kitching 1983a). Focusing attention again on the fauna, the most obvious difference from the European system is the presence of a considerable suite of predators, five of the nine species present. Another less striking but perhaps more significant difference is the concomitant reduction in diversity of the saprophage layer — four species, classing together the two species of mosquitoes, with principally filter feeding larvae. Turning from richness *per se* to evenness (*sensu* Pielou 1975) we find the situation summarized in Table 10.1. Both communities show similar measures of evenness, the Australian being dominated by helodid beetle larvae and the British by saprophagous chironomid larvae.

In addition to considerations of the effects of physicochemical factors, we may also ask whether or not we can draw any conclusions about the role of predators in determining community structure, given that the principal difference between the two communities is the presence of a

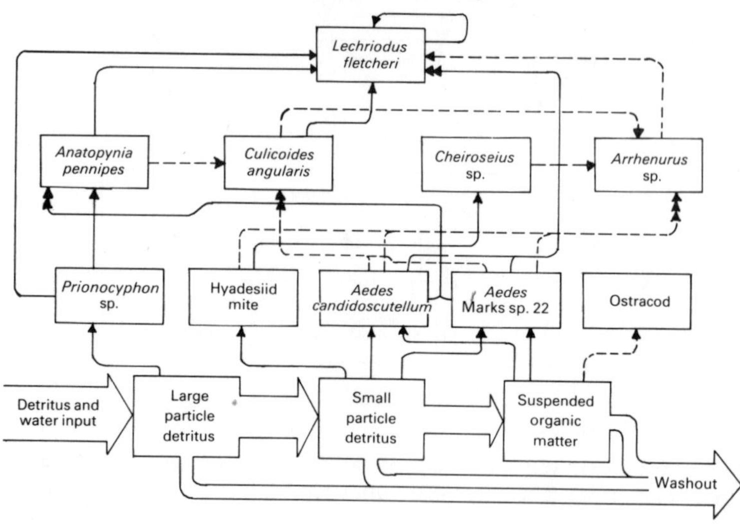

Fig. 10.2. Schematic representation of the food-web present in water-filled tree holes in subtropical box-forests of southeast Queensland (modified from Kitching & Callaghan 1982). Solid arrows represent known relationships; dashed arrows, putative ones.

Table 10.1. Comparison of the numbers of saprophages in selected tree holes in Wytham Woods, UK and Lamington National Park, Queensland, Australia.

	Wytham Woods[1]		Lamington NP[2]	
	No./l.	Frequency	No./l.	Frequency
Metriocnemus martinii	332.5	0.776	–	–
Dasyhelea dufouri	75.3	0.176	–	–
Myiatropa florea	6.5	0.015	–	–
Aedes sp.[3]	3.8	0.009	22.4	0.033
Prionocyphon sp.[4]	9.4	0.022	503.6	0.741
Hyadesiid mite	–	–	39.2	0.058
Ostracod	–	–	114.0	0.168
Richness (No. of species, *n*)	5		4	
Evenness (H/H_{max}[5])	0.43		0.58	

[1] Data extracted from Fig. 1 of Kitching (1971).
[2] Data extracted as means from six 2-monthly samples from 11 marked holes, previously unpublished.
[3] *Aedes geniculatus* in Wytham Woods; *A. candidoscutellum* and an undescribed species in Lamington National Park.
[4] *Prionocyphon serricornis* in Wytham Woods; *P.* n.sp. in Lamington National Park.

[5] After Pielou (1975). $H = -\sum^{n} p\ln p$ where p is the frequency of each of the n species.

suite of predators in one but not the other? In this regard two hypotheses come to mind.

(1) The simplicity of the primary consumer layer in the Australian community may be due to the presence of predatory species, the pressure from which effectively excludes saprophagous midge larvae from the community. In this regard the absence of any saprophagous chironomids from more southerly tree hole communities in North America has been attributed, tentatively, to the presence of predatory *Toxorhynchites* larvae in them (Carpenter 1982).

(2) The predominance of larvae of the helodid, *Prionocyphon* sp., may be due, firstly, to the absence of competitors at its trophic level (recalling it is a rare member of the British community) and its ability to withstand the predation pressures imposed. These two notions are of course complementary and may well operate simultaneously. Examination of our results over the years suggests that there is indeed a dynamic interaction between competition and predation. The cryptozoic habits of the helodid larvae, *Prionocyphon*, the larvae of which insinuate themselves among the interstices of the litter layer within the tree holes, put it in direct competition with other saprophages. It appears to be a slow-developing species with a two-year life cycle in the British situation (Kitching 1971) and these two factors have led to its observed rarity in the tree hole community there. In contrast, the very features that lead to its rarity in Britain have favoured the comparable species in Australia. The suite of predators present have removed potential competitors and the cryptozoic habits of larvae have allowed it to escape the full impact of predation itself. In the warmer ambience of Queensland the slower developmental rate that might be expected on the basis of their northern hemisphere congeners would not be a substantial disadvantage. These factors together reverse the European situation and elevate *Prionocyphon* to a dominant position in its trophic level.

10.2.2 Sea otters, kelp beds and aboriginal Aleuts — a keystone predator and its effects up and down food-chains

The northern sea otter *Enhydra lutis* once enjoyed a natural range that encompassed most of the subtidal, northern Pacific ocean from Japan to Mexico. In recent times, due to human exploitation of one sort or another this natural range has shrunk to portions of the Kurile, Commander and Aleutian Islands with an isolated, relict population off Southern California (Kenyon 1969; Estes & Palmisano 1974). Some reintroductions along the Pacific coast of North America have been made recently. Isolated populations in the Aleutians have provided ideal conditions for elucidating the role of the sea otter, a 'keystone' predator in the sense of Paine (1969), in influencing biological communities and, by its presence or absence, having major effects on species both below and above it in food-webs.

Estes and Palmisano (1974) carried out studies on the offshore animal and plant communities on Amchitka and Shemya, islands in the Rat and Near groups of the Aleutians. The study areas were physiographically similar and, *a priori*, might have been expected to support near-identical biotic communities. Amchitka had a population of 20–30 sea otters/km^2 which has existed at this level for some decades. In contrast, in recent memory, Shemya had received only occasional vagrant otters. The findings of these authors are summarized in Table 10.2. Basically the community with sea otters had low levels of sea urchin populations as a result of the otters' feeding activities. Healthy beds of macroalgae resulted with reduced numbers of epibenthic invertebrates such as mussels, limpets and chitons. The presence of flourishing macroalgal associations, in turn, increased the productivity of nearshore fishes and their predators, through increased richness of the detritus on which the fish food-web is based. On Shemya, on the other hand, lack of sea otters allowed large populations of sea-urchins to flourish with a consequent massive increase in grazing

Table 10.2. Comparison of nearshore marine communities of Amchitka (Rat Island group) and Shemya (Near Island Group), Aleutian Islands (extracted from Estes & Palmisano 1974 with some additional information from Simenstad *et al* 1978).

Feature examined	Amchitka	Shemya
Sea otter (*Enhydra lutis*) density	20–30/km^2	Vagrants only
Kelp (*Hedophyllum sessile* and *Laminaria longípes*)	Dense mats: <1% grazed	Heavily grazed — Average 50% of *H. sessile* plants, average 75% of *L. longipes* plants
Sea-urchins (*Strongylocentrotus* spp.)	8/m^2 average; 2–34 mm diameter	78/m^2 average; 2–86 mm diameter
Chitons (*Katherina tunicata*)	<1/m^2 average	38/m^2 average
Barnacles (*Balanus glandula* and *B. cariosus*)	Small and scarce 4.9/m^2 average	Dense populations 1215/m^2 average
Mussels (*Mytilus edulis*)	Small and scarce 3.8/m^2 average	Extensive beds 722/m^2 average
Rock greenling (*Hexagrammos lagocephalus*)	Abundant	Scarce or absent
Harbour seals (*Phoca vitulina*) (transect estimate)	Abundant; 8.1/km	Scarce or absent; 1.5–2.1/km
Bald eagles (*Halaeetus leucocephalus*)	Abundant	Scarce or absent

pressure on the macroalgae. This reduced primary productivity allowed substantial populations of epibenthos to establish themselves although associated detritus-based food-webs were depauperate.

Simenstad *et al* (1978) extended this work following close analysis of middens left by aboriginal Aleuts and their predecessors who inhabited these groups of islands in prehistoric times. These authors argue that the food-webs described by Estes and Palmisano (1974) represent two alternative stable state configurations for the offshore communities, the difference being accounted for by the presence or absence of the keystone predator, the sea otter. Simenstad *et al* (1978) identified both food-webs at different strata in Aleut middens using data of Dall (1877), Jochelson (1925), Desautels (1970) and others. These communities are illustrated in Fig. 10.3. Simenstad and his colleagues argue from these data that Aleut hunters, exploiting the sea otters to a greater or lesser extent, were responsible for driving the communities from one stable state to another. The time scale involved in this oscillation from one state to another is

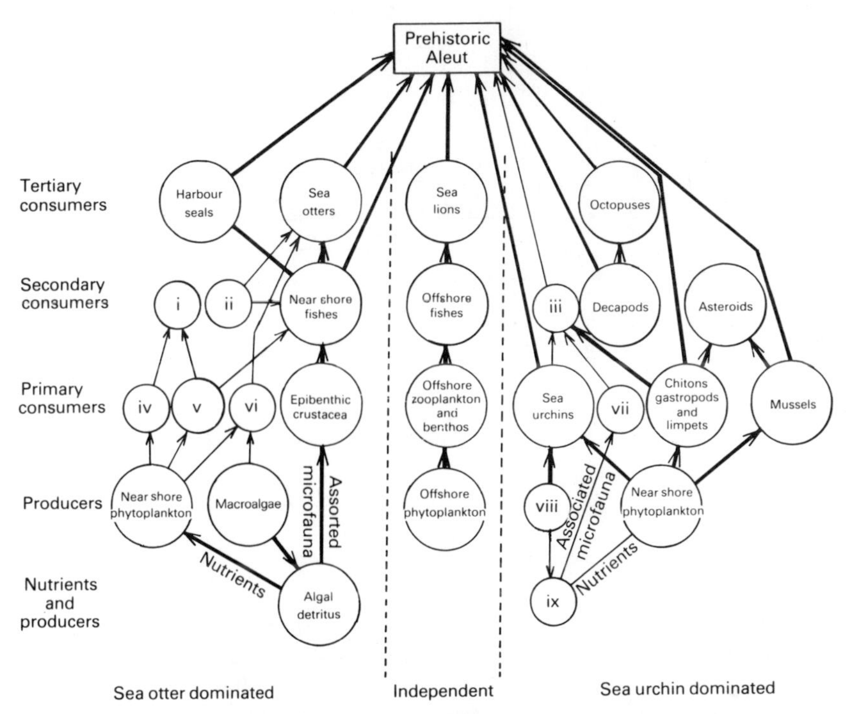

Fig. 10.3. Generalized marine food-web in the western Aleutian Islands (from Simenstad *et al* 1978). Sizes of circles are proportional to the size of the standing crops of the various components. (i) Asteroids. (ii) Decapods. (iii) Nearshore fishes. (iv) Mussels. (v) Chitons, gastropods and limpets. (vi) Sea urchins. (vii) Epibenthic Crustacea. (viii) Macroalgae. (ix) Algal detritus.

well illustrated by Fig. 10.4 (from Simenstad *et al* 1978) which shows the principal animal remains excavated from a midden on Amchitka Island. Before about 95 BC the midden 'community' was dominated by sea otters which then disappeared largely from the remains for about 70 cm of the deposit before reappearing in quantity. These authors argue that before the arrival of the Aleuts the sea otter-dominated community represented the stable state which had evolved following retreat of the Pleistocene glaciers. The Aleuts introduced major perturbation into the system, overcoming its inherent resilience (*sensu* Holling 1973) and driving it to the alternative stable state. Some long-term mechanism akin to 'switching' on the part of the Aleut hunters may have allowed recovery of the system and the switch back to the original state. Subsequent exploitation by Russian fur traders employing Aleut hunters reduced sea otter numbers over a wide area and, one must suppose, favoured the otter-less stable state until relaxation of hunting pressure in recent decades.

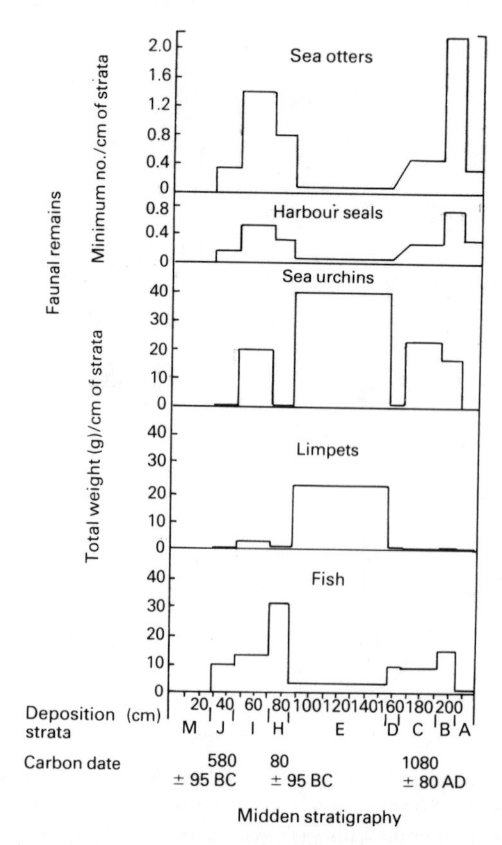

Fig. 10.4. Remains of the principal faunal elements found in an Aleut midden on Amchitka Island (from Simenstad *et al* 1978).

The Aleutian situation is that, from prehistoric to recent times, a keystone predator has controlled the fundamental structure of a large and complex community having profound effects on presence or absence, abundance, and/or feeding options of organisms from primary producers, through saprophages to top carnivores. The driving force in the situation described is the top predator involved, aboriginal people and their successors, but it is the more specialist sea otter, lower in the food-chain, that provides the mechanism for change.

The recent expansion in range and increase in abundance of Californian sea otters (e.g. Ostfeld 1982) has provided further opportunity and, in the face of misgivings by abalone fishermen, real need to assess the likely impact of the returning otters on the offshore community. Ostfeld (1982) has demonstrated that switching occurs in these situations with otters exploiting, first, sea-urchins and, later, crabs (*Pugettia producta*) and clams (*Gari californica*) responding, it seems, to changes in the relative abundance of these prey items. Again it would seem likely that the otters will alter the whole community from an urchin-dominated, low kelp situation to a different stable state with a greater preponderance of detritus-based food-chains (Pearse & Hines 1979). Epibenthic abalones may, indeed, decline (although there is little hard evidence for this) but a concomitant increase in detritus-based fish stocks is to be expected.

The fascinating and illuminating sea otter story may well have its parallel on the Eastern seaboard of the North American continent where kelp beds have been overgrazed by sea-urchins following overexploitation of the lobster *Homarus americanus* which probably acts in the role of keystone predator as does the sea otter in the Pacific (Mann & Breen 1972; Breen & Mann 1976; Mann 1977). It is curious to note the central role played in both Pacific and Atlantic systems by sea-urchins which, from a different perspective, might well be termed a keystone 'herbivore'!

10.2.3 Ants, lycaenids, homopterans and wattles — the blurring of definition

The ant genus *Iridomyrmex* contains upwards of 150 species in Australia with feeding habits ranging from predatory to herbivorous. In most species some mixed strategy of feeding is adopted with nectar, live prey and detrital matter making up the diet. One species, belonging to the *anceps* group, is abundant on the wattle bushes of the species, *Acacia irrorata*, in southeastern Queensland. Basically, foraging ants comb the foliage of wattles for prey. In this regard they are catholic, attacking and killing lepidopterous larvae, beetle larvae, spiders, and other species of ants among a wide range of other organisms. [See New (1979, 1983) for an indication of the wide range of insect species associated with related species of Australian *Acacia*.] In addition to these predatory activities the ants glean nectar from the wattles which have an abundance of extrafloral nectaries. Two groups of insects commonly encountered on the wattles are not preyed upon by the ants, namely Homoptera and the larvae of certain lycaenid butterflies. These the ants tend, removing from them secretions, anal in the case of the Homoptera and of various dorsal glands in the lycaenids. In return for this largesse, the ants protect its producers from at least a proportion of the predators and parasitoids which patrol the wattle in search of prey. The wattle bush then forms a microcosm comprising, for our present purposes, the bush itself, the *Iridomyrmex* ants, Homoptera, lycaenid larvae, other herbivores, predators and parasitoids. A traditional analysis of the interactions among populations of these organisms would examine the pairwise interactions indicated in the interaction matrix in Table 10.3.

The information contained in Table 10.3 is extracted largely from the work of Pierce (1983), Buckley (1983) and Kitching (1983b) together with a few additional inferences extracted from the wider literature. The nature of the *Acacia*-based community is further illustrated in Fig. 10.5 as a traditional Forrester diagram of the systems analyst [see Chapter 2 of Kitching (1983c) for further explanation of this technique] clearly identifying both material and information flows.

Turning in more detail then, to the role of the *Iridomyrmex* ants within the community we must focus on column five of Table 10.3. The *Acacia* bush itself provides nectar, through its extrafloral nectaries (EFN's), for a variety of animals associated with it. For *A. irrorata* both lycaenid larvae and the ants have been observed feeding from the EFN's. There are two hypotheses as to why the plant should provide 'handouts' in the form of this extrafloral nectar. The first is that it provides a diversion, turning the attention of would-be marauders from the reproductive parts of the plant where nectar is provided for potential pollinators and needs to be protected from the activities of nectar robbers. Alternatively, it is suggested that the EFN's, by maintaining an active population of foraging ants on the plants, perform an indirect defensive role directed against conventional herbivores. There is no direct evidence for *A. irrorata* to allow differentiation between these two nor is there any *a priori* reason why they need be considered mutually exclusive. A variety of authors provide data, from species other than *Acacia*, that the presence of ants reduces foliage damage (e.g. Tilman 1978; O'Dowd 1979) and other authors have demonstrated a reduction in damage to or robbing of flowers (Keeler 1977, 1980). Buckley (1983) demonstrates for *Acacia decurrens*, a species very similar to *A. irrorata* in growth form and leaf morphology, that the presence of ants does increase plant growth and seed set.

The *Iridomyrmex* ants have a close mutualistic association with the

Table 10.3. Interaction matrix for the 'community' of organisms in and around a wattle bush. + indicates that the presence of A increases the well-being of B; – that the presence of A decreases the well-being of B; and, 0 that the presence of A has no direct trophic effect on B. ± indicates that both positive and negative effects occur. Possible positive or negative intraspecific effects (the principal diagonal) are omitted from the table (indicated by X).

	A \ B	Acacia bush	Homoptera	Lycaenid larvae	Other herbivores	*Iridomyrmex* ants	Predators and parasites
		Effect of B on A					
Effect of A on B	Acacia bush	X	+	+	+	+	0
	Homoptera	–	X	+	–	+	+
	Lycaenid larvae	–	–	X	–	+	+
	Other herbivores	–	–	–	X	+	+
	Iridomyrmex ants	+	+	+	±	X	–
	Predators and parasites	0	–	–	–	+	X

homopterans which feed suctorially on the wattle bushes. Buckley (1983) demonstrated this association working with the membracid, *Sextius virescens* on *A. decurrens* and Pierce (1983) identified the same sort of interaction on *A. irrorata*. Buckley (1983) showed experimentally that ants removed honeydew droplets from the bugs, presumably using these as a source of energy, at the same time apparently defending the adult bugs from predation. It appears, however, that the bugs act rather like pseudo-EFN's. Buckley's work indicated that, although plants with ants alone performed better than those without, plants with ants and membracids performed less well than those with ants alone. In other words, the presence of the membracids modifies the basic +/+ interaction between plants and ants so its becomes, on balance, a +/– one.

As with the homopterans, the lycaenid larvae have a mutualistic relationship with the *Iridomyrmex* ants. Pierce (1983) clearly demonstrates that the ants protect the larvae and pupae from predators and some but not all parasitoids. In return the ants collect the secretions of the various dermal glands of the larvae (Kitching 1983b) from which they obtain both sugar-based and nitrogenous sources of energy. The larvae also benefit, indirectly, as the ants reduce competition by removal of competitors (other than homopterans). The complexities of the ant–lycaenid interaction are beyond the scope of this account but the coevolved mutualism has morphological, ecological, behavioural and chemical dimensions.

Larvae of *Jalmenus evagoras* of course are subject to predation and

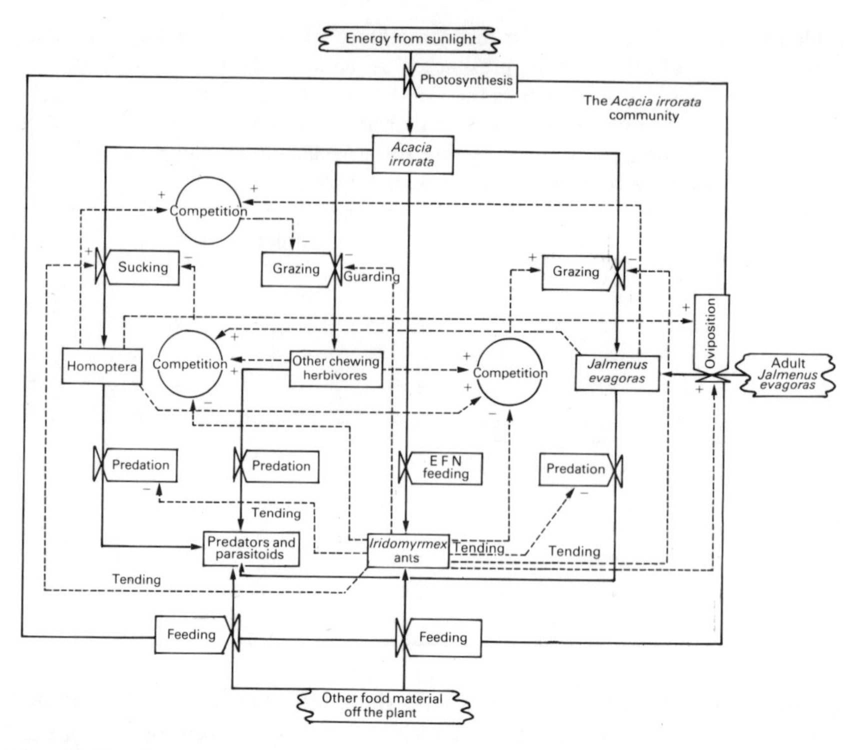

Fig. 10.5. Forrester diagram of the community centred on bushes of *Acacia irrorata*.

parasitism. About 60–66% of eggs are parasitized by a trichogrammatid wasp (*Trichogramma* sp.) (Pierce 1983). Some egg masses, especially those laid close to homopterans or other *J. evagoras* larvae and pupae, experienced far lower levels of attack, presumably due to close attendance by ants on these other organisms. A variety of other organisms may prey on eggs and, again, Pierce infers that the levels of attack are probably reduced through the presence of ants tending lycaenid larvae and pupae, homopterans and the plant EFN's. About 20–25% of lycaenid larvae are themselves parasitized by a braconid wasp (*Apanteles* sp.) and Pierce demonstrated that these rates were largely independent of the presence or absence of ants. In contrast, the presence of ants do reduce levels of predation *sensu stricto* by a range of natural enemies including mantids, larger ants, vespid wasps, coccinellids, reduviids and spiders. Pupae are also subjected to attacks by both parasitoids and predators and, in this instance, ants provide protection against both classes of natural enemy.

Completing the account, some reduction in numbers of 'other' herbivores probably occurs in response to the presence of large numbers of *Iridomyrmex* on the bush. Data presented by Pierce (1983) bear this out but, as she suggests, only longterm study will demonstrate just what feedback there is on plant performance. Again, the general nature of many of the predators recorded from the community suggests that these will have an impact on other herbivores as they do on the homopterans and lycaenids. Again, inasmuch as the presence of ants reduces the number of natural enemies on the bush, the ants will have an indirect positive effect on all the herbivores. However, in the case of those that are not in direct mutualistic interaction with the ants themselves, the net impact of ants on these herbivores, be it positive or negative, is hard to assess.

The key point is that organisms such as the *Iridomyrmex* ants that may be superficially designated as 'predators' play a role in the community which goes far beyond that simple designation. Certainly, the ants are predators, but they are also herbivores and mutualists and it is their overall role which defines their key position in the community. This is not an entirely novel observation, of course, but does point up the unfortunate if unavoidable legacy of population ecology to community ecology. Indeed it will remain an inevitable simplification except where we can assemble the degree of detailed biological knowledge available, at least by strong inference, for the *A. irrorata* community which, additionally, is a relatively simple example of a natural community.

Parenthetically, this case history also provides an additional instance of the two principal phenomena identified earlier: undoubtedly *Iridomyrmex* is a 'key' predator and its activities ramify both up and down the food-web of which it is a part.

10.3 Manipulated communities

Man has moved species about the world, either deliberately or inadvertently throughout, and indeed before, recorded history. Particular episodes of human expansion have had identifiable biological consequences apparent today. The British fauna retains the impact of the Romans in the form of fallow deer (*Dama dama*) and fat dormouse (*Glis glis*), the Spanish conquistadores left feral mustangs in the Americas, later European invaders of North America and, in particular Australasia, brought a suite of species, to those regions from their homelands and elsewhere. Elton's (1958) treatment remains the best general account of the impact of such exotics. Again though, the many case histories that can be cited are often anecdotal or, where based on numerical data, restricted to two-species treatments. Seldom can 'community-level' impacts be assessed both because of their intrinsic complexity and the lack of information about the pre-invasion state of the community.

Turning to predators as a subclass of the exotics that man has introduced into natural communities we might expect to gain special insight into the role of predators in communities in general, by examining the longterm consequences of these unintentional experiments. Sadly, this is seldom possible because of a dearth of information on the communities as they previously existed. One has but to reflect on the subtle and manifold interactions described in Section 10.2.3 for the simple community on *A. irrorata* to appreciate the difficulties of making judgements about rainforest communities in general, say, or even about the agroecosystems into which many of these predators have been placed. In addition the introduction of a particular species by man has been accompanied almost always by substantial concomitant habitat modification. Separating the impact of the exotic predator from that of the habitat change is usually impossible. Perhaps the only unequivocal system where predation alone was involved was the interaction between European seamen and the great auk *Pinguinus impennis* — sadly no longer available for study!

Two classes of community do allow some comment on this matter: those on oceanic islands and those that have been the intended or unintended targets of certain attempts at biological control. In the following brief treatments I can do little more than collate anecdotes and sparse data to pose questions and, possibly, stimulate further work, in the absence of more satisfactory studies at this time.

10.3.1 Oceanic islands

Two particular case histories are pertinent here and are presented, in part, because they represent recent advances in the field.

Ship rats on Lord Howe Island

Lord Howe is a Pacific Island under Australian jurisdiction situated some 232 km from the Australian coast on latitude 32°S. It was discovered, uninhabited, in 1788 by Lieutenant Ball of HMS *Supply*. It was visited frequently by mariners subsequently. By 1834 there was a small permanent settlement on the island and the island population grew slowly to the 1979 level of approximately 250. According to Recher (1974), at the time of its discovery Lord Howe Island possessed 15 species of terrestrial birds, 14 of which were endemic (five species and nine lower taxa). There were a small number of lizards but no snakes, mammals or amphibians. Little is known of the invertebrates but they included a peculiar flightless, endemic phasmid and about 70 species of land snail of which many are endemic (Smithers *et al* 1974). Almost from the point of discovery, man left his mark on the island in the form of introduced animals. Cats were present from about 1848 and pigs, goats and mice were also nineteenth century introductions. According to Recher (1974), however, it was in 1918 that the supply ship *Makambo* ran aground on the island releasing stock of the common ship rat *Rattus rattus* onto the island. The impact on the birds alone was, to say the least, dramatic. Table 10.4, taken from Recher's article, shows just how traumatic this was. No fewer than five of the 15 land birds were exterminated by the predatory activities of these rats over the next 20 years. In addition the rats contributed to the decline of the flightless woodhen *Tricholimnas sylvestris*, which presently survives at precariously low levels. The impact of the rats on other elements of the fauna is hard to assess as so little was known of that fauna beforehand. Certainly the giant phasmid, *Drycocelus australis* was exterminated on Lord Howe itself by the marauding rats although it may survive on the adjacent, ratless, islet known as Ball's Pyramid (Smithers *et al* 1974). The molluscan fauna has also suffered and the decline or even disappearance of many species of endemic land snail can be attributed, at least circumstantially, to the rats' activities. Other exotic predators, notably feral cats and the introduced Australian race of the owl *Ninox novaeseelandiae* have taken their toll: the first in direct predation on the woodhen in particular; the second, possibly, by interbreeding with the endemic race and swamping it such that its genetic integrity was lost. Man himself, of course, is the arch-predator and he destroyed the endemic gallinule and pigeon to serve as ships' fare before the end of the nineteenth century. The Lord Howe parakeet was destroyed by about 1870 as an actual or potential pest of crops (Recher 1974).

It is easy to account for these exterminations as the simple over-exploitation of the natural community by a 'superior' exotic predator but closer examination reveals the real complexity of the issue. Many of the species listed in Table 10.4, for instance, survived the introductions of cats and pigs as well as the presence of man before 1918. Why then did the ship rat have so much more of an impact than these other, potentially destructive species? Insectivorous song-birds have been introduced to the island, also, and several recent 'natural' invaders are established only because suitable habitat was manufactured for them by man's activities in agriculture and building. How do these other predators and competitors fit into the natural community and do some have more impact than others? We do not have answers to these questions but it could be postulated that some predators such as the ship rat have more impact than others because of their niche dimensions relative to those of other predators, both exotic and endemic.

Feral cats in New Zealand

Before Polynesian settlement the only terrestrial vertebrates present in New Zealand were birds, two species of bat, the tuatara *Sphenodon*

Table 10.4. History of the land birds of Lord Howe Island (modified from Recher 1974).

Species	Fate (as at 1974)	Direct agent of decline or demise
Woodhen (*Tricholimnas sylvestris*)	18–20 survive	ship rats, pigs, cats, man
White gallinule (*Porphyrio porphyrio albus*)	extinct, before 1844	man
Lord Howe pigeon (*Columba vitiensis godmanae*)	extinct, between 1853 and 1870	man
Green-winged pigeon (*Chalcophaps indica*)	common	–
Lord Howe parakeet (*Cyanoramphus novaezelandiae subflavescens*)	extinct, *c*. 1870	man
Lord Howe boobook owl (*Ninox novaeseelandiae albaria*)	extinct, date uncertain	unknown, possible *N.n.boobook*
Lord Howe kingfisher (*Halcyon sancta adamsi*)	abundant	–
Vinous-tinted thrush (*Turdus xanthopus vinitinctus*)	extinct, between 1919 and 1938	ship rats
Lord Howe warbler (*Gerygone insularis*)	extinct between 1919 and 1938	ship rats
Lord Howe fantail (*Rhipidura cervina*)	extinct between 1919 and 1938	ship rats
Lord Howe golden whistler (*Pachycephala pectoralis contempta*)	abundant	–
Lord Howe silvereye (*Zosterops tephropleura*)	abundant	–
Robust silvereye (*Zosterops strenua*)	extinct between 1919 and 1938	ship rats
Lord Howe starling (*Aplonis fuscus hullianus*)	extinct between 1919 and 1938	ship rats
Lord Howe currawong (*Strepera graculina crissalis*)	30–50 survive	unknown

punctatus and about 30 species of endemic geckos and skinks. The various waves of Polynesian invasion of the islands added the kiore *Rattus exulans* and the now-extinct Maori dog. European man, arriving at least 600 years after the first Polynesians, subsequently introduced 33 mammals, one reptile, 33 birds and innumerable insects and other invertebrates. These figures for post-European settlement introductions refer only to those species that established themselves (Bull 1969; Gibb & Flux 1973; Bull & Whitaker 1975; Salmon 1975).

Considering the mammals alone, all three species added to the fauna in pre-European times, the kiore, dog and man, were predators. These species, alone or in concert, are thought to have been responsible, either directly through hunting or, indirectly, through habitat modification, for the extinction of some 45 species of bird including all 24 species of a ratite order Dinornithiformes, the moas (Williams 1973; Bull & Whitaker 1975). Their 'sublethal' effects on other species are impossible to assess but the widespread use of fire, leading in many instances to the reversion of native forest to tussock grassland (Salmon 1975), must have led to major modifications in both range and abundance of large numbers of forest organisms.

European man, as elsewhere, began introducing animals and plants almost from the point of first contact. Cook himself released pigs and goats in New Zealand as early as 1773. Early settlers brought other domestic animals including dogs, cats and rabbits with them and by the early twentieth century the full range of mammals — marsupials, insectivores, lagomorphs, rodents, carnivores and ungulates — already alluded to were established. This veritable menagerie includes seven predators in addition to man — the ship rat *R. rattus*, the brown rat *R. norvegicus*, stoat *Mustela erminea*, weasel *M. nivalis*, ferret *M. putorius*, cat *Felis catus* and the insectivorous hedgehog *Erinaceus europaeus*. The impact of these predators on the native fauna has been much speculated upon but documented in only a few instances (e.g. Wodzicki 1950; Gibb & Flux 1973; King & Moody 1982). I propose to focus, in this account, on the impact of feral cats upon the communities into which they were introduced.

In examining the impact of cats on the New Zealand fauna, a number of key distinctions must be recognized: firstly, impact on offshore islands contrasted with that on the two principal islands and, secondly, impact in historical times as opposed to current effects. On the offshore islands there is little doubt that cats alone have been responsible for extinctions and consequent changes in community structure. Many authors recount the case of the Chatham Island wren *Xenicus lyelli* all known specimens of which were collected by the lighthouse keeper's cat in 1894. Karl and Best (1982) review other less well-known instances where cats have either exterminated or greatly reduced bird populations. For instance, Taylor (1979a) describes the devastation wreaked by cats and kiore on sooty tern (*Sterna fuscata*) colonies on Raoul Island in the Kermadec group. In contrast, on Ascension Island the terns continue to flourish in spite of an abundance of virtually monophagous cats. There are few rats on Ascension. In another paper (1976b), Taylor ascribes the extinction of Macquarie Island parakeets *Cyanorhamphus novaezelandiae* and banded rails *Rallus philippensis* to the combined activities of cats and wekas *Gallirallus australis*. Karl and Best (1982) in their own studies on Stewart Island estimated that 15.5% of the diet of feral cats was birds compared with

80% rats. Twelve of the 17 species of bird involved were native, with parakeets (*Cyanoramphus* sp.) and sooty shearwaters *Puffinus griseus* dominating the sample occurring in 15 and 12, respectively, of 229 scats examined. In contrast the near extinct kakapo *Strigops habroptilus* occurred in six scats.

As in other studies, Karl and Best found a confounding of the effects of cats and rats. They write (p. 291):

> "The relationship between cats, rats and birds is complex. Removing cats from an island may allow the rat population to increase, which in turn might be detrimental to birdlife. The effect may be direct through predation by rats on birds, or indirect through rats competing with birds for food."

Turning to cats on the major islands in recent times two studies are particularly instructive; that of Gibb *et al* (1978) carried out on a sheep property on the North Island, and that of Fitzgerald and Karl (1979) made in the Orongorongo Valley near Wellington. Gibb *et al*'s work examined the dynamics of a rabbit population within an 8.5 ha enclosure of cleared pastureland. A variety of avian and mammalian predators were active within the area. These included feral cats. Scat analyses made over a 6 year period indicated a diet dominated by rabbit (40–100%). Next most abundant were insect remains (0–41%) with the remainder of the diet made up of mice, rats, hedgehogs, stoats, lizards, birds and plant material. Birds varied as an item of diet from 0 to 17% and lizards from 0 to 0.2%. The birds involved were not determined. Basically what Gibb *et al* observed was a simplified community in which exotic animals and plants had almost totally replaced the original fauna. Only some of the insects, the lizard and, possibly, some of the birds were native fauna. However, to lay any of the mechanism for change at the door of the cat or any other predator (except man) has little justification. The clearing of the land, first by the burning activities of the Maoris and, subsequently by the agricultural practices of the European settlers, together with the introduction of large and small herbivores — sheep and rabbits in this study but, elsewhere in New Zealand, ranging from thar to chamois, hare to wapiti — both initiated the collapse of the original community and maintains the current one.

In contrast, the study of Fitzgerald and Karl (1979) was made in vegetationally undisturbed native forest. Again, estimates of the impact of cats was made using scat analysis. As in the previous study, exotic animals dominated the diet of the cats in the region. Over a 3-year period rats comprised between 20 and 62% of the diet, mice between 0 and 24%, rabbit between 10 and 40% and brush-tail possum *Trichosurus vulpecula* between 0 and 40%. Overall only 4.5% of the cats' diet was birds and less than 6% insects. Again, it is in these two minor categories that current direct impact on the natural community is occurring. The birds involved were largely exotic but included native pigeons and fantails. The insects were principally stenopelmatids ('wetas') and it is here that the major current impact of the cats on the natural community exists making up in one particular year 4.8% of the diet by weight. However, as in the agricultural ecosystem the current impact of feral cats on the endemic fauna could safely be said to be trivial.

Lastly, the much harder question of impact of the feral cats in earlier times must be examined. Wodzicki (1950) records a variety of small bird remains around cat dens while Gibb and Flux (1973) present data showing between 56 and 60% bird content from cat scats. These authors also point out that decline of native birds such as the laughing owl *Sceloglaux albifacies*, stitchbird *Notiomystis cincta*, kakapo, piopio *Turnagra capensis* and saddleback *Philesturnis carunculatus* before the introduction of mustelids may be due to the activities of cats. Again, though, they admit the involvement and added complexity of introduced rats. Karl and Best (1982) note the comments of early writers on the rapid decline shown by the flightless kakapo over most of the country following liberation of dogs, cats and rats.

The introduction and impact of feral cats in New Zealand, then, may be summarized as follows:

(1) Cats have been responsible, solely, for extinctions of some other vertebrate species on offshore islands.

(2) On the main islands, cats with other predators seem to have contributed to a general decline in native fauna and a consequent simplification of the communities concerned.

(3) There is a complex interaction between rats and cats wherever they occur and, on balance, the rats seem more damaging than the cats.

(4) Major community modification, other than on the offshore islands, is undoubtedly the consequence of habitat modification by man either directly or through the impact of introduced herbivores. Exotic predators play a minor role.

(5) In both agricultural and forest ecosystems, currently, feral cats prey predominantly on other exotic vertebrates. Impact on endemic species is minor and more significant, perhaps, upon insects than upon birds.

10.3.2 Biological control

Biological control is the process by which the natural enemies of some pest organism or object are introduced by man in an attempt to reduce that pest. Target organisms are commonly weeds, phytophagous pests of crops, predators or, even, biological products such as dung. The natural enemies involved may be parasites *sensu stricto*, parasitoids, predators, pathogens or, *vis-à-vis* dung, coprophages. This particular account concerns predators but, of course, many of the comments apply equally well to other potential controlling agents. The key point of importance, to be reiterated later, is that it is at the community level of ecological organization that the potential and actual impact of a controlling agent needs to be assessed.

Cane toads in Queensland

The cane or marine toad *Bufo marinus* is native to Guyana and Surinam from whence it was introduced, first around the Caribbean, then the Pacific and American Southeast and, ultimately to date, to subtropical and tropical Australia and Papua New Guinea. The complex web of introductions and reintroductions is traced fully by Easteal (1981). The reason for these introductions was the supposed efficacy of the toad to control certain boring beetles, pests of sugar cane. The cane toad is an omnivore, however, with a diet which includes earthworms, scarabs, wasps, centipedes, cicadas, skinks, small mammals, pet food and even, reportedly, cigarette ends (Pippet 1975; Tyler 1975; Niven & Stewart 1982). Accordingly, from the initial introductions in 1935 in North Queensland, the toad spread rapidly and, indeed, is still spreading, with a current distribution stretching from northern New South Wales throughout Queensland to the shores of the Gulf of Carpentaria in the Northern Territory. This range, of course, far exceeds the distribution of sugar plantations and, indeed, there is little evidence that the toads ever made any substantial impact on the sugar cane beetles (Fellows 1969; van Beurden 1978). The toad did, however, invade a wide range of natural habitats as well as the urbanized areas in which it appears to do particularly well.

That the toad has had substantial impact upon the natural communities which it has invaded is in little doubt but, as with earlier case histories, one must build anecdote upon anecdote, casual observation upon casual observation, in order to gain an idea of the *community*-level impact of the toad. Before and after studies at this level are in progress but, for the moment, the circumstantial picture is the only one available.

Following Covacevich and Archer (1975), there are four ways in which the toad appears to have direct impact on members of the community around it: firstly, it may be taken as prey by predators to which it subsequently proves fatally toxic; secondly, it may prove to be an acceptable food item for other predators; thirdly, it may prey upon members of the community; and, fourthly, it may compete with members of the community, displacing them from a particular locality or drastically reducing their levels of abundance there. Covacevich and Archer (1975) report eight species of reptilian predators which are known to have died from ingestion of toads. In particular it appears that the elapid snakes, red-bellied black snake *Pseudechis porphyriacus*, death adder *Acanthophis antarcticus* and brown snake *Pseudonaja textilis* may well have declined in number in areas of high toad abundance. These are all species for which frogs are an important dietary item. Fatal impacts on a few mammals and birds are also recorded.

An increasing number of predatory species, in contrast, are being recorded as exploiting adult and/or juvenile cane toads without any ill-effects. The keelback snake *Amphiesma mairii* and the freshwater tortoise *Elseya latisternum* do appear to feed on the whole animal frequently, being apparently immune to the toads' toxins (Covacevich & Archer 1975; T. Hamley & A. Georges personal communication). Birds and mammals appear to enjoy no such immunity but certain species have learnt to exploit parts of the toads' bodies while avoiding the toxic products of the parotid glands. In particular, the Australian native water rat *Hydromys chrysogaster* eats the stomachs of toads and there are reports that in certain regions of North Queensland these may form a substantial part of the rats' diet. A range of other species from fish to pheasant coucals *Centropus phasianinus* are known to feed occasionally on the toad or its tadpoles but, hereto at least, these do not comprise an important segment of their diets.

As early as 1936, Mungomery presented analyses of toad stomach contents. A wide range of invertebrates were included and subsequent work has confirmed the early suspicion that toads will feed on virtually anything small enough to pass between their jaws. It is as a competitor that even the qualitative impact of the toad is harder to assess. Probably, the toad is now the commonest terrestrial vertebrate in southeast Queensland and, as Covacevich and Archer (1975) write: "It is now common to find only *Bufo marinus*. . .in niches occupied by small native vertebrates in adjacent toad-free areas". The range of native species which may have been displaced includes frogs, skinks and, perhaps, even small mammals. We await definitive studies.

So, although lacking quantification, the toad has impacts up and down the food-web, as well as laterally. These are only the direct interactions; however, J. Covacevich (personal communication) suggests that with the decline in abundance of the red-bellied black snake, death adder and brown snake, a fourth elapid, the taipan *Oxyuranus scutellatus*, may have increased in numbers. If true, this reciprocal increase might well be the consequence of the original dietary overlap among these species: remove three of these, and the fourth may expand its realized niche taking in perhaps minor components of the niche spaces of the others.

Vertebrate predators and snakes, in particular, have a high 'apparency' to naturalists. The second-order effects of the cane toad in a natural community would necessarily be widespread among less visible members of the community as the effects of changed abundance of the toads' predators, competitors or victims, ramify through the food-web.

We may conclude, then, that the effects of the cane toad as a predator within the natural communities of northeastern Australia have been of the same kind we expect from other work on the role of indigenous predators (Section 10.2). The species, predictably, had very much greater impact than was intended by its original introducers and the consequences of these observations for biological control in general are outlined below.

Biological control: some general points

The misdirected impact of the cane toad on non-target organisms is particularly visible because of the size and familiarity of toad to the casual observer. Other vertebrate impacts, of stoats in New Zealand or mongooses in the Hawaiian Islands for instance, are similarly well known. I suspect that when invertebrate predators are introduced as controlling agents similarly wide impact other than on the target pest may occur: coccinellid beetles, chrysopid larvae or predacious mites, for example, are not known to be particularly selective in choice of prey. This may not matter, of course, if the predator concerned remains on the crop which is affected by the target pest. A degree of vagility, however, is one of the desirable features of an agent for biological control — a species that needs to be reintroduced to every fresh planting of the crop is not likely to be economically attractive to growers (although note the increasing use of 'tactical' biological control in schemes of integrated pest management). This vagility will also mean that such predators may well enter surrounding natural communities where their impact, although less visible, may be just as widespread and important as that of the cane toad on other vertebrates. Again, this is an area needing definitive studies but one where, *a priori*, one would expect to see major effects on occasion.

Parenthetically, it should be emphasized that the predicted impacts are likely to be much less when near-monophagous parasitoids rather than oligo- or polyphagous predators are used as controlling agents. To my knowledge little screening of such predators or parasitoids of invertebrate pests is done which involves indigenous members of adjacent communities.

Lastly, a long overdue point must be made. The practitioners of biological control frequently claim ecological principles for the scientific underpinning of their activities. This contention bears a brief closer examination. When a transfer of a species from its homeland to a target area is made, the predator or parasitoid concerned is being transferred from one coevolved complex community of animals and plants to another. To make *predictions* as to its likely success and general impact in the target community, detailed information about its role in the source community is required as is a profound understanding of the structure and functioning of the target community (or, even, communities). That such levels of knowledge are possible is demonstrated by some of the case histories described earlier in this chapter. In contrast current practice in biological control using predators or parasitoids follows one of two philosophies: either as many potential controlling agents as can be collected and cultured are released; or, the abiotic tolerances of a range of potential agents are examined so that they may be matched climatically with the target region. In both cases, ultimately, it comes down to 'suck-it-and-see' following release. Such processes draw little upon ecological theory in their execution beyond, perhaps, an incomplete, inchoate appreciation of the idea of the niche. The crucial point, however, is that almost certainly current practices are the most appropriate. As Geier and Clark (1978/79) have pointed out, pest control is not strictly a biological endeavour but comprises in addition, elements of economics, politics and sociology. Considering just the economic dimensions of biological control for present purposes, it seems to me that a truly ecological approach to biological control, including the necessary community-level studies at both ends, would be prohibitively expensive in both time and manpower. In contrast, the suck-it-and-see approach may be less elegant but, even with, say, a 5% chance of establishment and success in any particular species introduction, is almost certainly substantially cheaper. The only situation in which this would be arguable is where the possible conservation consequences of a potential controlling agent were paramount and, in these cases, the simple answer to most proposed introductions would be 'no' — not such a retrograde and conservative response given the growing sophistication of cultural and genetic methods for the control of pests.

10.3.3 Postscript on exotics

We have observed that some exotic predators will have major impacts on communities whereas others will remain at low levels having little impact. Why there should be this differential response is a matter for conjecture. I suspect it can be related to the breadth of the niche of the predator at the time of its introduction *relative* to the average breadth of the niches of the native predators already present in the community. The closer to the *r*-end of the habitat templet of Southwood (1977) the more likelihood there is of permanent establishment and major impacts especially if there is accompanying habitat modification going on (as is usually the case). Thus, the ship rat and the cane toad are broad-niched, quintessentially opportunistic species. However, if Mr G. W. Francis of Adelaide had had his way and the African secretary bird *Sagittarius serpentarius* had been introduced into Australia in the 1860s to become the bane of poisonous snakes there (Rolls 1969), I predict it would either have failed to establish or would have enjoyed the fate of the ostrich, a tiny relict population of which still roams the South Australian sand-hills following introduction in the late 1800s.

The longterm consequences of the presence of an exotic in a community are only likely to be calamitous when the community is simple and particularly vulnerable as on oceanic islands. Given resilience in the community then coevolution (perhaps 'polyevolution' is a more accurate term!) will act and the niches of both the exotic species and the indigenous species will be mutually modified and, ultimately, new equilibrium levels will be achieved. Whether or not all the original species will be present in the new equilibrium community is a moot point, of course, but we are not looking at an unnatural process: it is only the timing that may be unnatural. It is likely that the original, prehistoric, arrival of Australian boobook owls on Lord Howe Island, as storm-driven waifs perhaps, had as much impact on whatever community of animals existed there previously, as did the arrival of the ship rats following the beaching of the *Makambo*. The Australian communities containing the cane toad provide growing evidence of this process of accommodation and modification as the niches of some indigenous animals expand to include the cane toad as a prey item while those of others contract due to the 'adverse' effects of the toads' presence.

10.4 The future

This chapter has identified and emphasized the profound impacts that predators may have within communities, on other members of their own trophic level and on the level or levels above or below them. Different predators have different degrees of impact within the coevolved whole that is the community. In addition, it may be contended that labelling an organism with a single function such as predation or herbivory or parasitism is simplistic and on occasion, such as when attempting to assess the impact of exotic species, this may be disabling or at least grossly misleading.

A number of important avenues for further work have been identified either implicitly or explicitly within this account. Two general comments may be made. Firstly, there is need for much more information on the roles and occurrence of predators within communities. Questions about basic prey: predator ratios remain open (Pimm 1982) while questions about the more sophisticated notions of second-order effects or the impact of particular strategies of predation have seldom been discussed. The key methodological point is that to answer community-level questions, one must carry out community-level studies, rather than make inferences from population studies. This is not to denigrate population studies, of course — many of the accounts in this chapter draw in part at least on such inferences — but it is at the community level itself that our data base is sparse. Secondly, and not unrelated to the first point, we need to develop further an experimental approach to communities. Just as population ecology has moved from a basically descriptive phase to an experimental-theoretical state in recent decades, so must community ecology develop. In this regard, it may be advantageous to focus more attention on simpler natural communities such as those in delimited, discrete habitats such as dung pads, cadavers or, as demonstrated earlier, water-filled tree holes, and increase use of manipulated microcosms already used to effect by workers such as Beyers (1963, 1964) and Maguire (1971).

11 The Analysis of Competition by Field Experiments

Tony Underwood

11.1 Introduction

Controversy continues to rage about the relative importance of interspecific competition among species as a major determinant of the structure of natural communities. [See Chapter 6 and the arguments developed in Strong *et al* (1984b), and the issue of the *American Naturalist* (**122**, 1983) devoted to the topic.] Much of this debate is still a heated discussion of the recognition of patterns in nature that might be explained either as a result of competition in the past, or because of present day competitive processes. It is worth pointing out that such debate does not directly shed light on the problem of whether competition is actually important or not — it only generates heat about whether we can observe and document patterns of distribution and abundance of species that are consistent with competitive processes. For example, Simberloff (1978) and coworkers (Connor & Simberloff 1979; Strong *et al* 1979) have argued cogently that *post hoc* interpretations based on former competition among species are awry, and should be rethought, when the patterns of distribution of species on islands are not distinguishable from random colonization and occupancy of different islands. To rebut this line of reason with statements about its incorrectness (e.g. Diamond & Gilpin 1982; Gilpin & Diamond 1982) is to make no concrete progress in demonstrating the underlying importance of competition. Competitive interactions may or may not explain nonrandomness in species' distributions on islands. Indeed, intense competition may even have led to apparent randomness of distributions of species (although hypotheses to explain this are currently sparse). In both cases, the existence or nonexistence of nonrandom patterns does not, and cannot, demonstrate the importance or unimportance of competition as the process leading to the modern pattern.

The demonstration of patterns in space and time that are consistent with competitive interactions is obviously stronger evidence for the importance of these processes than would be the absence of such patterns. The patterns themselves are, however, of no value except to raise hypotheses about competition — they cannot (even where there is no argument about their existence) simultaneously be used as tests of the importance of competition. Dayton (1973) and Schroder and Rosenzweig (1975) have provided very convincing demonstrations of the need for well-planned field experiments to prevent totally spurious conclusions about processes in natural communities.

Correctly, therefore, there has been considerable investment of time, energy and ingenuity into the planning and carrying out of field experiments designed to eliminate alternative hypotheses, and thereby to support competition as an important process to explain present patterns of distribution and abundance of organisms. Several recent reviews have drawn attention to the great number of experiments on competition between organisms in the field (Birch 1979; Lawton & Hassell 1981; Lawton & Strong 1981; Connell 1983; Schoener 1983b). Experimental studies of competitive interactions in natural populations are scarcely new (Jackson 1981), and the reviews listed indicate a great abundance and diversity of studies on a wide range of organisms. What is not clear from these reviews is how good is the evidence. What sort of experiments are claimed to be successful in producing unequivocal evidence for competitive interactions? Are the available studies of such good quality that there is no longer any room for doubt about the widespread importance of competitive interactions? Connell (1983) has suggested that field experiments are often regarded as the *ne plus ultra* of ecological research, and has pointed out that poor experiments are often worse than useless because the results are often accepted without question. Schoener (1983b) opted for the position that experimental studies are complementary to observational studies (rather than the position that observations are useful for the proposal of explanatory models that can then be tested by properly designed experiments), but nevertheless concluded that field experiments had revealed much about interspecific competition.

I am less concerned about the outcome of experiments on interspecific competition than about their validity. Although the reviews already mentioned tend to the view that much is known, that many satisfactory experiments are pointing to some widespread advances and generalizations about the nature of interspecific competition, a more cautious question to ask is how good are field experiments on competition? This is not an idle question. There have been, in parallel with attempts to synthesize the current wad of experimentally acquired knowledge about competition, several reviews of the nature of biological experimentation, and the pitfalls and problems of much of the work. Dayton (1979), Dayton and Oliver (1980) and Underwood and Denley (1984) have pointed out serious deficiencies in the development of hypotheses and the lack of consideration of alternative viewpoints in the design, implementation and interpretation of many field experiments. Widespread problems with complex sets of experimentally acquired biological data have been revealed in the analyses discussed by Underwood (1981) and Hurlbert (1984). Is the experimental study of interspecific competition somehow better than other areas of ecological experimentation? How well will the currently available studies stand up when analysed by less personally involved historians of ecology at some point in the future when today's hypotheses and theory might be as interesting and relevant as Ptolemy's?

In the search for new theory today's is found wanting, will future ecologists view today's experiments excitedly, as an abandoned gold-mine to be reopened, or as the ideological equivalent of a used-car lot, littered with the rusty remains of vehicles of self-advancement, now abandoned by careless speedsters down the vanished highways of former ecological theory?

The present analysis is therefore of the experiments as experiments, not as contributors to theory. How well conceived have the experiments been? How might designs be improved? Are the available experimental studies as useful as recent reviews have suggested?

11.1.1 Sources of studies

I have benefited from the spade-work of Connell (1983) and Schoener (1983b), and have used their extensive reviews of experimental studies of competition as the source of material for the present purposes. Previous reviews have already selected from among the enormous number of papers available, using various criteria. Connell, in particular, emphasized the variance in quality of much of the available work, and excluded some papers that Schoener considered. All the studies listed by these two authors have not been covered, but, instead, the sample was limited to 57 papers published in *Ecology* (1960–1982), 16 in *Oecologia* (1975–1982 only), 10 in the *Journal of Animal Ecology* (1960–1982), 11 in the *Journal of Ecology* (1970–1982), and one paper in the *American Naturalist* (1981) (it was the only paper in that journal cited by Schoener as containing descriptions of experiments) — a total of 95 publications. The sample is by no means random, but it covers all the papers in Schoener from the journals and years indicated. This arbitrary number was examined to determine the hypotheses being tested, the nature and design of the experiments, the type of controls used, the degree of replication in space and/or time, the type of analysis of the data and the interpretations made by the authors. My selection biases the types of study to those currently in vogue with those arbiters of fashion who referee and edit the chosen journals, but may not be unrepresentative (and was considered to be largely representative by the other reviewers mentioned). Some 63 of the papers examined for the present work were also included in Connell's (1983) discussion; nine of the papers were also included in Birch's (1979) less extensive review. The decision to limit the sample to 95 papers (rather than any round, but still arbitrary, number) was determined by two things. First, the sample examined had sufficient evidence of some of the problems with field experimentation that a larger sample was not needed to gain some impression of the difficulties being experienced by the authors and, presumably, other readers. Secondly, as will be developed below, despair and tedium set in, in equal proportions, before I reached the larger individual papers in *Ecological Monographs* that were to make up 100 publications.

Several serious problems were encountered in the papers examined, including some studies that were partially incomprehensible, some that were not about competition, some that were not really experiments, and some that were not particularly in the field. Others were lacking in sensible controls and/or replication. Still others were designed to be confounded so that interesting comparisons, and crucial tests of important hypotheses were not possible, were invalid, or only indirectly interpretable. These topics are illustrated in the body of this chapter. Fortunately, however, some studies were of great value as indications of the successful application of hypothetico-deductive logic and simple principles of experimental design. These are also discussed below. Finally, some specific designs of experiments are described to illustrate procedures that may help eliminate some of the problems encountered here.

11.2 Non-experiments sometimes not about interspecific competition and not always in the field

11.2.1 Studies not about competition

In one of the studies examined, the experimental procedures did not really examine competitive interactions (regardless of the claims of the original author or the subsequent reviewers). Benke (1978) discovered that survivorship of early season species of dragonfly was unaffected by the arrival of later, smaller species. The late species were, however, deleteriously affected by the presence of the earlier species, because the early species ate the later arrivals. Benke (1978) considered this predation to be an extreme form of interference competition. I, like Connell (1983), wish to preserve the fundamental difference between predation and competition. The concept of competitive interactions is only important when the interactions occur as a result of a shortage of some critical resource required by the competing animals (Birch 1957). Predation by one species on another is not a competitive process, even if the prey might, if left untouched by the predator, compete with the predator for other food resources. In Benke's (1978) experiments, competition for food (small animals) had been hypothesized. The food resources were not in short supply in the enclosures and, as pointed out by Benke, dragonflies rarely die of shortage of food. Competitive interactions should not be invoked where resources are apparently adequate for all the consumers present.

11.2.2 Studies not in the field

Studies of organisms in the field are also problematic because it is often difficult to retain experimental densities or combinations of species in natural areas without the use of artificial enclosures (fences, cages, etc.). These installations may bring artefacts into the results that prevent sensible conclusions being drawn about the demonstrated competition and whether it occurs under natural conditions. Apart from noting that experimental enclosures might introduce diseases as artificial sources of mortality (leading to increased recognition of apparent competition where mortality is increased when densities are large) or may prevent the natural attrition of experimental organisms because predators are excluded from the experimental plots (thus making a competitive interaction appear more important than it might ever be in nature), no value judgements are offered about the accuracy of the match between experiments and nature. To do the experiments at all requires some interference with the natural train of events (see later for more discussion of proper controls in field experiments).

There are, however, investigations where laboratory experiments are appropriate for the questions being addressed, such as Wrobel *et al*'s (1980) experiments on salamanders. Such studies clearly should not be included in discussions of field experiments [although this study was included in Schoener's (1983b) list of field experiments].

Several studies have been made under conditions where the densities of the organisms investigated were raised well above naturally occurring densities. It is not clear from these papers that the ranges of densities investigated actually would occur in nature. Thus, Istock (1973) in a well-designed analysis of interactions among various species of corixids in a pond, increased the densities above those occurring naturally ("to overload the system somewhat at the beginning"), although the starting densities and weights used for stocking the experimental ponds were not obvious in the paper.

The analysis of competition between blue tits (*Parus caerules*) and great tits (*P. major*) is often done using nest-boxes instead of natural roosting or nest-sites (Dhondt & Eyckerman 1980; Minot 1981). In Minot's (1981) study, this presumably raised the densities of breeding pairs above those normally encountered when artificial nest-boxes are not available. Similarly, Werner and Hall (1977, 1979) in their study of interactions between freshwater fish, used densities somewhat greater than those occurring naturally. In experiments on intertidal grazing gastropods (Underwood 1978) I replaced dead animals at approximately fortnightly intervals through the experiments. This is entirely unnatural — adult snails and limpets would not appear rapidly in the system under natural, field conditions. In defence of these experiments, however, the mortality of the superior competitor *Nerita atramentosa* was very slight in cages with the limpet *Cellana tramoserica*; consequently, replacements of *Nerita* were very few, and could not have altered the conclusions from the experiments.

Finally, some of the classic experiments on barnacles by Connell (1961), which have deservedly been widely quoted as an excellent example of a successful field experiment on competition, were not done under natural conditions. To determine the nature of interactions between a high-shore species of British barnacle (*Chthamalus stellatus*) with the barnacle *Balanus balanoides*, that settled in great numbers lower on the shore, Connell mapped naturally settled *Chthamalus* and also moved stones with *Chthamalus* to various levels on the shore and removed all of the *Balanus* from one half of each stone, leaving the other half intact as a control. The stones were transplanted from higher levels, where *Chthamalus* had settled, to lower levels, because the barnacles did not settle lower than mean tide-level. Thus, the intense interference from *Balanus* reduced the densities of *Chthamalus* on all stones within and outside the observed natural range of distribution of the upper-shore species. *Balanus* did not eliminate the *Chthamalus* from stones at low levels (see also Underwood & Denley 1984). Connell's (1961b) great contribution was to demonstrate that competitive interactions could be investigated by manipulative experiments in the field. It does not, however, demonstrate the importance of competition as a determinant of the natural distribution of the upper species of barnacle, because competition with *Balanus* did not eliminate *Chthamalus* even in the unnatural situation where they were moved below their normal level of distribution on the shore.

All these papers clearly reveal much about competitive interactions between species. It is nevertheless arguable whether these particular studies should be used to determine the incidence and importance of competition in nature. They may not actually be about competition occurring in nature at natural densities of the participants.

11.2.3 'Natural experiments' interpreted as manipulative studies

General agreement appears to exist amongst the proponents of ecological field experiments that so-called 'natural' experiments are of little or no value in eliminating conflicting hypotheses about processes such as competition. [See the excellent discussion in Connell (1974).] A 'natural' experiment is one where the hypothesis concerns the effects of a putative competing species (species B) on the biology or demography of a target species (A). Such an investigation commonly consists of comparisons of the mortality, growth, migration, reproduction, etc., of species A in areas where the supposed competitor (B) is naturally absent, with the biology of species A in supposedly similar areas where the other species (B) is present. This type of comparison cannot eliminate the alternative hypothesis that the biology of species A differs between the two areas for reasons other than the presence of species B (and thus not as a result of any competitive interaction at all). The very nature of the 'experiment' makes it likely that noncompetitive hypotheses based on differences from one habitat to another are more important. The two types of area are, after all, always chosen *because* they are different (i.e. one set of study areas does not contain one species). For this reason, it would be more sensible if ecologists ceased to dignify such comparisons with the label 'experiment' and thus ceased to accord them similar ranking with experiments designed specifically to test hypotheses about competition. The latter are capable of distinguishing among alternative hypotheses that are confounded in the natural comparison.

In contrast, alternative hypotheses based on potential differences among the two sorts of habitat are easily eliminated in the common form of experimental manipulation [what Hurlbert (1984) called the "unnatural" experiment] where the potential competing species (B) is removed from some areas it shares with A, and the performance of A is then monitored in comparison with control, untouched situations.

This is well-known, and scarcely worth reiterating, except that the literature examined contained several natural experiments, that were accepted without question by some or all of the previous reviewers. For example, Gross and Werner (1982), in an excellent study of the colonization of established fields by various biennial plants, had numerous replicated plots sown with seed of each of four species. Pre-emptive competition from the established plants was suggested to be an important agent preventing such seeds from becoming established in some study areas. This was inferred partially from the fact that seeds germinated and survived better in a field where more bare ground was available than in two other fields. The fields were, however, chosen for study because they had not been cultivated for differing periods (1, 5 and 15 years). Competition was deduced from the comparative success of seeds in the three fields, and differences in germination and subsequent survivorship could just as easily have been due to *any* factors correlated with the differences in age (soil condition, nutrition, wind scour, etc.), not just the differences in cover of established plants. The potential competitors (the existing plants) were not manipulated in this part of the study. In a study of experimental introductions of teasel *Dipsacus sylvestris* into old-fields, Werner (1977) manipulated the colonization by the teasel, but did no experimental alterations of the existing plants, nor the litter on the ground, yet concluded that differences from one experimental field to another in the success of the invasion by teasels were due to pre-emptive competition by the existing plants, or because the resource of bare space was not available where extensive litter had formed under the existing coverage of plants. Again, the results of an otherwise perfectly good manipulative experiment that revealed patterns of difference from one field to another were explained by processes of competition inferred from a natural experimental comparison.

McLay (1974) provided a very clear example of a natural experiment for the supposed competitive interactions between duck-weed *Lemna perpusilla* and the weed *Potamogeton pectinatus* in a pond. Duckweed did not grow well over *Potamogeton*; according to McLay this was because of alterations of the pH of the water due to photosynthetic activity of the *Potamogeton* (and thus a form of chemical inhibition or competition). The major test of this hypothesized competition was during one summer when a dense front of duckweed uncharacteristically grew out over a bed of *Potamogeton*. The pH under the duckweed then altered, which was attributed to the shading of the lower weeds by the *Lemna*, decreasing their photosynthetic rate and thereby altering the pH to more favourable conditions for the duckweed. Again, there was no manipulation of abundance of either weed. The uncharacteristic growth of *Lemna* during that summer could well have been due to environmental conditions that also resulted in altered pH of the water, or simultaneously resulted in changes in the photosynthetic rate of *Potamogeton*, leading to altered pH, regardless of the presence of the duckweed.

Classic natural experiments were found in other papers examined. McClure (1980) contrasted the performance of two species of scale insects on hemlock by examining the insects on six plants that naturally had one species only, six with the other and six that naturally had both.

Some studies involved natural, nonexperimental comparisons in part of the work. For example, Davis (1973) suggested that golden-crowned sparrows *Zonotrichia atricapilla* were unaffected by competition from juncos *Junco hyemalis*, because the distribution of the sparrows did not alter during a period when juncos were naturally sparse. Such comparisons are equally suspect whether they claim to demonstrate or fail to demonstrate supposed competition. Similarly, Waser (1978) studied competition by plants for the resource of humming-birds (to act as pollinators). The flowering success of *Delphinium nelsoni* flowering before the presumed competitor (*Ipomopsis aggregata*) was compared with that of *Delphinium* flowering simultaneously with the *Ipomopsis*. If pollinators were sparse, the two species might be in competition for the attentions of the humming-birds, and thus flower less successfully than if the other species of plant were not also flowering. The flowering success of these two sorts of *Delphinium* may have differed for all sorts of reasons, including different times of the year, or any form of genetic difference in response to selection for flowering early or late in the season, etc. Again, this is a 'natural' experiment. Part of Morris and Grant's (1972) analysis of competition between rodents consisted of experimental removals of *Clethrionomys*, to determine the effects on *Microtus*. There was no replication, as noted by the authors (and see below), but they considered the experiment to have been repeated by examination of a population of *Microtus* after the natural die-off of *Clethrionomys* during one winter. This is obviously a natural experiment; winter conditions causing the deaths of *Clethrionomys* might have been advantageous to the subsequent well-being of *Microtus*. In all cases, the study of competition was considerably better served by subsequent, manipulative experiments that produced far less equivocal results.

One study, however, demonstrated a more subtle form of natural experiment, because they were wholly embedded in an otherwise valid experimental manipulation. Peckarsky and Dodson (1980) examined competition for prey among stoneflies (*Acroneuria* and *Megarcys*) in streams. Responses to prey were examined by recording migrations of the predators into and out of cages where prey were included or absent. Competition among the predators was apparently examined by comparing the colonization by other predatory species into cages that happened to contain a stonefly at the end of the previous experiments, with cages that had no predator inside. This is a 'natural' experiment because the presence or absence of stoneflies was not a result of any experimental manipulation. This is in no way different from a more obviously 'natural' experiment examining colonization into areas naturally containing some putative competitor compared with similar areas where that species was naturally absent. Rates of colonization into these cages may have differed in many ways, regardless of the previous occupancy by a putative competing predator. If other predatory species had different habitat preferences from those shown by the stoneflies, they would be *expected* to colonize different cages from those containing a stonefly. A better experiment would have been to remove stoneflies from half of the cages and then compare colonization into those with (controls) and those without (experimental) stoneflies. In this way, there would have been, on average, no potential intrinsic differences between cages containing a stonefly, and those without.

Thus, in the 95 papers examined, one was not really about competition, a further six different papers were arguably not about competition occurring in any natural manner under field conditions, and no less than eight included unmanipulated, correlative studies (so-called 'natural' experiments). Together, these studies constitute some 16% of the 95 papers examined. Uncritical acceptance of these papers as examples of successful field experiments about competition seriously overestimates the availability of such evidence.

11.3 Replication and controls in experiments

11.3.1 Lack of replication

There should be no further debate about the need for replication of experimental units, particularly in field studies where many variables are known to be uncontrolled, and many are obviously uncontrollable. It is inconceivable that only the desired, experimentally manipulated variable of interest will actually differ between any two pieces of the world (Connell 1974), particularly where these two pieces are arbitrarily chosen as control and experimental sites from among the possible sites available. Hurlbert (1984) has made an important contribution to this problem of the design of field experiments by pointing out the pitfalls of inadequate replication. In particular, he has drawn attention to the perils of 'pseudoreplication', where the incorrect level or type of experimental unit is replicated. Consider an experiment to investigate the effects on a target species (A) of removal of a hypothesized competing species (B) from one area (called the experimental area), by making comparisons with the ecology, survival, biology, behaviour, and so forth, of species A in the experimental area with that in an untouched area (called a control). It does not matter how many replicated readings, samples, traps, quadrats, etc., are actually monitored in each of these areas, the difference between the two plots is still confounded. Differences might be (as often hoped or simply asserted by experimenters) due to the experimental treatment (i.e. the removal of species B), or due to intrinsic differences between the plots at the start of, or during, the experiment (as is often denied, without any evidence, by the experimenters). The only sensible method for distinguishing between the two possible explanations for differences between the two experimental areas (control and experimental) would be to replicate the units, by having several experimental plots and several controls. In this manner, the potential intrinsic differences between plots can be considered as negligible, because they will have been averaged over the randomly chosen control plots, and averaged equally over the randomly chosen experimental plots. For further consideration of this, consult any sensible book about statistical procedures and random sampling [and such ecological advice as that in Eberhardt (1976) and Green (1979)]. Attempts to overcome the problems associated with lack of replication are sometimes made by testing for differences in various variables between the two study sites before experimental manipulation. This achieves nothing, even in cases where some considerable armoury of sophisticated statistical procedures is used, and many potentially important variables are examined. Lack of a significant difference in the rainfall, plant cover, numbers of each species, rate of attack by predators, etc., is not the same thing as *no difference* in these variables between the two areas. As Hurlbert (1984) has pointed out, such a lack of significant difference in any applied test is simply a result of inadequate sampling of the variables in the two areas. It is inconceivable that two randomly chosen areas will be identical in all the variables of interest or importance. Such prior tests are of *no* value in determining that two study sites will remain the same, except for the manipulated variable of interest, during the course of an experiment, particularly when the experiment may run for several weeks or years. The only experiments that can have any claim to being unequivocal are those that are replicated in space or time, by the inclusion of several independent control and experimental units into the design of the experiment. This is, of course, well known and should need no further discussion for present-day ecologists ["No one would now dream of testing the response to a treatment by comparing two plots, one treated and the other untreated" (Fisher & Wishart 1930)].

It comes as a surprise therefore to discover how many experiments on competition [15 papers (16%) of those examined] had no replication. Some experiments were simply done with no replication (Pontin 1960; Werner & Hall 1977, 1979; Peterson 1979; Bertness 1981a; Grace & Wetzel 1981). In some studies, there were replicates for some of the experimental treatments, but not for others. For example, Duggins (1980, 1981), in two totally separate series of experiments, had no replicates of the control plots (from which putative competitors were not removed). Menge (1972), in contrast, had data from several control plots, but no replicated areas of experimentally increased and decreased densities of starfish. In still other studies, there was pseudoreplication in various of the comparisons being made. For example, Hixon (1980) compared the effects of reef fish (*Embiotica lateralis* and *E. jacksoni*) by removing each species from one reef, and leaving a third, untouched reef as a control; this experiment was repeated at two depths. Several replicate counts of the numbers of fish of each species on each reef were then made. These are, however, pseudoreplicates in that the differences perceived between reefs are not unequivocally attributable to the experimental removals of the fish. Similarly, Montgomery (1981) removed one species of rodent (*Apodemus*) from one experimental grid, removed another species from a second grid, and left a third grid untouched as a control, despite the wide range of density of one of the species (*A. flavicollis*) throughout the study area (p. 130 of his paper). In this study, numerous replicated traps in rows, with replicated trap-points, traps, and nights of sampling were irrelevant to the validity of attributing differences between the three grids to the removals of species from two of them. Paine (1980) examined the effects of chitons (*Katharina*) on other grazing molluscs by removing chitons from a single experimental area, and comparing results with a single control plot. Replicated quadrats within each site were used to draw conclusions about the nature of interactions between these grazers. All that such pseudoreplicates can do is to indicate the existence of patterns of difference between experimental units (sites, grids, etc.). No amount of such replication can help to attribute the differences specifically to the experimental treatment.

In some studies [a further 11 papers (12%) of those sampled], there were replicates of the experimental treatments, but these were not considered as such in any comparisons among experimental treatments, or between experimental plots and controls. Thus, Haven (1973) repeated his experimental removals of each of two species of limpets in three separate experiments. The repeated experiments were not considered as replicates in any analysis. In this type of repeated, but unreplicated, experiment, differences among the repeats are usually ignored. Sometimes, differences among repeated experimental treatments must be attributed to various predetermined differences among the repeated sites or times of sampling. Lubchenco (1980) removed the thallus, or thallus and crust of the alga *Chondrus crispus* to determine its competitive effect on the distribution of other algae, *Fucus* spp. In the experiment, there was one plot cleared of the thallus, one cleared of thallus and crust, and a single control (untouched) plot. The experiment was repeated in two different sites (chosen originally because they were subject to different exposure to wave-action, and therefore the repeated experimental treatments were not actually replicates).

The excuses offered for lack of proper replication are as numerous and diverse as the types of experiments themselves. Some authors are, however, willing to concede that the degree of replication of their experiments was inadequate. Thus, Inger and Greenberg (1966) discussed the serious shortcomings of their experiments, and the possible effects of these on the results obtained. They were, however, working in a climate of considerable political uncertainty that allowed no subsequent experimentation. Montgomery (1981) lacked time to do better experiments. Morris and Grant (1972) considered the limitations of their experiments on rodents. Undoubtedly, lack of money or manpower must have limited the number of experimental plots that could be managed in some studies. Nevertheless, none of these causes of lack of replication can make up the serious deficiencies in unreplicated experimental work. For the purposes of the present analysis of experiments, those without replicates are useless. They do not form a proper test of any of the stated hypotheses about competition, because they are not designed in order to falsify many other hypotheses that could explain the patterns in the experimental data. In keeping with the aims of modern ecological experimentation, unreplicated studies do not really constitute valid experiments.

Some types of organisms are obviously much more difficult to study than others, because of mobility, size, rareness, size of territory, home-range or sparsity of natural resources. It is therefore more likely that investigation of these organisms (e.g. birds and mammals, in contrast to sessile marine invertebrates) will end up unreplicated. Examples of studies on small mammals and birds are those of Davis (1973), Chappell (1978), Minot (1981). It is difficult, however, to avoid the conclusion that lack of attention to the design of experiments and the need for replication are more important factors in the appalling state of much current experimentation that is the intrinsic difficulty of the subject matter. For example, replicated studies on small mammals have been done by Redfield *et al* (1977), Inouye (1981) and Joule and Jameson (1972). None of these authors reported any particular difficulty in achieving their experimental requirements. In contrast, many studies of rodents are unreplicated (Montgomery 1981; Abramsky & Sellah 1982) or are repeated with different treatments, in different places or at different times without any adequate replication in the repeats (e.g. DeLong 1966; Grant 1971; Price 1978; Holbrook 1979). Similarly, experiments on salamanders can easily be replicated as in the elegant studies by Hairston (1980b, 1981) and Wilbur (1972); there were no replicates in Jaeger's (1971) experiments, although they were repeated in different habitats. Competition between lizards was studied in replicated areas by Smith (1981); Tinkle's (1982) experiments on members of the same genera had no replication. There is usually no particular difficulty in manipulating replicate areas of a seashore as shown by replicated studies on limpets (Stimson 1970; Underwood 1978; Creese & Underwood 1982) and sessile invertebrates (Taylor & Littler 1982). Some studies of reef fish are replicated (e.g. Williams 1981) as are some studies of plants in oldfields (Allen & Forman 1976).

An overwhelming conclusion about field experiments on competitive interactions is that the authors apparently choose whether or not to include replicates in the experiments more or less at whim. This occurs despite the widely acknowledged need for replication in field experiments, and the lack of convincing justifications of experimental results when absence of replication leads to potential ambiguities of interpretations. If the simplest requirement for the design of an experiment had to be met by field ecologists, an inescapable consequence would be the harsh judgement that unreplicated studies are woefully inadequate as tests of meaningful hypotheses. If this evaluation (which is consistent with many other areas of experimental biology and other sciences) were used, unreplicated studies should be ignored in any assessment of the value of field experiments on competition. To do so here would eliminate some 28 (29%) of the 95 papers examined, whether in whole or in part as having inadequate levels of replication to allow meaningful conclusions to be drawn from some or all of the experimental work. This is not a dissimilar proportion of the total from that found by Hurlbert (1984) to contain pseudoreplication in some form or another (although the analyses are not entirely independent as some studies are common to both surveys).

11.3.2 Problems with controls

In addition to possible problems associated with experimental fences, cages, installations etc., some studies lacked sensible controls for parts of the manipulations. DeLong (1966), for example, examined interactions between house mice (*Mus*) and *Microtus californianus* on two grids. Both grids were then supplied with extra food, and *Microtus* were removed from one. In addition to the lack of replication, there were no controls for the experiment, and the results can only be used to infer processes where food supplies are artificially supplemented. Price (1978) made a detailed examination of habitat utilization by four species of rodents in a sampling area. Then, utilization of habitat was studied in experimental enclosures containing eight introduced individuals of a single species (Treatment 1 in her paper). She concluded that habitat use was less specialized than that in the intact community, because competitors were absent. As well as the intrinsic differences between the natural area sampled and the area used for the enclosures, and the possible artefacts of behaviour due to being in enclosures that might restrict the large scale movements of rodents, the experimental animals had also been trapped, handled and introduced into new, unfamiliar areas. Any of these three, uncontrolled variables, rather than the absence of other species, could explain the observed difference in behaviour compared with the natural community.

Lack of controls for the experimental disturbances of transplanted animals was found in several other studies. Pontin (1960, 1969) transplanted nests of ants (*Lasius niger* and *L. flavius*) to determine the nature of interactions with surrounding colonies. There were, however, no attempts to control for the effects of three types of experimental disturbances inevitable in such experiments. Firstly, there is the disturbance of the surrounding colonies due to digging out a space to insert a transplanted nest. Secondly, the transplanted ants are, presumably, seriously disturbed (and perhaps made more agitated?) by the physical disruption of the nest. Finally, the transplanted ants may behave differently as a result of being translocated to a new piece of habitat. Any of these disturbances, rather than just the presence of a putative competitor, could explain subsequent changes in surrounding nests. Controls for such disturbances are straightforward — comparisons could be made with areas in which nests were dug out and replaced (so that nests are disturbed but there is no translocation and no change of species); areas where a nest is removed but replaced with a nest of the same species (the nest is disturbed and translocated, but there is no change of species) and the usual experimental plot where a nest of another species is introduced (disturbed, translocated and a change of putative competitor). Using this protocol, the various experimental disturbances could be separated from the effects of the primary experimental variable — the introduction of a different species.

In contrast to very active animals, such as ants and rodents, it is not likely that experimental disturbance would seriously alter the behaviour, rate of predation etc., of sluggish animals, such as starfish (Menge 1972). Nevertheless, the lack of controls for disturbance confounds any comparison of experimental and other types of control areas.

Other experiments lacked different types of controls. Fowler (1981) removed several species of plants from experimental plots in oldfields, and Gross and Werner (1982) dug up the soil in some areas before planting seeds. In neither case was there a control for the disturbance of the surface of the soil as opposed to the removal, or absence, of competing species. Seeds in recently disturbed ground might germinate better than those in undisturbed plots, regardless of the presence of other plants. If this were the case in either of the cited studies, removal, or absence, of potential competitors would, incorrectly, appear to have caused greater germination.

In some studies, attempts to provide proper controls for the experimental manipulations failed. Thus, Kroh and Stephenson (1980; experiment 2) analysed interactions among plants, by examining single plants (species A) surrounded by six individuals of the same (A) or one of three other species (B, C, D). Control plants were isolated with no surrounding individuals. Unfortunately, the lack of surrounding plants caused increased stress on the control plants, because they were no longer protected from wind and other physical factors.

It appears that controls for various necessary manipulations are often lacking, or inadequate. This compromises the value of any experiment by allowing several possible interpretations to be made about differences among experimental treatments. Confounding variables must be removed from consideration, by the establishment of appropriate sets of controls, even though these may be more unwieldy than the experimental manipulation itself (Connell 1974). Controls to remove confounding variables are essential, even if this causes changes in the circumstances under which the hypothesis of competitive interactions is tested. If disturbance of some experimental animals is necessary to complete an experiment (for example, to introduce members of a competing species into some experimental areas), then control groups of similarly disturbed animals that have not had the competitor introduced will also be necessary, to ensure that differences between treatments are not due to handling and other types of disturbance, rather than the introduction of a competitor. If it is not possible to examine true controls (i.e. undisturbed animals) the nature of the experiment is such that the hypothesis being tested only applies to competitive interactions among disturbed (not necessarily *natural*) groups of organisms. This alteration of the hypothesis will, of course, reduce the generality of the results to those circumstances investigated. This, in turn, will limit the validity of extrapolations to entirely natural circumstances. More care in the establishment of proper controls for experimental procedures will make some experiments more complicated. It will, however, also increase the degree to which results of such experiments can be considered relevant to natural populations.

11.4 Confounded comparisons and other problems

11.4.1 Confounding in designs of experiments

Several different experimental designs have been used to investigate interspecific competition. Of these, two major types have been used extensively. The first is known as a replacement series, or reciprocal α design (de Wit 1960; de Benedictus 1974) and consists of a series of experimental plots, all at a fixed total density, but made up of different proportions of two species, ranging from a pure 'stand' of one through various mixtures to a pure 'stand' of the other. This design has been used commonly by plant ecologists. [See the extensive examples in Trenbath's (1974) review.] It has also been used in studies of animals [e.g. protozoa by Gill and Hairston (1972), freshwater bivalves by Mackie *et al* (1978)]. It is, however, often difficult to interpret the results of such experiments unambiguously, because any comparisons between experimental plots must involve two differences — the greater density of one species and the simultaneously reduced density of the other species. If species suffer from intraspecific in addition to interspecific competitive interactions (as they almost invariably will — see later), such experiments are of dubious value for interpreting processes of competition. In fact, some competition between two species might be completely masked by intraspecific effects of equal magnitude so that no change from treatment to treatment would be recorded when members of one species are replaced by members of another.

The alternative design, sometimes termed a 'mechanical diallel' (Putwain & Harper 1970) also makes comparisons between pure 'stands' of each species and mixtures of the two. Here, mixtures may be created by adding some number of another species to the individuals in a pure stand [a 'synthetic' design in Putwain and Harper (1970)] or pure stands may be created by removing members of one species from a mixture (Putwain and Harper's 'analytic approach'). In either case, the total density of all individuals (i.e. of both species combined) will be different between single species and mixed plots (Table 11.1b). Comparisons between such plots, provided common sense has been used, will be unconfounded. Any difference between members of species A at some chosen density when alone, and when members of species B are present can, presumably, only be due to the presence of species B. Care must be taken, however, to ensure that the experimental design is not inadvertently confounded. In several studies, the general design of a 'mechanical diallel' was followed, but densities of each species when alone were different from those when the two species were together. Thus, Bertness (1981a) compared rock pools with 200 *Clibanarius* (hermit crabs), pools with 200 *Calcinus* and pools with a mixture of 100 of each species. Jaeger (1971) and Werner and Hall (1977) created precisely the same difficulties in experiments with salamanders and fish, respectively. Haven (1973) removed limpets (*Acmaea digitalis* and *A. scabra*) to make experimental plots with single species, but caused confounded comparisons between plots because of very different starting densities of the two species (Underwood 1976). In these studies, comparisons between areas with a single species and with a mixture were therefore confused because the addition of a putative competitor was confounded with the simultaneous decrease in density of the target species (Table 11.1a).

Table 11.1. Confounded and valid designs of experiments to investigate interspecific competition. (a) Confounded comparisons. There are two differences between individuals of species A in treatments 1 and 3 — the intraspecific change of density of species A and the absence/presence of species B. (b) Unconfounded comparisons. There is only one difference between individuals of species A in treatments 1 and 3 — the absence/presence of species B.

	Experimental treatment		
	1	2	3
(a)			
Area (units)	5	5	10
Number of organisms	10 A	10 B	10 A + 10 B
Density of species A per unit area	2	–	1
Density of species B per unit area	–	2	1
(b)			
Area (units)	5	5	5
Number of organisms	10 A	10 B	10 A + 10 B
Density of species A per unit area	2	–	2
Density of species B per unit area	–	2	2

A more useful, and more validly interpretable design is of the kind illustrated in Table 11.1b. The hypothesis proposed is that there is competition between species A and B, and thus the presence of B will deleteriously affect members of A, and vice versa. A sensible null hypothesis to test is that the addition of some number of B to some number of A will have no effect compared with control groups of that number of A when on their own. This null hypothesis cannot be tested by the confounded design in Table 11.1a. It is not obvious what null hypothesis could be tested by that design. The unconfounded designs (as in Table 11.1b) have been used successfully by many authors (e.g. Wilbur 1972; Lynch 1978; Hairston 1980b, 1981; Peterson & Andre 1980; Wise 1981a, 1981b; Creese & Underwood 1982).

Other forms of confounding were inevitable in some designs of experiments. Grant (1971), for example, discussed the inherent confounding of his experimental manipulations of densities of rodents with differences in habitat from one enclosure to another and one time to another. Dhondt and Eyckerman (1980) demonstrated different patterns of utilization of experimental nest-boxes by blue tits before and after the openings of the boxes were made smaller, thus preventing the larger great tits from entering them. It is difficult to attribute this finding unequivocally to the removal of competition from great tits; the blue tits rarely used the boxes when the entry holes were large before the experiment began. The experiment is confounded because of the possibility that reducing the entry hole made the boxes more suitable for blue tits regardless of the concomitant exclusion of the larger species. Miles (1972) compared the germination of seedlings in the presence and absence of existing coverage of plants. Seeds of each species were planted in separate, experimentally bared plots. On covered plots, however, seeds of four species were scattered together. Differences in germination of any species between bared and covered plots might be due to the presence of seedlings of the other species, rather than the hypothesized pre-emption of space by older plants.

11.4.2 Other problems with experiments

An extremely common problem with experimental manipulations of active, wide-ranging animals is that of maintaining the treatment after the start of the experiment. Schroder and Rosenzweig (1975) and Abramsky *et al* (1979) have described the very rapid re-invasions of rodents after they had been removed from experimental grids. Pajunen (1982) described rapid invasions of adult corixids, *Callicorixa*, into pools supposed to contain only *Arctocorisa* (although there were no data of any kind in that paper). Sometimes, the competitors can be manipulated, but the resource is mobile. To determine any effect on the behaviour and distribution of ducks, Eriksson (1979) divided a lake into two with a net, and introduced fish into one half. The food of fish and ducks (adult and larval insects) could, however, move between the two halves, and may not have been reduced in abundance where fish were present.

Great spatial and temporal variations in numbers of a population also cause problems for small-scale or short-term experiments. Thus, in Tinkle's (1982) experiments on lizards, there were large changes in numbers of *Sceloporus undulatus* in the control area where two other species were present during the study. Such variations make it very difficult to detect any differences between areas with and without competitors, unless enormous numbers of replicate areas, or very long time courses of experiments are available. Such variations can, however, be expected for many organisms and should lead to variations in the intensity or importance of competitive interactions from time to time and place to place (Lynch 1978; Hils & Vankat 1982; Underwood *et al* 1983).

From this brief account of some of the problems with experimental procedures, it is clear that many studies of competitive interactions are going to lead to, at best, ambiguous results. Greater attention to the design of the experiments will remove obvious sources of confounding. Other problems are, however, unavoidable, although they might be detected by short term pilot studies in advance of committing large amounts of time and energy. As the whole purpose of experimental manipulations of hypothesized competitors is to produce unequivocal conclusions about the presence, intensity or nature of competition, many experiments must be considered far more preliminary than has been the case to date. It is doubtful that some of the confounded comparisons in the literature are of real value as demonstrations of competition as an important process. More care in the subsequent interpretation of such experiments might lead to better designs for future studies, rather than simple uncritical acceptance of the results regardless of their inherent difficulties.

11.5 Inter- and intraspecific competition?

11.5.1 Why study intraspecific competition?

There are four major reasons why interpretations of studies of interspecific competition would benefit from simultaneous examination of intraspecific interactions among the individuals of each of the competing species. First, the detection of intraspecific competition is a good starting basis from which to build hypotheses about interspecific competition, because the similarities of requirements for essential resources must be greater (or, at the very least, as great) among members of the same species than among members of different species. Thus, when resources are in short supply, intraspecific interactions must be occurring, if interspecific competition is also to occur. The point was raised by Darwin in *The Origin of Species*: "struggle for life most severe between individuals and varieties of the same species" (heading on p. 59 of the 1882 edition). This has been discussed by many authors (e.g. Pontin 1969) and was stated explicitly as criterion 3 in Reynoldson and Bellamy's (1970) discussion of necessary and sufficient conditions for demonstrating the existence of interspecific competition in natural populations. Yet it is often forgotten, and many experiments on interspecific competition are done without any investigation of the effects of members of the same species. Among other problems caused by this omission were the confounded comparisons in experiments described previously.

Secondly, intraspecific competition is clearly important in the development of theory about competitive interactions and subsequent development of general community theory. [For examples of Lotka–Volterra competition models, see Williamson (1972), Krebs (1978), Lawton & Hassell (1981) and Schoener (1983b).] Thirdly, the magnitude and intensity of interspecific competition detected in experiments will probably depend greatly on the starting densities of the species investigated, relative to their carrying capacities, because of intraspecific as well as interspecific interactions. [See the example of a nonlinear analysis in Schoener (1983), and the discussion in Connell (1983).] Fourthly, the outcome of interspecific competitive interactions, in terms of distribution and abundances of competitors, or in terms of structure of guilds or communities, will often depend on the relative importance of intraspecific in addition to interspecific interactions. Thus, an inferior competitor amongst intertidal limpets was able to coexist where intraspecific competition amongst members of the superior competitior was sufficiently intense to keep its density below the levels at which it could exclude the inferior species (Creese & Underwood 1982). Pontin (1969) discussed the regulatory role of intraspecific competition in returning competing species back to a stable equilibrium rather than a situation where the inferior species was excluded.

Sensible designs of experiments for the simultaneous analysis of inter- and intraspecific competition have been used by several authors (Wilbur 1972; Underwood 1978; Waser 1978; Seifert & Seifert 1979; Kroh & Stephenson 1980; Fonteyn & Mahall 1981; Brown 1982; Creese & Underwood 1982). Such experimental protocols are much more informative and reliable than indirect estimations of intraspecific interactions (e.g. Abrams 1981a).

11.5.2 Designs of experiments on inter- and intraspecific competition

The general design of experiments to investigate inter- and intraspecific competition between two species is illustrated in Table 11.2. Extensions to more than two species may be found in a number of studies on intertidal gastropods by Underwood (1978), Creese and Underwood (1982); salamanders by Wilbur (1972); cladocerans by Smith and Cooper (1982) and insects in pitcher plants by Seifert and Seifert (1979). Powerful statistical analyses of such experiments are available, particularly where orthogonal contrasts can be made simultaneously amongst several sets of treatments (e.g. Wilbur 1972; Underwood 1978; Creese & Underwood 1982). Experiments are described where intraspecific components of competition are investigated at only two densities of each species (Table 11.2a), and where several experimental densities are used (Table 11.2b). The advantage of the latter type of experiment is that the relationship between competitive effects and starting densities can be unravelled. [See the nonlinear plots in Schoener (1983b).] There are, however, problems where the natural, or otherwise relevant, densities of two species are not similar (see below).

Table 11.2. Designs of experiments to investigate intra- and interspecific competition between two species. (a) Two experimental intraspecific densities: both species at same density. (b) Three experimental intraspecific densities: both species at same density.

(a)

Density	Experimental treatment				
	1	2	3	4	5
Species A	10	20 (= 10A + 10A)	10	–	–
Species B	–	–	10	10	20 (= 10 B + 10B)
Comparison:	Intraspecific competition 2 vs 1 (A on A) 5 vs 4 (B on B)			Interspecific competition A's in 3 vs 1 (B on A) B's in 3 vs 4 (A on B)	

(b)

Density	Experimental treatment								
	1	2	3	4	5	6	7	8	9
Species A	10	20	30	10	10	20	–	–	–
Species B	–	–	–	10	20	10	10	20	30
Comparison:	Intraspecific competition 3 vs 2 vs 1 (A on A) 9 vs 8 vs 7 (B on B)				Interspecific competition A's in 5 vs 4 vs 1 (B on A) B's in 6 vs 4 vs 7 (A on B)				

Interpretations of interspecific interactions would benefit from simultaneous examination of intraspecific interactions as pointed out by Connell (1983). Relatively few studies of the total examined included any attempt to investigate effects of intraspecific increases in density on each species. This is again surprising given the long time-scale over which authors have urged consideration of intraspecific effects as a necessary prerequisite or corequisite to analyses of interspecific competition (Reynoldson & Bellamy 1970). The situation has changed very little since Pontin's (1969) statement that "this point [i.e. the importance of intraspecific interactions] continues to be overlooked in discussions of interspecific competition".

11.6 Symmetry and asymmetry in interspecific competition

11.6.1 Three types of symmetry

It has recently been fashionable to discover that competition between two species is rarely symmetrical — usually one species has greater effects on a second species than it experiences from the second species (e.g. Lawton & Hassell 1981; Connell 1983; Schoener 1983b). In terms of Lotka–Volterra theory, such asymmetry occurs whenever the competition coefficient (α_{ij}'s) are unequal. In the examples of experimental designs in Table 11.2, symmetry exists when the effect of species A on species B equals the effect of species B on species A. Lawton and Hassell (1981) concluded that asymmetrical competition was the norm for insects. Schoener (1983b) emphasized the importance of asymmetrical competition and suggested several possible processes to account for superiority by one species over another. I will not discuss here the importance of asymmetrical competition to the development of theory. Instead, I suggest that symmetry/asymmetry exists in three different forms, and each has intrinsic worth for understanding the role of competition among species in natural communities. First is the interspecific symmetry considered by previous workers (i.e. the degree of similarity of effect of each species interspecifically on the other species).

Of similar importance, however, is the degree of symmetry of interspecific versus intraspecific effects of individuals of a species. In the designs in Table 11.2, this symmetry exists only if the intraspecific effect of species A on species A is equal to the interspecific effect of species A on species B. This is important for predicting the effects of changes of numbers, fecundities, weights, survival, etc. of any species when new individuals of either species invade, arrive, recruit, or are experimentally introduced. Thus, if species A is asymmetrical by having a greater intraspecific competitive effect than its interspecific effect on B, an experimental increase in densities of species A into an area where A and B coexist will cause greater negative changes in members of species A than of species B (and could therefore actually have positive effects on species B if the resulting deleterious effects on A were of sufficient magnitude to reduce the overall resulting density of A in the area). For example, consider the resource of space on a rocky shore, and the situation when space in one area is occupied by members of species A and another area is occupied by members of species B. To make reliable predictions about future events in either area, when recruitments of further A individuals occur, will depend entirely on the relative competitive abilities of A on the occupant individuals of the species in each area. If A has a greater effect on existing A than on individuals of B, then recruitment of further A into each area will cause greater change where A occupies the space than where B occupies the space.

The third type of symmetry/asymmetry also involves comparisons of inter- and intraspecific competitive interactions. This is the comparative effect of members of species A on other A versus the effects of species B on species A. Knowledge of this form of symmetry would enhance predictive power about invasions or recruitments of species. Again, consider the resource of space on a rocky shore, and the situation when space in one area is occupied by members of species A. Future events in this area, if recruitments of further A or members of some other species B might occur, will depend entirely on the relative competitive abilities of the two species on the occupant members of species A. Thus, asymmetrical effects (A on A not equal to B on A) will cause different outcomes depending upon whether individuals of species A or B arrive first.

These three types of symmetry and asymmetry have not yet been considered for the complete analysis of any interaction, mostly because intraspecific effects of competing species are often ignored in many studies. They will not be considered further here, but, if one form of symmetry is important to the development of a complete understanding of competitive interactions, so are the other forms. Differences in the relative strengths of intraspecific interactions within two species might go a long way towards explaining asymmetrical interspecific competitive abilities, but have not been considered in this context in recent discussions (Lawton & Hassell 1981; Connell 1983; Schoener 1983b). One example is the inferior competitive ability of the limpet *Cellana* compared with the snail *Nerita* on rocky shores in New South Wales (Underwood 1978). *Cellana* also suffers much more intense intraspecific competition than does *Nerita* (Underwood 1976, 1978; Creese & Underwood 1982).

11.6.2 Experimental designs for the analysis of symmetry

The types of experiments outlined earlier (Table 11.2) would produce all the information necessary for complete analyses of all forms of symmetry. There is, however, a major problem when the densities of the competing species are not the same. Consider a simple example of competition between two species that are exactly equal (symmetrical) in their competitive effects (both inter- and intraspecifically). For the purposes of illustration, assume that members of both species consume all resources in proportion to their densities (thus, all food resource in an area is divided equally amongst all individuals, of whatever species, present in the area). Finally, consider that the weight of tissue of each individual increases during a given period of time by half the excess amount of food consumed over that required for maintenance (this is not realistic, but, to keep the example simple, no adjustments are made for changing weights of the animals). Similarly, if insufficient food is available for maintenance, tissue weight declines by half the deficiency of maintenance requirement minus amount consumed. Also, assume that densities of the two species are at some different number per unit area (but combined density represents the carrying capacity) when an experiment is planned. This situation is illustrated in Table 11.3, under experiment 1, where the experimental density is eight individuals of species A and five of species B. Each then gains exactly the maintenance requirements of food, and neither grows nor decreases in weight. Relative to the experimental plots where the other species has been removed, however, the animals in this plot will not weigh as much at the end of the experiment. The differences in weights between members of each species when alone and when together will be used by the experimenter as an estimate of the competitive effect of each species on the other. Apparent interspecific asymmetry will be inevitable in such an experiment. Note that the relative effect per individual of species A is to decrease the weight of individual B by 0.05 units (i.e. 0.40 divided by the 8 A in the mixed treatment). The effect of individual B is to reduce the weight of individual A by 0.03 units (0.16 divided by 5). This asymmetry is entirely artificial and occurs simply as the result of using different densities of the two species in the mixed treatment.

In experiments 2 and 3, each species is used at the same density (5 and 8 per enclosure in the two experiments; Table 11.3). Note that different relative effects of the two species are estimated in the two experiments: B has more effect on A when there were 5 A than when there were 8 A per enclosure. This is consistent with the non-linear analysis described by Schoener (1983b). In each of these experiments, however, the interspecific effects of each species on the other are symmetrical, as they should be from the situation modelled. Interspecific

Table 11.3. Symmetry and asymmetry in interspecific competition as a function of experimental design. Individuals of species A and B are equivalent (symmetrical) competitors needing 0.5 units of resource for maintenance. Each gains or loses weight as half the excess or shortage of resource compared with that needed for maintenance. The natural mean density in the area is 8A and 5B per unit area, and there are 6.5 units of resource available.

	Experimental density		
Experiment 1	8A	8A + 5B	5B
Consumption by individual A	0.81	0.50	–
Change of weight of A	+0.16	0	–
Change relative to 8A alone	–	−0.16	–
Consumption by individual B	–	0.50	1.30
Change of weight of B	–	0	+0.40
Change relative to 5B alone	–	−0.40	–
Experiment 2	5A	5A + 5B	5B
Consumption by individual A	1.30	0.65	–
Change of weight of A	+0.40	+0.08	–
Change relative to 5A alone	–	−0.32	–
Consumption by individual B	–	0.65	1.30
Change of weight of B	–	+0.08	+0.40
Change relative to 5B alone	–	−0.32	–
Experiment 3	8A	8A + 8B	8B
Consumption by individual A	0.81	0.41	–
Change of weight of A	+0.16	−0.05	–
Change relative to 8A alone	–	−0.21	–
Consumption by individual B	–	0.41	0.81
Change of weight of B	–	−0.05	+0.16
Change relative to 8B alone	–	−0.21	–

symmetry can only be properly detected when the same densities of the competitors are examined — because of the simultaneous effects of intraspecific competitive interactions among the individuals of each species.

It is not clear how much such confusion might have influenced conclusions about symmetry in previous analyses (Lawton & Hassell 1981; Connell 1983; Schoener 1983b), but all of these reviewers found a preponderance of asymmetry, but did not discuss the relative number of each species in the experiments. Some of the experiments (e.g. Peterson & Andre 1980; Williams 1981) certainly used different numbers of the competing species. Others did not (e.g. Brown 1982; Smith & Cooper

1982). Conclusions about the prevalence of symmetry and asymmetry in the literature are premature until the potential effects of experimental designs have been considered carefully. This does not seem to have been the case to date. As suggested earlier, the analysis of intraspecific competition in relation to interspecific competition is also made more difficult when different densities of two species are to be used to cause experimental mixtures.

In conclusion, symmetry, in three different forms, is important to allow realistic evaluation of the nature and importance of competitive interactions. Experiments must be more carefully designed with hypotheses about symmetry firmly in mind before any useful conclusions could be drawn about the existence of, or explanations for, symmetrical or non-symmetrical competitive abilities. Hypotheses about symmetry must, presumably, relate to the competitive effect per individual of competing species. Therefore, before any conclusion can be drawn from the literature, some manipulation of experimental results is necessary to correct for the numbers of each species, where densities of the competing species were different.

11.7 Conclusions

The major conclusion that emerges about experimental analyses of interspecific competition is a bewilderment that there are so many problems of design, implementation and interpretation in so many experiments. This does not include problems with such things as the statistical analyses (Underwood 1981; Hurlbert 1984) that have yet to be discovered. All in all, the message for future historians of ecology is 'do not look — spare yourselves'. After all, the bones of the principles of experimental design are at least 50 years old, but have not yet been fleshed out in the corpus of ecological investigation. This is despite the very recent development of ecological experiments on interspecific competition (Schoener 1983b). There are far too many problems with replication and controls, with unclear relationships between hypotheses and tests, and with confounded comparisons for us to be complacent. With Hurlbert (1984), I bemoan the poor state of experimentation, and suggest most forcefully that Schoener (1983b, p. 241) *was* premature to try his analysis, and that Connell (1983) was being excessively optimistic to assume that it was only when field experiments were *few* that the "tendency was to use this small amount of information uncritically".

One very clear conclusion is apparent — when evidence for competition is sought, seemingly any evidence will do. Before the trend to do experiments, competitive interactions were often invoked to explain many patterns in nature, without consideration of other equally tenable hypotheses. Now, regardless of the rigour, or lack of it, of an experimental design, competitive interactions are inferred from most experiments — even when other equally tenable hypotheses have not been eliminated by the experiment. A particularly obvious example was provided by De Long (1966, p. 481) who found a negative correlation between the densities of house mice and meadow mice in his sampling areas. He then manipulated the densities of *Microtus* "to see if the negative correlation between the densities of the two species was due to interference by *Microtus*, or was merely spurious". It could, of course, have been real — but due to other causes than competition. Other workers have not been so obvious about ignoring other possible models and hypotheses. If this state of affairs continues, there will have been little point in the experiments. Many will have been planned, carried out and interpreted solely as confirmations of existing hypotheses. [Aristotelian experimentation as defined in Medawar (1969), see also the thoughtful discussion of Dayton (1979).]

The present analysis confirms the worst fears of Hurlbert (1984). To paraphrase him, poorly designed and ill-conceived experiments on competition have literally flooded the literature during the last few years. This parlous situation can, of course, only come about with the active connivance of grant agencies, referees of papers, editors of journals, chairmen of scientific meetings, referees for promotion, selection committees for employment, compilers of reviews of the literature and the general ecological community. Why are we so afraid of constructive criticism about experimental work? Have we nothing new to learn about nature, and therefore know all there is to know about experiments? Perhaps it is time for experimental ecology to adopt a harsher judgement of poor work and reject it as inadequate, sloppy and open to numerous alternative explanations and conclusions than those reached about competition. Many of the problems could have been avoided by better designs of experiments, and therefore by more critical appraisal during the planning stage (by various of the categories of person listed previously). As a step to rid ecological theory of the encumbrance of dealing with inadequate experimental tests of former pieces of theory, and in the hope of moving towards an improved future set of experiments (rather than continued uncritical acceptance of the present examples followed by inadequacy becoming the norm for future work), the time is at hand to reject all unreplicated, uncontrolled, confounded, or otherwise inadequate experiments from further consideration. As a preliminary move towards this, papers that have offered sensible, competent or valuable description, discussion or interpretations about the design, planning and analysis of experiments on competition in the field have been listed (Table 11.4). Most of these papers probably also contain faults, but each has something to offer. This is obviously a personal choice amongst the papers sampled — many other papers are also valuable, but were not included in this

Table 11.4. Studies that provide useful experimental designs, discussion or interpretations of field experiments on competition.

Study organisms	Author
Desert plants	Friedman (1971); Friedman & Orshan (1974), Friedman *et al* (1977), Inouye (1980)
Old-field plants	Pinder (1975), Allen & Forman (1976), Gross (1980), Hils & Vankat (1982)
Heath plants	Miles (1974)
Prairie plants	Petranka & McPherson (1979)
Grassland plants	Putwain & Harper (1970)
Winter annuals	Raynal & Bazzaz (1975)
Jack-pine	Brown (1967)
Paramecium	Gill & Hairston (1972)
Pond-snails	Brown (1982)
Sea-pansy	Kastendiek (1982)
Fouling organisms	Sutherland (1978)
Anemones/barnacles, etc.	Taylor & Littler (1982)
Bivalves	Peterson & Andre (1980)
Marine gastropods	Creese & Underwood (1982), Underwood (1978)
Urchins/fish	Williams (1981)
Bees/flies	Morse (1981)
Beetles	Wise (1981a)
Spiders	Wise (1981b)
Salamanders	Hairston (1980b, 1981), Wilbur (1972)
Lizards	Smith (1981)
Rodents/fungus	Inouye (1981)
Rodents	Joule & Jameson (1972), Redfield *et al* (1977)

survey. I suggest these as the examples to be consulted early in the planning of future experimental work. I also urge all ecologists to read more about the design of experiments, and to consult cautionary reviews such as that by Hurlbert (1984).

The resources for which competition might occur and the mechanisms used by potential competitors to gain adequate amounts of resources are well reviewed by Connell (1983) and Schoener (1983b). These reviews and the present discussion confirm that competitive interactions can be investigated, interpreted and understood by use of manipulative field experiments. Many of the examples in Table 11.4 demonstrate this (although some demonstrated, equally validly, that competitive interactions were not important in some systems). The validity of the experimental procedures, however, depends entirely on the care with which the hypotheses being tested are defined. How closely the experimental conditions represent reality depends entirely on a clear understanding of the reality under investigation. The adequacy of controls for experimental manipulations is entirely determined by the experimenter. The need for replication in any field experiment is paramount — numerous sources of natural variation are already well-documented for most types of habitat and organisms. When questions are addressed about how widespread competition is in some ecosystems, more replication in space and time is needed. Otherwise, the question cannot be answered, even by the most elegant experiment at one site or time. The criticisms offered here are not to dissuade ecologists from attempting experiments, but to provide a framework in which improved methods for interpreting nature can be developed.

The papers listed in Table 11.4 discuss experimental designs, adequacy of controls and replication with respect to a wide variety of organisms, habitats and interesting hypotheses about competitive interactions and their role in structuring natural communities. There exists no more valuable guide to the methods available to overcome problems with field experiments.

Competition is undoubtedly important as an organizing force in the structure of communities and in the evolution of the organisms in many communities, as revealed by the syntheses of Connell (1983) and Schoener (1983b). Good, unequivocal evidence of this has been provided by many field experiments. The less valuable experiments are not needed to demonstrate the generality of competitive interactions in many types of environment.

Nevertheless, the prevalence of interspecific competition is not as well understood as suggested by the large number of experimental papers currently available in the literature. This situation will only be improved by a series of better experiments, not by further analysis of the existing ones. The smaller number of studies that have produced unequivocal results demonstrates that it is possible to do good field experiments on competition, and that such experiments still provide a powerful tool for unravelling complex biological communities. The instruments are, however, badly blunted by consistent, widespread, incorrect usage.

12 Ecological Succession

Derek J. Anderson

12.1 Introduction

Although some theorists undoubtedly would subscribe to Lawton's (1974) wry observation that "Ecology suffers from a surfeit of fascinating but apparently unrelated observations, superimposed upon an acute shortage of general theories", many ecologists would certainly argue that the concept of succession provides an exception to this generalization. The concept of the development of a biotic community or ecosystem over time has been a fundamental tenet of ecological theory; so much so that Margalef (1968) felt able to assert that "Succession in ecology occupies the same place as evolution in general biology".

Whether succession is defined as "the gradual change which occurs in vegetation of a given area of the earth's surface on which one population succeeds the other" (Tansley 1920) or more succinctly as "an ecocline in time" (Whittaker 1975), most ecologists are likely not only to accept that change in vegetational composition is a universal phenomenon but also that it is important and utilitarian in some contexts to predict the successional changes consequent upon such selective forces as grazing, logging or fire. In short, a comprehensive theory of succession is important both as a central tenet of ecological theory and as a potential management tool in the sphere of natural resource management.

Thus, as a general conceptual construct succession occupies a centrally significant place in the ecological literature. Curiously, however, although it has spawned a considerable volume of data and generalization, it has also generated much philosophical controversy; and surprisingly, for all that its obvious expression and ubiquity in real-life environments are accepted, an all-encompassing and definitive interpretation of the successional process seems still elusive.

12.2 Development of successional theory

As Whitmore (1982) and others have emphasized, the idea of ecological succession was well embedded in the writings of foresters dating from the eighteenth century, and the term was used expressly by Thoreau in 1863 in relation to the forests of New England. A little more than half a century later and on the opposite side of the same continent, Clements

(1916) presented his primary successional sequence with its component nudation, migration, ecesis, competition and reaction as modular processes leading to a regionally-characteristic, climatically determined, self-maintaining and regenerating, stable vegetation-type. The essence of this model of succession was that of directional change, with each successive stage (together forming the sere) being established as the result of the accumulated reactions on the habitat of earlier stages. This facilitation model was essentially deterministic, in that it was supposed that perturbation of any stage in the sequence would result in a steady rebuilding of that stage and a return to the trajectory leading to the regional monoclimax.

It is a familiar fact of ecological historiography that Clements's views were challenged then, as they are increasingly in the modern literature, by ecologists who saw either contrary, or no evidence for regional convergence, linear determinism or indeed the organismal structure Clements wished to impose on his monoclimax (e.g. Gams 1918; Gleason 1926; Du Rietz 1930). Many contemporary authors drew attention to the obviously different vegetation-types developed over contrasting soils and topographies within the one climatic zone as evidence for regional polyclimaxes, which then became widely accepted as a model of succession, particularly in the North American literature (e.g. Cain 1947). In essence the polyclimax theory distinguished a mosaic of edaphic, topographic, ecoclimatic, firing- or grazing-derived communities which with their self-perpetuating species populations were supposed to form the relatively stable endpoints of secondary successions.

Gleason's well known attack on Clementsian determinism was predicated on the view that the properties of vegetation depended absolutely on the properties of the individual plants it contained, and that no two vegetation samples were exactly alike either in quantitative or even qualitative composition. This 'individualistic' hypothesis was shared by some European workers (e.g. Ramensky 1930), although it was widely appreciated, as Gleason (1939) himself emphasized, that similar histories of migration and environmental development tend to produce samples of vegetation with a close species correspondence.

Apart from the organismal theory of vegetation (Chapter 1, Section 1.5), the 'facilitation' or 'relay floristics' model of succession which Clements espoused was not endorsed by several studies of old-field succession which demonstrated the existence of soil-stored seed of species represented in each of the later stages of succession (Beckwith 1954; Egler 1954; Drury & Nisbet 1973; Niering & Goodwin 1974). These authors stressed that most species appearing in successions developed on abandoned fields were mostly present at the outset as soil-stored seeds, rhizomes and regenerating roots. In the 'initial floristic composition' model, as Egler called it, successional changes in vegetation were the outcome of differential appearance, growth, reproduction and survival of species already present in the soil 'bank'. On this view of succession determinism is replaced by probabilism as a generating mode of changing composition; the successional 'niche' of any species is determined by its reproductive behaviour, its growth-form, growth potential and its ability to escape, tolerate or impose competitive selection.

In retrospect it may seem curious that so much emphasis was placed on deriving generalizations from successional trends which were perceived to operate at relatively large scales — that is to say, mostly at the association level — when in fact much of the observable data on successional phases provided particularly by foresters was garnered from much smaller spatial scales (e.g. Watt 1919). Whether the primary succession seemingly so successfully diagnosed by Cowles from his zonal studies of the Lake Michigan sand-dunes (Cowles 1899) created an intellectual mould too readily adoptable for uncritical mimicry is hard to assess, but there is no doubt from Walker's (1970) critical and revealing study that many 'hydroseral' sequences in Britain were given quite unwarranted status as putative successions on the basis of a supposed temporal connection between spatially contiguous vegetation zones on the open water–land interface of some lakes.

Apart from the central question addressed by Walker (1970), his approach serves to highlight both the inadequacy of much of the evidence accumulated to support putative successional sequences and the seeming intellectual need of the ecological community for both a simple and unifying model of the successional process.

The disparate nature of evidence brought to bear on successional theory requires little elaboration or emphasis. It has been derived, at a variety of spatial scales, from (i) the linking of spatial mosaics on adjacent sites to reproduce a postulated temporal sequence of phases in the same succession; (ii) detailed examination of small-scale vegetation mosaics overlying preserved fragments of antecedent phases; (iii) examination of stratified and preserved subfossil material, which from its mixed nature usually provides evidence of change on at least two spatial (local and regional) scales; (iv) direct historical evidence; and (v) experimental manipulation. Few theoreticians have seen fit to draw fully and simultaneously on this wide spectrum of evidence, and even present-day attempts to model the phenomenon of succession, with some notable exceptions, tend to emphasize but one dimension of this wide evidential canvas.

The intellectual need (as opposed to a recognition of the utility) for a generalized model of succession is evidenced in the modern literature in a number of ways, but two examples will suffice to illustrate the point that ideas generated three-quarters of a century ago are still potent as a basis for conceptualization.

John Curtis, who rediscovered the technique of ordination as an elegant method of summarizing variation in regional vegetation (Curtis & McIntosh 1951; Curtis 1959), and who clearly stood intellectually in line with Gleason as an ecologist who saw individual rather than collective species responses to environmental gradients (continua), nevertheless saw a need to ascribe *climax adaptation numbers* to species aligned along those gradients which were thought to be primarily reflecting successional trends. Quite apart from any methodological need for an appropriate weighting schema, there is implicit in the 'adaptation' term a notion that each species has a more or less precise niche in the successional scheme of things — an implication which seems more closely related to the facilitation model of the Clementsian school than the initial floristic composition model of the Gleasonian stream.

Following a quite different strategic approach, although one with quite clearly Clementsian antecedents, Odum (1969) has produced a functional model of succession by considering how community (organismal?) properties may emerge during the course of a succession:

> "Ecological succession may be defined in terms of the following three parameters. (i) It is an orderly process of community development that is reasonably directional, and, therefore, predictable. (ii) It results from modification of the physical environment by the community; that is, succession is community-controlled even though the physical environment determines the pattern, the rate of change, and often sets limits as to how far development can go. (iii) It culminates in a stabilized ecosystem in which maximum biomass (or high information content) and symbiotic function between organisms are maintained per unit of energy flow. In a word, the 'strategy' of long-term evolutionary development of the biosphere — namely, increased control of, or homeostasis with, the physical environment in the sense of achieving maximum protection from its perturbations".

Although to some authors such a model seems self-evidently incomplete as a model of succession (e.g. Drury & Nisbet 1973), the ecosystem attributes predicted on the basis of the model have been used widely as an exemplar of community properties emerging as succession proceeds towards more mature and complex stages. Not least, the Odum model has helped to nurture the view that structurally complex, *K*-selected, species-rich, climax vegetation is 'stable', a view which is consonant with arguments based on other criteria for a potentially pervasive 'balance of nature' (e.g. Vesey-Fitzgerald 1965; Costello 1969; McNaughton 1977). More recent contributions to the debate on successional theory have accepted a position somewhere between the original 'organismal' view of Clements and Gleason's polar 'individualistic' concept of vegetation. Drury and Nisbet (1973), for example, argue that most of the perceived phenomena of successional sequences can be ascribed to differences in colonizing ability, growth and survival of species pre-adapted to exist in environments distinguished by different suites of resource combinations. They argue that

> "The structural and function changes associated with successional change result primarily from the known correlations in plants between size, longevity and slow growth. A comprehensive theory of succession should be sought at the organismic level, and not in emergent properties of communities".

Such an approach — a preliminary comparison of ecophysiological attributes of species characterizing developmental or mature successional stages in both temperate and tropical forests — has been made by Bazzaz (1979) and Bazzaz and Pickett (1980), although it is clear from these reviews that more comparative data are required before the fine detail of species successional roles can be elaborated into a coherent predictive theory of succession.

Perhaps the nearest approach to such a coherent theory, developed in order to "analyze in simple terms the processes which control the structure and composition of vegetation" is that reported by Grime (1979). Grime's central thesis is that all plants exhibit one of three broad strategies in response to habitat *stress* (those environmental phenomena which reduce photosynthetic production) and habitat *disturbance* (the partial or total destruction of photosynthetically-derived biomass by biotic or physical agents) (Table 12.1).

Table 12.1. Strategies in response to habitat stress and habitat disturbance (from Grime 1979).

Intensity of disturbance	Intensity of stress: Low	Intensity of stress: High
Low	Competitors	Stress-tolerators
High	Ruderals	(No viable strategy)

Grime has expanded this basic model to illustrate how successional trajectories for environments differing in innate productivity, and the related development of biomass and sequence of life forms, might be predicted from this strategic consideration (Fig. 12.1).

In the most productive habitat (P_H), succession is characterized by a middle phase of intense competition in which competitive herbs, then competitive, shade-intolerant shrubs and trees come to dominate the vegetation. The climax phase is one in which stress-tolerance becomes more important as shading and nutrient-stress (induced by large, long-lived forest trees which cast deep shade and sequester nutrients) come to dominate the vegetation. In less productive habitats (P_I) the appearance

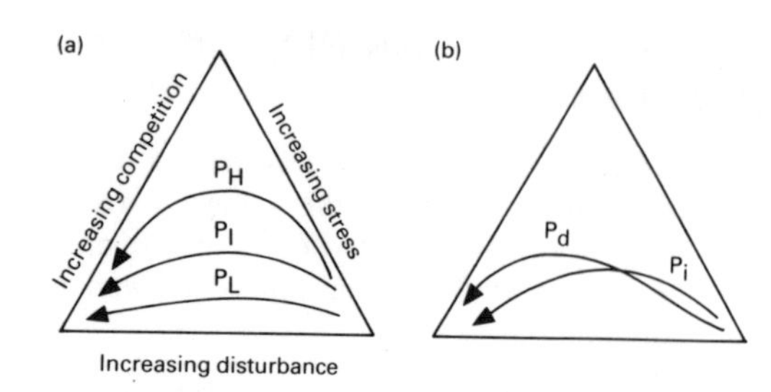

Fig. 12.1. Postulated trajectories of vegetation succession in areas with differing innate productivities. (a) In conditions of high (P_H), moderate (P_I) and low (P_L) potential productivity. (b) In conditions of increasing (P_i) or decreasing (P_d) productivity (after Grime 1979).

of highly competitive species is inhibited by the earlier onset of resource depletion and the stress-tolerant phase. Thus in Grime's analysis of successional mechanisms the predictive potential of his model is predicated on an appreciation of the innate productivity of particular seral habitats.

In contrast to Grime's approach Connell and Slatyer (1977) base their models of post-colonization succession on (i) species abilities to 'prepare the ground' for the advent of later successional species (facilitation), (ii) species abilities to colonize open sites and to maintain their captured space against all later would-be colonists (inhibition) or (iii) a collective ability to explore differentially low levels of resources (their tolerance model). Perhaps most significantly, Connell and Slatyer assert that, while they recognize some examples of the facilitation model and much evidence for the operation of the inhibition model, there exists little evidence in support of the tolerance model: a statement in stark contrast to Grime's emphasis on stress tolerance as a major component of his model.

A third major approach to the development of a coherent model of succession derives from a view of dynamic vegetational change as being probabilistic rather than deterministic, and thus being amenable to representation as a Markovian replacement process (Horn 1976). Given some background data on the gross 'adaptation' of a particular species assemblage, a table can be constructed which indicates the probability that a given plant will be replaced by one of its kin or by an individual of another species within a specified time period. The outcomes of these transition matrices are dependent on the fact that the proportional representation of a species in the stationary (stable) phase of the succession is independent of the initial vegetation composition in the sense that it depends only on the matrix of replacement probabilities. The biologically significant questions relate to the preparation of a matrix of replacement probabilities, which are dependent on a variety of idealized structures such as chronic and patchy disturbance, obligatory (facilitated) succession and the development of a competitive hierarchy.

A conceptually related, although essentially qualitative approach to the prediction of likely replacement sequences has been developed by Noble and Slatyer (1978, 1980) in which they seek to identify those attributes of a species which are essential to its active role in such sequences. The 'vital attributes' are grouped under three heads: (i) the method of arrival or (*sic*) persistence of a species at a site during and after disturbance, (ii) the ability of a species to establish and grow to maturity in a developing community and (iii) the time taken for a species to reach 'critical life stages', which include time to reproductive maturity, lifespan in an undisturbed community and the time period for all propagules to be lost from a community in those species which may become locally extinct. Using such an approach these authors derive a number of transition diagrams for various species 'types' — those that can dominate a particular post-disturbance phase — showing the sequence of life stages that a species will pass through as the result of a particular disturbance regime. The model has been employed to recreate successional sequences in which a few species rise to dominance seriatim (e.g. a Tasmanian example of a wet sclerophyll-rainforest system, cf. Jackson 1968), but in its original, qualitative form the model failed to reproduce satisfactorily the multi-dominant cycling phase complex of *Betula–Acer–Fagus* hardwood forest described by Forcier (1975).

Perhaps one of the most surprising points to emerge from this brief survey of the 'model' succession literature is that many land managers who presumably depend on an understanding of successional sequences for manipulative management see little reason to adopt either the models or the terminology of 'professional' ecological theory. For all their contributions to the successional debate, foresters' terminology seems curiously recalcitrant, with a persistent emphasis on such terms as 'gap-phase' and 'selection forest'. Range managers seem even further removed from the terminological mainstream, with their emphasis on species roles as invaders, increasers, decreasers, retreaters...and even neutrals (e.g. Vogl 1974). If so central a theoretical construct in community ecology as successional 'modelling' can be safely ignored by practical natural resource managers, perhaps it is fair to agree with White (1979) that "A need for a better perspective on the temporal dimension of communities is apparent".

12.3 Towards coherence in successional theory

12.3.1 The problem of definition

If we accept as axiomatic the notion that dynamic change, at least when judged against some relevant time-frame, is a universal characteristic of all vegetation, it is less readily self-evident that all such change is to be regarded as successional.

Tansley's (1920) circumscription "the gradual change which occurs in vegetation...on which one population succeeds the other" presumably cannot refer to successional progression in a monospecifically-dominated piece of vegetation, since it is presumably the replacement of species-populations rather than intraspecific age-class replacements that are the essential foundation of successional sequences. And if one looks at more modern prescriptions such as that of Lincoln *et al* (1982), wherein ecological succession is defined as "The gradual and predictable process of progressive community change and replacement, leading towards a stable climax community; the process of continuous colonization and extinction of species populations at a particular site...", it is not immediately obvious what the key diagnostic features of this fundamental ecological replacement might be. Must all individuals of a species have been replaced or supplanted before succession can be deemed to have occurred? Must the replacement process be totally and repeatably predictable? Are there in all successional sequences simple features which serve to identify if not diagnose a stable, climax community where it allegedly occurs?

These elementary but nonetheless fundamental definitional questions are further complicated by the notions that successional sequences can be themselves usefully classified as either primary or secondary, autogenic (endogenous, somatic) or allogenic (exogenous, external) in character. Whether or not such descriptors are essential to a fuller definition of succession, or rather are presumed to add predictive significance to an already recognized successional sequence, also becomes an important decision to resolve if the general phenomenon of succession is to be interpreted constructively. Colinvaux's (1973) tongue-in-cheek but nonetheless telling comment "From the beginning there was the problem of succession" encapsulates the need for care in dissociating folklore from science, and acceptable universalities from unwarrantedly extended local truths, in attempting this resolution.

I intend to approach this problem somewhat tangentially by accepting my own advice (Anderson 1971), namely by attempting to establish elements of a qualitative model as a basis for constructing an operationally effective definition and a coherently hierarchical interpretation of succession.

12.3.2 Patchwork quilt environments

Although it scarcely needed the advent of remote-sensing satellites to impress on field ecologists just how patchy is the mosaic of earthly environments within which biota are sorted into recognizable communities, such imagery has served to highlight, in a way descriptive prose never could, the essentially heterogeneous nature of the global surface. Despite many an innate desire and reported attempt to recognize homogeneity and uniformity within variously designated classes of landscape or environment, the reality is that a vast majority of global landscapes and environments are more or less spatially patterned in a wide variety of the parameters that are used to characterize them.

The scale or scales at which such spatial patterning can be perceived — the scales at which environmental scientists choose to recognize boundaries (whether discontinuities or transitions) between defined parameter states — are dependent in the main on geological history and the manner in which the products of that history have been moulded and redistributed by geophysical and geomorphological processes. The areal extent of the geological fabric, the nature, extent and intensity of the moulding processes, together with the pattern of atmospheric events, determine both the scale and amplitude of the patterned physical environment.

The essential features of the broad-scale patterning are readily recognized at both global and regional scales. Global patterns are familiar to us not only in terms of biogeographical classifications of the vegetation mantle but also in terms of the analytical attempts (e.g. Paterson 1956) to predict primary productivity from a latitudinally- and altitudinally-associated combinatorial matrix of physical constraints on plant growth and development. Inter- and intraregional patterning of environmental mosaics similarly have received analytical attention through a variety of approaches, but particularly those associated with various forms of gradient analysis, ordination and classification. [Greig-Smith (1983) provides an expanded summary account of these approaches.]

The heterogeneity of intraregional and community-scale environments has been exposed and to some extent categorized by a variety of statistical approaches, but those which have been predicated on scale-dependent analyses (e.g. Greig-Smith 1961; Noy-Meir & Anderson 1970) have served to emphasize the wide range of scales on which many factors of the physical environment are far from uniform.

Two important generalizations emerge from these elementary considerations. The first is that the global patterning of physical environments can be expected to impose upper limits on productivity and, within a given time-frame, biomass, in any particular habitat constrained roughly by its latitude, altitude and relative continentality. This primary constraint will therefore impose collateral limits on the range of growth-forms that can be accommodated, and therefore the potential diversity of physiognomic structure which will be developed in a particular area bounded by those limits.

Secondly, such evidence as we have points to the ubiquity and normality of patterned and heterogeneous, rather than uniform and constant, environments over a wide range of scales over the earth's surface. Patchwork quilt, patchy, harlequin, mosaic environments — however they may be terminologically characterized — are neither uncommon nor scale-invariant; rather, they are the basic warp of the environmental fabric within which we have to weave the process of ecological succession.

12.3.3 *Time and the impact of history*

It has been argued elsewhere (Osmond *et al* 1981) that a comprehensive explication of biological phenomena is both constrained by, and can only be achieved within, the context of an appropriate spatio-temporal continuum (Fig. 12.2). Equally well, it follows that any process which involves "continuous colonization and extinction of species populations at a particular site" can similarly be open to sensible interpretation only when an appropriate investigative time-frame and relevant spatial scales have been defined.

Given this premise, we can employ the concept of a time-frame in two particular ways. Initially, we can use it to help contrast those temporal changes which accord with the oscillatory period for effective regeneration (or regenerative replacement) of individual species populations with those (presumably longer) time scales that are associated with serial replacements of species populations. In other words the definitive identification of a successional sequence must depend in part on a recognition of the time scale of presumed cyclic regeneration phases of a particular successional stage relative to that required for a floristic shift in dominance within a successional sequence. Since the time scale of such regeneration phases may vary by at least two orders of magnitude (contrasting the 'turnover' period required for the regeneration of a population of ephemerals with that required for a long-lived tree population, say), it follows that an extant ecologist can in many instances seek hard evidence

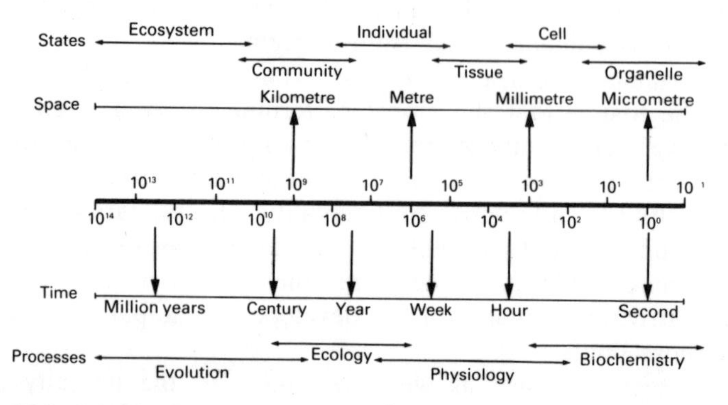

Fig. 12.2. Arbitrary scaling of state dimensions and relaxation times of processes in the biological continuum. Successional studies are broadly confined to the scales of $10^5 - 10^{10}$ (after Osmond *et al* 1980).

concerning successional change only from the historical record — itself a complex continuum ranging from the physicochemical residua of partially fossilized biota through to documented archival evidence relating to past biological communities.

Although the historical record — and particularly the palynological record — is frequently too short on detail to provide the fine-grained resolution we might cherish, regional-scale comparisons can provide evidence not only of vegetational change *per se*, but equally importantly, evidence relating to relative environmental stability or secular change which corresponds with such shifts in biotic composition. It is largely through this evidential avenue that we came to recognize the impact of environmental selective forces imposed not only by secular climatic change but also by anthropogenic cultural pressures, and particularly the latter as potent forcing functions of many so-called 'secondary' successions.

The historical record also tells us — given that it is sufficiently long in relation to major geological and climatic shifts or discontinuities — how long a regional environment has been in existence to provide for the floristic diversity and background against which modern selective processes have to operate. The historical origins of diversity are important to our cause not so much because of their evolutionary importance, but because it is this diversity from which a proximate cause of succession — the development of competitive hierarchies between species populations — must itself be selected.

In some instances the historical record — particularly when the sampling of that record is undertaken at particularly relevant scales — also serves to highlight the fact that some species populations can leave an indelible mark on the environments they inhabited long after those original populations have disappeared (e.g. D.J. Anderson 1965a). Since modern biota also frequently leave their imprimatur as marked modifications of the environments they themselves colonized (e.g. Charley & West 1975; Malik *et al* 1976), it seems fair to highlight the important generalization that an ecological site is itself likely to undergo a detectable change in its environmental characteristics. The site modifications are not only caused by environmental change but are also effected by virtue of its being occupied by biota which, through their innate physiologies and life-histories, are reactive and redistributive as well as exploitative. If the physical world is generally patchy the inhabited world is necessarily more richly textured still.

12.3.4 The development of competitive hierarchies

Although a comprehensive and universally agreed definition of succession may not be easy to achieve, a common element of most published definitions involves the concept of serial species replacements. In any given environment subject to reasonably constant selection, and assuming that the process of facilitation is not a consistent feature for all such seral stages, the most likely mechanism for a predictable replacement process involves later successional species out-competing earlier colonizing species, presumably by interception or sequestration of some important part of the resources available within the total environmental constraints which are operational at a particular site.

Any non-equitability of resource utilization by individuals in a community within a particular time-frame must lead to the development of differential dominance categories — dominants being defined as those individuals which have most impact on their neighbours while being themselves least affected by their own neighbours (cf. Greig-Smith 1957). The development of such dominance hierarchies — and their natural consequences, levels of suppression — are familiar both to experimental population ecologists and to practising foresters. An elementary *modus operandi* for the successional process therefore does exist, but its ubiquitous relevance as a singularly important successional mechanism must remain in question for at least three quite different reasons.

Firstly, if an earlier successional species population is to be subsequently and consistently dominated by a later seral species, their spatial intermingling must be such that these interactions can occur significantly often. In high latitude environments, with relatively short post-geological catastrophe histories and therefore relatively depauperate floras, such interactions are more likely to occur since the range of potential (growth-form) dominants is relatively small. It is therefore not surprising that the range of remarked dominants in temperate forests is itself small, since there is an order of magnitude difference in tree species diversity in temperate regions (< 10 species/ha) compared with species diversity in tropical rainforest (> 100 species/ha: Longman & Janik 1974). Notwithstanding the variety of reasons that have been put forward to account for this comparative species richness in rainforest, the important consequence in this context is that consistent spatial intermingling between the potential dominants and the suppressible species in species-rich vegetation is necessarily more unlikely. It is therefore not surprising to find that 'dominance' is seemingly much attenuated or even absent in tropical forests (cf. Colinvaux 1978), since it is in these ecosystems that pattern is at its most diverse (Whitmore 1982).

Secondly, there are in the modern world many instances where potentially dominant species are themselves suppressed by the imposition of continual environmental selection — as occurs for example in pasture-lands subjected to herbivorous selection or in vegetation subjected to periodic fire. Thus relatively species-depauperate but high density and physiognomically contrasted communities, within which dominance and suppression rankings might be repeatedly and therefore predictably established in the absence of external interference, may have such development thwarted by the impact of externally energetic selection. To put the point in another way, it is possible to see the development of dominance as a mechanism by which stored energy is allocated so as to most effectively intercept or sequester potentially shared resources. If that energetic allocation is reallocated or destroyed by the impact of externally energized selection, then innate dominance hierarchies will necessarily collapse.

Thirdly, it is a generalized feature of spatial pattern in those age-structured, perennial populations that have been analysed, that the greatest degree of patchiness is to be found in the younger elements of the population while the older cohorts exhibit progressively less pattern and eventually exhibit randomness (e.g. Greig-Smith & Chadwick 1965; Malik *et al* 1976). This finding in turn suggests that any forcing functions responsible for the biotic interactions underlying the successional process are likely to operate *before* the more obvious physiognomic dominance is perceived by an observer. This is based on the view that randomness is the spatial disposition of individuals "which results from a lack of response of any one individual to any other or to its environment" (Taylor *et al* 1978).

12.4 The future of successional theory

If there is one unequivocal conclusion that is derived directly from this brief survey and discussion it is that no one of the paradigms yet provided as an 'explanation' of succession seems really adequate as a univerally valid theoretical or utilitarian model of the phenomenon: White's (1979) gently phrased request for "a better perspective on the temporal dimension of communities" is clearly an unnecessarily modest call to arms.

Although it is not an original observation to suggest that theories on succession have been heavily context-sensitive and maybe for that reason over-generalized in their development and exposition, it seems likely that, irrespective of hemisphere, local geographical experience has made a considerable impact on those elements of a theory that have seemed critically important to their advocates. The fact that 'classical' successions based on experience gained in temperate forests were reported in the earlier literature from a range of equatorial rainforests is a reflection of the fact that many such successions were as often perceived on the basis of past experience as on the original analysis of hard-won local data. As Flenley (1979) was moved to remark after reviewing some relevant case histories, "equatorial successions in general have been more variable and less predictable than was formerly thought".

12.4.1 Development of dominance hierarchies and segregation

The fact that unequivocal successional sequences are not necessarily self-evident to critical observers in all vegetation types should give us cause both to pause and consider why this should be so. As one point of contrast, it is very much easier for a northern hemisphere, temperate zone botanist to 'know his plants', compared with his lower latitude colleague faced by the bewildering species diversity of an equatorial rainforest. The ease of recognizing a relatively species-poor flora stems from the fact that either that flora has had a short history of migration and/or evolution, or because an originally more diverse flora has been subject to intense and continued ecological selection sufficient to bring about significant species extinctions. Such differences in developmental background may well account for the major differences in dominance: diversity structures now widely recognized in plant communities (Fig. 12.3).

If we compare the contrasting dominance–diversity structures illustrated in Figs 12.3(a) and (c), it is not difficult to imagine a situation in which a few dominant species with relatively high importance values (Fig. 12.3a) will have more opportunities for recurring interspecific contacts. Recurrent contacts are likely to lead to better structured competitive hierarchies, and if such hierarchies are themselves repeated on a significant number of occasions throughout the regional environmental mosaic, then an emerging 'successional' sequence is likely to become apparent *at this scale.* By contrast, the situation reflected in Fig. 12.3(c) is more likely to lead to a wider range of interspecific contacts within a spatially comparable area; this in turn will reduce the likelihood of recurrent interspecific interference and thereby constrain the emergence of well-developed competitive hierarchies.

This initially simplistic comparison is subject to a number of further qualifications that are related to the operational scales of environmental (habitat) heterogeneity and particularly to the relevant scale(s) of regeneration niches (Grubb 1977). The point that requires emphasis here is that the pattern and scale of regeneration gaps or patches may well

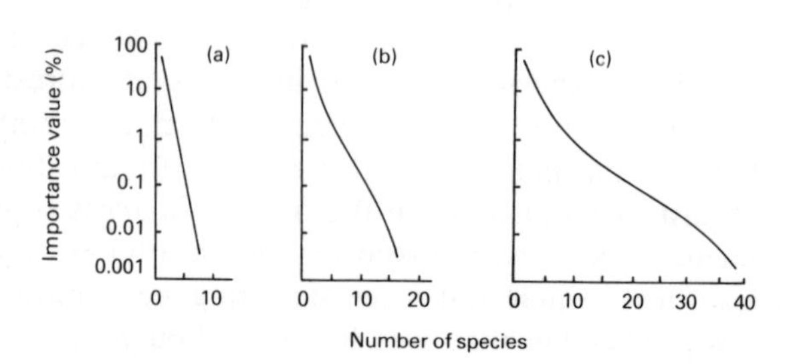

Fig. 12.3. Three principal types of dominance–diversity relationships recognized in vegetation. (a) Species-poor communities, (b) and (c) progressively more species-rich communities (after Whittaker 1965).

determine the population pattern (whether strongly or weakly segregated — cf. Pielou 1961) of a species, a pattern that is frequently apparent to an experienced observer. It should be appreciated, however, that different species also exhibit patterns relative to each other, which are not so readily self-evident and which may be analyzed only on the basis of appropriate tests of their relative spatial segregation (Pielou 1961). The interaction between selective forces operating to produce initially strong segregation (with consequential intraspecific thinning selection leading to a reduction of segregation in the pattern of adult individuals), compared with an initially lower level of segregation (followed by interspecific interference mediated by habitat selection, leading in turn to an increase in adult segregation) will have a powerful influence on whether or not unitary successional sequences become apparent in a particular context.

12.4.2 Environmental gradients, grain and ecological selection

Usually operating on a somewhat larger scale than the interactions discussed in the previous section are those environmental gradients within which regional diversity, relative productivity and species interactions are constrained (Section 12.3.2). Allowing for the seasonal, diurnal or continuous variation in some environmental factors which essentially reflect present latitudes, altitudes and relative continentality, it seems likely that the regional steepness of such gradients will in many instances be determined by the relative age of the landscape. Young, rejuvenated landscapes are likely to exhibit steep (in both senses) gradients, while older, more mature landscapes are likely to show more gradually assorted environmental variation. A young landscape which in turn has coarsely grained habitats and supports relatively few potential dominants may well be clothed eventually by an essentially monospecific forest. European beechwoods or the northern taiga probably fulfil such a prescription. Older landscapes in the main are likely to be less steeply graded and, because of the considerable redistribution of mantle material contingent on their age, are more likely to exhibit finer-grained habitats. Such circumstances argue for a mixed dominance in mature vegetation, which may or may not be maintained by modern selection processes.

Against this broad geographical background can be set the impact of further environmental selection imposed by regional soil development and redistribution (physical and chemical, particularly nutrient, gradients); selective logging, intermittent fire, selective herbivory (including seed predation), host-specific epidemic disease, vulcanism, wind-throw and lightning strike. Ecological selection clearly comes in many forms — the listing provided above is certainly not exhaustive — and varies in its longterm selective impact through its qualitative nature, frequency and intensity. What all these forms of selection achieve, however, is some modification, either diffuse or concentrated, of the species interactions which are set within a broad fabric of environmental history and heterogeneity. The matrix of potential outcomes — and the range of associated probabilities if they could be calculated adequately — are considerable. The fact that we can on occasion recognize a recurring pattern in such outcomes suggests that pattern generation is effectively dominated by relatively few of the range of selective processes operating in both natural and man-modified environments.

12.5 Conclusions

Although I have canvassed some of the seemingly essential components that need to be accommodated in any modern, coherent theory of ecological succession, I have deliberately eschewed any attempt to present a singular model of such a process for several reasons.

Firstly, although the concept of succession has been addressed by many leading investigators throughout the history of ecology — and many of these contributions have provided important insights — the literature is still remarkably short on wide-ranging examples of unequivocally successional sequences encompassed in comparable, precise terms. It seems likely that widely applicable conceptual generalizations will emerge only when they are based on a fuller evaluation of case studies which serve to contrast vegetation types that vary in species diversity, physiognomy, environmental history and present selective pressures.

Secondly, the definition and relevant scale of many putative successional sequences require critical evaluation before the evidence they offer can be justifiably incorporated into any developing, coherent theory. As one illustration of the importance of definition, the post-fire response of inflammable vegetation-types such as chaparral or sclerophyllous shrubland is regarded by some authors as 'resilience', while others regard it rather as a 'pyric succession'. Others still, depending on the scale and specificity of their relevant interest as either managers or theoretical modellers, might see the essential contrast as being between 'maintenance dynamics' or 'successional dynamics' (cf. Walker 1982). Clearly there needs to be more agreement as to what is properly encompassed in the sense and meaning of *succession* before much potentially relevant evidence can be properly evaluated and either incorporated into or dismissed from any potentially stable theoretical framework.

Thirdly, it seems that philosophical or pragmatic considerations still influence both the structure and the diagnosis of many successional trajectories described in the literature. Harper (1982) has been at pains to emphasize the conceptual gaps that exist between 'holists' and 'reductionists' in this regard and of course natural resource managers frequently present another angle to what is effectively a triad of approaches, while clearly embracing the view that a reductionist approach points the way to the attainment of precision, reality and generality in ecological science.

For these reasons among others, a universally accepted theory of succession is likely to emerge only when the plethora of present approaches and viewpoints are rigorously evaluated against the need to explain the processes by which present spatial patterns of organisms have been achieved, maintained and/or modified. It is the *process* — defined within an appropriate section of the spatio-temporal continuum — of Watt's (1947) *pattern and process* that should define the reference points within which successional dynamics are to be properly analysed and understood. Watt's seminal paper is still very relevant as a cardinal basis on which we might erect a more enduring and productive explication of the successional process than that which presently exists.

Part 5
Evolutionary Processes

Ecological processes discussed in Part 4 are all subject to selection in evolutionary time scale. The last part of this book focuses on evolutionary processes in communities from both specific and more generalized vantage points.

Chapter 13 reviews the development of arms races and covenants of restraint that develop between parasites and their hosts, and argues that the direction such relationships take is essentially dependent on the relative rates of evolution and generation-time of the two groups of organisms.

In Chapter 14 the authors paint this canvas more broadly still and advocate the view that community-generated evolution is likely to be detected most readily in rapidly evolving situations, many of which — such as weed systems and houseflies — are the products of human endeavour.

Thus the impact of such human endeavours, leading as they have done to an increasing awareness of the necessary and incessant interplay between evolution and ecology, forms a natural focus for the final Part and chapter of this book.

13 Arms Races and Covenants: The Evolution of Parasite Communities

Bill Freeland

13.1 Introduction

The shorter an organism's generation time, the faster its potential rate of evolution. The evolution of host defences against parasitism is only likely to match the evolution of parasite offences in situations where there is little difference between host and parasite generation times. Thus, short-lived hosts can be thought of as being involved in arms races (Gilbert 1971) which constantly generate selection favouring new host defences and new parasite offences. The evolution of defences by long-lived hosts is unlikely to match the faster evolving offences of parasites. Something other than arms races is necessary to account for the continued persistence of long-lived hosts.

Competition within and among parasite species may be a major force of selection acting on parasites of long-lived hosts, and may lead to the evolution of behavioural convenants between hosts and their parasites. Covenant is used in the sense of a binding agreement by two or more parties to do or keep from doing some specified thing. The agreement is binding because under most circumstances behavioural departures from the covenant are not selectively advantageous. Parasites may achieve their potential rates of evolution, but these occur within the constraints of the covenant.

The hypothetical existence of behavioural covenants between hosts and parasites assumes that parasitic communities exhibit structure defined by the effects of intra- and interspecific competition. This assumption cannot be tested as there has not been an adequate study of the structure of a natural parasite community. Failure to delineate structure in the species of ectoparasites or helminths infesting rats, or the mistletoes, or insects infesting trees, would not negate the hypothesis that parasite communities are structured. Structure is a characteristic of communities and need not be discernible in single taxonomic components of communities. Experimental laboratory evidence suggests that parasite communities are structured.

These hypotheses are developed following discussion of (i) the manner in which members of single parasite species and entire parasite communities are likely to generate selection for new host defences, (ii) the effects of new host defences on parasite communities, and (iii) the effects of host and parasite generation time on the evolution of parasite virulence, rates of resource use and modes of competition. The concepts of convenants and arms races are then discussed in relation to mutualistic interactions between organisms, and the structures of parasite communities.

13.2 Evolution of host defences

13.2.1 Parasites as selective agents

Common parasites with intermediate to high levels of virulence may select for new host defences. The larger the population of a parasite species the greater the probability of an individual host acquiring a member of that parasite species (Anderson 1979a). If acquisition of an individual parasite does not alter the probability of reinfestation of the host, then the larger the parasite population, the greater the parasite burdens carried by individual hosts.

The more individuals of a parasite species a host carries, the greater the likelihood of the host suffering fitness loss relative to other members of its population (e.g. Hall 1934; Willet 1956; Dobson 1965; Rhoades 1979). Anderson (1979) assessed parasite virulence using data on parasite dosage and host mortality for several parasite species. Crofton (1971) used the concept of lethal level as an index of a parasite's potential impact on individual host fitness. Lethal level was defined as the number of parasites required to kill the host. The use of mortality data, or lethal level, neglects host fitness losses that are expressed in other ways. Host fitness loss due to parasitism includes lowered probability of survival, impaired growth-rates, lowered access to mates and reductions in fertility (e.g. Arme 1968; Weatherly 1971; Kuris 1974; Shaw & Quadagna 1975; Freeland 1981). These modes of fitness loss are dosage dependent but may not be related to dosages required to kill hosts. The virulence of a particular host–parasite interaction is best regarded as dosage-dependent depression in the genetic fitness of individual hosts, irrespective of the mode of fitness depression. As natural selection is dependent on variance in genetic fitness (Fisher 1958), selection pressure leading to the evolution of new host defences is dependent on: frequent and significant depression of individual host fitness, and the occurrence of some novel, resistant host genotype.

Increases in parasite population size are likely to result in both increased frequency of host infestation and the potential for dosage-related decreases in the fitnesses of individual hosts. Thus parasites with large population sizes and high frequencies of host infestation have the potential for acting as agents selecting for new host defences.

Whether or not common parasites act as selection agents on host populations depends on the nature of the relationship between parasite dose and host response. Parasites of extremely low virulence, irrespective of dosage, or parasites that maintain low virulence by regulating the number of individuals infesting a host (Section 13.2.3), may frequently infest hosts and may not act as agents for the selection of new host defences. Parasites of intermediate levels of virulence may act to select new host defences if the level of virulence increases with increase in the size of parasite populations.

Parasites of exceptional virulence may be frequent enough to select for new host defences. Anderson (1979) concluded that parasites of exceptional virulence are necessarily rare. This conclusion is based on the suggestion that virulence is directly related to mortality, and that highly virulent parasites are likely to kill their hosts prior to their being able to reproduce and be transmitted to other hosts. As stated earlier, parasites may reduce individual host fitness to zero and affect full development, reproduction and transmission (e.g. insect parasitoids). Such parasites need not be rare, need not depress host population size (depending on host population growth-rate, host carrying capacity, age of host at infestation, etc.), and are capable of selecting for new host defences.

Loss of fitness by individual hosts may be aggravated by hosts being subject to infestation by members of several parasite species. Under natural circumstances, hosts are often subject to multi-species infestations that lead to additive or synergistic depression of individual host fitness (e.g. Adler 1954; Kates & Turner 1959; Al Dabagh 1961; Kilham & Oliver 1961; Box 1967; Nayak & Kelly 1969; Chowaniec *et al* 1972; Hein 1976; Whitlock & Georgi 1976). Less commonly, parasite species may inhibit each other's fitness-depressing effects on the host. The general conclusion can be drawn that as a parasite species increases in abundance, its impacts on fitnesses of individual hosts (and effects on host population size) are likely to increase due to interactions among parasite species. Because of multi-species infestations, the average impact that individuals of a parasite species have on individual host fitness is likely to be greater than it would be if it were the only parasite species infesting the host population.

Common parasites are likely to exert selection pressure on host populations unless they: (i) produce little or no additive or synergistic depression of host fitness when in infestation with other parasite species; or (ii) exhibit competitive exclusion of other parasite species; and (iii) have virtually no virulence capability; or (iv) limit virulence by effective regulation of the number of parasites of its own species infesting the host.

13.2.2 New host defences

Selective advantage is likely to accrue to host genotypes resistant to (i) particular common parasites of intermediate to high virulence or (ii) to a large number of parasite species irrespective of the individual species' virulences, population sizes or microhabitats on/in the host. If a parasite is rare its probability of encountering a host with a resistant genotype is extremely low and the resistant genotype is unlikely to gain any long-term parasite-related selective advantage. If the parasite is common but induces virtually no loss of host fitness, a resistant host genotype is unlikely to be selected for. A mutant host defence may be selected if it results in: (i) elimination of common, virulent parasites; (ii) reduction in the virulence of a common virulent parasite; or (iii) reduction in the growth rate of a parasite population. [The only advantage accruing from this would be a reduction in the probability of a resistant host's kin becoming infested (Hamilton 1964).] Janzen (1973) stressed that defences selected for by the activities of one parasite may incidentally eliminate one or more of other species. This is undoubtedly true. However, parasites exerting selection on host populations are likely to have a higher probability of being selected against by a new host defence than those that do not exert selection pressure.

Selective advantages may accrue to host genotypes that confer resistance to a large proportion of the total number of parasite species attacking the host population. This form of selection is independent of the frequencies of occurrence and host fitness impacts of individual parasite species. Parasite species eliminated by the new defence are those which together cause frequent and significant loss of individual host fitness (Janzen 1973b). Parasite species affected by such a defence are likely to include organisms from a variety of taxonomic categories and microhabitats in the host (Janzen 1973b).

Selective advantage based on resistance to several parasite species is most likely to be associated with a major addition to, or change in, host defence strategy. These changes are likely to be the types of change characteristic of organisms at the evolutionary roots of major taxonomic groups. For example, Ehrlich and Raven (1964) discuss the chemical defences of some plant families and the radiations of lepidopteran parasites that have followed radiations of the plant families. The acquisition of a new class of defences is likely to devastate a parasite community, and such phenomena are likely to be rare.

Host selective advantage based on resistance to single parasite species is likely to be associated with minor adjustments of already existing defences and can be expected to be more common than major changes leading to resistance to many parasite species.

Resistant host genotypes gain their selective advantage by lowering the probability of the host acquiring the infestation or by reducing the fitness depressing effects of the infestation. Reduction in the probability of acquiring an infestation may be achieved by: (i) the host being physiologically unsuitable for parasite survival (e.g. Rothman 1959; Van der Plank 1963); (ii) the host not providing a cue for a behavioural response by the parasite, and in consequence the parasite fails to infest an otherwise suitable host (e.g. Rogers 1963; Smyth & Haslewood 1963; Gilbert & Singer 1975; Schoonhaven & Jermy 1977); or (iii) host behaviour lowering the probability of the parasite encountering the host (e.g. Freeland 1976, 1977, 1979a, b).

A host may reduce the fitness depressing effects of an infestation by limiting: (i) the quantity and quality of host resources used by the parasites; (ii) the cost of tissue damage to the host in terms of disruptions to normal functions and cost of repair; (iii) the cost of any defence, and (iv) the cost of damage caused by a defence mechanism.

All these parameters are related to the number of parasites in an infestation. The size of a parasite infestation may be limited behaviourally (e.g. Freeland 1976, 1977, 1980), by the use of induced defences (e.g. Soulsby 1962; Dobson 1972; Janzen 1979; Rhoades 1979) or by slowing the rate of increase in the size of the infestation by slowing a parasite's rate of development and reproduction. Slowing a parasite's rate of development and reproduction may result in the additional advantage of lowering the probability of host's kin suffering massive infestations (Hamilton 1964).

As stressed by Sprent (1962), the use of a defence against a parasite may cost the host more than any fitness depressing effects the parasite is likely to produce. A host's failure to defend itself may be selectively advantageous in some cases.

13.2.3 Impact of new defences on parasite communities

The evolution of new host defences may lead to reduction in the species richness and average virulence of parasite communities, and result in parasite communities that do not exploit all potential microhabitats.

Complete host resistance to common parasites of high virulence is not an automatic correlate of selection for a new host defence. Whether or not a resistant genotype is selected for depends on the nature of competitive relationships among species constituting a parasite community. A common parasite species may have a superior interspecific competitive ability. A host genotype completely resistant to such a parasite may gain a temporary selective advantage. This advantage would be lost if previously outcompeted virulent species become more common.

Complete elimination of a common parasite species of appreciable virulence may not be of long-term selective advantage. A resistant host genotype may be of selective value if: (i) on elimination of the virulent common species, only avirulent parasites are released from competition; (ii) the new defence eliminates host fitness depression caused by a parasite's interactions with other less virulent species, or if (iii) removal of the parasite does not influence the abundance of other parasite species.

Alternatively, selective advantage may be gained by the resistant host genotype conferring control over the level of host fitness reduction, rather than eliminating the parasite. As mentioned earlier, this can be achieved by behaviourally or chemically limiting the size of the parasite infestation, slowing the rates of parasite development and reproduction or by reducing the amount of tissue damage done by the parasite. These kinds of effects have the advantage of maintaining a potentially highly competitive parasite (of low virulence) that prevents release of more virulent parasites. Maintenance of competitive low-virulence parasites may also protect the microhabitat of the parasite from colonization by another, potentially more virulent parasite from a different host species.

When selective advantage is based on host elimination (or reduction in the host fitness depressing effects) of a large number of parasite species, there is a lower probability of the resistant host genotype being subject to negative effects from competitively released parasites. As mentioned earlier, the particular species eliminated need not bear any relationship to the virulences or competitive abilities of individual parasite species. Highly competitive species need not be eliminated. If they have been, the few species remaining may lack the capability to immediately become more virulent. A major change in defence associated with loss of a large number of parasite species may also reduce the probability of the host being colonized by parasites from other species.

Elimination of parasite species is likely to result in the production of unexploited host microhabitats. Through time, some eliminated species will be replaced by colonization from parasites of other host species. Within a particular habitat or geographic area the number of potential colonists is limited. Long-term, repeated occurrence of new host defences and associated loss of parasite species may result in lowered probabilities of replacement of eliminated species. Because of this, not all potential parasite microhabitats are likely to be occupied. Microhabitats that are occupied may be occupied by more than one species. This pattern of vacancy and multiple occupancy of host microhabitats is likely to be reinforced by the rarity of many parasite species. Rarity of many parasite species may be due to massive density independent mortality in certain stages of parasites' life history, the activities of parasites and predators of the parasites themselves or poor quality of transmission between hosts.

None of these factors is likely to alter the discussed patterns of host–parasite evolution. Low frequencies of infestation (at any particular instant in time) associated with rapid transmission between hosts and short durations of infestation may be functionally equivalent to being common (e.g. Levins & Culver 1971).

Host selective advantage based on resistance to a single species of parasite, or to several parasite species, leads to a reduction in the number of species in parasite communities and in the average virulence of parasite communities, and results in parasite communities which do not exploit all host microhabitats.

13.3 Evolution of parasite strategies

13.3.1 Extent of host fitness depression

A parasite's virulence is a function of an interaction between the parasite's offences, host defences and competition among parasites. There is likely to be a continuum of parasite virulences ranging from total elimination of host fitness, through virtually no impact on host fitness, to the parasite having a positive effect on host fitness.

Abandonment of strategies that severely reduce host fitness may be selectively advantageous for some parasites. A highly virulent genotype with an effective means of transmission among hosts will initially infest a large number of hosts. Eventually, the host population will be depressed to the point where the mechanism of transmission is no longer effective. The parasite may persist at low frequencies of infestation or the parasite (or host and parasite) may become extinct. Low host population densities may result in selection favouring improved mechanisms of transmission for virulent parasite genotypes. This would result in either further reduction in host population density and frequency of parasite infestation, or extinction of the parasite, or host and parasite. If host population density was depressed and frequency of virulent parasite infestation thus reduced, selection may favour parasite genotypes with reduced impact on host fitness, and the intraspecific competitive ability to prevent colonization of a host by more virulent members of its species. At low host population densities, an extended period of residence in or on the host may significantly improve the probability of a parasite's propagules infecting other hosts.

Virulent parasite species reduce the fitnesses of individuals of other parasite species by reducing the availability of resources for growth and reproduction and by truncating life cycles prior to the production of propagules. This results in selection for improved interspecific competitive ability among other members of the parasite community. Thus both intra- and interspecific competition among parasites may lead to a reduction in the frequency of infestation of a virulent genotype.

Parasites of low virulence are most likely to evolve when the host is at a serious disadvantage in relation to the potential rate of evolution of a parasite. Among such hosts there is little likelihood that the appearance of a virulent parasite by mutation will be followed by the appearance of a resistant host genotype. A virulent parasite is more likely to be eliminated or reduced in frequency by the appearance of a highly competitive, less virulent genotype of the same parasite species, or more competitive genotypes of other parasite species. Thus although a long-lived host has little evolutionary potential to counter the occurrence of a virulent parasite, it may be defended by the competitive abilities and potentially faster evolutionary rates of other parasites of the host.

Highly virulent parasites are most likely to be selected for when there is little difference between the potential rates of evolution of host and parasite. If the probability of occurrence of mutant host genotypes resistant to virulent, common parasites approximately equals the probability of occurrence of less virulent, competitively superior mutant parasites, selection will frequently favour host resistance. Repeated selection for host resistance is likely to depress parasite fitness and select for parasites capable of overcoming host resistance, irrespective of impact on host fitness. This results in host–parasite arms races.

Virulent parasite strategies involving an abbreviated host longevity limit the duration of residence of parasites in or on the host. Because of this, selection for virulence depends on the parasites having the capacity to rapidly convert host resources into parasite biomass. This capacity may be based on rapid rates of propagule production in a live or dead host, or the parasite may be relatively large (e.g. some parasitoids). Short duration of infestation is likely to inhibit propagule transmission to additional hosts. This can be overcome by parasite reproduction in alternate biotic or abiotic environments, by the possession of extremely resistant propagules or by the production of highly mobile stages capable of locating and infesting hosts.

Although a slow rate of resource use over an extended period of time may have little impact on host survival [or may even enhance host survival (e.g. Vinson & Iwantsch 1980)], it does not preclude a parasite from reducing host fitness to zero. Fertility [e.g. parasitic castrators (Kuris 1974)] or access to mates (e.g. Freeland 1981) may be sufficiently impaired to greatly reduce host fitness. These kinds of parasites are highly virulent and may be most common in situations where host and parasite have similar rates of evolution.

13.3.2 Resource use by avirulent parasites

Avirulent parasites have rates of resource use dictated by resource availability in particular microhabitats, and the effects of resource removal on host fitness.

There is a limited amount of host resource available to parasites. Perhaps the only data giving some insight into the problem of resource availability to parasites are those on the effects of haemorrhage. A single loss of between 45 and 60% of blood volume usually results in the death of a wide variety of mammals (Schalm 1961). Medway *et al* (1969) provide data on the maximum safe volume of blood that can be taken from eleven species of domestic and laboratory mammals. These data suggest a mean of 9.98% (±0.64 SD) of blood volume (representing an average of 0.7% of body weight) as being the maximum safe volume. This is a relatively small proportion of total host resources and is safe only for single blood samples. Christie (1978) bled chickens daily, removing an average of approximately 17% of the blood volume per day. A substantial reduction in growth rate, and major changes in behaviour and haematological and biochemical parameters resulted. As little as approximately 1.5% of blood volume loss per day can have a significant effect on chickens (Sturkie & Newman 1951). Monthly withdrawal of 1.25–2 ml (approximately 0.05–0.08% of blood volume) from adult rats did not change a variety of haematological parameters but did result in a lowered rate of weight gain (Cardy & Warner 1979). It appears that relatively minor losses of host tissue can have significant effects on host fitness and the quality and quantity of resources available to parasites.

Among virtually avirulent parasites, selection is likely to favour mechanisms increasing parasite reproductive output without (i) increasing use of host resources or (ii) increasing disruption of host functions to levels that depress host and parasite fitnesses. These can be achieved by reductions in parasite body size and/or by an increase in the parasite's efficiency in the use of resources. Efficiency of resource use could be improved by the appearance of mutant enzymes of greater substrate specificity etc., or by elimination of redundant biochemical pathways. In the highly predictable environment of parasites, the possession of some biochemical pathways may be superfluous. Zemonhof and Eichon (1967) grew mutant *Bacillus subtilis*, defective for a particular biochemical pathway, and normal *B. subtilis*, together in continuous culture, using media adequate for the growth of the defective mutants. In all cases, the mutants outcompeted the normal strain. Parasitic nematodes and platyhelminthes often exhibit abbreviated energy metabolism (Read 1970). These abbreviations may reflect an absence of genetic capacity to produce the missing pathways, or the genetic capacity may exist, but is not used (Read 1970). Either of these mechanisms is likely to lead to efficient use of energy and nutrients and increased rates of development and reproduction without depressing individual host fitness.

The rates at which host resources can be used and the impact of resource removal on the host are likely to depend on the microhabitat and resources used by a parasite. Daily losses of 1.5% of blood volume, 1.5% of brain tissue or 1.5% of faecal matter in a rectum, are likely to have very different impacts on parasite and host. *Trichosomoides crassicauda*, a nematode living in the bladder of rats, has a relatively low nutrient availability, an extremely low rate of reproduction (Smith 1946; van der Gulden 1967), and removal of all or most available nutrients (not including those of host tissues) is unlikely to have a great impact on host fitness. Removal by *Hymenolepis diminuta* (a tapeworm) of all simple carbohydrate in the small intestine of a rat may be disastrous to both host and parasite. Thus, there is likely to be a negative relationship between fitness value of a microhabitat to a host, and the rate at which avirulent parasites exploit that microhabitat. Resources used by a parasite may be the result of host defences forcing a parasite from a microhabitat of high host fitness value, to one of low host fitness value. Alternatively, a microhabitat shift by a parasite may allow a pattern of resource use to become less virulent, and so enhance parasite fitness.

13.3.3 Competition among parasites

Intraspecific competition among parasites is likely to result in hosts maintaining a certain number of parasites of a single species. This is an evolutionary consequence of interactions between the level of fitness loss experienced by the host, resource availability, resource requirements of individual parasites and the consequences of kin selection (Hamilton 1964) among parasites.

Intraspecific competition results in lowered survival, growth, and reproduction of individual parasites (Duszynski 1972; Moqbel & Wakelin 1979; Stewart *et al* 1980). Intraspecific competition is likely to select for any mechanism that enables a parasite to prevent or limit additional infestation of its host, or eliminate or reduce infestations in the host being colonized. Mechanisms include physical interactions, the excretion of allelochemicals and the induction or manipulation of host defences to a parasite's advantage (e.g. Holmes 1961, 1962; Louch 1962; Damian 1964; Rigby & Chobota 1966; Schad 1966; Tomiyama *et al* 1967; Reed 1970; Gordon 1973; Kazacos & Thorson 1975; Smith 1976).

In many natural situations, colonization of a host by a parasite is likely to be accompanied or followed by colonization by closely related individuals of the same species. The best studied example is that of the mouse, *Mus musculus*, and the tapeworm, *Hymenolepis nana* (Heyneman 1961, 1962a, b, c; Ito 1978, 1980). *H. nana* can infect mice in two ways: by the mouse eating an arthropod carrying the cysticercoid stage, or by the mouse ingesting an egg derived from its own or another mouse's faeces.

If a mouse ingests a cysticercoid in an arthropod, the resulting tapeworm lays eggs. Eggs are passed in the faeces and can infest the mouse with the parent tapeworm, and other mice. Ingestion of a cysticercoid confers resistance against colonization of the host by additional cysticercoids. This does not prevent colonization of the mouse by tapeworms originating from eggs (Ito 1978). If a mouse ingests eggs, it becomes resistant to colonization by either eggs or a cysticercoid (Ito 1978).

Wild mice live in social units composed of one or two adult males, and several adult females and immature offspring (P.K. Anderson 1964, 1965, 1970; Petras 1967a, b; Berry 1968; Selander 1970). If a mouse social unit is colonized by a cysticercoid, eggs from that tapeworm may colonize all members of the social group. All individuals in the mouse social unit would then exhibit resistance to colonization by eggs or cysticercoids. If a mouse social unit is colonized by an egg or eggs, the unit as a whole is likely to be colonized by eggs from the resultant tapeworm, and exhibit resistance to colonization by eggs and cysticercoids from outside the host group.

H. nana is unusual in its ability to infest the definitive host (mouse) by direct transmission of eggs. It is this short circuiting of the normal tapeworm life cycle that appears to be responsible for the difference in the types of host resistance induced by eggs and cysticercoids (Chandler & Read 1961). Host social groups provide several tapeworm habitats in close proximity. Colonization of these habitats by a resident tapeworm's offspring is selectively beneficial to the tapeworm but is likely to result in competition among the offspring and between parent and offspring. However, the pattern of resistance generated by *H. nana* is such that a genetically related group of hosts is likely to maintain a 'clone' of tapeworms that collectively prevent invasion by unrelated *H. nana*. Other examples of parasites limiting the number of conspecifics infesting a single host via manipulation of host defences include bacterial resistance to superinfection by phage (Lewis 1977), *Nematospiroides dubius* in mice (Behnke & Parish 1979), *Nippostrongylus brasiliensis*, *Strongyloides ratti* and *Trichinella spiralis* in rats (Ogilvie & Jones 1971; Moqbel & Denham 1977; Bell *et al* 1979) and many more.

The number of individuals of a parasite species living in a host is likely to result from kin selection acting on parasites such that the fitness of kin groups is optimized, rather than individual selection acting to promote the fitness of individual parasites. A theoretical mechanism by which number and size of individuals are likely to be determined in relation to efficiency of resource use and resource availability is discussed by Harris (1981). The only additional parameter necessary when considering parasites is that the combined reproductive output of the kin group needs to be optimized in relation to duration of infestation and probability of transmission to other hosts.

Use of different parts of a host by individuals of different parasite species reduces the potential for members of one parasite species to competitively eliminate members of other species from a host, and does not necessarily reduce interspecific competition for host resources.

Interspecific competition among parasites is likely to lead to: (i) an increase in the rate of parasite use of host resources; (ii) potentially decreased availability of resources to individual parasites; (iii) possible decreases in parasite longevity; and (iv) a probable increase in the level of fitness depression experienced by the host. These effects are likely to decrease individual parasite fitness and select for competitive dominance of one parasite species by another, or stable coexistence among parasite species.

Members of a parasite species occupying the same microhabitat or the same part of a host as members of another species have opportunities for competing via non-specific modification of the habitat to make it unsuitable for the other species (Behnke *et al* 1977; Howard *et al* 1978; Christie *et al* 1979), and for interfering with members of the other species using allelochemical interactions (Read 1970), by manipulation of inducible host defences (De Vay & Adler 1976; Schad 1966) and by purely physical means (e.g. Holmes 1961, 1962, 1971).

Coexistence within a host by members of two or more parasite species is often associated with the use of differing host parts. However, in these situations the individual parasites are still in competition for a limited resource (the host). Competitive interactions among parasites in different parts of a host are limited to manipulation of inducible host defences or other manifestations of the host resource budget (Janzen 1973b).

Perhaps the best competitive strategy available to a parasite is to prevent members of other parasite species from colonizing its host. Such a strategy is most likely to be effective when members of the competing species occupy the same microhabitat, when the host has defences that a parasite can turn against a competing species (e.g. the immunological systems of birds and mammals) or when a parasite makes use of an allelochemical. These mechanisms may exclude a competitor without the host or resident parasite undergoing significant fitness losses. Competitive success in these situations often favours members of the parasite species first colonizing a host (e.g. Rigby & Chobota 1966; Moqbel & Wakelin 1979; Stewart *et al* 1980).

In situations where the host does not have defences suitable for manipulation by the parasite (e.g. lacking in specificity, not penetrating to all parts of the host) and/or where the parasites live in different parts of the host, interspecific competition may be mediated via manipulation of the host resource budget. The mere presence of a parasite is likely to constitute an alteration in the pattern of host resource budgeting. Members of some parasite species are likely to benefit from changes in the pattern of distribution of host resources produced by the presence of other species. Similarly, parasites can be expected to have been selected to manipulate host resources such that members of competing species experience lowered resource availabilities. Manipulation of host resources may take the form of scramble competition, or chemical or physical diversion of host resource towards or away from certain host functions or parts.

Parasites may compensate for competition-induced fitness losses by adopting a more enhanced rate of use of host resources. It may be an effective strategy for an avirulent parasite to adopt when in competition with virulent parasites, or when its host is otherwise debilitated. Enhanced rates of use of host resources may also be employed by competing members of virulent species. Any advantage from this strategy depends on the parasite gaining more resource than it would if it did not change its pattern of resource use and/or the enhanced rate of resource use resulting in reduced resource availability to, and possible elimination of, the competitor. By using this strategy, a parasite may succeed in producing some propagules prior to host death.

An accelerated rate of parasite development or reproduction leading to host death may have a severe impact on the fitnesses of members of competing species of avirulent parasites. The rate of use of host resources by an avirulent parasite is likely to have evolved such that the rate and duration of reproduction are optimized to ensure the greatest probability of transmission to other hosts (Section 13.2.3). Rates of resource use greater than optimum are likely to reduce host fitness. Depressed host fitness may lead to reduced duration of parasite reproduction and possible reduced rates of reproduction. These effects may lower parasite fitness below that of individuals that do not alter their strategy of resource use in the presence of members of other avirulent species.

It may be possible for a parasite (virulent or avirulent) to manipulate a host resource budget without altering either its rate of resource use or its impact on host fitness. A parasite may lower the supply or availability of resources in a microhabitat such that members of competing species cannot colonize or survive in a host.

Selection resulting from avirulent parasites competing via the host resource budget may favour the manipulation of the pattern of a host resource budget (as described above) or, compensation for competitive fitness loss by: (a) more efficient conversion of host resources into parasite propagules, or (b) reduction in the rate of use of host resources and associated extension of parasite longevity.

Compensation mechanism (a) may take the form of more efficient biochemical pathways, the elimination of biochemical pathways, or a reduction in parasite body size (reduced growth and maintenance requirements). Mechanism (a) may be selected for even if levels of propagule production achieved by these means are not greater than those achieved in the absence of interspecific competition. Levels of propagule production need only be significantly greater than those of other genotypes when faced with competition.

Compensation mechanism (b) is less likely to be selected. It has the disadvantages of slowing the rate of reproduction and increasing the age at which sexual maturity is attained. These changes automatically result in a lowered rate of population growth and fitness for the genotype. For reduced rates of resource use to be selected for, interspecific competition is likely to have to be frequent, and result in massive fitness losses by individual parasites. This mechanism has the disadvantage of fitness benefits to the parasite depending on host survivorship. Many kinds of host are not likely to have probabilities of individual survival sufficient to allow for selection of mechanism (b).

13.4 Behavioural covenants

13.4.1 Coinciding selective interests

The occurrence of a behavioural covenant (defined as binding agreements by two or more parties to do, or keep from doing, some specified thing), between a host and its parasites, depends on a coincidence of host and parasite selective interests in regulating parasite burdens (number of species and individuals per host) and limiting fitness losses by individual hosts. Whether or not coincidence of interest occurs depends on the potential rate of evolution of the parasite in relation to that of the host.

Complete coincidence of selective interest occurs when the parasite has a greater potential evolutionary rate than the host (Fig. 13.1). In these circumstances, host and parasites have a vested interest in limiting fitness loss by the host (Section 13.2.3). Increased host fitness loss due to intra- and interspecific parasite competition results in losses of fitness by the parasites. Selection is likely to result in limited loads of individual

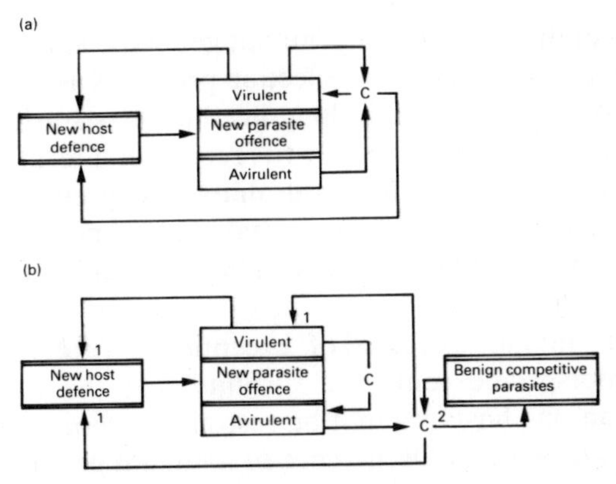

Fig. 13.1. Interactions between hosts and parasites lead to arms races (a) (little difference between host and parasite potential rates of evolution) or behavioural covenants (b) (parasites with potentially faster rates of evolution than hosts). The evolution of both types of host–parasite relationship is the product of the evolution of new host defences, new parasite offences (virulent and avirulent) and the results of intra- and interspecific competition among parasites. Arrows indicate the major directions in which selection pressure is exerted, with high (no number), medium (2) or low (1) probability of selection leading to the phenomenon indicated, and C denotes the occurrence of intra- and interspecific competition.

species, and either a limited number of parasite species infesting individual hosts, or limited fitness losses by multiple-infested hosts. Thus it is possible for hosts with long generation times to have complete coincidence in selective interest with their parasites in controlling parasite-induced reductions in host fitness.

The long-term maintenance of host–parasite covenants depends on there being no selective advantage to either party in behaving in ways contrary to the covenant. New host defences that disrupt covenants are likely to be infrequent among long-lived hosts. Even if a new defence mutation did occur, selection leading to its incorporation within a host population depends on frequent depression of individual host fitness by parasites. Any violation of the covenant of restraint is most likely to occur due to the occurrence of some mutant parasite genotype.

Selection favouring any mutant parasite genotype is reflected in an increased relative abundance of the genotype and a probable increase in parasite population. This automatically leads to potential increases in parasite-induced loss of individual host fitness due to increased doses of the parasite received by hosts, and increased frequency of interspecific parasite interactions in hosts. Both these effects are likely to depress the fitness of mutant genotypes and limit their rates of population increase. Excessive virulence resulting in host death may limit the rate of transmission of a mutant. A virulent parasite is likely to result in selection leading to improved interspecific competitive ability among other members of the parasite community. In short, any mutant genotype appearing in a parasite population is subject to the same selection pressures as are other genotypes and species. Among parasites of long-lived hosts, parasites of exceptional or enhanced virulence are not likely to receive long-term selective advantage. Selection is more likely to favour mutants expressing enhanced intra- or interspecific competitive ability, greater efficiency of resource use or enhanced transmission in the absence of high virulence. These mechanisms do not disturb the covenant.

A 'super' mutant parasite of exceptional virulence, having great intra- and interspecific competitive ability and the capacity for rapid transmission among hosts, cannot be restrained from short-term devastation of a host population. Occurrence of a mutant genotype incorporating all these characteristics would seem unlikely. The probability of occurrence of 'super' mutants may decrease over long periods of selection favouring efficient parasites which lack biochemical pathways that could otherwise be subject to change via mutation. The advent of a 'super' mutant exerts selection pressure on the host population, on members of the mutant's species not carrying the mutation, and on other parasite species. Among long-lived host species, these selection pressures are likely to result in the evolution of enhanced competitive, 'host-protecting' abilities among the nonvirulent members of the parasite community (Section 13.2.3).

The behavioural covenant of restraint between long-lived hosts and their parasites is protected from disruption by: (i) the probable low fitness value of most virulent genotypes; (ii) the low probability of 'super' mutant genotypes; (iii) the intra- and interspecific competitive abilities of parasites, and (iv) the potentially rapid rates of evolution of species in parasite communities.

Hosts with short generation times may have a coincidence of fitness interest with some of their parasites; i.e. those with potential rates of evolution significantly greater than that of the host. Parasites with rates of evolution greater than their host are endangered by virulent parasites. Selection may act to produce competitive mechanisms that protect host and parasite from virulent parasites.

These expectations imply that the average virulence of species in parasite communities is likely to decline with increasing host generation time. Increasing host generation time is also likely to be associated with decreasing variance in the virulence of species in parasite communities.

13.4.2 *Parasite community structure*

The number of parasite species infesting a host species is likely to change through time. Temporal change in parasite species number is likely to follow a pattern of increasing species number (due to acquisition of species from other hosts) punctuated by drops in species number associated with the occasional occurrence of new host defences. As discussed earlier, the abundance and virulence of many parasite species are likely to be decreased by new host defences. Parasite species number, relative abundances and virulences will also be influenced by the evolutionary outcomes of intra- and interspecific competition among parasites.

The structure of parasite communities probably varies according to the relative rates of evolution of host and its parasites. Among hosts with long generation times, new defences against parasitism are likely to be rare and the structure of parasite communities primarily determined by interactions among the species of parasite. Such communities are characterized by (i) few abundant parasite species; (ii) individuals of parasite species relatively evenly distributed among infested hosts; (iii) interspecific competition having produced coexistence of species without fitness loss to host or parasites; (iv) clearly defined (though subject to change through time) competitive relationships among species, and (v) more potential microhabitats for parasites than there are parasite species to occupy them (Section 13.2.3). As stated previously, parasites in these communities are likely to be relatively avirulent. Both host and parasite are likely to conform to behavioural covenants of mutual restraint, defined by an absence of selective advantage in doing otherwise.

The greater the likelihood of a new host defence (i.e. the shorter the host generation time), the less likely the structure of a parasite community is to be determined by evolutionary and ecological interactions among the parasites. Parasites having generation times equivalent to that of their host are likely to be virulent and generate marked temporal and spatial fluctuations in populations of host and parasites. Spatial and temporal variation in parasitic selection pressure on host populations results in infrequent benefits to hosts carrying particular resistance genes. Rapid rates of host evolution, and temporal and spatial variation in selection pressure due to particular parasite species, may result in host populations polymorphic for levels and mechanisms of resistance to particular parasites. Temporal and spatial variation in the abundance of parasites is unlikely to result in consistent selection pressures for the evolution of avirulent competitive relationships among particular parasite species. Avirulent mechanisms leading to competitive success at one time or place may be inappropriate to another. Competition via rates of exploitation of host resources is likely to provide a means for parasites to exert intra- or interspecific competitive dominance.

Parasite communities infesting hosts with short generation times may be characterized by wild fluctuations in the abundance of all parasite species, frequent overdispersion of individuals of a species among hosts and varying competitive relationships based on rates of exploitation of the host. Behavioural covenants of restraint between parasite and host are likely to be less common than in parasite communities infesting hosts of long generation times. Parasites normally using low virulence strategies may become virulent when in competition with more virulent species.

Similar kinds of host (e.g. species of trees) living in the same geographic area are unlikely to have parasite communities with similar numbers of species. This is due to the unpredictable occurrence of parasite mutation, new host defences and interactions among parasites.

13.4.3 Mutualism

I have discussed mechanisms evolved to protect a host and parasite from fitness losses incurred by both parties when a host is invaded by a second parasite species. In these cases the host–parasite relationship can be viewed as mutualistic.

Fitness benefits to a host from a mutualism based on shared nutritional advantages may be lost because of nutrient-caused disruption of host–parasite covenants. A nutritional mutualism is likely to introduce quantitative and/or qualitative changes in the resource base exploited by parasites. Quantitative increases in nutrient availability to a host automatically increases nutrient availability to the parasite community. Non-virulent parasites are likely to respond to increased nutrient supply with faster rates of growth, reproduction and transmission among hosts. The resultant increase in single and multiple species burden may depress individual host fitness. Parasite dosages may exceed levels controllable by the parasites' mechanisms of inter- or intraspecific competition. The value of a mutualism to a host's fitness may be further reduced if the host's enhanced nutritional status attracts parasites not usually attacking the host. Unless a mutualism's nutritional benefits to a host are translated into additional anti-parasite defences (e.g. an enhanced growth-rate, allelochemicals), or a parasite turned mutualist is capable of protecting its host from excessive parasitism, selection is unlikely to favour the evolution of nutrition-based, mutualistic relationships.

If a nutrition-based mutualistic relationship between host and parasite confers a qualitative change in resources available to parasites, it may result in: changes that favour the reproduction and transmission of non-mutualist parasites and consequences similar to those resulting from quantitative changes, or changes that impair parasite reproduction and/or survival. Qualitative changes in resource availability to parasites may cause loss of species from, or reductions in the abundances of species in, parasite communities. These changes may be due to the inability of parasites to deal with the qualitative change, or the mutualism may allow the host to use a new set of foods or habitats that impair parasite reproduction and/or transmission.

Qualitative changes in host nutrient status leading to loss or reduction in the abundance of parasites need not be advantageous to a host. Loss of an avirulent parasite may leave a host vulnerable to a more virulent, previously rare parasite. A changed host resource status may make a previously avirulent parasite's rate and manner of resource use more virulent. Alternatively, the parasite may increase its rate of resource use such that virulence is enhanced. The effects of loss (and changed abundances) of parasite species are compounded if the qualitative resource changes, or new foods and habitats used by a host, result in host vulnerability to a set of parasites that the host (and its parasite community) has never experienced. Colonization by a new set of parasite species may result in total disruption of behavioural covenants.

For a nutritional mutualism between a host and a parasite to evolve, it is necessary for the mutualism to protect the host from the consequences of any disruption to behavioural covenants between host and parasites. Alternatively, the mutualism may alter host resources and habitat to the extent that the host may become virtually parasite free.

13.5 Summary

The existence of arms races or convenants of restraint between host and parasite is dependent on the relative rates of evolution of host and parasite. Arms races between host and parasite develop when host generation time is similar to that of its parasite or parasites. Covenants of restraint evolve when the parasite's potential rate of evolution greatly exceeds that of the host. These patterns are based on the following propositions:

(1) Common parasites with intermediate to high levels of virulence may select for new host defences.

(2) Selective advantage is likely to accrue to host genotypes resistant to particular common parasites of intermediate to high virulence or, to a large number of parasite species irrespective of the individual species' virulences, population sizes or microhabitats on or in the host.

(3) The evolution of new host defences leads to a reduction in the species richness and average virulence of parasite communities, and to parasite communities that do not exploit all potential microhabitats.

(4) The effect a parasite has on the fitness of an individual host is an evolved function of the relative potential evolutionary rates of host and parasite.

(5) Avirulent parasites have rates of resource use dictated by resource availability in particular microhabitats, and the effects of resource removal on host fitness.

(6) The number of individuals of a parasite species in a host is the result

of kin selection optimizing the fitness of kin groups.
(7) Use of different parts of a host by individuals of different parasite species reduces the potential for members of one parasite species to competitively eliminate members of other species from a host, without necessarily reducing competition for host resources.

The selective value of virulent parasite genotypes parasitizing long-lived hosts is restrained by reduction of host longevity due to virulence, the effects of intra- and interspecific competition among parasites, selection acting to promote parasite genotypes competitively superior to the virulent mutant, and to the rare occurrence of new host defences against parasitism. The majority of virulent mutants are unlikely to be of selective value to individual parasites. Potentially rapid rates of parasite evolution are likely to be expressed within the confines of covenants of restraint and involve enhanced parasite efficiency in the use of resources, enhanced mechanisms of parasite competition and the possible evolution of host–parasite mutualisms protective of host fitness losses due to infestation by virulent parasites, or virulent combinations of parasite species.

The evolution of behavioural covenants between a host and its parasites results in parasite communities whose structure is largely determined by evolutionary and ecological interactions among the parasites. Such communities are likely to have few abundant parasite species and individuals are likely to be distributed relatively evenly among infested hosts.

Host–parasite arms races result in parasite communities in which the evolution of new host defences is of relatively frequent occurrence. These communities are likely to be characterized by dramatic temporal and spatial changes in the abundance of parasite species, and by over-dispersion of parasite individuals among hosts.

14 Evolution in Communities

Tony Bradshaw and Martin Mortimer

14.1 Introduction

When ecologists study any plant or animal community or compare any two species they are studying the products of evolution. As a result there is always the temptation to come to conclusions about the path that evolution has taken and even about the mechanisms involved. Indeed, modern ecological literature is full of speculative generalizations about evolutionary processes. This is an acceptable intellectual exercise, but its limitations must be realized.

Evolution, precisely defined, is net change in the gene frequencies of a population. The mechanism as far as we can infer is due to natural selection acting on genetic variation. The cumulative result of this process is to cause the variety of living organisms. If we look at this variety we are not looking at the processes by which it has arisen, only at the results. If we want to see the processes, which is what this chapter is all about, we must pay attention to genetic variability and the way gene frequency in populations can be changed by selection.

To look at the latter critically we can either look at changes in populations while they occur, in experiments or in nature, or we can look at situations where the changes have just occurred and where the conditions that caused the change can be presumed because they still exist.

In both cases we may be able to measure differences or changes in the populations directly, because the genes concerned have effects which are clear cut and not affected by the environment. Otherwise we may need to raise the material under standard conditions to remove the effects of differential environments. With concomitant breeding experiments, we will then know whether genuine genetic differences occur. Similarly we can carry out reciprocal transplant experiments to test performance more critically in different environments.

While much of this may not be perfect evidence, about which recently there has been much discussion (e.g. Quinn & Dunham 1983), it is more likely to yield reasonable evidence about evolution than comparisons made between established species, because with these we can only guess at the evolutionary processes that may have occurred. Even if we carry out typical ecological experiments on these species, although we shall know more about how they differ, again we will only be able to guess at the path of evolution that has produced them. Model making is better because it can test evolutionary hypotheses, but again it is imperfect, because it can only test the processes and parameters that are fed into the model, and it is always possible that in the real world others occur. We will therefore attempt to use the more critical evidence in this chapter.

14.2 Basic mechanisms

Evolution depends on the existence of genetic variation and selective effects of the environment. If both variation and selection are present, the mechanism can be very potent. Simple calculations show that selection pressures of only 0.2, due to a relative fitness of 80%, can radically change gene frequencies over only a few generations (e.g. Falconer 1981).

There are now several excellent examples looking at changes in progress in populations, such as industrial melanism in the moth *Biston betularia*, heavy metal tolerance in grasses, and herbicide resistance in agricultural weeds (Bishop & Cook 1981; LeBaron & Gressel 1982), where the nature of the genetic variability and the selective factors have been properly determined. These show that evolution can be rapid and local, and sensitive to subtleties of the environment. They also show that evolution can readily reverse if environmental pressures demand it, as in the case of industrial melanism.

The power of selection can equally be demonstrated, if further evidence is needed, by the innumerable plant and animal breeding programmes that have given us our complex and productive array of breeds and cultivars. But these programmes, and many laboratory selection experiments, have also shown that there is a limit set by the availability of genetic variation. If there is no appropriate genetic variation there can be no evolutionary change, however potentially powerful are the selective forces of the environment. This has recently been demonstrated for natural processes of evolution, explaining why only some plant species evolve metal tolerance (Bradshaw 1984). It also supports the idea that at any one point in time most species are static from an evolutionary point of view because they have run out of appropriate variation, a condition of *genostasis*.

In any specific situation therefore it is important to establish whether the appropriate variation is available. The word 'appropriate' here is very important. Genes on the whole have independent effects. It is therefore no use assessing the variability of a population in relation, for instance, to selection for increased reproductive output, by assessing its variability in respect of particular isozymes. The variability of the character in question must be assessed directly. In particular the heritability of this variability must be determined, because this decides whether it can be affected by selection. Heritability is the ratio of genetically determined variability to total (including environmentally determined) variability. Obviously it is only the genetically determined variability that can be affected by selection. A discussion of genetic variation in relation to ecological characteristics and the ways in which it can be measured are the subject of two recent reviews (Bradshaw 1984; Lawrence 1984).

However, it may still be difficult to assess precisely the genetic variability that is available, because there are processes of recombination of tightly linked genes and of mutation by which the appropriate variation may eventually appear but not immediately and not within the material being examined. Two good examples are the delayed appearance of industrial melanism in the Rosy Minor moth (*Miana literosa*) (Kettlewell 1973) or the highly sporadic occurrence of warfarin resistance in rats (Berry 1977). So in a geological timescale, there is usually a wealth of diversity and subtle adaptation that may not be immediately apparent, enabling evolutionary mechanisms to produce remarkable changes in plant and animal species.

That pervasive evolutionary processes are constantly occurring can be seen at the level of local population, geographical race and species. It is most obvious at the level of the population where there are two or more populations that have diverged from one another, because each acts as a reference point for the other. Changes in one population over time are more difficult to observe, if only because of the need for continuity of observation. However, where these sorts of observations are available, as in the reversal of melanism in *Biston betularia* (Cook *et al* 1970) the evidence is very convincing. There are excellent reviews of the differences that can be found between populations (e.g. Clausen 1951; Mayr 1963; Grant 1971). These provide overwhelming and critical evidence for the power of evolution.

Two crucial points arise from this wealth of evidence. Firstly, such evolutionary divergence can take place on a very local geographical scale, because of the power of selection in overcoming gene flow between the populations being subject to different selection. This gene flow would otherwise tend to prevent their gene frequencies becoming different. As a result we do not have to think that divergence can only take place between populations that are *allopatric* — isolated from each other. It can also take place in populations that are *parapatric* — adjacent to each other and able to exchange genes. A good example is in the cliff populations of the grass *Agrostis stolonifera* (Fig. 14.1). Whether or not it can take place in populations that are *sympatric* — coincident and having free exchange of genes — is not so clear, because many examples thought to be sympatric turn out, on careful examination, to be parapatric on the actual scale at which selection interacts with gene flow. This is clearly so in the case of the hawthorn fly/apply maggot (*Rhagoletis*) which has evolved the apple-eating race within the population range of the original

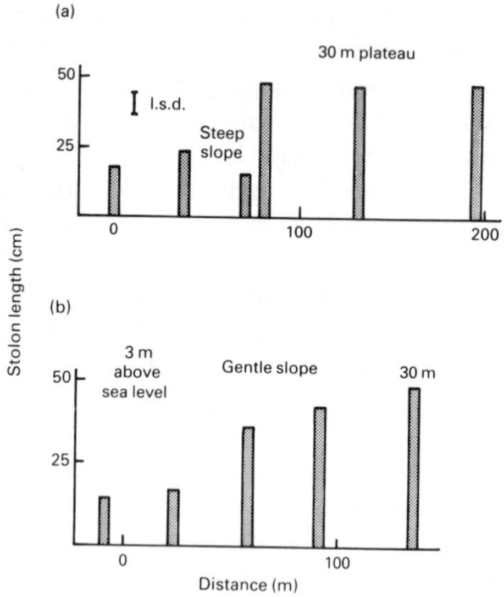

Fig. 14.1. The stolon length in cultivation of populations of the grass *Agrostis stolonifera* taken from two transects across exposed sea cliffs showing how localized population differences can be if the environment changes sharply. (a) Sudden change transect. (b) Gradual change transect (from Aston & Bradshaw 1966).

hawthorn race, in apparently sympatric, but really parapatric, populations associated with individual trees, aided by genes for host preference (Bush 1975). The same seems true for other cases of apparent sympatric evolution, for example, in the codling moth.

The second important point is related to fitness. It is obvious that the evolutionary process must lead to increased 'fitness', i.e. 'adaptation'. Adaptation is the state for which fitness is the measure. Because we are invoking a genetic determination of survivorship, there must be a direct relationship between natural selection and increased survivorship. But of course it does not lead to perfect adaptation, because 'evolution' is a process of 'tinkering' — adding to or modifying that which already exists (Jacob 1977). Therefore we can expect to find curious compromises and modifications where an existing character is converted, as best it can be, to a new function. Perhaps we should use the term 'abaptation' (Harper 1982).

Nevertheless evolution allows a species to move into new habitats, and the range of habitats colonized by a species, its *ecological amplitude* or its *niche diversity*, will, to a large extent, be determined by its evolutionary flexibility. Certainly, for instance, the species that colonize metal-contaminated sites would be precluded from them if they did not possess the necessary evolutionary flexibility to develop metal-tolerant populations.

14.3 Evolution at the level of the community

Many of the examples of evolution we have looked at so far have nothing to do with life in a community. Heavy metal tolerance and herbicide resistance in plants arise as the result of interaction between species and their simple physical environment. The same can readily be found in animals, for example, in adaptation to climate, as in *Drosophila pseudo-obscura* (Dobzhansky 1941) and *Rana pipiens* (Moore 1949).

However, as has been amply displayed throughout this book, there are profound community factors affecting the performance and survival of individual species, populations and individuals. At the extreme, we have seen how interspecific competition, which is a property of communities, can be so powerful that it leads to the extinction of one component in a mixture. The crucial point is that any such factor which causes reduction in performance or survivorship will cause evolution, because it can cause selection if appropriate genetic variability is present that can be differentially selected.

Since community-generated environmental forces can be so powerful, it would seem that they must be an important component of the evolutionary process. Darwin (1859) certainly envisaged this when he described natural selection as "silently and sensibly working, whenever and wherever opportunity offers, at the improvement of each organic being in relation to its *organic* and *inorganic* conditions of life" [italics added]. Inorganic conditions must be taken to be *physical* and *chemical* factors. Organic conditions would seem to be *biotic* factors, the effects of other living organisms, although Darwin does not actually define what he means. Biotic factors, of course, can give rise to physical and chemical conditions, but if they have resulted from organisms living in communities, they are relevant. The range of species that occur in a community will be primarily determined by history and by external physical and chemical factors. But beyond this it is clear that community-generated factors can have a crucial role. They cannot change the basic genetic processes which generate variability in populations, but they can certainly influence its fate. However, community-generated factors are complex. As a result, in a community there is a complicated hierarchy of selective, and therefore, evolutionary, processes.

Firstly, there are the essentially internal physical and chemical effects generated by the community — affecting the general local environment. Then there are the specific effects of particular living organisms, acting individually or jointly, on each other. Not only can these effects be complex by nature — the previous chapters have made this amply clear — but they are generated by populations of living organisms that can change in abundance and activity and can have profound effects on the type of selection that they will cause.

But populations of living organisms can also change in genetic constitution. If we are considering that one species may be evolving in relation to the effects on it of another, we must also remember that the second species can itself evolve in relation to the first. This, which is true coevolution as defined by Janzen (1980a), can lead to complicated evolutionary pathways. One particular consequence is that any evolutionary advantage gained by the first species may be nullified by the second, with the additional implication that a species may have to evolve if it is to keep its position in the community, because the latter is itself not static, aptly described as the 'Red Queen effect' by Van Valen (1973).

We are therefore dealing with a new and exciting part of evolutionary biology, with many possible new principles whose relevance to reality are not yet altogether understood. We suggest that because such community-dominated evolution is so important it should be distinguished as 'synevolution', analogously to synecology. This follows the idea of a remarkable early worker in this field, Sinskaia (1931), who introduced the term 'synecotype' in a long forgotten review. Introducing 'synevolution' has the advantage that it is a term covering all aspects of evolution where community factors are involved, leaving 'coevolution' to be limited to the process of reciprocal genetic change that occurs in populations of ecologically interacting species. [See Futuyma and Slatkin (1983) for recent discussion.] But despite such an early start, the surface of the problem has only been scratched, and there is scope and need for a great deal of further critical work on populations actually exposed to community-induced conditions.

14.4 Evolution in relation to physical factors produced by the community

Although they may be inorganic in strict terms, the physical and chemical conditions of the environment produced by the community give rise to perhaps the simplest sort of evolution that is community related. This can manifest itself in many different ways. The products can be tested in simple transplant experiments into a standard environment or into a series of control environments.

In plant communities, tall growing vegetation produces a very distinctive light profile for other plants. So, on the one hand, we have species that are light demanders, and shade intolerant, being selected to be as tall as this vegetation; for instance, tormentil (*Potentilla erecta*) populations associated with the tall growing grass *Molinia caerulea* are nearly twice as tall under a common environment as those from the adjacent dwarf *Agrostis–Festuca* grassland (Watson 1969). On the other hand, we have species that are shade tolerant and cannot in any way match the height of the community, evolving various adaptations to shade conditions; for instance, woodland populations of cocksfoot (*Dactylis glomerata*) have a lax growth form (Turesson 1925); similar populations of the goldenrod (*Solidago virgaurea*) have both distinctive light intensity–photosynthesis response curves and physiological plasticity adapted to low light intensities (Björkman & Holmgren 1963) (Fig. 14.2).

Some aspects of the physical environment of a community are less easy to reproduce experimentally but their evolutionary effects are very spectacular. The most remarkable example is in the early Russian work on the cruciferous weed of flax (*Camelina sativa*) and other crops (Sinskaia & Beztuzcheva 1931). Overlain on a simple geographical differentiation is a specific differentiation in adaptation to the particular crops with which the weed coexists. The populations that have become obligate weeds of flax, by being harvested and subsequently sown with the flax, mimic the flax almost completely in height, time of ripening and seed morphology. They have even lost the plasticity of growth form possessed by the more common generalist populations. This mimicry relates even to the particular flax variety with which the *Camelina* is associated.

Sinskaia (1931) called these populations that have evolved in relation to other plants 'synecotypes', and distinguished examples that are 'mimics' like *Camelina* from others that are 'differentials' which have evolved a growth form to evade the crop, e.g. *Spergularia vulgaris* and *Lepidium sativum*.

Recently a considerable amount of attention has been given to the effects of general community characteristics on evolution, particularly on life history parameters, as a result of MacArthur and Wilson (1967), who distinguished between r and K selection after the two terms in the logistic

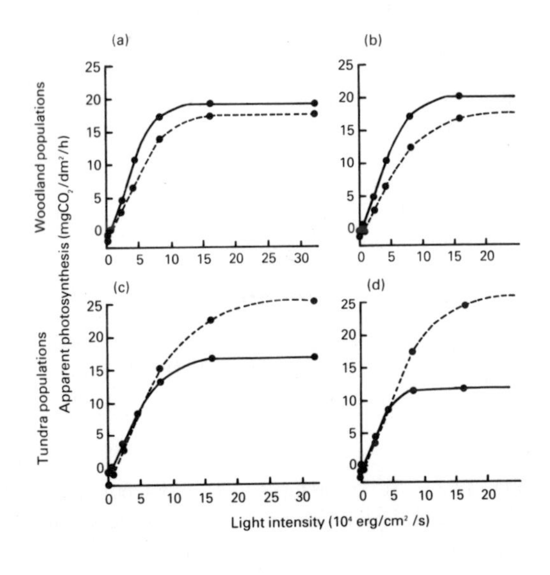

Fig. 14.2. The response of photosynthesis to light intensity in populations of goldenrod *Solidago virgaurea* (upper two from woodland, lower two from tundra) after pretreatment with high light intensity (broken line) or low light intensity (unbroken line), showing the remarkable adaptation of each population to the light regime of its environment of origin: (a) Hallands Väderö, (b) Nörreskov, (c) Beskades, (d) Rönneberga, (from Björkman & Holmgren 1963).

equation. In situations where there is no community and no competitors, it is suggested that selection will be for reproduction, whereas in situations that are saturated there will be selection for survival and persistence. In colonizing species earliness of reproduction particularly will be favoured (Lewontin 1965). Whilst providing a useful conceptual framework, there have been several critical appraisals of this concept (e.g. Stearns 1977; Itô 1980; Begon 1985).

r selection favours	*K* selection favours
rapid development	slower development
high maximum rate of increase	greater competitive ability
early reproduction	delayed reproduction
small body size	large body size
single reproduction	repeated reproduction
many small offspring	fewer larger offspring

The acid test for such ideas on synevolution is to see what happens in practice. There is certainly excellent evidence in the shortlived grass *Poa annua* (Law *et al* 1977). The populations from open (disturbed) habitats are early and profusely flowering; the populations from closed (stable) habitats are late flowering and put much more emphasis on vegetative growth, and live longer (Fig. 14.3). The same is found in goldenrods (*Solidago* spp.) (Abrahamson & Gadgil 1973). The populations from disturbed sites invest heavily in seed production; woodland and pasture populations do not. Since population differences must arise from genotype differences, it is interesting that in populations of *Taraxacum officinale*, clear differences can be found between genotypes in their life history characters (Table 14.1) (Solbrig & Simpson 1974). It is significant that genotype A was the most frequent genotype in populations from dry disturbed habitats and genotype D the most frequent in wet undisturbed habitats.

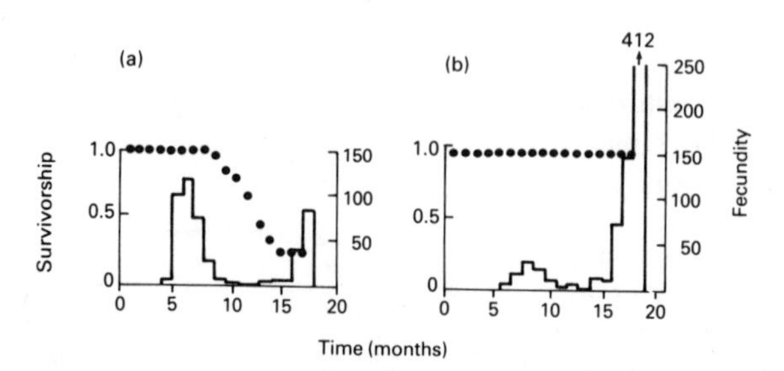

Fig. 14.3. Survivorship (dots) and fecundity (histogram) for two populations of *Poa annua* derived from (a) a disturbed environment, and (b) a stable environment (from Law *et al* 1977).

Table 14.1. Number of inflorescences per plant, number of seeds per inflorescence, mean seed weight, and average number of seeds for two dandelion biotypes in a greenhouse experiment (from Solbrig & Simpson 1974).

	Biotype A	Biotype D
Average number of inflorescences/plant	0.24	0.04
Average number of seeds/inflorescence	105.5	204.3
Average seed weight (mg)	0.315	0.437
Total number of seeds/plant	25.3	8.2

14.5 Evolution at the same trophic level — competitive ability

In fact in the last example we have been considering, we have slipped without realizing it from synevolution in relation to general physical effects of communities to synevolution in relation to specific biological effects produced by other organisms, because this is what *K* selection is all about. In a community, as previous chapters have shown, growth usually proceeds until it becomes limited by the supply of essential resources. For animals it is usually food. For plants it may be light, water or nutrients — which crudely are also forms of food. It was Wallace rather than Darwin who emphasized that food supply was the crucial factor limiting population growth and that consequently any variation leading to enhanced ability to obtain food would be selected for. What is interesting is the paucity of critical evidence to back this obviously important principle. It is true that every comparison between species in nature, or better, in competition experiments, which shows that one is better than the other under a given set of conditions, is indirect evidence of such evolution. But it does not give clear evidence of selection acting on relevant genetic variation. The differences found could indeed be due to something else.

However, genetic variability in competitive ability (i.e. relative success in mixtures) was first shown in rice, among different genotypes (Sakai 1955, 1961), although the heritability was low (0.12). Sakai considered competitive ability as a character in its own right, but subsequent work with rice indicated that the differences were due to differences in growth form and therefore light interception, and that the outcome of competition experiments depends on the conditions under which they are carried out since these can affect the relative advantage of any particular growth form (Jennings & Aquino 1968).

From this it is a short step to selection experiments in which a mixture of genotypes is maintained for several generations. This can readily be achieved in a species like rice by constructing a segregating population by hybridization of two distinct parents and maintaining it over a number of generations. In such a mixture the gene for dwarf was progressively eliminated, most rapidly in dense stands at high nitrogen levels, in only a few generations (Jennings & Herrera 1968).

Similar evidence has been obtained in *Drosophila* and *Tribolium*. There is abundant evidence that competitive ability is heritable (e.g. Lerner & Ho 1961; Mather & Cooke 1962). In relation to this, various experiments have been carried out to see if experience of competition between populations of two species has any effect on the subsequent performance of those populations in competition. If *Drosophila simulans* is mixed with *D. melanogaster*, *D. simulans* is successful; but in subsequent generations the *D. simulans* eliminates *D. melanogaster* more rapidly (Moore 1952). When houseflies are put into competition with blowflies, the latter are inferior; but after 25 generations of repeated exposure of the species to each other, the blowflies evolve to be superior and to eliminate the houseflies in competition (Pimentel *et al* 1965) (Fig. 14.4). This was confirmed by test competition experiments. However, unlike the plant examples, little is known as to the nature of the changes that have taken place. It is possible that it is due to differences in resource (i.e. food) acquisition, but it is also possible that it is due to some other interaction, for example, chemical antagonism. The changes require analysis in the ways suggested in Chapter 11.

If we move away from direct evidence for changes in competitive ability we can find in plants a number of pieces of evidence which show just how much local differentiation of populations (in these cases over distances of only 10 m or so) leads to profound superiority of native material compared with alien of the same species in the face of the total competition provided by the community. By reciprocally transplanting genotypes of sweet vernal grass *Anthoxanthum odoratum* among different plots of the Park Grass fertilizer/lime experiment at Rothamstead, Davies and Snaydon (1976) were able to show that, in competition with the whole community, the survival and growth of the alien compared with that of the native populations was on average only 50% even after 12 months only. The half-life of the alien was 8 months, that of the natives 24 months.

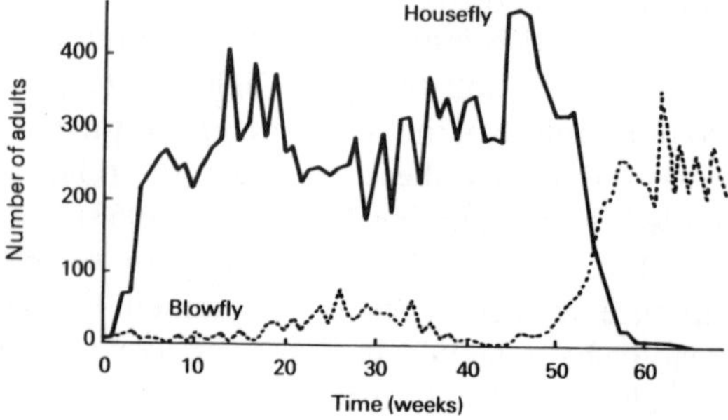

Fig. 14.4. Changes in competing populations of houseflies and blowflies in a 16-celled cage (from Pimentel *et al* 1965).

These populations had already been shown to have evolved a complete complex of characters by which each appeared to be better adapted to the conditions of their native communities. Because the plots were artificial the evolution must have taken place in less than 60 years.

Perhaps the extreme of this sort of evidence is that of *Ranunculus repens* from adjacent woodland and grassland. Its relative survival in reciprocal transplant experiments between the two habitats is given in Table 14.2 (Lovett Doust 1981). It is clear that the importance of the differences which occur among populations within species can only be appreciated when the populations are put into competition. This can only readily be done with perennial plants. But the conclusion, that competitive ability can be drastically changed by evolution in natural conditions over only a few generations, is of universal relevance.

14.6 Evolution at the same trophic level — coexistence of species

When two or more species occur together at the same trophic level in a mixture, evolution of enhanced competitive ability in one species may occur, implying that the species concerned takes an increased share of the resources available. But we have already seen (Chapter 6) that there is an important type of ecological interaction in which, by diversity of resource utilization, each interacting species in a mixture can succeed without

Table 14.2. Selection coefficients (1−performance of alien/performance of native) against alien ramets of *Ranunculus repens* when reciprocally transplanted into woodland and grassland (from Lovett Doust 1981).

Measure of growth	Coefficient* of selection against aliens	
	Woodland site	Grassland site
(a) Total production of daughter ramets	0.071	0.784
(b) Number of ramets produced by each stoloniferous ramet	−0.004	0.460
(c) Ramets producing stolons (%)	0.250	0.670
(d) Total length of stolon tissue produced	0.786	0.802
(e) Ramets producing flowers (%)	0.0	0.285
(f) Mean birth rate of leaves	0.064	0.161
(g) Mean death rate of leaves	0.116	−0.008
(h) Mean number of leaves	0.072	0.159
(i) Mean turnover of leaves	−0.247	0.207

* A negative selection coefficient indicates that aliens had a greater value for a given measure of growth than native ramets.

having a negative effect on the other(s). Such niche diversification, or annidation (Ludwig 1959), is now seen to be a crucial and widespread phenomenon. It may well explain much of the coexistence of species in natural communities, since one of its crucial consequences is that each species reacts more to its own density than to that of the other species. As a result fitness is frequency dependent in a way leading to stable mixtures; each species performs best when at low frequency in the mixture. There would appear to be positive interactions between the species because each will usually do better in the mixture than expected from its growth alone at the same density.

Several authors have explored what would happen theoretically if, in a mixture of interacting species, there was genetic variation within the species in resource utilization. Despite earlier suggestions, most models now indicate that there would be evolution of niche diversity leading to stability (e.g. León 1974; Lawlor & Smith 1976). This form of synevolution has been called 'coevolution', in which two (or more) species (or populations) evolve in relation to each other. However, exactly what occurs will depend on the variety of resources available and the number of competing species (Roughgarden 1976) and genetic and phenotypic variance (Slatkin 1980). A good review is provided by Slatkin and Maynard Smith (1979).

There are some good examples showing the occurrence of genetic variability within species leading to positive interactions and therefore coexistence of genotypes, although the best examples involve coexistence of species. In a comparison of 22 laboratory strains of *Drosophila melanogaster* it was found that many performed better in mixed culture than in pure culture (Lewontin 1955). Careful analysis of changes in population cages containing mixtures of different chromosome arrangements in *Drosophila persimilis* (Spiess 1957) argues similarly for positive interactions, leading to frequency-dependent fitness; both Klamath and Mendocino chromosome arrangements when combined with Whitney reached stable mixtures containing a frequency of Whitney of about 0.65.

Similar studies have been made in *Drosophila melanogaster.* The Esterase 6 locus shows marked interactions between the fast and slow alleles so that whatever the starting frequency in mixtures the equilibrium value returns to 0.3 for Esterase 6F. The alleles were always found to be most fit when at low frequency (Kojima & Yarbrough 1967; Yarbrough & Kojima 1967).

In plants similar frequency dependent fitness leading to stable equilibria has been found for the genes at the S locus for seed coat colour in populations of the lima bean (*Phaseolus lunatus*) (Harding *et al* 1966). There was always a negative correlation between fitness and frequency.

Although studies on a mixture of varieties may seem a little arbitrary because they can involve the combined effects of several genes, they have a particular interest in that they show what the combined effect of several genes can achieve. When mixtures of different varieties of flax and linseed (*Linum usitatissimum*) are made, startling increases in fitness of the varieties compared with their performance in pure stands, can be found (Khan *et al* 1975) (Fig. 14.5).

As already argued, the corollary of this is that each component of the mixture does best when it is at low frequency. The degree of positive interaction was found to depend on the particular combination of varieties, implying that there is perhaps a specific as well as a general 'combining' ability among such genotypes. Similar interactions have been found in mixtures of varieties of *Lolium perenne* although positive interactions were less common (Rhodes 1968).

In these two instances some indication could be obtained as to the reasons why the varieties concerned performed better when in the presence of another variety rather than itself. There was always a way in which the components differed in their use of the environment, in *Linum* in time, and in *Lolium* in space.

So there is plenty of evidence for genetic variation and differentiation which allows individuals to coexist in communities by differences in resource utilization. However, there are plenty of other examples where this has not been found (Harper 1977). But such differentiation is unlikely to be common and well developed unless it has actually been selected for. Tests between randomly chosen genotypes are unlikely to show it, a point we will return to. But the crucial point is that variability does exist.

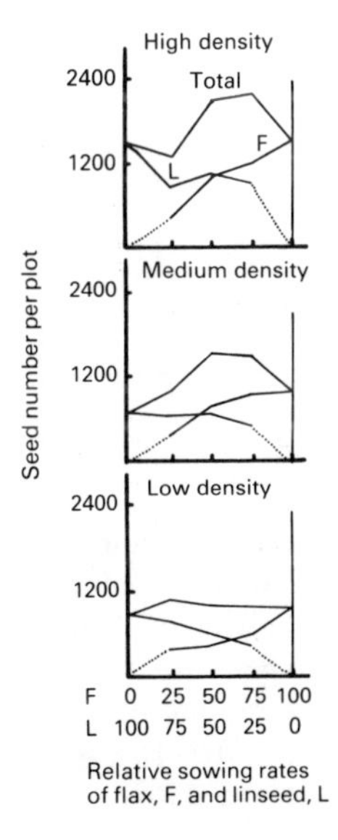

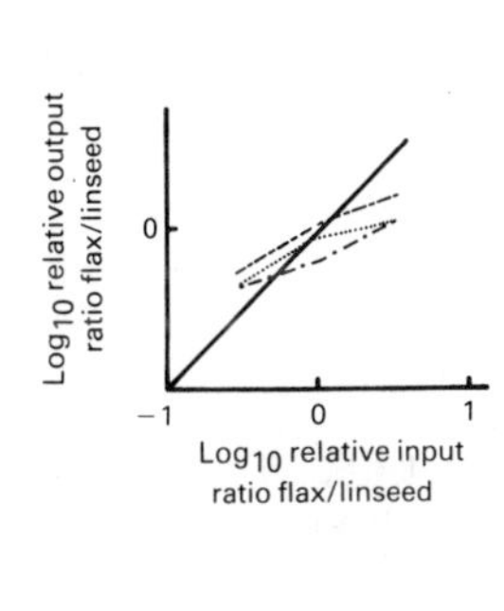

Fig. 14.5. Replacement series graphs for seed number per plot of flax and linseed varieties grown in mixtures at different densities, expressed as raw data (left) and ratio diagrams (right) (density:·····low; –·–medium; –––– high.), showing that each component does best when it is at low frequency in the mixture, and that flax is favoured by high density (from Khan *et al* 1975).

Various selection experiments have been carried out in an attempt to induce the evolution of differences in resource utilization experimentally. We have already seen that marked improvement in competitive ability of one species occurred in an experiment involving mixtures of houseflies and blowflies (Fig. 14.4). However, in one population of the two species mixture, a balanced situation evolved in which the two species remained in reasonable balance over a very long period (Pimentel *et al* 1965). This is good evidence that evolution of a coexistence mechanism is possible. The other good example is in *Drosophila melanogaster* (Seaton & Antonovics 1967). Two populations marked by 'wild' and 'dumpy' were put together for five generations without being allowed to interbreed. At the end of this time marked improvements could be detected in the ability of each population to coexist with the other (Fig. 14.6).

However, when other investigators have looked for the same coevolutionary changes in experimental populations, they have not necessarily been able to find them. Perhaps this is because the distinctive genotypes required for niche differentiation are commonly destroyed by recombination (Futuyma 1979). But this is not a very satisfactory explanation since other sorts of adaptive changes are readily evolved. More investigation is required.

This leads us, then, to the evidence of such coevolution in nature. There are, of course, plenty of examples of coexisting species pairs, where the coexistence appears to be due to resource partitioning because of divergence in niche utilization, which may be revealed by character displacement (Chapter 6). But many of these appear to be of ancient origin

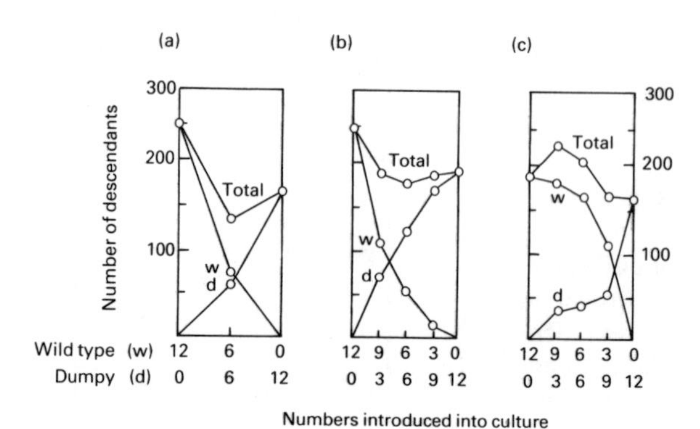

Fig. 14.6. The relationship between the numbers of fertilized female *Drosophila* (wild type and 'dumpy') introduced into a culture and the number of descendants produced after one generation, showing evidence for evolution of coexistence. (a) populations with no previous experience of mixed culture; (b) wild type with selected 'dumpy' (after five generations of mixed culture with wild type); (c) 'dumpy' with selected wild type (after five generations of mixed culture with 'dumpy') (from Seaton & Antonovics 1967).

and could be the result of allopatric evolution in response to differing selective pressures in the different countries of origin. To be convincing we need to find character displacement occurring only where the species overlap, and for these examples to be, as far as possible, due to recent overlap and genuine competition, so we can be certain that effects other than overlap have not been the cause. At the same time the actual manifestation must be clear — it must be heritable and able to be distinguished from clinal changes (Arthur 1982). As a result some putative examples, such as the rock nuthatches (*Sitta neumayer* and *S. tephronota*), may not be good evidence (Grant 1972) (see Chapter 6, Fig. 6.8). Similar problems occur in wider studies on character displacement in whole guilds of species (Strong *et al* 1979). However, further analysis does suggest that there are cases which are incontrovertible (Hendrickson 1981).

On the Galapagos Islands there is the remarkable series of finches, made famous by Darwin for their adaptive radiation. What is of interest to us are the particular inter-relationships of one group, the ground finches (*Geospiza*). There are three very similar but distinct species. When two of the species (*G. fuliginosa* and *G. fortis*) occur separately on different islands (Crossman & Daphne) their beak depths are more or less the same, but where they occur together on the same islands (Charles & Chatham) the species have different beak depths and even the ranges of variation do not overlap. When a third species occurs (*G. magnirostris*) on Abingdon and Bindloe, the species differences are modified so that the ranges of all three species do not overlap (Lack 1947) (Fig. 14.7). All this can be related to changes in the size of food items in the diets of the species. It is difficult not to believe that this is a clear case of simple coevolution.

This type of analysis has been extended to give it a community-wide basis for all the terrestrial breeding bird species of a region, for example,

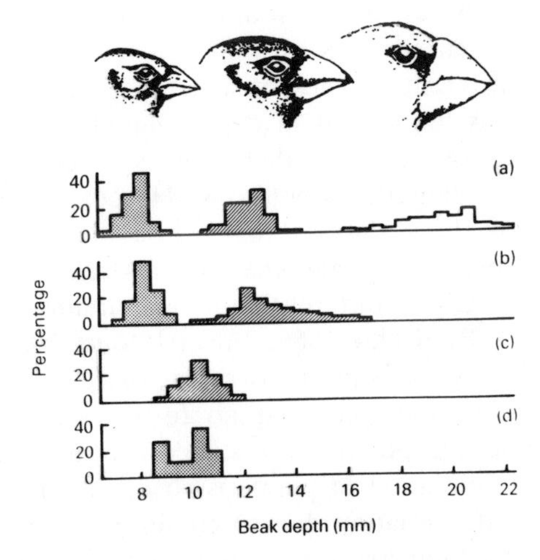

Fig. 14.7. Frequency distributions of beak depths in the genus *Geospiza* (left and stippled histogram *G. fuliginosa*, middle and shaded histogram *G. fortis*, right and open histogram *G. magnirostris*) on the Galapagos Islands, showing character displacement on those islands where two or more species coexist (a) Abingdom and Bindloe. (b) Charles and Chatham. (c) Daphne. (d) Crossman (from Lack 1947).

for the Tres Marias Islands off the coast of Mexico. However, the displacement in *Geospiza* does not then necessarily match what happens on a community-wide basis. There must be other local factors at work causing changes in the individual populations; but this is only to be expected since evolution is a strictly local phenomenon in origin.

The detailed evidence provided by mud snails (*Hydrobia*) in Northern Denmark (Fenchel 1975b) (Chapter 6, Fig. 6.10) is very tantalizing. Here, because of glaciation, we have a series of populations which cannot have coexisted for more than 150 years. It is difficult to explain the regularity of the changes, the character displacement, except as an outcome of coevolution leading to resource partitioning, even if the experimental evidence on feeding is not yet completely clear.

The real test for partitioning of resources is performance in mixtures, as we have already seen. In *Anolis* lizards, on all the Lesser Antilles islands with one species, male *Anolis* have average adult jaw sizes around 10 mm, whereas on all the islands where there are two species one has jaws of about 7 mm and one of about 15 mm (Schoener 1969b). Recently Pacala and Roughgarden (1982b) have tested the implied niche separation (different jaw sizes mean different prey sizes) by mixtures put into experimental enclosures. When *A. wattsi* was introduced into *A. gingivinus*, to which it is very similar, it had a very negative effect on growth and caused *A. gingivinus* to alter its perch height. However, when introduced into *A. bimaculatus*, from which it is much more different, no effects were detectable. However, while we can presume such proven niche differentiation has evolved as a result of species interaction, we cannot be certain because we are dealing with an event that took place some considerable time in the past.

It is therefore interesting that the only evidence for evolution of resource partitioning in plants is not only based on performance in mixtures, but must also have involved a situation clearly of recent origin. When samples of white clover (*Trifolium repens*) were removed from different patches of a field dominated by one of four grass species, it was found that in experimental conditions the clover always grew best in association with the grass species with which it had previously been growing (Turkington & Harper 1979). The most remarkable feature of these data is that white clover, because of its stolon growth, is very mobile and that the material came from a single field. However, very local differentiation in stoloniferous perennials is well known (Aston & Bradshaw 1966; Lovett Doust 1981). It would appear here that coevolution is probably taking place almost on a genotype/genotype basis. It is a great pity that no further evidence is yet forthcoming for plants on a population basis, because direct experimental testing of niche differentiation in plants is so much simpler than in animals. Evolutionary differentiation due to community effects, such as that described earlier for *Camelina sativa*, needs investigating for resource partitioning.

14.7 Evolution at the same trophic level — coexistence of genotypes

The possibility of coevolutionary changes as a result of interactions between genotypes of different species raises the interesting possibility of effects arising from interactions between genotypes of the same species. Because of their essentially greater similarity to each other than to genotypes of other species, it is between genotypes of the same species that the greatest interactions should occur. The effects of density on the individuals of both plant and animal species is very obvious. It therefore follows that there should be strong selection pressure for niche divergence between genotypes. The mathematical possibility of this has been demonstrated very clearly by Antonovics (1978). It will, of course, be limited by the genetic exchange inevitable between members of a single species, but the genetic mechanisms which enable a discrete polymorphism would overcome this. We have already seen evidence for frequency-dependent fitness associated with individual major genes, and with distinct chromosome arrangements. We have used this as evidence for the existence of genetic variation from which populations differing in resource utilization can be evolved. But this variability may indeed be an inbuilt characteristic of populations, produced by the selection which arises from the intense competition between individuals.

The example which lends remarkable support to this argument is the bulk hybrid population of barley analysed by Allard and Adams (1969). This population, produced originally by intercrossing a number of distinct cultivars, was shown after the elapse of 18 generations to have retained a remarkable range of variability manifested as different homozygous genotypes (barley is inbreeding) retained in the population, against what would have been expected from theory. When eight of these genotypes were extracted from the population and grown in pair mixtures in all possible combinations, in nearly all cases they showed positive interactions, i.e. their performance in mixture was superior to that in pure stands at the same density (Fig. 14.8). This can only be interpreted as being due to niche differentiation of the sort already shown occurring in *Linum* (Fig. 14.5).

When pair mixtures of a random set of different barley cultivars were made by the same authors, positive interactions were not found. It has to be concluded that the situation in the bulk hybrid populations arose by coevolution, as a result of the continued interactions experienced by the genotypes segregating out of the hybrid population. This would not have occurred in the random pairs of cultivars. Although it is true that a self-fertilizing species allows the formation of genotypes differing in several genes, it seems reasonable to believe that this kind of coevolution should occur within, as well as between, species.

It is certainly found between the two sexes of sorrel, *Rumex acetosella*

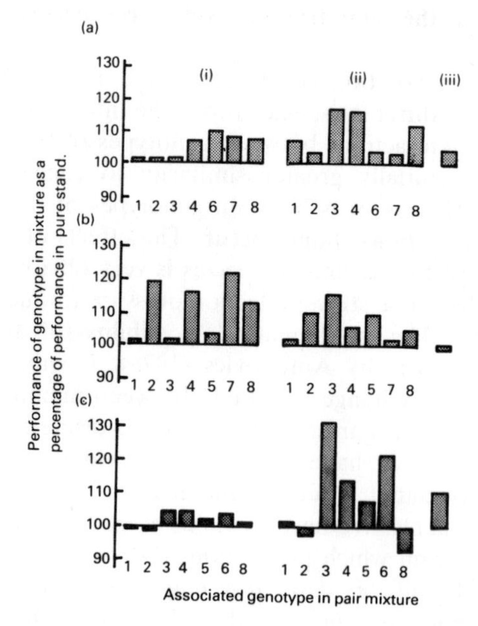

Fig. 14.8. Examples of the positive interactions in pair mixtures found between eight genotypes abstracted from a stable polymorphic bulk hybrid population of barley *Hordeum vulgare* [illustrated by three genotypes: (a) Genotype 5 (b) Genotype 6 (c) Genotype 7] suggesting coevolution of genotypes within a single population. (i) Effect of associate on genotype. (ii) Effect of genotype on associate. (iii) Yield relative to all eight genotypes when in pure stand (from Allard & Adams 1969).

(Putwain & Harper 1972) and leads to maintenance of the sex ratio in the lengthy vegetative phase of the species. Coevolution in sexual differentiation, however, is outside the scope of this chapter.

More vague evidence for the possibility comes from what is commonly called 'character release', greater variability occurring in the allopatric populations of species pairs, originally pointed out by Van Valen (1965). In the absence of the effects of the other species, it is suggested that natural selection is encouraging diversity within each species itself because each species is able to occupy a wider range of habitats when the other species is not present. Niche diversity is only possible when the alternative niches are not already occupied. There are several possible examples; perhaps the best is that in *Geospiza* sp. (Grant *et al* 1976). Here the heritability of the variation was proven, thus removing the possibility that it is merely a phenotypic effect. But this character release could equally be due to the allopatric populations having a wider geographic distribution in the absence of the other species and therefore being exposed to a wider range of different local selection pressures.

14.8 Evolution across trophic levels — pathogens and parasites

Of all the complexities that exist in communities of organisms, one very distinct type of interaction is that which acts across trophic levels. This process is manifested in a myriad of interactions, both ecological and evolutionary, between hosts and their predators, parasites and pathogens. The immediate consequences of herbivory, parasitism and disease in communities are readily apparent to the ecologist. As a result there is an extensive array of arguments that the ultimate effect of these events is evolution through lowered fitness and exposure to selection. Haldane (1957), for example, suggests that in the development of modern man "infectious diseases have exerted some of the strongest pressures".

Ecologically there is a dichotomy in the events that occur across trophic levels. On the one hand 'prey' may be consumed in entirety, yet on the other hand it may be only damaged. What is crucial to evolution is what precludes extinction of the host by the predator, herbivore or parasite. Removal of part or whole of the host should play an important part in determining the type of defence mechanism that is evolved which determines the co-occurrence or otherwise of both constituents.

Four hypotheses have been advanced in explanation for the co-occurrence. Van Valen's (1973) 'Red Queen hypothesis' proposes that a constant rate of extinction of species implies a constant rate of deterioration of the environment for the interacting species as they evolve. The species therefore maintain themselves and escape extinction by constant evolution. Maynard Smith (1976), however, rejects this on the grounds that the zero-sum assumption — the increase in fitness in any one species is exactly balanced by the sum of the losses of fitness of all the others — is rarely likely to be met, because nature is not so meticulous. Co-occurrence is achieved because nature is sloppy. As a result, selection for compensatory alterations in fitness will proceed at reduced rates unless major changes in species or environment occur that dramatically alter the framework of interspecific interactions. Rosenzweig (1973) subscribes to the Van Valen hypothesis, suggesting that in exploiter–victim situations evolutionary rates will be balanced because the fixation rate of mutations enhancing fitness will be proportional to the degree of vulnerability. Thus in predator–prey interactions, predator efficiency will evolve faster than prey resistance if an inefficient predator is chasing already resistant prey, and vice versa. Against all this Simberloff (1978) and Gould (1980) propose that substantial evolutionary changes are not the result of deterministic causes but occur by historical and evolutionary accidents. Can we make any sense out of the problem?

If we start with the particular case of host–parasite interactions, we find that the persistence of pathogens and parasites and their hosts over geological time is well documented (Pirozynski 1976). The reasons for this apparent stability between interacting species have, however, not been easy to see. Strong (1977) has summed up the question eloquently.

> "I am...in a lowland tropical rainforest, an appropriate place to contemplate plant–parasite interactions. It is almost constantly wet here and never very cold; microbial growth is not restrained by weather. Every plant bears fungal colonies or epiphyllae, and the decay from bacterial digestion begins in foliage within hours after it is cut or falls from plants. Yet the forest is healthy. None of the 500 or so plant species within walking distance is apparently free of microbial activity. The burns, blazes, spots, and lesions of infection are tiny and rare on all but the oldest leaves, even on plants that retain leaves for many months. This is the ultimate enigma of host–parasite interactions. What are the mechanisms and mechanics of the equilibrium between hosts and parasites? That the balance is unstable we see in the case of plagues, pest outbreaks, and disease epidemics in agricultural plants; that it is not simple we see in the plethora of diseases suffered by every plant species."

Perhaps the solution to these questions can come better from studies of evolution during epidemic disequilibria when evolutionary change must occur, than from situations where there are long-term equilibria when evolutionary change must be unlikely. But an explanation of equilibrium situations is, of course, our ultimate aim.

When host and pathogen persist in an ecosystem in dynamic balance, endemic host–pathogen interactions must constantly occur. A classic example is southern maize rust (*Puccinia polysora*) in Central America and northern South America. This rust is commonly found infecting individual maize plants throughout its natural range, yet there is no historical evidence that extensive epidemics of rust on maize in its native Latin American habitat have occurred in colonial or recent times (Borlaug 1972). Similar data are also available for wild barley (*Hordeum spontaneum*) and mildew (*Erisyphe graminis*), and wild oats (*Avena sterilis*) and crown rust (*Puccinia coronata*) in Israel (Wahl *et al* 1978). In both these cases considerable genetic diversity characterizes both the host plant and the pathogen populations (Fig. 14.9), even though the hosts are predominantly autogamous species (Allard 1965).

Browning (1974) considers the several mechanisms which can determine the success of the pathogen (Table 14.3). In natural populations, genetic variation for many, if not all, of these different mechanisms can be found, despite the views of early investigators. In *Avena sterilis*, genes conferring nonspecific (horizontal) resistance (Van der Plank 1968) give protection against stem rust (*Puccinia graminis*) (Sztejnberg & Wahl 1969), while genes conferring specific (vertical) resistance to individual crown rust (*Puccinia coronata*) races also occur (Browning 1974). Tolerance, where the hosts are susceptible but show less damage than true susceptible ones, is an equally important mechanism with high

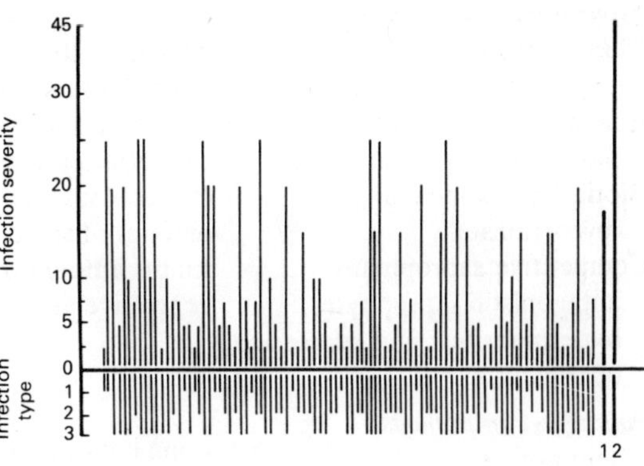

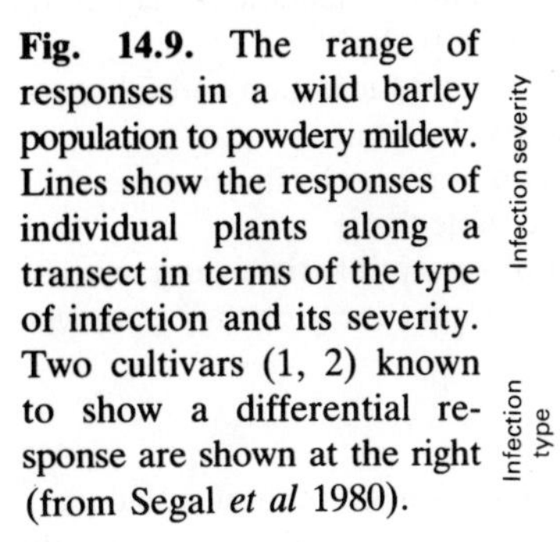

Fig. 14.9. The range of responses in a wild barley population to powdery mildew. Lines show the responses of individual plants along a transect in terms of the type of infection and its severity. Two cultivars (1, 2) known to show a differential response are shown at the right (from Segal *et al* 1980).

Table 14.3. Mechanisms determining the success of a pathogen (from Schmidt 1978)

Components of mechanism	Effect of disease development
Host components	
Immunity	Inoculum has zero potential on non-host species
Specific genetic resistance	Some genotypes of the pathogen have zero potential on resistant genotypes of a host species
General genetic resistance	May not limit initial amount of disease, but may limit colonization or production of inoculum and therein subsequent amount of disease
Tolerance	Does not limit amount of disease, but when amount of disease is not directly related to the amount of inoculum then a tolerant species can conceivably reduce amount of inoculum, i.e. tolerant species may also possess general disease resistance
Limited infection sites	Limits the amount of susceptible tissue similar to the correction factor $(1-x)$ when x is near 1; however, this limitation is present throughout the epidemic
Physiological age	Limits amount and distribution of susceptible tissue and subsequent colonization and disease development
Numbers of species	Reduces the probability that inoculum will contact a susceptible host
Horizontal dispersion of the host plants	Reduces the probability that inoculum will contact a susceptible host
Vertical dispersion of the host plants	Reduces the probability that inoculum will contact a susceptible host
Natural succession	Restricts the increase of pathogen populations by changing the host species

Environmental components	
Climatic factors such as temperature, moisture, etc.	May limit the amount of infection and conditions latent and infectious periods
Edaphic factors such as nutrients, moisture, etc.	May limit the amount of infection and subsequent rate of disease development
Biotic factors including mycorrhizae	Protects susceptible species by limiting the amount of susceptible tissue
Competitive saprophytes, antagonistic saprophytes, hyperparasites, and succession of organisms	Limits amount of susceptible tissue and restricts vegetative and reproductive growth of pathogens
Pathogen components	
Latent period	Long latent periods limit amount of inoculum
Infectious period	Short infectious periods limit amount of inoculum (usually not a limiting factor)
Genetic potential (for infection)	Limits number of successful infections as indicated for specific genetic resistance of the host
Requirements for a wound as an infection site	Limits amount of susceptible tissue
Dependence on insects for production, release, dissemination, or deposition of inoculum	Limits amount of susceptible tissue and probability that inoculum will reach the infection site if those insects are few or inefficient

heritability in cultivated oats (Simons 1972; Simons *et al* 1978). It is therefore significant that from ecological studies both *Hordeum* and *Avena* have relatively high fitness in relation to these diseases in natural populations (Browning 1974). A formal genetical analysis has yet to be carried out (Barrett 1984).

The importance of some of the mechanisms outlined in Table 14.3 is seen when studying recent epidemics. A common cause of increases in pathogen incidence to epidemic proportions is host uniformity. Operationally this usually means a genetical, uniform, susceptibility to a limited range of pathogen genotypes. Such an event may be man-made, as in hybrid maize and *Helminthosporium maydis* (Ullstrup 1972) where the genetic base of the maize has been restricted in the course of plant breeding, or natural as in the case of the American chestnut (*Castanea dentata*) and blight fungus (*Endothia parasitica*) (Jaynes *et al* 1976) where the necessary genes did not occur. In instances where the necessary genes do occur, we begin to investigate coevolutionary responses. A major step was the early discoveries of oligogenic resistance in host plants (Biffen 1905). This led to the isolation of physiologic races in pathogens (Stakman & Piemeisel 1917) and finally the recognition of duality in the definition of host and parasite phenotypes: a resistant host defines an avirulent parasite phenotype, a virulent parasite identifies a susceptible host phenotype. The genetic basis of this interaction (Flor 1946, 1947) led to the formulation of the 'gene for gene' hypothesis, or the matching gene theory (Gallun & Khush 1980). What is envisaged is that the products of a resistance gene interact with the products of a specific gene in the parasite to determine host resistance and parasite avirulence. In cultivated plants at least it is commonly due to multiple alleles (Day 1974; Barrett 1984). Such genes became widely used in plant breeding. The result has been a whole series of 'boom–bust' cycles in both agriculture and horticulture (Fig. 14.10). A plausible reason for these cycles is that the selection pressure exerted by a genetically uniform and resistant host can be met by a simple evolutionary response in the parasite, but that then Van Valen's hypothesis is fulfilled because the host itself cannot evolve.

But in more natural situations coevolution is more likely. The deliberate introduction of myxoma virus for the control of the rabbit (*Oryctolagus cuniculus*) in Australia (Fenner 1965) is a remarkable example. The rabbit populations in Australia before 1950 were so damaging that the myxoma virus was introduced as a control measure. The initial effects in 1951 were dramatic (Fig. 14.11), almost certainly because of the narrow genetic base from which the host populations were originally derived (24 individuals) and because the two populations had not previously been exposed to each other. Mortality of the rabbits was 99.8%, yet in the outbreak in the following year mortality dropped to 90%, and in subsequent years to less than 40%. Careful analysis showed that two important changes occurred. Firstly the virulence of the myxoma

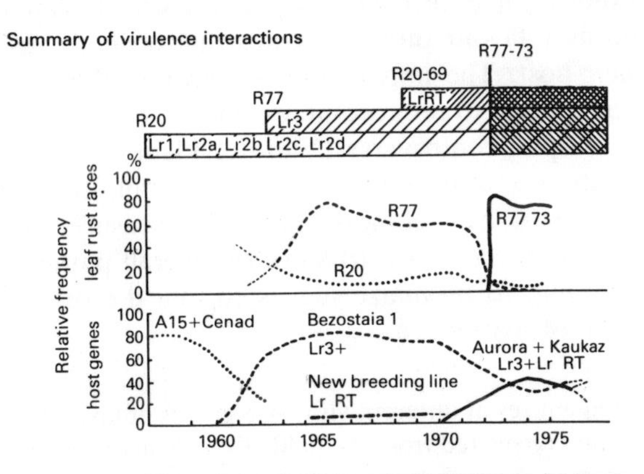

Fig. 14.10. 'Boom–bust' cycles and the evolution of the wheat-leaf rust system in Roumania. The diagram shows the changes that have occurred in the acreage of different wheat cultivars sown, the relative frequency of leaf rust (*Puccinia recondita*) races and a summary of the rust races present and the host genes in cultivars.

Lr: host leaf rust resistance gene. R: leaf rust race. (R20: virulent to resistance genes Lr 1, Lr 2a, Lr 2b, Lr 2c, Lr 2d. R77: virulent to Lr3. R20-69: virulent to LrRT. R77-73: virulent to Lr 3 + LrRT.) (from Ionescu-Cojocarn & Negulescu 1976).

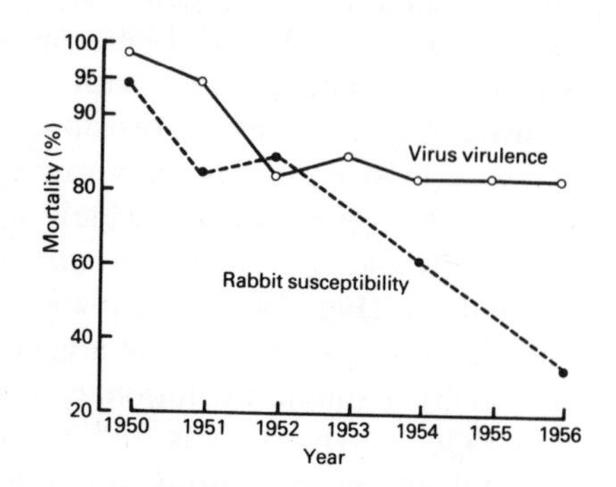

Fig. 14.11. Changes in the virulence of myxoma virus and in the susceptibility of *Oryctolagus cuniculus* in Australia (from Burnet & White 1972).

declined (Table 14.4) causing lowered rabbit lethality. Secondly there was an increase in host resistance. Fenner and Ratcliffe (1965) suggest that selection in the myxoma was not for reduced virulence *per se* but for increased survival time of the infected host. Myxoma strains which result in rapid host death are less likely to be transmitted via mosquitoes (which feed on living rabbits) than those which only slowly lead to mortality. The speed of appearance of these less virulent forms in the early stages of the interaction then enabled sufficient turnover in the rabbit population for selection of resistance (Sobey 1960). Interestingly the myxoma did not evolve towards a benign state although the potential was present.

There are several important implications in this example. Firstly the evolutionary response of both host and parasite can be rapid. Secondly natural selection may favour those parasites that do not cause the rapid extinction of their host. The implication is therefore that virulence of a pathogen must be related to reduced fitness measured over the whole life cycle. However, despite controversy (Van der Plank 1968; Nelson 1973) there is a virtual absence of information on the overall fitness of virulence mutations. An attempt by Leonard (1977) to measure the effects of specific alleles for virulence and avirulence on overall pathogen fitness in *Helminthosporium maydis* on maize suggested that the relative fitness of virulence was 70% of avirulence.

Table 14.4. The frequencies of different virulence grades of myxoma virus (to *Oryctolagus cuniculus*) recovered from the field (from Fenner 1965).

Grade of severity	I	II	IIIA	IIIB	IV	V
Mean survival time of infected test rabbits (days)	≤13	14–16	17–22	23–28	29–50	–
Case mortality rate (%)	99.5	99	>90	<90	60	≤30
Australia						
1950–1951	100	–	–	–	–	–
1958–1959	0	25	29	27	14	5

A third implication is that genetic diversity increases in both coevolving populations and stabilizes the interaction. A new genotype in a pathogen will evoke by selection a new resistant genotype in the host. This could go on *ad infinitum*. However, it may be that each gene in the pathogen for virulence has a cost. Models examining the outcome of selection in initially genetically homogeneous populations (Person *et al* 1976; Groth & Person 1977) suggest that there will be a limit to the number of virulence genes that will accumulate in the pathogen or parasite through natural selection. This limit occurs when the advantage of extending the range of host genotypes is countered by the loss of fitness in individual host–parasite interactions in the enlarged host population open to infection. The overall fitness of the pathogen population will be related to the number of genes for resistance in the host population. Interestingly this is a situation analogous to that discussed earlier of the evolution of coexisting genotypes within a single population, because it is niche divergence among the genotypes.

The disease stabilization commented on by Schmidt (1978) therefore seems to be the outcome of a simple process of selection in the host for resistance, together with selection in the pathogen for virulence, but not too much. Genetic diversity is an obviously important cause but is only one component bearing on a complex evolutionary process in which all components of fitness (Table 14.3) need to be evaluated. Functional diversity — the consequence of a gamut of processes — remains a little explored area of exciting promise (Barrett 1984). Reluctantly now, we have to conclude that there is a shortage of data to evaluate Person *et al*'s conclusions in practice (Day 1978; Burdon 1982) and their appropriateness to the hypotheses for coevolution mentioned above.

14.9 Evolution across trophic levels — grazers and predators

Nowhere is the array of potential patterns of synevolution so vast and complex as in grazers and predators. Grazer–host and predator–prey interactions are ecologically different in a number of ways from the host–parasite interactions we have just considered. [See Thompson (1982) for a detailed discussion of them.] Pathogens and parasites interact with their host for the duration of their life cycle, but with grazers and predators the interaction time is often relatively short. As a result, in parasite interactions selection commonly acts on biochemical and physiological traits, but in grazer and predator interactions selection leads to grazer or predator avoidance on the one hand and increased predator efficiency on the other, and therefore may involve a wide range of morphological, anatomical and behavioural characteristics.

This may then well lead to coevolution. In the antagonist interactions there will first be evolution by the host leading to simple reduction in damage. This will in turn impose evolutionary pressures on the predator, which may lead to an evolutionary arms race (Chapter 13) or, as we shall see, to interactions which are mutualistic. As was argued earlier over coexistence at the same trophic level, and as Janzen (1980a) points out, coevolution amongst species across trophic levels cannot necessarily be shown simply by the demonstration of mutual congruence of sets of traits. Coevolution involves the specific coordination of the gene pools of two species and the fitness changes associated with them. If such coevolution is to be inferred where it is impossible to follow experimentally the genetics of the interaction, detailed ecological experimentation is necessary to demonstrate fitness relationships between the species, often by mimicking the selection pressures which are presumed to have occurred. In studies of grazer–host and predator–prey interactions it has often been the only method employed. But even then we would issue the caveat we gave at the beginning of this chapter.

If we consider first the grazer and predator side of the relationship, we might expect that evolution would produce two traits common to both grazers and predators (Chapter 8), firstly evolution for a mixed diet (polyphagy) and secondly the evolution of an ability to learn which prey to eat and which to avoid — both associated with a low degree of specialization. Selection for polyphagy may come about for two reasons, firstly to maximize the balance of nutrients in the diet and secondly for energetic reasons. The time spent searching for the next meal of the same host species may not justify ignoring other species as potential food sources. Equally clearly these two are not related.

Yet unequivocal evidence for the form and intensity of selection favouring mixed diet is not easy to find. Ecological studies point to its importance: Barnes (1965) showed that the grasshopper *Melanophus sanguinipes* had improved survivorship (67%) on a diet mixture of alfalfa (*Medicago sativa*) and barnyard grass (*Cynodon dactylon*) in comparison to the two alone (alfalfa, 24%; barnyard grass, 0%). Similarly limpets, sea-urchins and parrotfish have higher fitness on mixed rather than specific diets (Lowe & Lawrence 1976; Kitting 1980; Lobel & Ogden 1981). Also, predators display host switching according to the relative abundance of hosts (Murdoch & Oaten 1975). Selection for a mixed diet to ensure dietary intake of a spectrum of nutrients, or avoidance of a food which may have deleterious effects if eaten in bulk, however, have not been seriously investigated.

Critical data also need to be gathered on variability in the activity of individual grazers and predators and its genetic control. The importance of this level of detail is underlined by studies which show that food choice is influenced by a learning response in offspring either from parents or by trial and error. Learning constitutes one mechanism by which grazers and predators may avoid toxic and indigestible foodstuffs. Whilst it is a phenotypic response, it clearly has a genetic basis (McClean & DeFries 1973) but the direct evidence of its evolution in nature is not available.

On the host and prey side of the relationship, however, there is clearer evidence for genetic variation, and related evolution, in mechanisms of defence. At a very simple level, it has been known for many years that populations of grass species found in heavily grazed pastures are always more prostrate and densely tillered than populations of the same species occurring in other habitats. To determine directly the effect of grazing, Harris and Brougham (1970) subjected a mixture of two cultivars of *Lolium* to different intensities of sheep grazing. After 5 years of lax grazing, the population was dominated (>80%) by the tall cultivar, but under intense grazing only 12% of this cultivar persisted. Of considerable significance was the analysis of the genotypes surviving within each cultivar under this grazing regime. They showed a noticeable shift, measured in terms of vegetative characteristics towards the genotypes that naturally dominate under heavy grazing. Similarly, in clover (*Trifolium repens*) and trefoil (*Lotus corniculatus*) populations, some plants are cyanogenic (capable of producing hydrogen cyanide) and some are not, the result of segregation of genes at two separate loci. In *Lotus*, both small vertebrates and invertebrates show a clear preference to eat acyanogenic genotypes (Jones 1962) and slugs may distinguish between the genotypes in clover (Dirzo & Harper 1982). In natural populations a high frequency of the cyanogenic genotypes is found in habitats where there is a high frequency of these herbivores.

However, because both grazers and predators can develop a search image (Tinbergen 1960), the selection pressures which are exerted on prey can be complicated. Predation intensity commonly increases in relation to prey availability, giving rise to density-dependent selection. If there are genotypes which are distinguishable by the predator or grazer, then those which occur at low frequencies in a population, and are well dispersed, may be less likely to suffer predation than those occurring at high frequencies, so that there is frequency-dependent selection. This will be enhanced where predators respond to the relative abundance of differing prey types by switching food preference. This is particularly important because it may lead to the maintenance of polymorphisms in populations — the occurrence in the same population of two or more genetically distinct forms (or morphs) of a species. Predators that form a search image for the most frequent form will continue to predate that form until it diminishes in abundance to a certain low frequency. At this point, choice of prey is switched to the more common form. If this occurs not only are rare forms likely to be maintained in the population, but also novel forms are also likely to be rapidly incorporated into the population. This form of selection with its stabilizing properties has been termed apostatic selection by Clarke (1962).

Convincing evidence for apostatic selection has come from studies of fish predation which have been reanalysed by Clarke (1962). Silverside (*Antherina laticeps*) are coral fish that display bright coloration. Snappers (*Lutianus griseus*) prey on them and moreover can learn to associate colour with different degrees of palatability. From experiments of Reighard (1908), Clarke was able to show that there was a tendency for snappers to take prey of the particular colours with which they were most familiar. This predatory behaviour indicates that they would take the colour type most common in the population, selecting in a frequency-dependent way. Figure 14.12 shows this process when corixid water bugs *Sigara distincta* are preyed upon by rudd, *Scardinius erythrophthalmus*. These bugs display colour variation to achieve matching with environmental backgrounds. Analysing experiments in which bugs with different degrees of crypsis were offered to rudd, Clarke showed that the most cryptic and least cryptic were taken more frequently than expected when common, and less frequently than expected when rare.

Predation can, in some instances, cause remarkable evolutionary changes, of which mimicry is a good example. There are two forms, Batesian — a palatable and harmless species gaining advantage through resemblance to a distasteful species, and Mullerian — where already protected species gain mutual advantage by resembling one another. The morphological change is often coloration but may also involve body shape. Mimicry has been recognized as an ecological phenomenon for a long time, but the earliest experimental tests of mimicry were those of Brower (1958, 1960). These involved offering starlings (*Sturnus vulgaris*) artificial mealworms. These were painted green and either made unpalatable (through use of quinine), to act as models, or left palatable to act as mimics. Other palatable ones were painted orange. The experiments showed firstly that the starlings could associate colour with a distasteful experience, secondly that the mimic could gain advantage from the presence of the unpalatable model, and thirdly that the selective advantage or the mimetic form was frequency dependent. Thus the mimics which are common in a population gain less advantage than when they are rare.

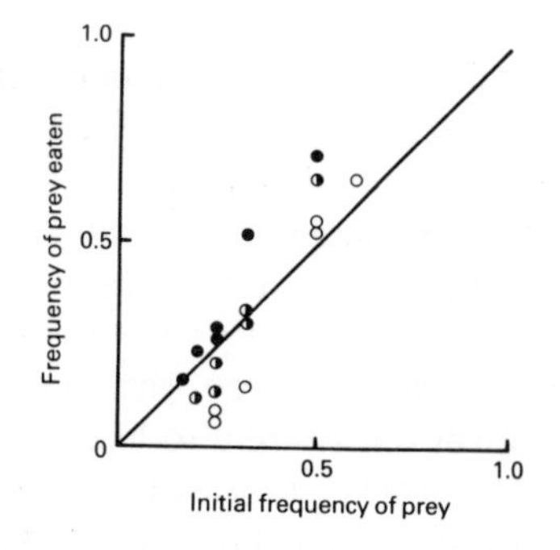

Fig. 14.12. Frequency-dependent predation of the water bug *Sigaria distincta* by rudd, *Scardinius erythrophthalmus*. All morphs are overpredated when they are common and underpredated when rare. Morph colour of *Sigaria*: ○ most cryptic — dark brown; ◑ intermediate; ● least cryptic — light brown. (from Clarke 1962).

O'Donald and Pilecki (1970) have shown that the selective advantage gained depends on the degree of unpalatability and the abundance of the model. Where partial palatability occurs a mimetic polymorphism may evolve that is frequency dependent. The equilibrium situation depends upon the relative abundances and the palatabilities of the models (Turner 1984).

Nevertheless it follows from all this that there is a limit to the numbers of the palatable species, which is determined by the numbers of the model species it mimics. As a result any increase in number will depend on the palatable species mimicking other distasteful species that occur in its area. This can lead to complex evolutionary situations with fine adjustments, well shown in the detailed analysis of swallow-tailed butterflies by Clark and Sheppard [described in Ford (1971)]. In *Papilio dardanus* and *P. memnon* the females are usually highly variable, being mimetic of a whole range of distasteful butterfly species. When different models occur in the same area, the different appropriate mimics also occur, sympatrically. They differ from one another in a series of characteristics — body and wing colour, wing tails and pattern of forewing: the mimic is only successful if it resembles the model. Genetic analysis has shown that many of these characters are controlled by single loci which are multiallelic, and that five of these loci are tightly linked on the same chromosome to give a 'supergene'. This is important since it ensures that the individual complex mimicry patterns are maintained over generations and that genes governing them do not segregate yielding non-mimetic recombinants. What is interesting is that the effects of the supergene for a particular form in one geographic race of *Papilio* can be quite different when put into another race, because the operation of the 'gene' depends on a particular set of accompanying modifiers in the rest of the genotype.

In Mullerian mimicry the situation is somewhat different. Here all members of the group (of species) gain by having characters in common when any member is attacked. Thus selection tends to promote uniformity among Mullerian mimics, removing individuals furthest from the mean. It therefore suppresses the likelihood of polymorphisms.

Both forms of mimicry clearly result from the role of avian predators as agents of selection and provide crucial examples of synevolution. But it is wrong to consider them as examples of coevolution since the interaction and its responses is unidirectional — no changes are presumed in the predator. However, there is no reason why coevolution should not occur in some situations (Gilbert 1983).

Defence, as previously described, is due to unpalatability to predators and may be a result of chemical changes. A more acute form of chemical defence is seen in the range of toxins that host plants possess (Feeney 1976). Because these are powerful, they are potentially able to exert strong evolutionary pressures on the predator or grazer. Dolinger *et al*

(1973) found that the intensity of predation experienced by lupins in Colorado was inversely related to the number of different alkaloids present in the plant tissue. Species which were lightly attacked showed wide variation in the proportional complement of the different alkaloids, whereas severely attacked species had no variability. We can argue from this that variability in the lupin population is maintained because the proportion of predators adapted to rare alkaloid combinations will be low and hence these rare plant phenotypes will be selected for. The situation seems similar to that already seen in host–pathogen evolution.

But is there any direct evidence for such a coevolutionary effect, on a single grazer or predator species? In pineleaf scale insects (*Nuculaspis californica*) on ponderosa pine (*Pinus ponderosa*) Edmunds and Alstad (1978) have shown that individual pine trees differ in their defence mechanisms (monoterpene complements) towards specific subpopulations of the insect. When the insects are transplanted amongst host trees their fitness drops dramatically on 'alien' hosts, indicating that they are genetically adapted to their original host. Selection pressures exerted in this way will favour novel and rare resistant forms (different complements of monoterpenes) in the pine populations and therefore maintain polymorphisms in both the host and the grazer. Coevolution in grazers and their hosts can clearly occur, at a very simple level.

But can we see evidence of genuine coevolution at a higher level? In a remarkable review of evolution of butterflies and plants Erhlich and Raven (1964) assembled the wealth of indirect evidence that suggests that the evolution of these two groups has been a remarkable process of evolution and counter-evolution. In this the availability of the appropriate variability, which we discussed earlier, can be presumed to have had a profound effect, because the counter-evolution has often occurred only in one particular butterfly (or plant) group. But for detailed evidence we must turn to the work of Gilbert (1975) in *Heliconius* butterflies and *Passiflora* vines. The larvae of *Heliconius* butterflies are the major herbivores of the vines and in their turn form the basis of an intricate and elaborate associated food-web. The vines have defensive traits against herbivory, one of the most important being leaf shape. The butterflies detect hosts on which to lay eggs by sight and hence novel leaf shapes in the vine may enable escape from the predator, even by evolving leaves which appear to have eggs of the butterfly already laid on them (Fig. 14.13). Among sympatric species of *Passiflora* leaf shape variation is enormous (Gilbert 1975) yet *Heliconius* species are locally specific to particular *Passiflora* species (Benson *et al* 1975; Benson 1978). Coevolution between butterfly and vine appears to be the consequence of a sophisticated arms race between predator and prey involving a range of defences — changing leaf shape, toxic compounds, and butterfly egg mimicry are invoked in the vines and combatted by the butterflies.

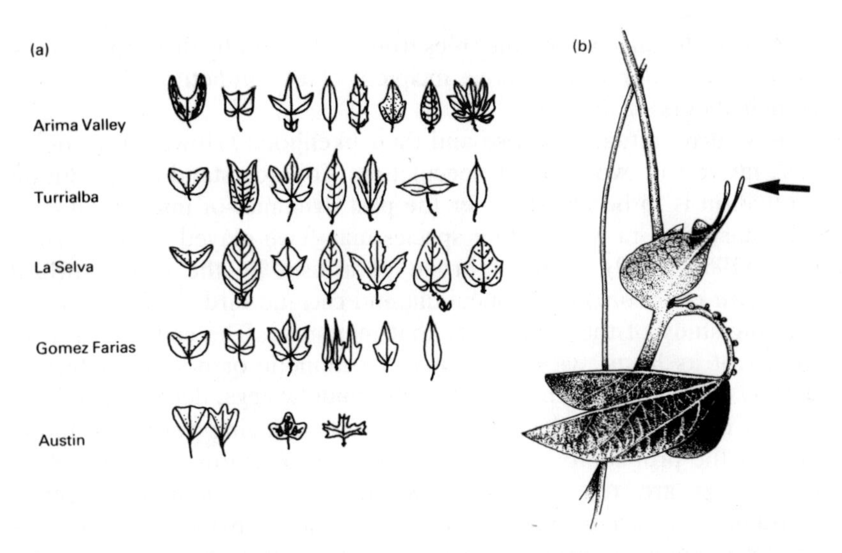

Fig. 14.13. (a) The variation in leaf shape amongst sympatric species of *Passiflora* at five locations. (b) Egg mimicry in leaves of *Passiflora cyanea*. The arrow shows the stipular outgrowth in the form of a mimetic egg (from Gilbert 1975).

14.10 Evolution across trophic levels — mutualistic interactions

An important consequence of these antagonistic interactions amongst communities is that they can lead to mutualistic interactions which can be considered to be the end point of synevolution. The arguments for the evolution of mutualism are straightforward. Although it causes damage, the predator/grazer may increase the fitness of the prey/plant while, by providing food, the latter obviously helps the former. Frugivores and seed dispersal, nectar feeders and pollination are obvious examples. Many authors have discussed their ecological and putative evolutionary relevance (Janzen 1983). But the ecological conditions leading to them and the actual evolutionary mechanisms by which they develop remain unclear. Where a predator, parasite or grazer benefits by limiting its attack, so that it does not lead to selection for an escalation of defences or avoidance, the circumstances may become reinforced if the host gains benefit. But the precise consequences of such increased mutual dependence leading to highly coevolved systems need further investigation. The mutualism between *Pseudomyrmex* ants and *Acacia* trees (Janzen 1966), however, does provide striking evidence of evolution of mutualism. The ants feed at the nectaries on the petioles and highly proteinaceous Beltian bodies at the tips of young leaflets and shelter in enlarged stipular thorns. This investment is recouped, since survivorship of the acacias themselves is increased, as the ants defend the trees from herbivores by their pugnacious behaviour, and moreover remove incipient plant competitors by grazing close neighbours of the tree.

How such mutualisms arise and their likelihood (Howe 1984) are of intense interest in evolutionary speculation. To understand the evolution of mutualism is to be able to infer the past dynamics of interaction over evolutionary time in which other species may have played a crucial role. Smith's (1968, 1979) study of brood parasitism by the giant cowbird (*Scaphidura oryzivora*) on oropendula and cacique birds in Panama is a fascinating study of the complexity that may arise. The cowbird lays, in the nests of its hosts, eggs which may be mimetic or not. Host female birds may discriminate or not against non-mimetic eggs, depending on the proximity of colonies of hymenopteran insects (stinging wasps and biting bees) and the probability of attack by botflies (*Philornis* species). Non-mimetic eggs are rejected from the nests of discriminating female oropendula and cacique birds if nests are near to hymenopteran nests where for unknown reasons botflies are absent. Away from the wasp and bee colonies, however, botflies heavily parasitize oropendula and cacique chicks. Female botflies lay eggs or larvae on chicks which are likely to die if they bear more than 10 of these ectoparasites. Oropendula and cacique females with nests away from hymenopteran colonies, however, accept non-mimetic cowbird eggs. Under botfly attack, the presence of cowbird chicks in the nest increases the probability of host chick survival, as the cowbird chicks eat the botfly larvae from the bodies of the host chicks. The probability of successfully rearing a chick to the fledgling is nearly threefold greater if another chick in the nest is a cowbird rather than a sibling. Clearly this complex mutualism is made up of several antagonist interactions — botflies and oropendulas and caciques; hymenopterans and botflies; and cowbirds and oropendulas and caciques — which in sum total in certain circumstances confers advantage to both avian antagonists. But like so many examples, the actual evolutionary processes which have led to this situation have yet to be demonstrated.

Fascinating as the ant–acacia and cowbird–oropendula–cacique mutualisms are, Howe (1984) has questioned whether we should accept these apparently tightly linked non-symbiotic mutualisms as anything more than rarities in practice and anomalous in theory. In complex communities the potential for many more such interactions exists than actually do occur. Diffuse selection arising from many interacting species in a diverse community, changing selective regimes over evolutionary time as a consequence of normal ecological processes (such as succession and disturbance) and variation in temporal and spatial attributes of populations, are suggested as powerful forces altering the intensity of specific pair-wise interactions amongst species. All these promote selection for general rather than narrow mutualisms amongst species.

14.11 Conclusions

This approach to evolution in communities has been essentially analytical rather than synthetic, because we believe that, at the moment, the only way forward is by identifying all the separate components that can make up evolution in communities. What is essential is to realize that any one of these components can act on its own, or in concert with other factors. There is a terrible tendency in biology for concepts about a complex process to be dominated, at any one time, by one particular idea. Nature does not work in this fashion; it is a complex web of interacting processes.

As a result we will not attempt to come to conclusions about which of the processes we have discussed is the most important. The crucial conclusion, however, is that since community-generated interactions can be very powerful, leading not just to the reduction of a species, but to its elimination, community-generated evolution is an important part of evolution as a whole, and should be investigated further. Those species which cannot so evolve are likely to be eliminated. However, if we see a powerful process we must not necessarily expect to find its effects always manifested as evolutionary change. Firstly, change may be restricted by problems of pleiotropy — genetic changes that are adaptive in one direction that may be accompanied by other effects that are deleterious. Secondly, and perhaps more importantly, although natural selection can obviously be very powerful, it can only operate if the necessary genetic variation is present, and it can be argued that usually such variation has, temporarily at least, been used up, i.e. has reached effective fixation. So it may be best to look for community-generated evolution in new situations in which the evolutionary process has not been exhausted. Certainly some of the best examples that we have examined, such as in weed populations of *Camelina sativa*, rabbits and myxomatosis, barley bulk hybrids, and houseflies, are new evolutionary changes. As a result, although community effects may be most highly developed in complex communities such as a tropical rainforest, the best place to study their evolutionary consequences may be in invading animals, weed and crop populations and man-made habitats. Darwin, by his use of the evidence of cultivated animals and plants, has given us the mandate. But whatever we choose, we must remember that evolution can be very rapid, and that in communities it is not only the species that is being studied that may change but the others as well. What we are looking for may therefore not only be elusive but also complex.

15 Living with Rainforest: the Human Dimension

Peter D. Dwyer

"My search is further.
There's still to name and know
beyond the flowers I gather
that one that does not wither —
the truth from which they grow."

(from *The Forest* by Judith wright)

15.1 Introduction

Many people live within the wet, forested, tropical regions of the world (e.g. Padoch & Vayda 1983). Some have done so from time immemorial. Others have colonized more recently. There are people who derive all their needs from the tropical forest, altering it only slightly. Other people obtain their livelihood in the tropics by eliminating the forest. This occurs where forest is deflected to grassland in the course of slash and burn agriculture or where the means of production is the extraction and export of timber (Clarke 1976; Webb 1977). Between these extremes there are many ways of living with rainforest.

This chapter is one attempt to understand ways in which subsistence societies live with rainforest. It commences by developing a framework within which human subsistence modes may be analysed as ecological systems. It then moves to interpreting the natural history of some rainforest peoples within the context of that frame and provides a tentative classification for the subsistence modes discussed. Last, it considers aspects of the evolution of human subsistence patterns.

15.2 Companion foods and compatible schedules

To satisfy their need for an adequate diet people move through time and space. Models intended to describe or predict food-getting behaviour should take these contexts for behaviour as primary. [Here, all human behaviour directed at obtaining food, from visiting 'fast-food' emporia to gathering mongongo nuts, is labelled 'food-getting'. Most people who study the ecology of people use the term 'foraging' to describe food-getting behaviour of 'hunters and gatherers' (Lee 1979, pp. 116–9). I like to think that 'foraging' is also what people do in supermarkets.] The inquiry strategies advocated by adherents to 'optimal foraging theory' may, if used cautiously, be valuable (e.g. Pyke *et al* 1977; Smith 1979; Winterhalder 1980; Winterhalder & Smith 1981). Their methodology is founded in assumptions about optimal behaviour and they ask questions about the effectiveness and efficiency of different food-getting practices. They are enthusiastic in their search for behavioural modes that show particular organisms, including people, are suited to particular environments. They ask, for example, if the choice of food items can be explained in cost–benefit terms and, in testing such questions, employ energy as a convenient 'currency' for their equations (e.g. Hawkes *et al* 1982). The intention has merit but is oversimplifying. People cannot live on calories alone. The diet must also include a suite of nutrients and the proportions in which these occur must fall within certain limits (e.g. Steele & Bourne 1975; Arlin 1977; Hegsted 1978; Diener *et al* 1980). An assumption that people routinely behave as optimizers or rational strategists is too strong for many anthropologists (e.g. Sahlins 1976). In this chapter, I provide a framework within which human food-getting behaviour may be described. The framework allows for multiple currencies (e.g. energy, protein, vitamins). My account borrows from optimization models but softens them; it accepts less rigid assumptions and to that extent sacrifices precision to realism and generality (cf. Levins 1966, 1970).

Within a given environment there are many potential food resources. Here I define a *food resource* as a species or part of a species of plant or animal that can be eaten by an animal. Some attributes of food resources qualify as environmental givens; their empirical content is independent of the particular animal species that utilize the potential foods. These attributes are *dispersion* and *particle size* and each has a spatial and temporal dimension; i.e. *supply* is the temporal analogue of dispersion in space and *lasting properties* the temporal analogue of particle size. For each of these four attributes potential food resources may be ranked on a scale that runs from less to more patchy. That is, for each attribute a given food resource takes a single 'state' on this scale. For the same absolute quantity of a given resource patchiness increases as dispersion shifts from even to clumped, as supply shifts from continuous to intermittent, as particle size shifts from small to large and as lasting properties shift from brief to long.

A second set of attributes for food resources is definable with reference to the impact they have for the particular animals using the foods. These attributes are *abundance*, *time to capture or harvest an item* (measured after encounter and hence independent of abundance), *nutrient mix* and *preparation (or handling) time.* They do not qualify as environmental givens; each assumes empirical content only relative to a particular user. That is, for any food resource, each species, whether it be ants, antelope, pipits or people, will 'assess' abundance, time to capture, nutrient mix and preparation time in a different way. That assessment will flow from the impact, for the user, of the way in which the attribute concerned is manifest. Impact may be recorded as a cost–benefit function. Thus, for each of these four attributes, potential food resources may be ranked on a scale that runs from higher to lower cost–benefit ratio. Cost–benefit ratio decreases as abundance increases, time to capture decreases, nutrient mix improves and preparation time decreases.

Table 15.1 lists the attributes of food resources described above and provides a generalized scheme for comparing different foods according to the states taken by those attributes (i.e. less to more patchy, higher to lower cost–benefit ratio). For animals other than people, predictions about food-getting behaviour may be derived directly from the tabulated information on the assumption that the attribute states of particular foods will limit options among other potential food resources. People confront similar constraints but, for them, it is necessary to first allow that they manipulate many of the resources they use.

There are many ways in which people manipulate food resources. Some manipulations do not seem to be intentional. For example, Hynes and Chase (1982) described 'domicultures' from Cape York, Australia, where dispersion and local abundance of some otherwise wild plant foods were altered because, for generations, people had discarded fruit or seeds at camp sites. The people's behaviour had produced groves of fruit or seed bearing trees at particular places. Other manipulations are a consequence of deliberate behaviour; netting, trapping, animal husbandry and gardening are examples. Manipulations, whether deliberate or

Table 15.1. A classification of the attributes of food resources.

Attribute	Attribute state	
(1) Environmental givens		
	less patchy	*more patchy*
dispersion	even	clumped
supply	continuous	intermittent
particle size	small	large
lasting properties	brief	long
(2) Impacts		
	higher cost	*lower cost*
abundance	uncommon	common
time to capture	long	brief
nutrient mix	poor	good
preparation time	long	brief

otherwise, are technical adjustments that alter the status of particular food resources *vis-à-vis* the people using those foods. They may be placed in two broad classes. First, people may manipulate the patchiness of food resources. They may assume a level of control over dispersion, supply, particle size and lasting properties. In general, they are likely to increase clumping, make supply more regular either at a particular time or across time, increase particle size and, through techniques of storage, increase lasting properties (cf. Table 15.1). When people manipulate patchiness they are likely to be controlling access to a resource by making access more predictable. The manipulated resource will contribute more to diet without any necessary change in the cost per unit of nutrient derived from it. Secondly, people may directly manipulate the cost–benefit functions associated with food resources. They may assume a level of control over abundance, time to capture, nutrient mix and preparation time. In general they are likely to reduce cost–benefit ratios by increasing abundance, reducing time to capture or, through techniques of processing, reducing preparation time (cf. Table 15.1). For societies that lack the requisite knowledge of nutrients and nutrient balance, alterations to nutrient mix will usually be long-term and arise as incidental (and unrecognized) consequences of other, more immediate, manipulations. Where nutrient mix is improved the manipulations concerned have reduced cost; where nutrient mix is shifted toward an inferior state then cost has increased. Manipulations that reduce cost per unit of nutrient may result in increased representation of the manipulated resource in diet and/or increase the time available for obtaining other food resources where these latter might, for example, supply nutrients that are missing from the manipulated resource.

The above treatment of manipulative practices differs from conventional statements. I have grouped the practices according to their consequences for particular food resources or, more specifically, for key attributes of those resources. It is more usual for anthropologists, archaeologists, geographers and others who study the ecology of people to discuss manipulative practices according to perceived similarities in the actions performed by people. Thus, animal domestication, plant management (including agriculture), hunting techniques, traditional 'conservation' practices, etc., may be isolated as separate expressions of the manipulative capacities of people. Literature relevant to the topic has been diffused and misses the important point that functional consequences of manipulative practices may be of the same kind for very different sorts of food resources and despite different actions by people. Students of material culture may know, for example, that netting techniques increase predictability of supply of certain game animals but may be neither impressed by, nor interested in the fact that rules governing the frequency or location of hunting may achieve the same result. The framework developed above highlights these sorts of similarities. It allows comparison of food resources independent of the actions of people; that is, food resources may be regarded as a single set, all of whose attributes may be ranked on the scales of less to more patchy and higher to lower cost–benefit ratio irrespective of whether they are manipulated. A manipulated resource simply changes rank relative to its 'wild' analogue. A classification of potential resources that has, *a priori*, accommodated extant manipulative practices allows predictions, about food-getting behaviour, that are not possible when those practices are grouped according to similarities in the actions of people. The framework suggests 'rules of assembly' which will result in an adequate diet. It suggests, first, how choices among potential foods are mutually constraining and, secondly, ways in which people pattern food-getting behaviour within the context of these constraints. A generalized statement of these rules follows.

The subsistence modes of people may be ranked from fine-grained to coarse-grained. The terms 'fine-grained' and 'coarse-grained' have been used to characterize ways in which the resources used by particular animal species are distributed; they refer, respectively, to relatively homogeneous and relatively heterogeneous patterns (MacArthur & Connell 1966). This usage encounters difficulties for species with the capacity to either accept or ignore potential resources found within particular environments. Co-occurring populations of the one species might utilize sets of resources that have different patterns of distribution. To accommodate this difficulty the terms fine-grained and coarse-grained are used here to characterize the way in which people respond to the particular environments they occupy; the terms are not used to specify patterns of distribution for all the resources people could use. Thus, in the *fine-grained subsistence mode* people respond to environment as though it were relatively homogeneous; they are likely to eat small quantities of a diversity of food resources. In the *coarse-grained subsistence mode* people respond to environment as though it were relatively heterogeneous; they are likely to take a few food resources in relatively large quantities.

The array of potential food resources in an environment will constrain possible expressions of subsistence mode. These constraints are likely to be least where the diversity of potential resources is greatest. In environments of relatively low predictability, such as deserts and savannahs, the distribution of resources will tend to be patchy in time and space. Here people may have little choice but to opt for relatively coarse-grained subsistence modes. In environments, such as rainforests, that are rich in resources and show relatively high predictability, different patterns of resource distribution will co-occur and people may opt for either fine-grained or coarse-grained subsistence modes. Thus, rainforests potentially afford people with many subsistence options.

An extreme of fine-grained subsistence may be characterized as a group of people who choose to forage from a set of resources $a_{1,2...i}$ that take the following states: even dispersion, continuous supply, small particle size, brief lasting properties, low abundance, brief time to capture, good nutrient mix (i.e. each item contains all essential nutrients in the proportions required for an adequate diet) and brief preparation time. If, for such people, the number of items encountered and consumed in available foraging time (T_A) satisfied everyone's dietary needs then the strategy of taking all encountered items from the resource set '*a*' would result in an adequate diet.

Environments, within which people obtain food, include many potential food resources, and people do not eat everything that can be eaten. In addition, none of the actual food resources used supply all the nutrients needed. The preceding model of fine-grained subsistence will alter in as much as people have available and use a different set of food resources, $b_{1,2...i}$, that provide useful amounts of some nutrients but are deficient in others. For constant T_A, accepting items from this set will result in nutrient imbalance that cannot be satisfied by the original set '*a*'. Dietary needs will be met in available foraging time only if there is a compensatory shift to other sets of food resources, $c_{1,2...i}$, $d_{1,2...i}$, etc., that contain useful amounts of the nutrients which are missing from resources of the set '*b*'.

The magnitude of compensatory changes of the sort described will increase in as much as exploited resources from the '*b*' set take *any* of the following states relative to those of the '*a*' set: less even dispersion, less continuous supply, larger particle size, longer lasting properties, greater abundance, poorer nutrient mix, longer time to capture, longer time to prepare. The magnitude of the necessary shift will increase further in as much as a few resources from the '*b*' set are accepted as 'staples' for the particular nutrients they supply in abundance.

The states taken by the attributes of any major food resource impose constraints upon the space and time within which other food resources may be obtained. Thus, when people obtain a high proportion of one sort of nutrient from one or a few resources that take any of the states listed above, so they restrict their options among other resources that could supply the missing nutrients. To achieve an adequate diet they must limit their choice to those particular resources whose attributes are expressed such that sufficient quantities of the needed nutrients can be obtained from within the now limited spatial and temporal domains. For example, if a group of people meet virtually all their calorific needs from a highly clumped plant food then the area available to them for satisfying other nutrient needs is limited. If, in addition, preparation of the energy source is time demanding then the time available for satisfying other nutrient needs is also limited. 'Capture' of these other nutrients must occur within an area and an interval that can be 'defined' in relation to properties of the calorie-rich food. Only food resources that can yield sufficient quantities of the other nutrients in the area and time available can contribute to an adequate diet. People's options among all potential foods containing the relevant nutrients are, therefore, limited.

The simple conclusion is that some foods make good companions while others do not. The critical criteria that make some foods good companions have to do with the ways in which the attributes of those foods are expressed. For this argument it does not matter whether the foods are manipulated. The interlocking nature of choices from potential food resources arises irrespective of the manipulative practices of people. If the manner in which a group of people obtains its calories limits their choice of protein foods to those that have certain properties then it is the presence of these properties which is crucial; it does not matter whether the properties are a result of manipulations.

Thus far, the framework I have developed has not emphasized the social dimension of human food-getting. It needs an added component. When a group of people obtain nutrients from a single food resource they engage in a particular set of activities. I shall define the set of activities for each sort of food resource as a *procurement system* (cf. Flannery 1968). I have shown that choices among potential food resources are mutually constraining; that the use of a particular food to obtain nutrients of one sort will limit options among potential foods that contain other nutrients. An adequate diet results from particular combinations of food resources and, for people, this means a variety of potentially different procurement systems must be brought into viable articulation. People accomplish this by organizing their food-getting behaviour in particular ways in time and space; they assign task-specific places, task-specific times, task-specific roles and task-specific group sizes. These structuring principles are social adjustments to food-getting. Taken together, they may be labelled *scheduling* (cf. Flannery 1968). The ways in which scheduling is expressed in different societies will depend on the variety of procurement systems that are to be brought into harmony and on the magnitude of differences between those systems. The now deceased anthropological dictum that 'the men shall hunt and the women shall gather' is one way this idea has been portrayed (Lee & DeVore 1968; Dahlberg 1981). At the same time the dictum highlights the fact that, by scheduling their activities, people may increase subsistence options. People regularly share the food they obtain and so different individuals in a group can contribute in different ways to the dietary needs of everyone. While it is true that companion foods need compatible schedules it is also true that scheduling is a way to make very different sorts of food resources companionable. This will become evident in the accounts of rainforest people given below.

15.3 The natural history of rainforest people

Here, ways in which some human societies live with rainforest are described. The descriptions are placed within the context of the preceding section. The large and very scattered literature on the ecology of rainforest people is not reviewed but, rather, information from some societies is used to fulfil two aims. These aims are: to provide concrete support for the framework outlined above, and to show that rainforest subsistence modes may be ordered on the scale of fine-grained to coarse-grained. This ordering of subsistence modes is cross-cut by a tentative classificatory scheme. The latter is represented by four categories that serve as subheadings below. Not all activities of the people discussed within each category satisfy the definition offered for the category. Further, within each category a continuum of subsistence modes occurs or is possible.

15.3.1 Conformers: who are

'Conformers' are of the rainforest. Ideally, like any other species of animal living within the rainforest, 'conformers' are an integral part of the environment. And, as is the case for other animals, for very long periods their impact does not significantly or permanently alter the composition or the structure of the forest. In essence, 'conformers' may be regarded as members of the community of species with which they interact. Their manipulations are few and their scheduling regimes are influenced directly by seasonality or by using one food as a 'staple'.

There are no recent or extant societies of rainforest people who fit this idealist definition of 'conformers'. Here the category covers many different expressions of subsistence, which might, under a different typology, have been labelled 'hunters and gatherers'. 'Conformers' include those human societies whose food-getting practices entail the least manipulation of resources. Pre-eminent among them are the Tasaday of Mindanao, Philippines; or, more precisely, the Tasaday before about 1960 when they met a hunter named Dafal and were taught new ways of obtaining food. When the Tasaday were 'discovered' by the western world they numbered about 26 individuals, lived in caves, owned no weapons and used a small array of stone tools (Yen & Nance 1976). They were popularized as a relict people who showed ancient adaptive modes. In fact, like the !Kung San of the Kalahari, they may have been refugees (Yen 1976a). For present purposes their history does not matter; it is the way they lived with rainforest that is instructive.

Before they met Dafal the Tasaday apparently gathered a diversity of small plant and animal foods; seeds, fruit, leaves, palm hearts, crabs, fish, tadpoles and grubs (Yen 1976b). Yams, which were small in size and not numerous, may have come nearest to qualifying as a staple. The people practised minimal management of these tubers by replanting cut crowns. They did not eat mammals nor, indeed, large game of any sort and they ignored readily available large packages of starch in the form of caryota palms. The Tasaday ate, in effect, from a nutrient rich 'fruit bowl'. Their diet would have seemed familiar to Australian bushwalkers who carry an energy-rich and protein-rich mix of nuts, dried fruit, muesli, chocolate and, sometimes, a 'sweetener' (e.g. jellybeans) which they call 'scroggin'. On long, energy-demanding, walks 'scroggin' is very sustaining; a little goes a long way. Yen (1976b) remarked on the small amounts eaten by the Tasaday and wondered how, before Dafal, they would have managed without caryota starch. They managed because they had chosen to eat some of the finest foods available. Because these were present only as small morsels scattered through the forest they may have been time-consuming to gather but the return of diverse nutrients per unit weight would have been high. The Tasaday responded to rainforest in a fine-grained way. They selected small items of food which were rich in nutrients, highly dispersed and not amenable to storage; their foraging was not unduly constrained by seasonal variation and the demands upon their time, up to harvesting, were on-going and regular.

After Dafal arrived the diet of Tasaday changed significantly. He showed them how to process and store starch from caryota palms. He encouraged them to perceive and respond to plant resources in a less fine-grained way. Had this been all he did, Dafal may have enhanced quantity and calories in their diet at considerable cost to quality and protein. There is more than three times as much protein per unit (wet) weight in fresh yams than in caryota palm starch. But Dafal's wisdom was greater than that which motivated the 'green revolution'. He balanced the shift to caryota starch with a shift to larger animals; he showed the people how to trap birds, monkeys and deer and how to store meat by smoking it. In this his contribution was not merely to simultaneously adjust plant and animal foods, but to accommodate the people to very different scheduling constraints. The processing of palm starch requires sustained inputs of effort from small groups of people. They cannot, at the same time, seek out small and widely dispersed foods which compensate for protein costs incurred by the new activity. To balance both diet and time they must have available larger packages of protein rich food which can be obtained at relatively low cost. Trapping mammals is one way to satisfy this need.

The foregoing account of Tasaday diet differs from that provided by Yen (1976b). My interpretation may help answer Endicott's (1979) question concerning forest-living Batek of Malaysia: "Why do the Batek hunt small game almost exclusively, and not hunt larger game?" The game was arboreal and terrestrial birds and mammals up to the size of monkeys and gibbons. These were obtained by blowgun or by digging. Wild pig, wild cattle, deer, rhinoceros and elephant were available in the forest. They were not hunted by Batek but were hunted by their neighbours who were gardeners. Relative to Tasaday, wild yams were a staple for Batek. The latter had achieved a balance of dietary components and time utilization similar to the post-Dafal situation of the Tasaday. The significant difference between the two groups is that Tasaday combined the time-intensive task of processing palm starch with trapping game where Batek combined the time-extensive task of gathering wild yams with hunting game. For Batek, a shift to even larger parcels of protein food and, hence, the commitment of more time to this activity might have necessitated complementary changes in the exploitation of plant resources. Intermittent hunting for very large game animals may be well matched to subsistence gardening but not to the on-going collection of wild, though managed, yams.

The diet of Batek, as described by Endicott (1979), appeared to be plant dominated. Surrounded by numerous large animals the people ignored them. This conflicts with the opinion of Hawkes *et al* (1982) that people who forage 'optimally' will, where possible, concentrate their effort on animal foods. They have described food-getting behaviour of the Aché of Paraguay. Aché were once forest-living hunters and gatherers. At the time of the study they resided at a mission where they grew crops. Hawkes *et al*'s data are from several hunting trips made into the forest. On these trips the variety of plant and animal foods obtained was great. Indeed, with the peculiar addition of honey, it was remarkably similar to the inventory of *wild* foods utilized today by people from the New Guinean Highland Fringe (cf. Hyndman 1982). The manner in which Aché utilized palms (*Cocos romanazoffiana*) is of special interest. Palms were relatively common but many were ignored. On most occasions when they were used processing was minimal — the fibre was sucked dry and discarded on the spot or sometimes taken to camp and cooked. "On rainy days when there was little or no hunting, or after an early stop, husbands and wives might exploit palms as a team..." (Hawkes *et al* 1982, p. 385). Now processing was more elaborate because starch was extracted by pounding palm fibre.

Palms are a resource that can provide many calories but they require a considerable input of time. To some extent the Aché seem to have treated palms as a supplementary, perhaps emergency, food. Hawkes *et al* (1982) found that palm starch ranked low in terms of calories returned for effort expended after encounter, and explained its place in the diet on the basis of this criterion. There may have been other complications. Had processing of palm starch been more regular or more fussy, people's hunting activities would have been curtailed. In the forests where they hunted, two species of peccary were available. These were large (20–30 kg) and fat. Coati, paca and armadillo (3.5–7.5 kg) were probably fat too, but monkeys were sometimes ignored by Aché 'because they are not fat' (Hawkes *et al* 1982, p. 391). By weight, the contribution of peccaries to diet was considerable (25% of mammals) and one of the two species was obtained by shotgun. Peccaries are like 'mobile palms', with very high energy yields but with an advantage over real palms because they contain lots of protein. They are a self-contained companion food; one procurement system provides large amounts of two primary nutrients. Real palms must be complemented with real meat which, for forest wanderers, may render scheduling arrangements awkward.

A final example is instructive. Radcliffe-Brown (1964, pp. 36–40) described the food-getting activities of forest-living Andaman Islanders. Through the rainy season (mid-May to September) people lived at one place. Groups of 2–5 men hunted, with the primary object of obtaining wild pigs. Other game was taken if encountered and some honey and plant foods were gathered. Women gathered plant food from the forest and performed maintenance tasks at camp. Radcliffe-Brown wrote that women contributed little to the total food supply in this season. After the rainy season there was a period of unsettled weather to November, a cool-dry season to February and a hot-dry season to May. Foraging activities changed progressively through these months and people dispersed to temporary camps. Hunting for pig gradually declined and by March pigs might be ignored, or killed and abandoned. In this season they were not fat enough to eat. Men and women increasingly worked together as gatherers, taking a wide range of small plant and animal foods. Food was plentiful with some types more abundant at one period than at another; this variability was due to changes in seasonal availability. Toward the close of the hot-dry season people began to congregate at productive breadfruit (*Artocarpus*) trees where both men and women collected ripe fruit. The flesh of the fruit was eaten and the seeds processed and stored. These seeds supplemented diet through the rainy season.

In this ethnographic case a primary shift from coarse-grained to fine-grained food-getting was reflected, simultaneously, in settlement pattern, food particle size and the way men and women scheduled activities. If Radcliffe-Brown was correct in saying women contributed very little quantity in the rainy season then their contribution may well have been important in terms of quality. That is, in the rainy season a mix of coarse-grained and fine-grained foraging occurred with men and women contributing in different ways. Women, perhaps, supplied essential vitamins and minerals that were absent from the energy and protein rich pigs. Stored breadfruit seeds may have had a similar dietary role. Relative to the other cases cited above, the forest-living Andamans were 'Aché-like' through the wet season and 'Tasaday-like' (pre-Dafal) through the hot-dry season.

15.3.2 Copiers: who do

'Copiers' mimic the rainforest. They manage their major resources, both plant and animal, often in subtle ways. The thrust of their manipulations is to increase access to wanted resources by making these more predictable. They assume a level of control over the patchiness of key resources; their primary manipulations are not, in the first instance, motivated by reducing cost–benefit ratios. The connections between the diverse manipulatively-based procurement systems of 'copiers' give coherence to their subsistence mode. For them, scheduling assumes a significance not observed in 'conformers'. For the latter people differences of seasonal availability or particularistic constraints that arise from using one food as a 'staple' influence scheduling regimes. For 'copiers', scheduling integrates the exploitation of an array of resources each of which has been managed as a separate entity and each of which, therefore, assumes a different pattern in time and space. To a lesser or greater extent 'copiers' adjust to rainforest by patterning their manipulations in a way that is analogous to the pattern of rainforest itself; that is, a multitude of simultaneous processes, connected acts, interdependent places. The totality of their subsistence activities is a microcosm of the processes in space and time which make up the greater environment within which they live. Because 'copiers' create and manage the environments they occupy their relationship to the wider community of organisms is to impose a different and artificial dynamism upon it (e.g. Dwyer 1984).

This definition of 'copiers' is influenced by Geertz's (1963) classic description of swidden gardens as analogues of rainforest. That description has been recently dissected with care and found wanting (see *Human Ecology*, Vol. 11, No. 1, 1983). I want to retain Geertz's metaphor by expanding its focus out from the specificity of the swidden to a totality of food-getting activities for groups of people who combine some gardening with a variety of other procurement systems. With one proviso, societies of the Highland Fringe of New Guinea, particularly of the southern and wetter slopes, exemplify 'copiers'. It seems that through recent millenia the subsistence orientation of many Highland Fringe societies has deflected increasingly to gardening; the people have borrowed from the elaborated horticultural practices of true Highland societies. If it were possible to strip away history and picture Highland Fringe societies at an earlier era then they would better fulfil my idealist definition of 'copiers'. My own experience among Etolo of the Southern Highlands Province will serve as illustration (Kelly 1977; Dwyer 1982, 1983, 1985).

Root crops and sago starch were the major carbohydrate foods of Etolo. The people also ate a bright red, oil rich, sauce made from fruit of the screw-pine *Pandanus conoideus*, bananas and about 30 different vegetables which they grew. From forest they gathered fungi, fronds of tree ferns, *Pandanus* nuts in some years, occasional palm hearts and, intermittently, the fruit of many plants. Plant management practices were diverse. Thirteen varieties of *P. conoideus* were grown in orchards which varied in size from just a few plants to several hundred plants. Moist slopes exposed to the sun were preferred sites. Pruning and plant propagation, protection of ripening fruit from predators, manipulations to delay ripening and general maintenance were on-going tasks.

Etolo named at least 16 varieties of sago palm (*Metroxylon*) that they had established as isolated palms or large groves in swamps, on the banks of streams that were prone to flood, in awkward dips of landscape that had become soaks and in and alongside diversion channels dug for the purpose. The palms were large and processing was elaborate; the pith was chopped from the trunk, pulverized and then washed, beaten and filtered in troughs. Family groups worked together on the venture and much of the starch was wrapped in leaves and stored for long periods in mud.

Different sorts of gardens were prepared for each of the three root crops. These were often located in different places because the drainage needs of sweet potato (*Ipomoea*), taro (*Colocasia*) and yams (*Dioscorea*) were different. Taro was sometimes grown in communal gardens a hectare or more in size, at other times a few plants were maintained at a seepage on the bank of a mountain stream. The latter sites were reused over several years with new plantings following immediately upon harvest. Most bananas and vegetables were grown in one or another of these three sorts of gardens but there were a few exceptions. Small plots of watercress (*Nasturtium officinale*) were sometimes established away from gardens in parts of streams that could be easily fenced off from pigs. Lowland pitpit (*Saccharum edule*) was often grown in young *Pandanus* orchards where it yielded each year until the growing screw-pines threw too much shade. Near living places, bamboo, palm lilies (*Cordyline*), tobacco and some other plants were grown; of these only bamboo provided food and then only occasionally.

Domestic pigs provided Etolo with about one-quarter of the protein they received from animal sources but they were not the only animal resources manipulated by people. Trapping for mammals in long lines of deadfalls set in low fences, snaring cassowaries, poisoning fish in small dams using derris that had been grown for the purpose and incubating beetle larvae in specially felled palms were techniques by which people increased either the predictability of supply, the size of the haul or both. Large traps for wild pigs were sometimes built where sago palms had been recently processed and were intended, in part, to protect palms in which beetle larvae were being incubated. Management of the eggs of megapodes and of the larvae of wood-boring beetles was achieved by regulating access to mounds or suitable trees respectively while, for some small and medium-sized mammals and for large orb web spiders (*Nephila*), availability was enhanced as a direct result of gardening, orcharding and processing sago. When dens of the terrestrial cuscus *Phalanger gymnotis* were found they were covered by special traps that could capture a succession of occupants. Secondary growth, 10–15 years old, was the best place to set tree traps for another cuscus, *P. interpositus*; the activities of people had both encouraged the habitat used by the species and made the habitat physically suited to a particular form of trapping.

Etolo subsistence was based in a variety of very different procurement systems that required much manipulative input from people. The primary systems differed from one another in their dispersion across the landscape and in their operative time scales. Most sago palms were at altitudes below 900 m, most gardens between 1000 to 1200 m and lines of mammal traps in advanced secondary forest or primary forest from 1000 to 1400 m. The sago palms were individually owned but took 15 or more years to mature. The lines of traps were operated by family groups for up to 6 months of each year. In sweet potato gardens nearly all crops had ceased to yield by 10 or 11 months but bananas and sugar cane continued to 15 or 18 months, lowland pitpit to 2 or 3 years and, *P. conoideus*, if it had been planted, took several years to bear but then continued to produce for a generation or more. The manipulations of people gave a new dynamism to the landscape. New gardens were regularly cleared from advanced regrowth as abandoned ones slowly reverted to forest. A scatter of orchards, some a few decades old, interrupted the regrowing forest. At the bend of a stream the forest was pushed back to accommodate, for a few years, a small plot of taro or, for generations, a stand of sago. In their day to day food-getting, people moved between many places whose positions they had established. They moved too within a shifting temporal frame that accommodated both the immediacy of early producing vegetables in recently planted gardens and the relative plasticity and diffuseness of lowland pitpit and the orchards where it grew. Through daily and seasonal scheduling people established connections in time and space between the diversity of their procurement systems. Men, women and sometimes children were often assigned different tasks. The size of work parties altered for different tasks. Some procurement systems were emphasized at one time of the year, others at different times (Dwyer 1982). These latter changes did not seem to be environmentally necessitated; rather, they were social adjustments to food-getting. Through the year there was a shift from low to high reliance upon sago starch and another from trapping to hunting as the primary means of obtaining game mammals. These two shifts did not coincide. There were also differences between families in their commitment to different combinations of procurement systems. Families that were most reliant upon sago starch tended to hunt mammals more than they trapped them; families that were least reliant upon sago starch tended to trap mammals more than they hunted them (Dwyer 1985). The associations were influenced by the altitudinal zones where the different activities took place.

Ellen (1982, p. 170) wrote that "much confusion has been caused... by assuming ostensibly predominant subsistence techniques to represent total subsistence strategies...". He commented "...it is characteristic of systems involving swiddening that it is only part of a broader and more complex strategy linking together a range of techniques" (p. 171). For 'copiers' these remarks are apposite. Ellen's (1978) work among the Nuaulu of Seram demonstrated this clearly. In broad outline Nuaulu subsistence was not unlike that of Etolo or, indeed, of some other, though certainly not all, societies that have been classed as 'swidden cultivators'. Within New Guinea, Wopkaimin, Mianmin, Gadio Enga, Sanio Hiowe and Umeda may all be classed as 'copiers' (Dornstreich 1973, 1977; Gell 1975; Hyndman 1979, 1982; Morren 1974, 1977; Townsend 1974). They all make much use of 'non-domesticated' resources though, as Ellen (1975) has made clear for Nuaulu and as described for Etolo, a large component of the 'non-domesticated' resources used by 'copiers' has been 'tamed' either by overt manipulation or through unintentional alteration of habitat.

The New Guinean societies classed as 'copiers' differ in many ways. The major carbohydrate food may be sweet potato, taro or sago; some groups keep very few pigs, others keep more. Within this variety Etolo subsistence is relatively coarse-grained. The visual dominance of their gardens, even the sound of trees being felled, conceals the diversity of smaller scale management practices that are integral to their food-getting. It is these smaller scale ventures I want to highlight; a tiny patch of taro in a seepage, a few *P. conoideus* in a forest clearing, some lowland pitpit planted nearby, a ditch built to irrigate a sago palm, beetle larvae growing fat in a felled palm or a trap permanently set above a cuscus den. This list is lengthened if other Highland Fringe societies are examined. Wopkaimin subsistence, for example, is finer-grained that that of Etolo; notable examples of small scale management not found among the latter people are small groves of breadfruit trees (*Artocarpus*) and of nut pandans (*P. julianetti*), a diversity of techniques, including the promotion of flowering trees, for obtaining birds and, in secondary forest, the preparation of special mounds of litter where eggs and adults of two species of skink can be subsequently collected. Or again, for Etolo, observing the important vegetable acanth spinach (*Rungia klossii*) as a weed in regrowth, an overabundance of the self-propagating and edible climbing curcurbit (*Trichosanthes*) in newly planted gardens or aibika (*Hibiscus manihot*) lingering in long-abandoned gardens suggests an ancient potential for their easy management.

15.3.3 Clingers: who don't

'Clingers' do not live with rainforest. They live alongside it, at the interface with another environment. They combine exploitation of faunal resources located in the nonrainforest environment with the management of plant resources at or just within the rainforest boundary. The forest itself is, typically, rather unimportant for food-getting but provides materials for shelter, fuel and the paraphernalia of subsistence. 'Clingers' are coarse-grained in their response to environment either selecting a mix of plant and animal foods of large particle size or using techniques that clump their major plant foods and increase the haul of hunted animals. Because animal and plant foods are often located in different places which may be far apart, scheduling, usually expressed by assigning task-specific roles, is important.

There are three major sorts of 'clingers'; riverine people, people who live near the ocean including populations on tropical islands and people at the interface of rainforest and savannah. Examples from South America are discussed below. In the New Guinean and Pacific regions, populations living along the Sepik River and on some islands combine intensive cultivation of taro or yams with the exploitation of faunal resources from aquatic environments (river, lagoon, ocean) while on the upper reaches of the Fly River sago palms and riverine faunal resources are staples (Malinowski 1922; Williams 1936; Ohtsuka 1977; Johannes 1980; Rhoades 1981). To varying extents the societies concerned husband pigs. Techniques for 'fishing' are highly elaborated and the people may concentrate effort on large or clumped resources or use a mix of strategies to obtain large hauls.

The Miskito of coastal Nicaragua employ a complex of food-getting practices yet conform, in essentials, to the definition of 'clingers'. Nietschmann (1973) described the ecology of these people some years after changes resulting from emergence of a cash economy had been initiated. Portrayal of a 'traditional' subsistence mode is, therefore, rather speculative. The complexity of Miskito food-getting arose, in large part, from seasonal influences that altered availability of key resources. Growing root crops, particularly manioc (*Manihot esculenta*), and 'ocean fishing' for turtles were predominant themes in terms of both effort and contribution to diet. Because turtles were difficult or dangerous to obtain during periods of very wet weather [June, July and October–December (Nietschmann 1973, p. 166)] people shifted to fishing in sheltered lagoons and on rivers and increased their hunting of land mammals at these times. That the Miskito turned almost 'a blind eye' to food resources of the tropical forest which flourished near them is shown by the facts that (i) 88% of the area they exploited comprised ocean, lagoon, rivers and creeks, (ii) 87% of animal foods (by butchered meat weight) came from aquatic environments and (iii) about 85% of land-derived animal foods were from habitats highly modified by people (e.g. plantations and secondary forest). The coarse-grained theme of their subsistence is emphasized by the facts that (i) returns for effort in gardening were extremely high (approximately 30 : 1), (ii) a few large animals — green turtle, white-lipped peccary and white-tailed deer — contributed 87% of the butchered weight of all hunted animals, (iii) some hunting on land took advantage of game that was marooned on islands by rising flood waters, (iv) relatively long-term storage was possible for both manioc (in the ground) and turtles (tied up in the shade or held captive in races), (v) large and long-lasting dugout canoes and several sorts of intricate harpoons were manufactured and (vi) task specific roles played an important part in patterning subsistence (i.e. men and women had different gardening tasks with, traditionally, women doing more work than men and, among the men, different individuals tended to specialize on turtling with nets, turtling with harpoons and hunting mammals on land).

The ecology of Amazonian riverine societies has been described by many (e.g. Goldman 1963; Lathrap 1970; Meggers 1971; Ross 1978; Johnson 1982). The riverine people are swidden gardeners and hunters. Maize and manioc are their major crops. The fauna they pursue are fish and the larger aquatic or semi-aquatic reptiles, birds and mammals. Fishing techniques are elaborate. When hunting does occur on land the search is often concentrated along river banks. Seasonal flooding influences availability of foods and the people adjust to this by switching between different resources. Eggs of turtle and caiman are important foods at some times of the year. The lack of attention paid to the rainforest itself was captured by Goldman (1963, p. 44) who wrote "The orientation of the Cubeo is toward the river and not toward the forest. Whereas the forest is undifferentiated terrain, the rivers are known to every turn and outcrop of rock or other feature. The river is the source of ancestral powers, of benefits as well as dangers. The forest is a source mainly of dangers". The contrast between this attitude and the joyous reverence shown to *their* forest by the Mbuti pygmies of the Congo is profound (cf. Turnbull 1962).

Finally, among the Gé and Bororo of central Brazil there were relatively fine-grained 'clingers' who exploited rainforest and savannah environments (Turner 1979; Maybury-Lewis 1967). These people made their villages at rainforest margins, established swiddens within the forest and hunted and gathered from rainforest, savannah and the ecotonal zone between the two. For some Gé–Bororo societies, gardens accounted for about half the food eaten (Turner 1979, p. 149). Hunting and gathering were oriented primarily toward savannah where tapir, deer, rhea, honey, termites, roots, nuts, fruit and palm shoots were available and were important foods. Gathering was often associated with long distance 'trekking' where groups of people remained away from villages for several weeks at a time. Although some people lived near rivers, it seems that at least traditionally, they did little fishing, lacked knowledge of trapping, poisoning and preserving techniques and used rafts, not canoes, for transport on water (Maybury-Lewis 1967).

15.3.4 Controllers: who won't

'Controllers' live where the rainforest was. They eliminate the forest and make their living as gardeners and animal husbanders on land that once was forest. Their impact is conspicuous, either long-term or permanent. Remnants of forest that remain within their domain have usually been altered both structurally and floristically as a result of their activities. The manipulations of 'controllers' have the effect of bringing a number of separate procurement systems under one roof. It is hardly necessary to look beyond the gardens and the well regulated behaviour of domestic animals to see the primary sources of their food. If additional procurement systems are of significance then these too will probably be found close to where the people reside. For 'controllers' the thrust of their manipulations is the reduction of cost–benefit ratios. This is achieved by emphasizing a few food resources and, particularly, by locating different procurement systems alongside each other. For 'controllers', therefore, scheduling can be less intricate than for 'copiers'. The way in which the former people coordinate different procurement systems in time and space means that scheduling regimes, though often strict, may be structurally simple. It is in this, particularly, that 'controllers' reduce cost–benefit ratios. Their subsistence mode is very coarse-grained because, to exaggerate, all their foods have been clumped at one place. The primary domain occupied by 'controllers' becomes, in effect, a community that is largely produced by and dependent upon their activities.

Highland societies of New Guinea are 'controllers'. Sweet potato is the staple and gardening and pig husbandry the major tasks (e.g. Brookfield & Brown 1963; Clarke 1971; Waddell 1972; Sillitoe 1983). Taro gardens and orchards of nut *Pandanus* are subsidiary procurement systems. Across much of the central highlands, forest has been replaced by grasslands and stands of *Casuarina* trees. The latter have been planted in maturing gardens; they enrich soil and provide both fuel and building material. In these societies gardening is intensive. Techniques of tillage, composting and drainage for sweet potato and of irrigation for taro are widespread (Clarke & Street 1967; Waddell 1975; Spriggs 1978, 1982; Wood & Humphrey 1982). Crop diversity is high and, on most days, 95% or more of food may come from gardens. A considerable surplus production is directed to maintaining pigs.

The pigs kept by New Guinean Highlanders were often killed on special occasions — to mourn a death, celebrate a marriage or, every few years, hold a pig feast. [See, for example Rappaport (1968) on the Tsembaga Maring. Note that Maring subsistence is finer-grained than that of true Highlanders; the subsistence base is broader and more diffuse than occurs at the higher altitudes.] This implies that pigs are not of great significance as protein-providers; their protein yield is too erratic to meet day to day needs. Where then, do Highlanders obtain their protein and what is the role of pigs?

Many authors have commented on the apparently low protein intake of New Guinean Highlanders (Hipsley & Clements 1950; Oomen 1961, 1970; Hipsley & Kirk 1965; McArthur 1977). Norgan *et al* (1974) showed for one population, at Lufa, that diet was adequate. Nearly 80% of the daily intake of protein (43.2 g and 47.1 g for female and male, respectively) came from garden foods with sweet potato contributing about half this amount. Pigs contributed about two-thirds of the protein from animal sources but the supply was not regular and variation between individuals was high. At the time of this study store foods (e.g. rice and tinned fish) were significant and traditional constraints on the killing and consumption of pigs were probably relaxed. Were supplementary protein sources available to Highlanders? Hunting for medium-sized mammals was probably not of nutritional significance (Dwyer 1983) but two categories of food resources may have been useful. First, a variety of seasonally available larger or clumped foods that are rich in protein, e.g. fungi, *Pandanus* nuts and flying foxes. These foods, combined with 'unplanned' attrition from pig herds, may have allowed some patterning of supplementary protein sources through a year but, if important, this would imply diversification of subsistence beyond the strict definition of 'controllers'. Secondly, many species of insects and of small frogs, lizards, birds and mammals would have been available for casual collection and immediate eating by people engaged in a normal daily round. Access to these latter foods would be an incidental byproduct of overt food-getting practices. The foods might be 'invisible' to research workers and, given Highlander's perceptions of roles (Lawrence & Meggitt 1965; Brown & Buchbinder 1976), most likely to be eaten by women and children who might need them most.

When people interrupt their normal round of food-getting to engage in time-demanding social activity they need to have on hand, or be able to tap, some special resource that provides energy and important nutrients in less time than is usual. These special resources should exhibit many of the following states relative to the sorts of foods in normal use: either small particle size and highly clumped or large particle size and short time to capture, good nutrient mix, high abundance for at least short periods and short preparation time. If the resource in question lasts only briefly then its availability will dictate timing of the social activities; if it is long lasting then the people may time the activities as they please. Domestic pigs in the New Guinea Highlands have many of the properties listed above and I see their primary role as foods that facilitate festivals. Other foods with a similar role are bogong moths in the southern highlands of Australia, bunya nuts in southeastern Queensland and *Pandanus* nuts at very high altitudes of New Guinea (Flood 1980; Bowdler 1981). Each of these foods is characterized by an excellent mix of nutrients. In particular, protein levels are so high that during the periods the foods are eaten, people receive far more protein than they need. But it is the high fat (or lipid) content that makes these foods admirable for festivals. They are protein-rich packages of energy that allow people, for short periods, to defer from their usual subsistence mode to fulfil special social needs.

15.3.5 Overview: mobility, population and change

The above account has stressed foods eaten and patterns of food-getting by rainforest peoples. Categorization as four primary modes was merely a scaffolding; no simple typology will capture the rich variety of human ecological systems. The symbiotic association between Bantu agriculturalists and Mbuti net-hunters is a case in point (Turnbull 1965; Tanno 1976). Here, coarse-grained 'controllers' exchange garden products throughout the year for smoked and fresh meat obtained, during a short dry season, by much finer-grained 'conformers'. The latter people intensify their hunting to obtain the surplus meat.

Within each of the categories of rainforest people there is a range of subsistence modes that may be ordered from fine-grained to coarse-grained. Between them there is a trend from 'conformers' through 'copiers' to 'controllers' for the grain of subsistence to increase. This trend is not unidimensional. Finer-grained modes among 'copiers' are not, *per se*, coarser than the coarsest-grained modes among 'conformers'. Grain may be increased in several ways; e.g. when a few foods are taken as staples, when the dispersion of primary foods is highly clumped or when management is intensive. The differences between societies that have been emphasized in this essay are, first, the sorts of interactions that link choices people make among potential food resources, secondly, the varying degrees to which procurement systems are manipulated and, thirdly, the differing roles of scheduling in patterning subsistence modes. In broad outline the patterning of choice gave coherence to the ecology of 'conformers', the connections between diverse manipulated procurement systems were important for 'copiers' and the organization of procurement systems under a relatively unified scheduling regime characterized 'controllers'. To an extent 'clingers' were anomalous. Indeed, this category may be envisaged as cross-cutting all others — we could have, as it were, 'clingers' who were 'conformers', 'copiers' or 'controllers' of the dual environment with which they lived.

Patterns of mobility and population density hold much fascination for ecologists who study people. They have not been stressed in this essay. In part, they are concomitant upon subsistence mode (Jochim 1981, pp. 148–151). Thus, 'conformers' are able to be more mobile than rainforest people of other categories and, should they cause local depletion of foods or alter their diet seasonally, mobility may be obligatory. Where people manipulate resources in ways that fix the location of those resources so they restrict opportunities for movement. Manipulations that control supply or increase clumping, particle size, lasting properties or the abundance of foods are likely to be associated with either highly patterned movement or increased sedentism. By contrast, some manipulations that reduce the time to capture or prepare food resources may be portable. Within each broad category of rainforest people an increase in the grain of subsistence tends to be associated with increased sedentism. 'Copiers', for example, combine daily and seasonal movement patterns between places where various procurement systems are located with shifts of settlement at more or less regular intervals. In fact, these latter shifts reinforce the analogy between 'copiers' and the pattern of rainforest. Among these people, those whose response to environment is finer-grained tend to shift their settlements more frequently than those whose response is coarser-grained; compare, for example, Wopkaimin and Mianmin who may shift every year with Etolo who, traditionally, shifted every 3 or 4 years (Kelly 1977; Morren 1977; Hyndman 1979). Again, for coarser-grained 'clingers' the intensity of gardening may fix the site of settlements for relatively long periods but the need to obtain faunal resources at a distance may dictate mobility and at least short-term absences for the hunters. Increased sedentism, when linked to intensification of a few management practices, is generally associated with higher population density though the causal pathway remains obscure. In rainforest environments 'conformers' and 'copiers' typically have low absolute population densities ($< 2/km^2$); for the latter people population density in currently utilized areas may be several times higher. 'Clingers' span a wide range of densities while 'controllers' attain the highest densities (e.g. to $100/km^2$ in the highlands of New Guinea).

The colonization of rainforest by people and the subsequent colonization of those people by others has continued for millenia. The fragility of the rainforest environment, so recently appreciated by ecologists, has long been the experience of societies who lived there (e.g. Davis 1977; DéAth 1980; Waiko & Jiregari 1982). Subsistence modes have not remained static and, from recent history, even in the absence of overt coercion or threats of annihilation, there has been pattern to the change. Tasaday deferred to Dafal and took up trapping and processing of palm starch. Mbuti accepted produce from the gardens of Bantu as staples. Aché grew crops under the guidance of missionaries. Gé and Bororo became keen fishermen when riverine Brazilians came onto their land. Etolo increasingly experiment with the intensive gardening techniques used by their highland Huli neighbours. New Guinean Highlanders and Miskito have developed cash cropping. 'Conformers' seem to defer to 'copiers', 'copiers' to 'controllers'. More generally, people who freely adopt subsistence modes to which they have been exposed usually shift to modes of coarser-grain than their own. The reverse is rarely the case. Why is this so and can it offer insight into past changes that were not motivated by the practices, preaching and persuasion of neighbours or invaders?

15.4 The evolution of food-getting: a speculation on the unit of change

In the course of her or his life each person carves out an ontogenetic trajectory. There are rules that constrain the form of these trajectories. There is also scope for transgression and error, for reaching beyond operative constraints to express latent or new information. In form and in fate this uncommitted information is analogous to puns in language.

The model for cultural evolution used here is an analogue of ideas developed by epigeneticists and others (Wilden 1972; Gould 1977; Riedl 1978; Bateson 1979; Løvtrup 1981; Rachootin & Thomson 1981; Wiley & Brooks 1982). They have argued that new or latent information arising within the regulatory or ontogenetic domain of biological systems may contribute more to change of form than comparable information arising from structural genes. If this is true for biological systems it may apply with greater force for cultural ones. Here I take the ontogenetic errors and transgressions of people, from their birth to their death, to be the raw material of cultural change. Like puns they reshuffle conventional modes of expression. They may be accidental or highly motivated and everyone tries them at one time or another. They may be ignored or captured by, and incorporated into, the sphere of human activity from which they spring. Their capture requires replication through imitation and testing in a public arena. Their incorporation means they have decayed into convention itself and alters the domain for further change (Dwyer 1979). Puns are monstrous. They are helpless if they abuse limits of acceptability. They are hopeful when they do not infringe the harmony of language. By accepting ontogenetic 'puns' as the basis for an account of change, individual actors are assigned a fundamental role as initiators of change. Not all accounts of cultural evolution have met this requirement. [cf. Friedman and Rowlands (1977) who also offer an epigenetic perspective on cultural evolution.] In what follows I consider the role of ontogenetic 'puns' for the evolution of human food-getting. The discussion opens with four scenarios.

(1) A person digs for an awkwardly placed yam. The tuber breaks

away leaving the stem in the ground. This has not happened before. Months later the same person obtains another tuber at the same place. This time he takes care to leave the stem planted.

(2) A recent fall of trees lies down a steep slope. Crossing the trees a woman slips and drops some tubers she had gathered earlier. Some months later she observes a healthy yam vine in the overgrowing clearing; it climbs on a wild banana and is larger than those she had collected the tubers from. Several days pass and the woman digs another yam that is near a different clearing. She transplants the crown of this yam into that clearing.

(3) A family group gathers yams from a small patch of natural regrowth. They comment that vines exposed to the light yield larger tubers. As they replant crowns one person breaks away some overhanging branches.

(4) Pleased with the yams they have collected people eat them at once and then replant the crowns. A youth remarks that one tuber was extremely bitter and discards its crown.

The list of similar scenarios could be continued. A regular feeding site gives way to a grove of fruit trees. The area around a shelter harbours a scatter of self-propagated food plants. Someone uses a stick to prize pith from an obstinate palm and the starch falls away from the fibre. The acrid taste of some seeds is relieved because they are left overnight in a downpour. A plant stem discarded in a pool 'stuns' a few small fish. A large haul of insect larvae is collected from a palm trunk that, fortuitously, had been covered by a fall of leafy branches.

From the repetition of acts as simple as these people commence to fix, reposition and modify the locations of food resources or to alter the quality of those resources. An implication from the examples mentioned is that alterations to specific procurement systems might not be difficult to initiate. Such alterations could, however, lead to increased use of the manipulated resource with potentially adverse effects on nutrient balance. Thus, a new manipulative practice may be difficult to incorporate into the subsistence repertoire of a group of people unless coincident and compensatory adjustments are made in other procurement systems. It might not be hard, for example, to fall upon and refine techniques that produced more starch than usual from large palms but, unless these techniques were matched by changes in the way protein was obtained, their use could cause serious debilitation. Yet the people would not know this in advance. They would find out only through trial and error. Groups that regularly exploited the bountiful feast from large palms at the expense of protein intake would show reduced viability. The need for an adequate diet would act as a major constraint on the adoption of seemingly advantageous manipulations. This would slow the rate at which human food-getting behaviour altered through time.

The sorts of changes indicated above occur because the idiosyncratic behaviour of one individual gives a result that happens to appeal to many. What is the appeal? In the first instance the ontogenetic 'puns' described make access to, or some quality of, a resource more predictable. The location, supply or taste of the food, on the next occasion it is chosen, is guaranteed. Hazard or risk is, or appears to be, reduced. For these reasons the 'puns' are repeated and, eventually, become entrained. Reductions in cost are, initially, a secondary consideration. In fact, the new behaviour might be repeated even though some costs had increased.

Incorporation of new food-getting behaviour into the everyday repertoire of a group of people shifts their subsistence mode. Through time they will progressively fix the location of resources, increase usage of a few resources and adjust scheduling regimes to blend a reduced variety of more intricate procurement systems. As they do so their response to environment will have altered from finer-grained to coarser-grained. Tentative experiments in plant management may give way to gardening. The casual taming of young wild animals for later consumption may slide almost unnoticed into husbandry. Or, to change focus, pig-keeping may be replaced by battery chickens, MacDonald's farm by his hamburgers. Through time people have chosen from the diversity of what were once natural resources and progressively harnessed their selection to hearth and to home. Step by step they have abandoned nature. It has all been possible because 'punning' is unavoidable, tempting and delicious to copy — especially delicious when to copy is to gain control.

Many accounts of the evolution of subsistence focus on the transitions to agriculture on the one hand and animal domestication on the other (e.g. Ucko & Dimbleby 1969; Cohen 1977; Harris 1977). Speculations about transition are varied. They include intensification prompted by environmental change, encouraged by a need to support a growing population or as a result of borrowing from 'more knowledgeable' neighbours. Explanations of the last sort are 'diffusionist'; they may, for example, postulate a few core areas of the globe where people, living in extraordinarily favourable environments, had the time on their hands to first 'invent' and then 'export' the new ways (e.g. Sauer 1969). Each of these three models may contain a grain of truth. But the grain is particularized. At times and in particular places people will have responded to various pressures and there may well have been some 'innovative centres'. Within prehistoric New Guinea, for example, inundation of the Arafura Sea after 17 500 BP, was associated with expansion of rainforest up the mountains and onto the plain (Hope & Hope 1976; Hope *et al* 1983). At higher altitudes people may have responded to the changing environment by intensifying a broad range of plant management practices. At lower altitudes they may have promoted stands of sago palm within rainforest where, previously, it had been associated with the forest edge (cf. Rhoads 1982). The latter change would have been easily accomplished by holding stands of palms where they were as the forest itself surged southwards onto the plain. In both cases the alterations to subsistence mode may have occurred more because individual people attempted to continue doing what they had always done rather than because the environment forced change upon them.

The models referred to do not depict underlying units of change. In this they fail to describe a general process that can both accommodate and specify many different products, each of which has had its own history (cf. Rosen 1982). Subsistence evolution has not been unidirectional (Ellen 1982, p. 259). Refugees and castaways, for example, must generalize on former subsistence modes for they cannot carry all their manipulative skills with them and may be too few to adhere to former scheduling arrangements. Different responses are needed for different environments and people, in general, have responded capably. They have devised a variety of subsistence modes suited to a variety of environments. There is no sound reason for asserting that in most places people's existence was less than favourable or that they had little time to innovate; that view, in any case, is rather ungenerous. The evolutionary model sketched here focuses on ontogenetic 'puns' as the units of change. It is not inconsistent with the models mentioned above. Borrowing from one's neighbours is merely unimaginative 'punning' and, under pressure, people may well 'pun' more frequently — the peculiar circumstances and opportunities accelerate the rate of change, the mechanism is the same.

Over the millenia of human history, subsistence modes have tended to shift from finer-grained to coarser-grained. Why has this been so? One reason is proposed above. Alterations to food-getting behaviour that have increased predictability have been replicated. There is a second reason. Each time an ontogenetic 'pun' is built into a food-getting system it alters irrevocably people's perception of their environment. In doing so it prescribes a different domain of potential 'puns'. In the last analysis it is people who both perceive and define resources as acceptable or otherwise. This dimension of human food-getting has lain dormant through the present essay. It is crucial. People distinguish between the acceptable and the unacceptable. They clothe themselves in the acceptable, they internalize it. As this happens the boundary between Nature and Culture is repositioned though, for the participants, it seems to have sharpened (cf. Douglas 1966; Lévi-Strauss 1966). The sorts of ontogenetic 'puns' that are accepted are those whose psychological impact is to connect food and food-getting to Culture and disconnect them from Nature. Slowly but inevitably, in the minds of women and men, Nature itself has been driven into the wilderness.

The history of human subsistence has been an on-going reification of resources. What people have chosen to eat and the ways they have chosen to get it have become, for them, cultural. The rest is Nature. Food itself "...is accorded meaning; it is consumed, altered, displayed, distributed and destroyed" (Ellen 1982, p. 253). Resources exist in a state of perpetual redefinition; they are idealized and institutionalized. All of us are caught up in an ecological tautology wherein an ever-contracting resource base is perceived as ever-expanding; where Nature, which once was the source of our succour, is reduced to a Commons and a commodity. All over the world this, surely, has become the fate of rainforest — a commodity for those of us whose notions of food-getting have all but broken their links with Nature. For the others, of whom I wrote, only rainforest may sustain their living.

References

Abou Akkada A. R. & Howard B. H. (1960) The biochemistry of rumen protozoa. 3. The carbohydrate metabolism of *Entodinium*. *Biochem. J.* **76**, 445–451. 7.2.3

Abrahamson W. G. & Gadgil M. (1973) Growth form and reproductive effort in goldenrods (*Solidago* Compositae). *Amer. Nat.* **107**, 651–661. 14.4

Abrams P. (1975) Limiting similarity and the form of the competition coefficient. *Theor. Popul. Biol.* **8**, 356–375. 6.5.2

Abrams P. (1980) Some comments on measuring niche overlap. *Ecology* **61**, 44–49. 6.7.2

Abrams P. (1981a) Alternative methods of measuring competition applied to two Australian hermit crabs. *Oecologia* **51**, 233–239. 11.5.1

Abrams P. (1981b) Competition in an Indo-Pacific hermit crab community. *Oecologia* **51**, 240–249. 6.8

Abrams P. (1983) The theory of limiting similarity. *Ann. Rev. Ecol. Syst.* **14**, 359–376. 6.5.2

Abrams P. A. (1984) Recruitment, lotteries, and coexistence in coral fish. *Amer. Natur.* **123**, 44–55. 6.8

Abramsky Z. & Sellah C. (1982) Competition and the role of habitat selection in *Gerbillus allenbyi* and *Merionoes tristrami*: a removal experiment. *Ecology* **63**, 1242–1247. 11.3.1

Abramsky Z., Dyer M. I. & Harrison P. D. (1979) Competition among small mammals in experimentally perturbed areas of the shortgrass prairie. *Ecology* **60**, 530–536. 11.4.2

Adler S. (1954) The behavior of *Plasmodium berghi* in the golden hamster *Mesocricetus auratus* infected with visceral leishmaniasis. *Trans. roy. Soc. trop. Med. Hyg.* **48**, 431–440. 13.2.1

Al Dabagh M. A. (1961) Synergism between coccidia parasites (*Eimeria mitis* and *E. acervalina*) and malaria parasites (*Plasmodium gallinaceum* and *P. juxtanucleare*) in the chick. *Parasitology* **51**, 257–261. 13.2.1

Al-Hasan R. H., Coughlan S. J., Pant A. & Fogg G. E. (1975) Seasonal variations in phytoplankton and glycollate concentrations in the Menai Straits, Anglesey. *J. Mar. Biol. Assoc. U.K.* **55**, 557–565. 7.3.3

Allard R. W. (1965) Genetic systems associated with colonizing ability in predominantly self-pollinated species. In *The Genetics of Colonizing Species* (eds Baker H. G. & Stebbins G. L.) pp. 49–76. Academic Press, New York. 14.8

Allard R. W. & Adams J. (1969) Population studies in predominantly self-pollinating species. XII. Intergenotypic competition and population structure in barley and wheat. *Amer. Natur.* **103**, 621–645. 14.7

Alldredge A. L. (1976) Discarded appendicularian houses as sources of food, surface habitats and particulate organic matter in planktonic environments. *Limnol. Oceanogr.* **21**, 14–23. 7.3.3

Allee W. C., Emerson A. E., Park O., Park T. & Schmidt K. P. (1949) *Principles of Animal Ecology*. Saunders, Philadelphia. 3.1

Allen E. B. & Forman R. T. T. (1976) Plant species removals and old-field community structure and stability. *Ecology* **57**, 1233–1243. 11.3.1, 11.7

Amadon D. (1973) Birds of the Congo and Amazon forests: A comparison. In *Tropical Forest Ecosystems in Africa and South America: A Comparative Review* (eds Ayensu E. S., Duckworth W. D. & Meggers B. J.) pp. 267–277. Smithsonian Inst. Press, Washington, D.C. 5.5.1

Anderson D. J. (1965a) Studies on structure in plant communities. I. An analysis of limestone grassland in Monk's Dale, Derbyshire, *J. Ecol.* **53**, 97–107. 12.3.3

Anderson D. J. (1965b) Classification and ordination in vegetation science: controversy over a non-existent problem? *J. Ecol.* **53**, 521–526. 1.5

Anderson D. J. (1971) Hierarchies and integration in ecology. *Proc. Ecol. Soc. Aust.* **6**, 1–6 preface, 12.3.1

Anderson G. R. V., Ehrlich A. H., Ehrlich P. R., Roughgarden J. D., Russell B. C. & Talbot F. H. (1981) The community structure of coral reef fishes. *Amer. Natur.* **117**, 476–495. 6.8

Anderson J. M. (1975) Succession, diversity and trophic relationships of some soil animals in decomposing leaf litter. *J. Anim. Ecol.* **44**, 475–495. 7.2.2

Anderson J. M. & Bignell D. E. (1980) Bacteria in the food, gut content and faeces of the litter-feeding millipede *Glomeris marginata* (Villers). *Soil Biol. Biochem.* **12**, 251–254. 7.2.2

Anderson P. K. (1964) Lethal alleles in *Mus musculus*. Local distribution and evidence for isolation of demes. *Science* **145**, 177–178. 13.3.3

Anderson P. K. (1965) The role of breeding structure in evolutionary processes of *Mus musculus* populations. In *Mutation in Population. Proc. Symp. Mutational Process* (ed. Hončariv R.) pp. 17–21. Academia, Prague. 13.3.3

Anderson P. K. (1970) Ecological structure and gene flow in small mammals. *Symp. Zool. Soc. Lond.* **26**, 299–325. 13.3.3

Anderson R. M. (1979a) Parasite pathogenicity and the depression of host population equilibrium. *Nature* **279**, 151–152. 13.2.1

Anderson R. M. (1979b) The influence of parasitic infection on the dynamics of host population growth. In *Population Dynamics* (eds Anderson R. M., Turner B. D. & Taylor L. R.) pp. 245–281. Blackwell Scientific Publications, Oxford. 9.3, 9.3.1

Anderson R. M. (1981) Population ecology of infectious disease agents. In *Theoretical Ecology* (ed. May R. M.) (2nd edn) pp. 318–355. Blackwell Scientific Publications, Oxford. 8.3.3

Anderson R. M. & May R. M. (1978) Regulation and stability of host–parasite population interactions. I. Regulatory processes. *J. Anim. Ecol.* **47**, 219–247. 9.3.1

Anderson R. M. & May R. M. (1979) Population biology of infectious diseases: Part 1. *Nature* **280**, 361–367. 9.3.1

Anderson R. M. & May R. M. (1980) Infectious diseases and population cycles of forest insects. *Science* **210**, 658–661. 8.3.3

Anderson R. M. & May R. M. (1981) The population dynamics of microparasites and their invertebrate hosts. *Phil. Trans. roy. Soc. Lond.* **B291**, 451–524. 9.3

Anderson R. M. & May R. M. (1982) Coevolution of hosts and parasites. *Parasitology* **85**, 411–426. 9.3, 9.3.2

Anderson R. M. & Michel J. F. (1977) Density-dependent survival in populations of *Ostertagia ostertagi*. *Intern. J. Parasitol.* **7**, 321–329. 9.2.5

Andrewartha H. A. & Birch L. C. (1954) *Distribution and Abundance of Animals*. University of Chicago Press, Chicago. 6.8

Antonovics J. (1978) The population genetics of mixtures. In *Plant Relations in Pastures*. (ed. Wilson J. R.) pp. 233–252. CSIRO, Melbourne. 14.7

Arai H. P. (1980) Migratory activity and related phenomena in *Hymenolepis diminuta*. In *Biology of the Tapeworm* Hymenolepis diminuta (ed. Arai H. P.) pp. 615–637. Academic Press, New York. 9.2.1

Arlin M. (1977) *The Science of Nutrition*. MacMillan, New York. 15.2

Arme C. (1968) Effects of the plerocercoid larva of a pseudophyllidean cestode, *Ligula intestinalis*, on the pituitary gland and gonads of its hosts. *Biol. Bull.* **134**, 14–25. 13.2.1

Armstrong R. A. (1982) The effects of connectivity on community stability. *Amer. Natur.* **120**, 391–402. 4.2.2, 4.2.3

Arnold S. J. (1982) The microevolution of feeding behavior. In *Foraging Behavior* (eds Kamil A. C. & Sargent T. D.) pp. 409–454. Garland STPM Press, New York. 6.6.3

Arrhenius O. (1921) Species and area. *J. Ecol.* **9**, 95–99. 4.3.1

Arthur W. (1982) The evolutionary consequence of interspecific competition. *Adv. Ecol. Res.* **12**, 127–188. 14.6

Ashmole N. P. (1968) Body size, prey size and ecological segregation in 5 sympatric tropical terns (*Aves*: Laridae). *Syst. Zool.* **17**, 292–304. 6.4

Askew R. R. (1980) The diversity of insect communities in leaf-miners and plant galls. *J. Anim. Ecol.* **49**, 817–829. 5.2

Aston J. & Bradshaw A. D. (1966) Evolution in closely adjacent plant populations. II. *Agrostis stolonifera* in maritime habitats. *Heredity* **21**, 649–664. 14.2, 14.6

Atkinson W. D. & Shorrocks B. (1981) Competition on a divided and ephemeral resource: a simulation model. *J. Anim. Ecol.* **50**, 461–471. 8.3.1

Ausmus B. S. (1977) Regulation of wood decomposition rates by arthropod and annelid populations. *Soil Organisms as Components of Ecosystems. Ecol. Bull.* (Stockholm) **25**, 180–192. 7.2.2

Ayling A. M. (1981) The role of biological disturbance in temperate subtidal encrusting communities. *Ecology* **62**, 830–847. 9.3.4

Bach C. E. (1980a) Effects of plant density and diversity on the population dynamics of a specialist herbivore, the striped cucumber beetle, *Acalymma vittata* (Fab). *Ecology* **61**, 1515–1530. 8.2.2

Bach C. E. (1980b) Effects of plant diversity and time of colonisation on a herbivore–plant interaction. *Oecologia* **44**, 319–326. 8.2.2

Bailey T. G. & Robertson D. R. (1982) Organic and caloric levels of fish feces relative to its consumption by coprophagous reef fishes. *Mar. Biol.* **69**, 45–50. 7.3.2

Baldwin R. L., Lucas H. L. & Cabrera R. (1969) Energetic relationships in the formation and utilization of fermentation end-products. In *Physiology of Digestion and Metabolism in the Ruminant. Proc. Third Intern. Symp.* (ed. Phillipson A. T.) pp. 319–334. Oriel Press, Newcastle upon Tyne. 7.2.3

Barbehenn K. R. (1969) Host–parasite relationships and species diversity in mammals: an hypothesis. *Biotropica* **1**, 29–35. 9.3.2

Barnes O. L. (1965) Further tests of the effects of food plants on the migratory grasshopper. *J. econ. Entomol.* **58**, 475–479. 14.9

Barrett J. A. (1984) Genetics of host–parasite interactions. In *Evolutionary Ecology* (ed. Shorrocks B.) pp. 275–294. Blackwell Scientific Publications, Oxford. 14.8

Barrow J. H. (1955) Social behavior in freshwater fish and its effect on resistance to trypanosomes. *Proc. Natl. Acad. Sci.* **41**, 676–679. 9.3

Bateson G. (1979) *Mind and Nature: A Necessary Unity*. Dutton, New York. 15.4

Bauer O. N. & Hoffman G. L. (1976) Helminth range extension by translocation of fish. In *Wildlife Diseases* (ed. Page L. A.) pp. 163–172. Plenum Press, New York. 9.3.2

Bazzaz F. A. (1979) The physiological ecology of plant succession. *Ann. Rev. Ecol. Syst.* **10**, 351–371. 12.2

Bazzaz F. A. & Pickett S. T. A. (1980) Physiological ecology of tropical succession: a comparative review. *Ann. Rev. Ecol. Syst.* **11**, 287–310. 12.2

Beard J. S. (1946) The mora forests of Trinidad, British West Indies. *J. Ecol.* **33**, 173–192. 2.5

Beattie A. J. & Culver D. C. (1981) The guild of myrmecochores in the herbaceous flora of West Virginia forests. *Ecology* **62**, 107–115. 5.2

Beck L. & Friebe B. (1981) Verwertung von Kohlenhydraten bei *Oniscus asellus* (Isopoda) und *Polydesmus angustus* (Diplopoda). *Pedobiologia* **21**, 19–29. 7.2.2

Beckwith S. L. (1954) Ecological succession on abandoned farm lands and its relation to wildlife management. *Ecol. Monog.* **24**, 349–376. 12.2

Beddington J. R., Free C. A. & Lawton J. H. (1978) Characteristics of successful natural enemies in models of biological control of insect pests. *Nature* **273**, 513–519. 8.3.3

Beever J. W. III. (1979) The niche-variation hypothesis: an examination of assumptions and organisms. *Evol. Theory* **4**, 181–191. 6.6.3

Begon M. & Mortimer M. (1981) *Population Ecology*. Blackwell Scientific Publications, Oxford. 8.3.1

Begon M. E. (1985) A general theory of life history variation. In *Behavioural Ecology* (eds Sibley R. M. & Smith R.H) pp. 91–97. Blackwell Scientific Publications, Oxford. 14.4

Behnke J. M., Bland P. W. & Wakelin D. (1977) Effect of the expulsion phase of *Trichinella spiralis* on *Hymenolepis diminuta* infection in mice. *Parasitology* **75**, 79–88. 13.3.3

Behnke J. M. & Parish H. A. (1979) *Nematospiroides dubius*: arrested development of larvae in immune mice. *Exp. Parasitol.* **47**, 116–127. 13.3.3

Bell R. G., McGregor D. D. & Desbommier D. D. (1979) *Trichinella spiralis*: mediation of the intestinal component of protective immunity in the rat by multiple, phase-specific antiparasite responses. *Exp. Parasitol.* **47**, 140–157. 13.3.3

Bell R. H. V. (1970) The use of the herb layer by grazing ungulates in the Serengeti. In *Animal Populations in Relation to Their Food Resources* (ed. Watson A.) pp. 111–124. Blackwell Scientific Publications, Oxford. 8.4.3

Bell R. H. V. (1971) A grazing ecosystem in the Serengeti. *Sci. Amer.* **224**(1), 86–93. 8.4.3

Bell R. H. V. (1982) The effect of soil nutrient availability on community structure in African ecosystems. In *Ecology of Tropical Savannas* (eds Huntley B. J. & Walker B. H.) pp. 193–216. Springer-Verlag, Berlin. 8.4.3, 8.5

Belovsky G. E. (1984) Moose and snowshoe hare competition and a mechanistic explanation from foraging theory. *Oecologia* **61**, 150–159. 6.7.2, 6.8

Benke A. C. (1978) Interactions among coexisting predators — a field experiment with dragonfly larvae. *J. Anim. Ecol.* **47**, 335–350. 11.2.1

Benson W. W. (1978) Resource partitioning in passion vine butterflies. *Evolution* **32**, 493–518. 14.9

Benson W. W., Brown K. S. & Gilbert L. E. (1975) Coevolution of plants and herbivores: passion flower butterflies. *Evolution* **29**, 659–680. 14.9

Berenbaum M. (1981) Patterns of furanocoumarin distribution and insect herbivory in the Umbelliferae: plant chemistry and community structure. *Ecology* **62**, 1254–1266. 8.2.4

Berk S. G., Brownlee D. C., Heinle D. R., Kling H. J. & Colwell R. R. (1977) Ciliates as a food source for marine planktonic copepods. *Microb. Ecol.* **4**, 27–40. 7.3.3

Berry R. J. (1968) Epigenetic polymorphism in wild populations of *Mus musculus*. *Genet. Res.* **4**, 193–220. 13.3.3

Berry R. J. (1977) *Inheritance and Natural Selection*. Collins, London. 14.2

Bertness M. D. (1981a) Competitive dynamics of a tropical hermit crab assemblage. *Ecology* **62**, 751–761. 11.3.1, 11.4.1

Bertness M. D. (1981b) Conflicting advantages in resource utilization: the hermit crab housing dilemma. *Amer. Natur.* **118**, 432–437. 6.8

Bethel W. M. & Holmes J. C. (1973) Altered evasive behavior and responses to light in amphipods harboring acanthocephalan cystacanths. *J. Parasitol.* **59**, 945–956. 9.3, 9.3.3

Bethel W. M. & Holmes J. C. (1977) Increased vulnerability of amphipods to predation owing to altered behavior induced by larval acanthocephalans. *Can. J. Zool.* **55**, 110–115. 9.3, 9.3.3

Beyers R. J. (1963) The metabolism of twelve aquatic microecosystems. *Ecol. Management* **33**, 281–306. 10.4

Beyers R. J. (1964) The microcosm approach to ecosystem biology. *Amer. Biol. Teacher* **26**, 491–498. 10.4

Biehl C. C. & Cody M. L. Ecological segregation by body size: tests, interpretation, and misinterpretation of field data. *Evolution*, submitted for publication. 6.5.1

Biffen R. H. (1905) Mendel's laws of inheritance and wheatbreeding. *J. agric. Sci.* (*Camb.*) **2**, 109–128. 14.8

Birch L. C. (1957) The meanings of competition. *Amer. Natur.* **91**, 5–18. 11.2.1

Birch L. C. (1979) The effects of species of animals which share common resources on one another's distribution and abundance. *Fortschr. Zool.* **25**, 197–221. 11.1, 11.1.1

Bishop J. A. & Cook L. M. (eds) (1981) *Genetic Consequences of Man Made Change*. Academic Press, London. 14.2

Björkman O. & Holmgren P. (1963) Adaptability of the photosynthetic apparatus to light intensity in ecotypes from exposed and shaded habitats. *Physiol. Plant.* **16**, 889–914. 14.4

Blackman G. E. & Rutter A. J. (1954) Biological flora of the British Isles. *Endymion nonscriptus* (L.) Garcke. *J. Ecol.* **42**, 629–638. 2.6

Blaustein A. R., Kuris A. M. & Alió J. J. (1983) Pest and parasite species-richness problems. *Amer. Natur.* **122**, 556–566. 8.2.3

Bocock K. L. (1963) Changes in the amount of nitrogen in decomposing leaf litter of sessile oak (*Quercus petraea*). *J. Ecol.* **51**, 555–566. 7.2.2

Bocock K. L., Gilbert O., Capstick C. K., Twinn D. C., Waid J. S. & Woodman M. J. (1960) Changes in leaf litter when placed on the surface of soils with contrasting humus types. I. Losses in dry weight of oak and ash leaf litter. *J. Soil Sci.* **11**, 1–9. 7.2.2

Booth R. G. & Anderson J. M. (1979) The influence of fungal food quality on the growth and fecundity of *Folsomia candida* (Collembola: Isotomidae). *Oecologia* **38**, 317–323. 7.2.2

Borlaug N. E. (1972) A cereal breeder and ex-forester's evaluation of the progress and problems involved in breeding rust resistant forest trees: moderator's summary. *U.S. Dep. Agric. Misc. Publ.* **1221**, 615–642. Washington, D.C. 14.8

Boucher D. H. (1982) The ecology of mutualism. *Ann. Rev. Ecol. Syst.* **13**, 315–347. 6.8

Boucot A. J. (1983) Area-dependent-richness hypotheses and rates of parasite/pest evolution. *Amer. Natur.* **121**, 294–300. 9.2.3

Bourlière F. (1973) The comparative ecology of rain forest mammals in Africa and tropical America: some introductory remarks. In *Tropical Forest Ecosystems in Africa and South America: A Comparative Review* (eds Meggers B. J., Ayensu E. S. & Duckworth W. D.) pp. 279–292. Smithsonian Institute Press, Washington D.C. 5.5, 5.5.1

Bournier A. (1977) Grape insects. *Ann. Rev. Ent.* **22**, 355–376. 8.2.4

Bowdler S. (1981) Hunters in the highlands: aboriginal adaptations in the Eastern Australian Uplands. *Archaeology in Oceania* **16**, 19–111. 15.3.4

Bowman R. I. (1961) Morphological differentiation and adaptation in the Galapagos finches. *Univ. Calif. Publ. Zool.* **58**, 1–302. 6.4, 6.6.2

Box E. D. (1967) Influence of *Isospora* infections on patency of avian *Lankesterella* (*Atoxoplasma*, Garnham, 1950). *J. Parasitol.* **53**, 1140–1167. 13.2.1

Bradley D. J. (1972) Regulation of parasite populations: a general theory of the epidemiology and control of parasitic infections. *Trans. roy. Soc. trop. Med. Hyg.* **66**, 697–708. 9.3

Bradley D. J. (1974) Stability in host-parasite systems. In *Ecological Stability* (eds Usher M. B. & Williamson M. H.) pp. 71–87. Chapman & Hall, London. 9.2.5

Bradley D. J. (1982) Epidemiological models — theory and reality. In *Population Dynamics of Infectious Diseases, Theory and Applications* (ed. Anderson R. M.) pp. 320–333. Chapman & Hall, London. 9.3.1

Bradshaw A. D. (1984) The importance of evolutionary ideas in ecology — and vice versa. In *Evolutionary Ecology* (ed. Shorrocks B.) pp. 1–25. Blackwell Scientific Publications, Oxford. 14.2

Bradshaw A. D. & Chadwick M. J. (1980) *The Restoration of Land: The Ecology and*

Reclamation of Derelict and Degraded Land. Blackwell Scientific Publications, Oxford. 1.5

Breedlove D. E. & Ehrlich P. R. (1968) Plant–herbivore co-evolution: lupines and lycaenids. *Science* **162**, 671–672. 1.7

Breen P. A. & Mann K. H. (1976) Changing lobster abundance and the destruction of kelp beds by sea urchins. *Mar. Biol.* **34**, 137–142. 10.2.2

Bremermann H. J. & Pickering J. (1983) A game-theoretical model of parasite virulence. *J. theor. Biol.* **100**, 411–426. 9.3

Brereton J. Le Gay & Kikkawa J. (1963) Diversity of avian species. *Aust. J. Sci.* **26**, 12–14. 4.4.1

Briand F. (1983) Environmental control of food web structure. *Ecology* **64**, 253–263. 4.2.3

Briand F. & Cohen J. E. (1984) Community food webs have scale-invariant structure. *Nature* **307**, 264–267. 4.2.3

Broekhuizen S. & Kemmers R. (1976) The stomach worm, *Graphidium strigosum* (Dujardin) Railliett & Henry, in the European hare, *Lepus europaeus* Pallas. In *Ecology and Management of European Hare Populations* (eds Pielowski Z. & Pucek Z.) pp. 157–171. Polish Hunt. Assoc., Warsaw. 9.3.2

Brookfield H. C. & Brown P. (1963) *Struggle for Land: Agriculture and Group Territories among the Chimbu of the New Guinea Highlands*. Oxford Univ. Press, Melbourne. 15.3.4

Brooks D. R. (1980a) Allopatric speciation and non-interactive parasite community structure. *Syst. Zool.* **29**, 192–203. 9.2.2, 9.2.3

Brooks D. R. (1980b) Testing hypotheses of evolutionary relationships among parasitic helminths: the digeneans of crocodilians. *Amer. Zool.* **19**, 1225–1238. 9.2.3

Brower J. VZ. (1958) Experimental studies of mimicry in some North American butterflies. Part II. *Battus philenor* and *Papilio troilus*, *P. polyxenes* and *P. glaucus*. *Evolution* **12**, 123–136. 14.9

Brower J. VZ. (1960) Experimental studies of mimicry. IV. The reactions of starlings to different proportions of models and mimics. *Amer. Natur.* **94**, 271–282. 14.9

Brown J. H. (1975) Geographical ecology of desert rodents. In *Ecology and Evolution of Communities* (eds Cody M. L. & Diamond J. M.) pp. 315–341. Belknap Press, Cambridge, Massachusetts. 5.5.3, 6.8

Brown J. H. & Davidson D. W. (1977) Competition between seed-eating rodents and ants in desert ecosystems. *Science* **196**, 880–882. 5.2

Brown J. H., Reichman O. J. & Davidson D. W. (1979) Granivory in desert ecosystems. *Ann. Rev. Ecol. Syst.* **10**, 201–227. 8.5

Brown K. M. (1982) Resource overlap and competition in pond snails: an experimental analysis. *Ecology* **63**, 412–422. 11.5.1, 11.6.2, 11.7

Brown P. & Buchbinder G. (eds) (1976) *Man and Woman in the New Guinea Highlands* Special Publ. No. **8**. American Anthropological Association, Washington, D.C. 15.3.4

Brown R. T. (1967) Influence of naturally occurring compounds on germination and growth of jack pine. *Ecology* **48**, 542–546. 11.7

Brown V. K. (1982) The phytophagous insect community and its impact on early successional habitats. *Proc. 5th Int. Symp. Insect–Plant Relationships, Wageningen*, pp. 205–212. PUDOC, Wageningen. 8.4.3

Brown W. L. & Wilson E. O. (1956) Character displacement. *Syst. Zool.* **5**, 49–64. 6.6.1

Browning J. A. (1974) Relevance of knowledge about natural ecosystems to development of pest management programs for agro-ecosystems. *Proc. Amer. Phytopath.* **1**, 191–199. 14.8

Brylinsky M. (1977) Release of dissolved organic matter by some marine macrophytes. *Mar. Biol.* **39**, 213–220. 7.3.3

Bucher E. H. (1982) Chaco and Caatinga — South American arid savannas, woodlands and thickets. In *Ecology of Tropical Savannas* (eds Huntley B. J. & Walker B. H.) pp. 48–79. Springer-Verlag, Berlin. 8.5

Buckley R. (1983) Interaction between ants and membracid bugs decreased growth and seed set of host plant bearing extrafloral nectaries. *Oecologia* **58**, 132–136. 10.2.3

Bull P. C. (1969) The smaller placental mammals of Canterbury. In *The Natural History of Canterbury* (ed. Knox G. A.) pp. 400–419. A. H. & H. W. Reed, Wellington, New Zealand. 10.3.1

Bull P. C. & Whitaker A. H. (1975) The amphibians, reptiles, birds and mammals. In *Biogeography and Ecology in New Zealand* (ed. Kuschel G.) pp. 231–276. W. Junk, The Hague. 10.3.1

Burchfield R. W. (ed.) (1972) *A Supplement to the Oxford English Dictionary*, Vol. **1**. Clarendon Press, Oxford. 3.1

Burdon J. J. (1982) The effect of fungal pathogens on plant communities. In *The Plant Community as a Working Mechanism* (ed. Newman E. I.) pp. 99–112. Blackwell Scientific Publications, Oxford. 14.8

Burnet M. & White D. O. (1972) *Natural History of Infectious Disease* (4th edn). Cambridge University Press, Cambridge. 14.8

Burrough R. J., Bregazzi P. R. & Kennedy C. R. (1979) Interspecific dominance among three species of coarse fish in Slapton Ley, Devon. *J. Fish Biol.* **15**, 535-544. 9.3.2

Bush A. O. & Holmes J. C. (1983) Niche separation and the broken-stick model: use with multiple assemblages. *Amer. Natur.* **122**, 849–855. 9.2.1, 9.2.5

Bush A. O. & Holmes J. C. (1986) Intestinal helminths of lesser scaup ducks: pattern of associations. *Can. J. Zool.* (in press). 9.2.1

Bush G. L. (1975) Modes of animal speciation. *Ann. Rev. Ecol. Syst.* **6**, 339–364. 8.2.4, 14.2

Bush G. L. & Diehl S. R. (1982) Host shifts, genetic models of sympatric speciation and the origin of parasitic insects. *Proc. 5th Int. Symp. Insect–Plant Relationships, Wageningen*, pp. 297–305. PUDOC, Wageningen. 8.5

Butterworth E. W. (1982) A study of the structure and organization of intestinal helminth communities in ten species of waterfowl. PhD thesis, University of Alberta, Edmonton. 9.2.1, 9.2.4

Cain S. A. (1947) Characteristics of natural areas and factors in their development. *Ecol. Monogr.* **17**, 185–200. 12.2

Cameron G. N. & LaPoint T. W. (1978) Effects of tannins on the decomposition of Chinese tallow leaves by terrestrial and aquatic invertebrates. *Oecologia* **32**, 349–366. 7.2.2

Cammen L. M. (1980) The significance of microbial carbon in the nutrition of the deposit feeding polychaete *Nereis succinea*. *Mar. Biol.* **61**, 9–20. 7.3.3

Camp J. W. & Huizinga H. W. (1979) Altered color, behavior and predation susceptibility of the isopod *Asellus intermedius* infected with *Acanthocephalus dirus*. *J. Parasitol.* **65**, 667–669. 9.3.3

Cardy R. H. & Warner J. W. (1979) Effect of sequential bleeding on body weight in rats. *Lab. Anim. Sci.* **29**, 179–181. 13.3.2

Carey A. B., McLean R. G. & Maupin G. O. (1980) The structure of a Colorado tick fever ecosystem. *Ecol. Monogr.* **50**, 131–151. 9.3, 9.3.4

Carney W. P. (1969) Behavioral and morphological changes in carpenter ants harboring dicrocoeliid metacercaria. *Amer. Midl. Natur.* **82**, 605–611. 9.3.3

Carpenter S. R. (1982) Stemflow chemistry: effect on population dynamics of detritivorous mosquitoes in treehole ecosystems. *Oecologia* **53**, 1–6. 10.2.1

Carroll E. J. & Hungate R. E. (1954) The magnitude of the microbial fermentation in the bovine rumen. *Applied Microbiology* **2**, 205–214. 7.2.3

Carter M. A. (1968) Thrust predation of an experimental population of the snail *Cepea nemoralis* (L.). *Proc. Linn. Soc. Lond.* **179**, 241–249. 9.3.3

Case T. J. (1979) Character displacement and coevolution in some *Cnemidophorus* lizards. *Fortschr. Zool.* **25**, 235–282. 6.5.2

Case T. J. (1982) Coevolution in resource-limited competition communities. *Theor. Popul. Biol.* **21**, 69–91. 6.5.1, 6.5.2, 6.6.3

Case T. J. (1983) Sympatry and size similarity in *Cnemidophorus*. In *Lizard Ecology: Studies of a Model Organism* (eds Huey R. B., Pianka E. R. & Schoener T. W.) pp. 297–325. Harvard University Press, Cambridge, Massachusetts. 6.5.1

Case. T. J., Faaborg J. & Sidell R. (1983) The role of body size in the assembly of West Indian bird communities. *Evolution* **37**, 1062–1074. 6.5.1

Caswell H. (1976) Community structure: a neutral model analysis. *Ecol. Monogr.* **46**, 327–354. 6.3

Caughley G. (1970) Eruption of ungulate populations, with emphasis on Himalayan thar in New Zealand. *Ecology* **51**, 53–72. 1.7

Caughley G. & Lawton J. H. (1981) Plant–herbivore systems. In *Theoretical Ecology* (ed. May R. M.) (2nd edn) pp. 132–166. Blackwell Scientific Publications, Oxford. 8.3.2, 8.4.3

Chandler A. C. & Read C. P. (1961) *Introduction to Parasitology*. Wiley, New York. 13.3.3

Chappel M. A. (1978) Behavioral factors in the altitudinal zonation of chipmunks (*Eutamias*). *Ecology* **59**, 565–579. 11.3.1

Charley J. L. & West N. E. (1975) Plant-induced chemical patterns in some shrub-dominated semi-desert ecosystems of Utah. *J. Ecol.* **63**, 945–963. 12.3.3

Chesson P. L. & Warner R. R. (1981) Environmental variability promotes coexistence in lottery competitive systems. *Amer. Natur.* **117**, 923–943. 6.8

Chiang H. C. (1978) Pest management in corn. *Ann. Rev. Ent.* **23**, 101–123. 8.2.4

Choi C. I. (1972) Primary production and release of dissolved organic carbon from phytoplankton in the western North Atlantic Ocean. *Deep-Sea Res.* **19**, 731–735. 7.3.3

Chowaniec W., Westcott R. B. & Congdon L. L. (1972) Interaction of *Nematospiroides dubius* and influenza virus in mice. *Exp. Parasitol.* **32**, 33–46. 13.2.1

Christian R. R. & Wetzel R. L. (1978) Interaction between substrate, microbes and consumers of *Spartina* detritus in estuaries. In *Estuarine Interactions* (ed. Wiley M. L.) pp. 93–113. Academic Press, New York. 7.3.3

Christie G. (1978) Haematological and biochemical findings in an anaemia induced by the daily bleeding of ten-week-old cockerels. *Br. vet. J.* **134**, 358–365. 13.3.2

Christie P. R., Wakelin D. & Wilson M. M. (1979) The effect of the expulsion phase of *Trichinella spiralis* on *Hymenolepis diminuta* infection in rats. *Parasitology* **78**, 323–330. 13.3.3

Claridge M. F. & Wilson M. R. (1982a) Insect herbivore guilds and species-area relationships: leafminers on British trees. *Ecol. Ent.* **7**, 19–30. 8.2.3

Claridge M. F. & Wilson M. R. (1982b) Species-area effects for leafhoppers on British trees: comments on a paper by Rey *et al. Amer. Natur.* **119**, 573–575. 8.2.3

Clarke B. (1962) Balanced polymorphism and the diversity of sympatric species. *Syst. Assoc. Publ., Taxonomy & Geography* **4**, 47–70. 6.3, 14.9

Clarke W. C. (1971) *Place and People: an Ecology of a New Guinea Community*. Australian National University Press, Canberra. 15.3.4

Clarke W. C. (1976) Maintenance of agriculture and human habitats within the tropical forest ecosystem. *Human Ecol.* **4**, 247–259. 15.1

Clarke W. C. & Street J. M. (1967) Soil fertility and cultivation practices in New Guinea. *J. trop. Geogr.* **24**, 7–11. 15.3.4

Clausen C. P. (ed.) (1978) *Introduced Parasites and Predators of Arthropod Pests and*

Weeds: a World Review. Agriculture Handbook **480**. United States Department of Agriculture, Washington, D.C. 8.3.3

Clausen J. (1951) *Stages in the Evolution of Plant Species*. Cornell University Press, Ithaca. 14.2

Clements F. E. (1916) Plant succession: an analysis of the development of vegetation. *Publ. Carnegie Inst.* **242**. 12.2

Clifford H. T. & Stephenson W. (1975). *An Introduction to Numerical Classification*. Academic Press, New York. 4.4.2

Cody M. L. (1968) On the methods of resource division in grassland bird communities. *Amer. Natur.* **102**, 107–147. 6.5.8, 6.8

Cody M. L. (1974) *Competition and the Structure of Bird Communities*. Princeton University Press, Princeton. 5.5, 6.8

Cody M. L. (1975) Towards a theory of continental species diversities: bird distributions over Mediterranean habitat gradients. In *Ecology and Evolution of Communities* (eds Cody M. L. & Diamond J. M.) pp. 214–257. Belknap Press, Cambridge, Massachusetts. 5.5.2, 5.5.3

Cody M. L. & Diamond J. M. (eds) (1975) *Ecology and Evolution of Communities*. Belknap Press, Cambridge, Massachusetts. 5.4

Cohen J. E. (1966) *A Model of Simple Competition*. Harvard University Press, Cambridge, Massachusetts. 4.3.2

Cohen J. E. (1970) A Markov contingency table model for replicated Lotka–Volterra systems near equilibrium. *Amer. Natur.* **104**, 547–560. 6.5.4

Cohen J. E. (1978) *Food Webs and Niche Space*. Princeton University Press, Princeton. 4.2.3, 10.2

Cohen M. N. (1977) *The Food Crisis in Prehistory: Overpopulation and the Origins of Agriculture*. Yale University Press, New Haven. 15.4

Colinvaux P. A. (1973) *Introduction to Ecology*. Wiley, New York. 12.3.1

Colinvaux P. A. (1978) *Why Big Fierce Animals are Rare: an Ecologist's Perspective*. Princeton University Press, Princeton. 12.3.4

Colwell R. K. & Fuentes E. R. (1975) Experimental studies of the niche. *Ann. Rev. Ecol. Syst.* **6**, 281–310. 6.7.1

Colwell R. K. & Futuyma D. J. (1971) On the measurement of niche breadth and overlap. *Ecology* **52**, 567–576. 6.7.2

Colwell R. K. & Winkler D. W. (1984) A null model for null models in biogeography. *In Ecological Communities: Conceptual Issues and the Evidence* (eds Strong D. R., Simberloff D., Abele L. G. & Thistle A. B.) pp. 344–359. Princeton University Press, Princeton. 6.5.1

Combes C. (1968) Biologie, ecologie des cycles et biogéographie de digènes et monogènes d'amphibiens dans l'est des Pyrénées. *Mem. Mus. natn. Hist. nat., Paris. Ser. A. Zool.* **51**, 1–195. 9.3

Connell J. H. (1961a) The effects of competition predation by *Thais lapillus*, and other factors on populations of the barnacle *Balanus balanoides*. *Ecol. Monogr.* **31**, 61–104. 10.1

Connell J. H. (1961b) The influence of interspecific competition and other factors on the distribution of the barnacle *Chthamalus stellatus*. *Ecology* **42**, 710–723. 11.2.2

Connell J. H. (1974) Ecology: field experiments in marine ecology. In *Experimental Marine Biology* (ed. Mariscal R.) pp. 21–54. Academic Press, New York. 11.2.3, 11.3.1, 11.3.2

Connell J. H. (1975) Producing structure in natural communities. In *Ecology and Evolution of Communities* (eds Cody M. L. & Diamond J. M.) pp. 460–490. Belknap Press, Cambridge, Massachusetts. 6.8

Connell J. H. (1978) Diversity in tropical rainforests and coral reefs. *Science* **199**, 1302–1310. 4.3.4, 9.3.4

Connell J. H. (1980) Diversity and coevolution of competitors, or the ghost of competition past. *Oikos* **35**, 131–138. 8.3.1, 9.2.2

Connell J. H. (1983) On the prevalence and relative importance of interspecific competition: evidence from field experiments. *Amer. Natur.* **122**, 661–696. 11.1, 11.1.1, 11.2.1, 11.5.1, 11.5.2, 11.6.1, 11.6.2, 11.7

Connell J. H. & Slatyer R. O. (1977) Mechanisms of succession in natural communities and their role in community stability and organization. *Amer. Natur.* **111**, 1119–1144. 12.2

Connell J. H. & Sousa W. P. (1983) On the evidence needed to judge ecological stability or persistence. *Amer. Natur.* **121**, 789–824. 8.3.2, 8.4.3

Connor E. F., Faeth S. H., Simberloff D. & Opler P. A. (1980) Taxonomic isolation and the accumulation of herbivorous insects: a comparison of introduced and native trees. *Ecol. Ent.* **5**, 205–211. 8.2.3, 8.2.4

Connor E. F. & McCoy E. D. (1979) The statistics and biology of the species-area relationship. *Amer. Natur.* **113**, 791–833. 8.2, 8.2.1, 8.4.2, 9.2.3

Connor E. F., McCoy E. D. & Cosby B. J. (1983) Model discrimination and expected slope values in species-area studies. *Amer. Natur.* **122**, 789–796. 4.3.1

Connor E. F. & Simberloff D. (1979) The assembly of species communities: chance or competition? *Ecology* **60**, 1132–1140. 6.5.4, 11.1

Cook L. M., Askew R. R. & Bishop J. A. (1970) Increasing frequency of the typical form of the peppered moth in Manchester. *Nature* **227**, 1155. 14.2

Coop R. L. (1982) The impact of subclinical parasitism in ruminants. In *Parasites — Their World and Ours* (eds Mettrick D. F. & Desser S. S.) pp. 439–450. Elsevier Biomedical Press, Amsterdam. 9.3

Cornell H. (1974) Parasitism and distributional gaps between allopatric species. *Amer. Natur.* **108**, 880–883. 9.3.2

Cornell H. V. (1985) Local and regional richness of cynipine gall wasps on Californian oaks. *Ecology* **66**, 1247–1260. 8.2.3

Costello D. F. (1969) *The Prairie World*. Thomas Y. Crowell, New York. 12.2

Covacevich J. & Archer M. (1975) The distribution of the cane toad, *Bufo marinus* in Australia and its effects on indigenous vertebrates. *Mem. Qld. Mus.* **17**, 305–310. 10.3.2

Cowles H. C. (1899) The ecological relations of the vegetation on the sand dunes of Lake Michigan. Part I. Geographical relations of the dune floras. *Bot. Gaz.* **27**, 95–117; 167–201; 281–308; 361–391. 1.4, 1.5, 12.2

Cram E. B. (1931) Developmental stages of some nematodes of the Spiruroidea parasitic in poultry and game birds. *Tech. Bull, U.S. Dept. Agriculture, Washington, D.C.* **227**, 1–27. 9.3.3

Crawley M. J. (1983) *Herbivory — the Dynamics of Animal–Plant Interactions*. Blackwell Scientific Publications, Oxford. 1.7, 8.3.3, 8.5

Creese R. G. & Underwood A. J. (1982) Analysis of inter- and intra-specific competition amongst intertidal limpets with different methods of feeding. *Oecologia* **53**, 337–346. 11.3.1, 11.4.1, 11.5.1, 11.5.2, 11.6.1, 11.7

Crofton H. D. (1971) A model of host–parasite relationships. *Parasitology* **63**, 343–364. 13.2.1

Cromack K. Jr, Todd R. T. & Monk C. D. (1975) Patterns of basidiomycete nutrient accumulation in conifer deciduous forest litter. *Soil Biol. Biochem.* **7**, 265–268. 7.2.2

Crombie A. C. (1946) Further experiments on insect competition. *Proc. Roy. Soc. B* **133**, 76–109. 1.7

Crompton D. W. T. (1973) The sites occupied by some parasitic helminths in the alimentary tract of vertebrates. *Biol. Rev.* **48**, 27–83. 9.2.1

Crossley D. A. Jr (1977) The role of terrestrial saprophagous arthropods in forest soils: current status of concepts. In *The Role of Arthropods in Forest Ecosystems* (ed. Mattson W. J.) pp. 49–56. Springer-Verlag, New York. 7.2.2

Crowell K. L. (1962) Reduced interspecific competition among the birds of Bermuda.

Ecology **43**, 75–88. 6.6.3

Cumming D. H. M. (1982) The influences of large herbivores on savanna structure in Africa. In *Ecology of Tropical Savannas* (eds Huntley B. J. & Walker B. H.) pp. 217–245. Springer-Verlag, Berlin. 8.4.1, 8.4.3

Cummins K. W. (1975) Macroinvertebrates. In *River Ecology* (ed. Whitton B. A.) pp. 170–198. Blackwell Scientific Publications, Oxford. 7.3.1

Curio E. (1976) *The Ethology of Predation*. Springer-Verlag, Berlin. 6.6.3

Curtis J. T. (1959) *The Vegetation of Wisconsin*. University of Wisconsin Press, Madison. 1.5, 12.2

Curtis J. T. & McIntosh R. P. (1951) An upland forest continuum in the prairie-forest border region of Wisconsin. *Ecology* **32**, 476–496. 12.2

Dahlberg F. (ed.) (1981) *Woman the Gatherer*. Yale University Press, New Haven. 15.2

Dall W. H. (1877) *On the Succession in the Shellheaps of the Aleutian Islands*. Government Printing Office, Washington D.C. 10.2.2

Dallinger R. & Wieser W. (1977) The flow of copper through a terrestrial food chain. I. Copper and nutrition in isopods. *Oecologia* **30**, 253–264. 7.2.2

Damian R. T. (1964) Molecular mimicry: antigen sharing by parasite and host and its consequences. *Amer. Natur.* **98**, 129–149. 13.3.3

Damuth J. (1981) Population density and body size in mammals. *Nature* **290**, 699–700. 8.4.1

Darwin C. R. (1859) *The Origin of Species by Means of Natural Selection or the Preservation of Favoured Races in the Struggle for Life*. Murray, London. 11.5.1, 14.3

Daubenmire R. F. (1947) *Plants and Environment* (1st edn). Wiley, New York. 1.5

Davidson D. W. (1977) Foraging ecology and community organization in desert seed-eating ants. *Ecology* **58**, 725–737. 5.3.1

Davidson D. W. (1978) Size variability in the worker caste of a social insect (*Veromessor pergandei mayr*) as a function of the competitive environment. *Amer. Natur.* **112**, 523–532. 6.6.3

Davies M. S. & Snaydon R. W. (1976) Rapid population differentiation in a mosaic environment. III. Measures of selection pressures. *Heredity* **36**, 57–66. 14.5

Davis J. (1973) Habitat preferences and competition of wintering juncos and golden-crowned sparrows. *Ecology* **54**, 174–180. 11.2.3, 11.3.1

Davis S. H. (1977) *Victims of the Miracle: Development and the Indians of Brazil*. Cambridge University Press, Cambridge. 15.3.5

Davison E. A. (1982) Seed utilization by harvester ants. In *Ant–Plant Interactions in Australia* (ed. Buckley R. C.) pp. 1–6. W. Junk, The Hague. 7.2.1

Day P. R. (1974) *Genetics of Host–Parasite Interactions*. Freeman, San Francisco. 14.8

Day P. R. (1978) The genetic basis of epidemics. In *Plant Disease: an Advanced Treatise* (eds Horsfall J. G. & Cowling E. B.) Vol. 2, pp. 263–285. Academic Press, New York. 14.8

Dayton P. K. (1971) Competition, disturbance, and community organization: the provision and subsequent utilization of space in a rocky intertidal community. *Ecol. Monogr.* **41**, 351–389. 9.3.4

Dayton P. K. (1972) Toward an understanding of community resilience and the potential effects of enrichments to the benthos of McMurdo Sound, Antarctica. In *Proc. Colloquium on Conservation Problems in Antarctica* (ed. Parker B. D.) pp. 81–95. Allen Press, Lawrence, Kansas. 9.3.2

Dayton P. K. (1973) Two cases of resource partitioning in an intertidal community: making the right prediction for the wrong reason. *Amer. Natur.* **104**, 662–670. 11.1

Dayton P. K. (1979) Ecology: a science and a religion. In *Ecological Processes in Coastal and Marine Systems* (ed. Livingstone R. J.) pp. 3–18. Plenum Press, New York. 11.1, 11.7

Dayton P. K. & Oliver J. S. (1980) An evaluation of experimental analyses of population and community patterns in benthic marine environments. In *Marine Benthic Dynamics* (eds Tenore K. R. & Coull B. C.) pp. 93–120. University of South Carolina Press, Columbia, South Carolina. 11.1

De'Ath C. (1980) *The Throwaway People: Social Impact of the Gogol Timber Project, Madang Province*. Monograph **13**, Institute of Applied Social and Economic Research, Boroko, Papua New Guinea. 15.3.5

De Bach P. (1964) *Biological Control of Insect Pests and Weeds*. Chapman & Hall, London. 8.3.3

De Bach P. (1974) *Biological Control by Natural Enemies*. Cambridge University Press, Cambridge. 8.3.3

De Benedictus P. A. (1974) Interspecific competition between tadpoles of *Rana pipiens* and *Rana sylvatica*: an experimental field study. *Ecol. Monogr.* **44**, 129–141. 11.4.1

de la Cruz A. A. & Poe W. E. (1975) Amino acids in salt marsh detritus. *Limnol. Oceanogr.* **20**, 124–127. 7.3.2

De Long K. T. (1966) Population ecology of feral house mice: interference by *Microtus*. *Ecology* **47**, 481–484. 11.3.1, 11.3.2, 11.7

Dempster J. P. (1983) The natural control of populations of butterflies and moths. *Biol. Rev.* **58**, 461–481. 8.3.2

Derenbach J. B. & Williams P. J. Le B. (1974) Autotrophic and bacterial production: fractionation of plankton populations by differential filtration of samples from the English Channel. *Mar. Biol.* **25**, 263–269. 7.3.3

Desautels R. (1970) United States Atomic Energy Commission, Research Development Report AT**(29-2)-20.** 10.2.2

De Vay J. E. & Adler H. E. (1976) Antigens common to hosts and parasites. *Ann. Rev. Microbiol.* **30**, 147–168. 13.3.3

De Wit C. T. (1960) *On Competition*. Versl. Landbouwk. Onderzoek No. **66**, 8, Wageningen, The Netherlands. 11.4.1

Dexter R. W. (1944) Ecological significance of the disappearance of eel-grass at Cape Ann, Massachusetts. *J. Wildl. Mgmt.* **9**, 173–176. 7.3.2

Dhondt A. A. & Eyckerman R. (1980) Competition between the great tit and blue tit outside the breeding season in field experiments. *Ecology* **61**, 1291–1296. 11.2.2, 11.4.1

Diamond J. M. (1970) Ecological consequences of island colonization by Southwest Pacific birds. I. Types of niche shifts. *Proc. Natl. Acad. Sci.* **67**, 529–536. 6.6.2

Diamond J. M. (1973) Distributional ecology of New Guinea birds. *Science* **179**, 759–769. 5.5.3

Diamond J. M. (1975) Assembly of species communities. In *Ecology and Evolution of Communities* (eds Cody M. L. & Diamond J. M.) pp. 342–444. Belknap Press, Cambridge, Massachusetts. 5.1, 6.5.1, 6.5.4

Diamond J. M. & Gilpin M. E. (1982) Examination of the "null" model of Connor and Simberloff for species co-occurrences on islands. *Oecologia* **52**, 64–74. 11.1

Diamond J. M. & Marshall A. G. (1977) Niche shifts in New Hebridean birds. *Emu* **77**, 61–72. 6.6.2

Diamond J. M. & May R. M. (1981) Island biogeography and the design of natural reserves. In *Theoretical Ecology* (ed. May R. M.) (2nd edn) pp. 228–252. Blackwell Scientific Publications, Oxford. 4.3.1

Diener P., Moore K. & Mutaw R. (1980) Meats, markets, and mechanical materialism: the great protein fiasco in anthropology. *Dialectical Anthropology* **5**, 171–192. 15.2

Dirzo R. & Harper J. L. (1982) Experimental studies on slug-plant interactions. IV. The performance of cyanogenic and acyanogenic morphs of *Trifolium repens* in the field. *J. Ecol.* **70**, 119–138. 14.9

Dobson C. (1965) The relationship between worm population density, host survival and the growth of the third-stage larva of *Amplicaecum robertsi*: Sprent and Mines, 1960, in the

mouse. *Parasitology* **55**, 183–193. 13.2.1
Dobson C. (1972) Immune response to gastrointestinal helminths. In *Immunity to Animal Parasites* (ed. Soulsby E. J. L.) pp. 191–222. Academic Press, New York. 13.2.2
Dobzhansky Th. (1941) *Genetics and the Origin of Species* (2nd edn). Columbia University Press, New York. 14.3
Dogiel V. A., Polyanski Yu. I. & Kheisin E. M. (1964) *General Parasitology*. (Engl. transl. by Z. Kabata). Oliver & Boyd, Edinburgh. 9.2.3
Dolinger P. M., Ehrlich P. R., Fitch W. L. & Breedlove D. E. (1973) Alkaloid and predation patterns in Colorado lupine populations. *Oecologia* **13**, 191–204. 14.9
Dornstreich M. (1973) An ecological study of Gadio Enga (New Guinea) subsistence. PhD thesis, Dept. Anthropology, Columbia University, New York. 15.3.2
Dornstreich M. (1977) The ecological description and analysis of tropical subsistence patterns: an example from New Guinea. In *Subsistence and Survival: Rural Ecology in the Pacific*. (eds Bayliss-Smith T. & Feacham R. G.) pp. 245–271. Academic Press, London. 15.3.2
Douglas M. (1966) *Purity and Danger: an Analysis of Concepts of Pollution and Taboo*. Routledge & Kegan Paul, London. 15.4
Dritschilo W., Cornell H., Nafus D. & O'Connor B. (1975) Insular biogeography: of mice and mites. *Science* **190**, 467–469. 9.2.3
Drury W. H. & Nisbet I. C. T. (1973) Succession. *J. Arnold Arboretum* **54**, 331–368. 12.2
Duggins D. O. (1980) Kelp beds and sea otters: an experimental approach. *Ecology* **61**, 447–453. 11.3.1
Duggins D. O. (1981) Interspecific facilitation in a guild of benthic marine herbivores. *Oecologia* **48**, 157–163. 11.3.1
Dunham A. E., Smith G. R. & Taylor J. N. (1979) Evidence for ecological character displacement in western American catostomid fishes. *Evolution* **33**, 877–896. 6.6.1
Dunsmore J. D. (1981) The role of parasites in population regulation of the European rabbit (*Oryctolagus cuniculus*) in Australia. In *Proceedings of the First Worldwide Furbearer Conference* (eds Chapman J. A. & Pursley D.) pp. 654–669. Worldwide Furbearer Conference, Frostburg, Maryland. 9.3
Du Rietz G. E. (1930) Vegetationforschung auf soziationsanalytischer Grundlage. *Abderhalden, Handb. biol. Arbeitsmeth.* **11**, 293–480. 1.4, 12.2
Duszynski D. W. (1972) Host and parasite interactions during single and concurrent infections with *Eimeria nieschulzi* and *E. separata* in the rat. *J. Protozoology* **25**, 226–231. 13.3.3
Dwyer P. D. (1979) Animal metaphors: an evolutionary model. *Mankind* **12**, 13–27. 15.4
Dwyer P. D. (1982) Prey switching: a case study from New Guinea. *J. Anim. Ecol.* **51**, 529–542. 15.3.2
Dwyer P. D. (1983) Etolo hunting performance and energetics. *Human Ecology* **11**, 145–174. 15.3.2, 15.3.4
Dwyer P. D. (1984) From garden to forest: small rodents and plant succession in Papua, New Guinea. *Aust. Mammal.* **7**, 26–39. 15.3.2
Dwyer P. D. (1985) Choice and constraint in a Papua New Guinean food quest. *Human Ecology* **13**, 49–70 15.3.2
Eadie J. M. & Howard B. H. (1963) Rumen ciliate protozoology. In *Progress in Nutrition and Allied Sciences* (ed. Cuthbertson D. P.) pp. 57–67. Oliver & Boyd, Edinburgh. 7.2.3
Easteal S. (1981) The history of introductions of *Bufo marinus* (Amphibia: Anura); a natural experiment in evolution. *Biol. J. Linn. Soc.* **16**, 93–113. 10.3.2
Eberhardt L. L. (1976) Quantitative ecology and impact assessment. *J. Env. Manag.* **4**, 27–70. 11.3.1
Eckhardt R. C. (1979) The adaptive syndromes of two guilds of insectivorous birds in the

Colorado Rocky Mountains. *Ecol. Monogr.* **49**, 129–149. 5.2

Edmunds G. F. Jr & Alstad D. N. (1978) Coevolution in insect herbivores and conifers. *Science* **199**, 941–945. 14.9

Edwards C. A. (1974) Macroarthropods. In *Biology of Plant Litter Decomposition* (eds Dickinson C. H. & Pugh G. J. F.) pp. 533–554. Academic Press, London. 7.2.2

Edwards J. (1983) Diet shifts in moose due to predator avoidance. *Oecologia* **60**, 185–189. 8.4.3

Egler F. E. (1942) Vegetation as an object of study. *Phil. Sci.* **9**, 245–260. 1.4

Egler F. E. (1951) A commentary on American plant ecology, based on the textbooks of 1947-1949. *Ecology* **32**, 673–695. 1.4

Egler F. E. (1954) Vegetation science concepts. I. Initial floristic compositon — a factor in old-field vegetation development. *Vegetatio* **4**, 442–417. 12.2

Ehrlich P. R. & Raven P. H. (1964) Butterflies and plants: a study in coevolution. *Evolution* **18**, 586–608. 13.2.2, 14.9

Ekman S. (1947) Uber die Festigkeit der Marinen Sediments als Faktor der Tierverbreitung, ein Beitrag zur Associations-Analyse. *Zool. Biol. Univ. Uppsala, Sweden* **25**, 1–20. 7.3.2

Ellen R. F. (1975) Non-domesticated resources in Nuaulu ecological relations. *Soc. Sci. Information* **14**, 127–150. 15.3.2

Ellen R. F. (1978) *Nuaulu Settlement and Ecology: An Approach to the Environmental Relations of an Eastern Indonesian Community*. Verhandelingen van het Koninklijk Instituut voor Taal-, Land en Volkenkunde 83, Martinus Nijhoff, The Hague. 15.3.2

Ellen R. F. (1982) *Environment, Subsistence and System: the Ecology of Small-Scale Social Formations*. Cambridge University Press, Cambridge. 15.3.2, 15.4

Elton C. S. (1927) *Animal Ecology*. Sidgwick & Jackson, London. 1.6, 5.1

Elton C. S. (1942) *Voles, Mice and Lemmings: Problems in Population Dynamics*. Oxford University Press, London. 1.7

Elton C. S. (1958) *The Ecology of Invasions by Animals and Plants*. Methuen, London. 2.6, 4.2.1, 10.3

Elton C. S. & Miller R. S. (1954) The ecological survey of animal communities: with a practical system of classifying habitats by structural characters. *J. Ecol.* **42**, 460–496. 1.1, 1.6, 1.7

Enders F. (1976) Size, food-finding and Dyar's constant. *Environ. Entomol.* **5**, 1–10. 6.6.3

Endicott K. (1979) The hunting methods of the Batek Negritos of Malaysia: a problem of alternatives. *Canberra Anthropology* **2**(2), 7–22. 15.3.1

Eriksson M. O. G. (1979) Competition between freshwater fish and goldeneyes *Bucephala clangula* (L.) for common prey. *Oecologia* **41**, 99–107. 11.4.2

Errington P. L. (1946) Predation and vertebrate populations. *Quant. Rev. Biol.* **21**, 144–177. 9.3.1

Esch G. W. (1971) Impact of ecological succession on the parasite fauna in centrarchids from oligotrophic and eutrophic ecosystems. *Amer. Midl. Natur.* **86**, 160–168. 9.2.1

Esch G. W., Campbell G. C., Conners R. E. & Coggins J. R. (1976) Recruitment of helminth parasites by bluegills (*Lepomis macrochirus*) using a modified live-box technique. *Trans. Am. Fish. Soc.* **105**, 486–490. 9.2.5

Esch G. W., Gibbons J. W. & Bourque J. E. (1975) An analysis of the relationship between stress and parasitism. *Amer. Midl. Natur.* **93**, 339–353. 9.2.1

Estes J. A. & Palmisano J. F. (1974) Sea otters: their role in structuring nearshore communities. *Science* **185**, 1058–1060. 10.2.2

Fager E. W. (1968) The community of invertebrates in decaying oak wood. *J. Anim. Ecol.* **37**, 121–142. 2.5

Fahey T. J. (1983) Nutrient dynamics of aboveground detritus in lodgepole pine (*Pinus contorta* ssp. *latifolia*) ecosystems, south eastern Wyoming. *Ecol. Monogr.* **53**, 51–72. 7.2.2

Falconer D. S. (1981) *Introduction to Quantitative Genetics* (2nd edn). Longmans, London. 14.2

Feder H. M. (1966) Cleaning symbiosis in the marine environment. In *Symbiosis* (ed. Henry S. M.) Vol. 1, pp. 327–380. Academic Press, New York. 9.3.2

Feeney P. (1976) Plant apparency and chemical defense. In *Biochemical Interaction between Plants and Insects* (eds Wallace J. & Mansell R.) pp. 1–40. Plenum Press, New York. 14.9

Feinsinger P. (1976) Organization of a tropical guild of nectarivorous birds. *Ecol. Monogr.* **46**, 257–291. 5.2, 5.5.1

Fellows A. G. (1969) Cane beetles and toads. *Vict. Nat.* **86**, 166. 10.3.2

Fenchel T. (1970) Studies on the decomposition of organic detritus derived from the turtle grass *Thalassia testudinum*. *Limnol. Oceanogr.* **15**, 14–20. 7.3.2

Fenchel T. (1975a) The quantitative importance of benthic microflora of an arctic tundra pond. *Hydrobiology* **46**, 445–452. 7.3.3

Fenchel T. (1975b) Character displacement and coexistence in mud snails (Hydrobiidae). *Oecologia* **20**, 19–32. 6.6.1, 14.6

Fenchel T. & Kofoed L. H. (1976) Evidence for exploitative interspecific competition in mud snails (Hydrobiidae). *Oikos* **27**, 367–376. 6.6.1

Fenner F. (1965) Myxoma virus and *Oryctolagus cuniculus*: two colonising species. In *The Genetics of Colonising Species* (eds Baker H. G. & Stebbins G. L.) pp. 485–501. Academic Press, New York. 14.8

Fenner F. & Ratcliffe F. N. (1965) *Myxomatosis*. Cambridge University Press, Cambridge. 14.8

Ficken R. W., Ficken M. S. & Morse D. H. (1968) Competition and character displacement in two sympatric pine-dwelling warblers (*Dentroica*, Parulidae). *Evolution* **22**, 307–314. 6.6.1

Fish D. & Carpenter S. R. (1982) Leaf litter and larval mosquito dynamics in treehole ecosystems. *Ecology* **63**, 283–288. 10.2.1

Fisher R. A. (1958) *The Genetical Theory of Natural Selection* (2nd edn). Dover, New York. 13.2.1

Fisher R. A., Corbet A. S. & Williams C. B. (1943) The relation between the number of species and the number of individuals in a random sample of an animal population. *J. Anim. Ecol.* **12**, 42–58. 4.3.2

Fisher R. A. & Wishart J. (1930) The arrangement of field experiments and the statistical reduction of the results. *Tech. Comm. No.* **10**, pp. 1–23. Imperial Bureau of Soil Science, London. 11.5.1

Fisher S. G. & Likens G. E. (1973) Energy flow in Bear Brook, New Hampshire: An integrative approach to stream ecosystem metabolism. *Ecol. Monogr.* **43**, 421–439. 7.3.1

Fitzgerald B. M. & Karl B. J. (1979) Foods of feral house cats (*Felis catus* L.) in forest in the Orongorongo Valley, Wellington. *N.Z. J. Zool.* **6**, 107–126. 10.3.1

Fitzpatrick J. W. (1980) Foraging behavior of Neotropical tyrant flycatchers. *Condor* **82**, 43–57. 5.2, 5.5.1

Flannery K. V. (1968) Archeological systems theory and early Mesoamerica. In *Anthropological Archaeology in the Americas* (ed. Meggers B.) pp. 67–87. Anthropological Society of Washington, Washington. 15.2

Flemming T. H. (1973) Numbers of mammal species in North and Central American forest communities. *Ecology* **54**, 555–563. 5.6

Flenley J. (1979) *The Equatorial Rain Forest: a Geological History*. Butterworth, London. 12.4

Flood J. M. (1980) *The Moth Hunters*. Australian Institute of Aboriginal Studies, Canberra. 15.3.4

Flor H. H. (1946) Genetics of pathogenicity in *Melampsora lini*. *J. agric. Res.* **73**, 335–357. 14.8

Flor H. H. (1947) Inheritance of reaction to rust in flax. *J. agric. Res.* **74**, 241–262. 14.8

Fonteyn P. J. & Mahall B. E. (1981) An experimental analysis of structure in a desert plant community. *J. Ecol.* **69**, 883–896. 11.5.1

Forcier L. K. (1975) Reproductive strategies and the co-occurrence of climax tree species. *Science* **189**, 808–810. 12.2

Ford E. B. (1971) *Ecological Genetics* (3rd edn) Chapman & Hall, London. 14.9

Ford J. (1971) *The Role of the Trypanosomiases in African Ecology: A Study of the Tsetse-fly Problem*. Clarendon Press, Oxford. 8.4.3, 9.3.2, 9.3.4

Forrester D. J. (1971) Bighorn sheep lungworm pneumonia complex. In *Parasitic Diseases of Wild Mammals* (eds Davis J. W. & Anderson R. C.) pp. 158–173. Iowa State University Press, Ames. 9.3

Foster J. & Kearney D. (1967) Nairobi National Park game census, 1966. *E. Afr. Wildlife J.* **5**, 112–120. 8.4.3

Foster J. & McLaughlin R. (1968) Nairobi National Park game census, 1967. *E. Afr. Wildlife J.* **6**, 152–154. 8.4.3

Fowler C. W. & MacMahon J. A. (1982) Selective extinction and speciation: their influence on the structure and functioning of communities and ecosystems. *Amer. Natur.* **119**, 480–498. 8.5

Fowler N. (1981) Competition and coexistence in a North Carolina grassland. II. The effects of the experimental removal of species. *J. Ecol.* **69**, 883–896. 11.3.2

Frankenberg D. & Smith K. L. Jr (1967) Coprophagy in marine animals. *Limnol. Oceanogr.* **12**, 443–450. 7.3.2

Freeland W. J. (1976) Pathogens and the evolution of primate sociality. *Biotropica* **8**, 12–24. 13.2.2

Freeland W. J. (1977) Blood sucking flies and primate poly-specific associations. *Nature* **269**, 801–802. 13.2.2

Freeland W. J. (1979a) Primate social groups as biological islands. *Ecology* **60**, 719–728. 9.2.3, 13.2.2

Freeland W. J. (1979b) Ecology and social organization of mangabeys, *Cercocebus albigena*, in two localities. *Folia primat* **32**, 108–124. 13.2.2

Freeland W. J. (1980) *Cercocebus albigena*: movement patterns and ecological variables. *Ecology* **61**, 297–303. 13.2.2

Freeland W. J. (1981) Parasitism and behavioral dominance among male mice. *Science* **213**, 461–462. 13.2.1, 13.3.1

Freeland W. J. (1983) Parasites and the coexistence of host animal species. *Amer. Natur.* **121**, 223–236. 9.2.1, 9.2.3, 9.2.4, 9.3, 9.3.2

Friedman J. (1971) The effect of competition by adult *Zygophyllum dumosum* Boiss. on seedlings of *Artemisia herba-alba* Asso in the Negev desert of Israel. *J. Ecol.* **59**, 775–782. 11.7

Friedman J. & Orshan G. (1974) Allopatric distribution of two varieties of *Medicago laciniata* (L.) Mill. in the Negev desert of Israel. *J. Ecol.* **62**, 107–114. 11.7

Friedman J., Orshan G. & Ziger-Cfir Y. (1977) Suppression of annuals by *Artemisia herba-alba* in the Negev desert of Israel. *J. Ecol.* **65**, 413–426. 11.7

Friedman J. & Rowlands M. J. (1977) Notes toward an epigenetic model of the evolution of 'civilization'. In *The Evolution of Social Systems* (eds Friedman J. & Rowlands M. J.) pp. 201–276. Duckworth, London. 15.4

Fritz R. S. (1982) Selection for host modification by insect parasitoids. *Evolution* **36**, 283–288. 9.3.3

Fuller M. E. (1934) The insect inhabitants of carrion: a study in animal ecology. *Bull. Coun. Scient. Indust. Res., Aust.* **195**, 1–62. 10.1

Futuyma D. J. (1979) *Evolutionary Biology*. Sinauer, Sunderland, Mass. 14.6

Futuyma D. J. (1983) Evolutionary interactions among herbivorous insects and plants. In *Coevolution* (eds Futuyma D. J. & Slatkin M.) pp. 207–231. Sinauer, Sunderland, Mass. 8.2.4

Futuyma D. J. & Mayer G. C. (1980) Non-allopatric speciation in animals. *Syst. Zool.* **29**, 254–271. 8.2.4

Futuyma D. J. & Slatkin M. (eds) (1983) *Coevolution*. Sinauer, Sunderland, Mass. 14.3

Gallun R. L. & Khush G. S. (1980) Genetic factors affecting expression and stability of resistance. In *Breeding Plants Resistant to Insects* (eds Maxwell F. G. & Jennings P. R.) pp. 64–85. Wiley-Interscience, New York. 14.8

Gams H. (1918) Principienfragen der Vegetationsforschung. *Vierteljahrsschr. Naturf. Ges. Zurich* **67**, 132–156. 12.2

Gatz A. J. Jr (1979) Community organization in fishes as indicated by morphological features. *Ecology* **60**, 711–718. 6.5.3

Gauch H. J. Jr (1982) *Multivariate Analysis in Community Ecology*. Cambridge University Press, Cambridge. 1.5, 4.4.2

Gause G. (1934) *The Struggle for Existence*. Haefner, New York. 1.7, 6.1, 6.7.2

Geertz C. (1963) *Agricultural Involution: the Process of Ecological Change in Indonesia*. University of California Press, Berkeley. 15.3.2

Geier P. W. & Clark L. R. (1978/1979) The nature and future of pest control: production process or applied ecology? *Prot. Ecol.* **1**, 79–101. 10.3.2

Gell A. (1975) *Metamorphosis of the Cassowaries: Umeda Society, Language and Ritual*. Athlone Press, London. 15.3.2

George A. S., Hopkins A. J. M. & Marchant N. G. (1979) The heathlands of Western Australia. In *Heathlands and Related Shrublands. Ecosystems of the World* (ed. Specht R. L.) pp. 211–230. Elsevier, Amsterdam. 4.3.1

Gere G. (1956) The examination of the feeding biology and humificative function of Diplopoda and Isopoda. *Acta. Biol. Hungr.* **6**, 257–271. 7.2.2

Getz W. M. & Pickering J. (1983) Epidemic models: thresholds and population regulation. *Amer. Natur.* **121**, 892–898. 9.3.1

Gibb J. A. & Flux J. E. C. (1973) Mammals. In *The Natural History of New Zealand* (ed. Williams G. R.) pp. 334–371. A. H. & A. W. Reed, Wellington. 10.3.1

Gibb J. A., Ward C. P. & Ward G. D. (1978) Natural control of a population of rabbits, *Oryctolagus cuniculus* (L.), for ten years in the Kourarau enclosure. *D. S. I. R. Bull., Wellington* **223**, 1–89. 10.3.1

Gibson C. W. D. (1980) Niche use patterns among some Stenodermini (Heteroptera: Miridae) of limestone grassland, and an investigation of the possibility of interspecific competition between *Notostira elongata* Geoffroy and *Megaloceraea recticornis* Geoffroy. *Oecologia* **47**, 352–364. 8.3.3

Gibson C. & Visser M. (1982) Interspecific competition between two field populations of grass-feeding bugs. *Ecol. Ent.* **7**, 61–67. 8.3.3

Gilbert L. E. (1971) Butterfly plant coevolution: has *Passiflora adenopoda* won the selectional race with heliconine butterflies? *Science* **172**, 585–586. 13.1

Gilbert L. E. (1975) Ecological consequences of a coevolved mutualism between butterflies and plants. In *Coevolution of Animals and Plants* (eds Gilbert L. E. & Raven P. H.) pp. 210–240. University of Texas Press, Texas. 14.9

Gilbert L. E. (1983) Coevolution and mimicry. In *Coevolution* (eds Futuyma D. J. & Slatkin M.) pp. 263–281. Sinauer, Sunderland, Mass. 14.9

Gilbert L. E. & Singer M. C. (1975) Butterfly ecology. *Ann. Rev. Ecol. Syst.* **6**, 365–397. 13.2.2

Gill D. E. & Hairston N. G. (1972) The dynamics of a natural population of *Paramecium* and the role of interspecific competition in community structure. *J. Anim. Ecol.* **41**, 137–151. 11.4.1, 11.7

Gilpin M. E. & Diamond J. M. (1982) Factors contributing to non-randomness in species co-occurrences on islands. *Oecologia* **52**, 75–84. 6.5.4, 11.1

Gleason H. A. (1917) The structure and development of the plant association. *Bull. Torrey Bot. Club* **44**, 463–481. 1.5

Gleason H. A. (1926) The individualistic concept of the plant association. *Bull. Torrey Bot. Club* **53**, 7–26. 1.5, 9.2.2, 12.2

Gleason H. A. (1939) The individualistic concept of the plant association. *Amer. Midland Nat.* **21**, 92–110. 12.2

Glover P. E., Trump E. C. & Wateridge L. E. D. (1964) Termitaria and vegetation patterns on the Loita plains of Kenya. *J. Ecol.* **52**, 367–377. 2.3

Goeden R. D. (1971) Insect ecology of silverleaf nightshade. *Weed Science* **19**, 45–51. 8.2.4

Goeden R. D. (1974) Comparative survey of the phytophagous insect faunas of Italian thistle, *Carduus pycnocephalus*, in southern California and southern Europe relative to biological weed control. *Environ. Entomol.* **3**, 464–474. 8.2.4

Goeden R. D. & Ricker D. W. (1968) The phytophagous insect fauna of Russian thistle (*Salsola kali* var. *tenuifolia*) in southern California. *Ann. Ent. Soc. Amer.* **61**, 67–72. 8.2.4

Goldman I. (1963) *The Cubeo: Indians of the Northwest Amazon*. University of Illinois Press, Urbana. 15.3.3

Goodall D. W. (1952) Quantitative aspects of plant distribution. *Biol. Rev.* **27**, 194–245. 4.3.1

Gordon H. McL. (1960) Nutrition and helminthosis in sheep. *Proc. Aust. Soc. Anim. Prod.* **3**, 93–104. 9.3

Gordon H. McL. (1973) Epidemiology and control of gastrointestinal nematodes of ruminants. *Adv. Vet. Sci. Comp. Med.* **17**, 395–437. 13.3.3

Gould S. J. (1977) *Ontogeny and Phylogeny*. Belknap Press, Cambridge, Massachusetts. 15.4

Gould S. J. (1980) Is a new and general theory of evolution emerging? *Paleobiology* **6**, 119–130. 14.8

Grace J. B. & Wetzel R. G. (1981) Habitat partitioning and competitive displacement in cattails (*Typha*): experimental field studies. *Amer. Natur.* **118**, 463–474. 11.3.1

Grant P. R. (1969) Experimental studies of competitive interaction in a two-species system. I. *Microtus* and *Clethrionomys* species in enclosures. *Can. J. Zool.* **47**, 1059–1082. 6.7.1

Grant P. R. (1971) Experimental studies of competitive interactions in a two-species system. III. *Microtus* and *Peromyscus* species in enclosures. *J. Anim. Ecol.* **40**, 323–350. 6.7.1, 11.3.1, 11.4.1

Grant P. R. (1972) Convergent and divergent character displacement. *Biol. J. Linn. Soc.* **4**, 39–68. 6.6.1, 14.6

Grant P. R. (1975) The classical case of character displacement. In *Evolutionary Biology* (eds Dobzhansky T., Hecht M. K. & Steere W. C.) Vol. 8, pp. 237–337. Plenum Press, New York. 6.6.1

Grant P. R. & Abbott I. (1980) Interspecific competition, island biogeography and null hypotheses. *Evolution* **34**, 332–341. 6.5.1

Grant P. R., Grant B. R., Smith J. N., Abbott I. J. & Abbott L. K. (1976) Darwin's finches: population variation and natural selection. *Proc. Natl. Acad. Sci.* **73**, 257–261. 14.7

Grant P. & Schluter D. (1984) Interspecific competition inferred from patterns of guild structure. In *Ecological Communities: Conceptual Issues and the Evidence* (eds Strong D. R., Simberloff D., Abele L. G. & Thistle A. B.) pp. 201–233. Princeton University Press, Princeton. 6.5.1, 6.6.1

Grant V. (1971) *Plant Speciation*. Columbia University Press, New York. 14.2

Green R. H. (1979) *Sampling Design and Statistical Methods for Environmental Biologists*. Wiley, New York. 11.3.1

Greenfield M. D. & Karandinos M. G. (1979) Resource partitioning of the sex communication channel in clear wing moths (Lepidoptera: Sesiidae) of Wisconsin. *Ecol. Monogr.* **49**, 403–426. 6.8

Greig-Smith P. (1957) *Quantitative Plant Ecology* (1st edn). Butterworth, London. 1.8, 12.3.4

Greig-Smith P. (1961) Data on pattern within plant communities. I. The analysis of pattern. *J. Ecol.* **49**, 695–702. 1.8, 12.3.2

Greig-Smith P. (1979) Pattern in vegetation. *J. Ecol.* **67**, 755–779. 2.3, 2.5

Greig-Smith P. (1983) *Quantitative Plant Ecology* (3rd edn). Blackwell Scientific Publications, Oxford. 1.8, 2.2, 2.6, 4.3.1, 4.4.2, 12.3.2

Greig-Smith P. & Chadwick M. J. (1965) Data on pattern within communities. III. *Acacia-Capparis* semi-desert scrub in the Sudan. *J. Ecol.* **53**, 465–474. 12.3.4

Griffiths R. P., Caldwell B. A. & Morita R. Y. (1982) Seasonal changes in microbial heterotrophic activity in subarctic marine waters as related to phytoplankton primary productivity. *Mar. Biol.* **71**, 121–127. 7.3.3

Grime J. P. (1979) *Plant Strategies and Vegetation Processes*. John Wiley, Chichester. 12.2

Grime J. P., Macpherson-Stewart S. F. & Dearman R. S. (1968) An investigation of leaf palatability using the snail *Cepaea nemoralis* L. *J. Ecol.* **56**, 405–420. 2.2

Gross K. L. (1980) Colonization by *Verbascum thapsus* (mullein) of an old-field in Michigan: experiments on the effects of vegetation. *J. Ecol.* **68**, 919–927. 11.7

Gross K. L. & Werner P. A. (1982) Colonizing abilities of "biennial" plant species in relation to ground cover: implications for their distributions in a successional sere. *Ecology* **63**, 929–931. 11.2.3, 11.3.2

Groth J. V. & Person C. O. (1977) Genetic interdependence of host and parasite in epidemics. *Ann. N.Y. Acad. Sci.* **287**, 97–106. 14.8

Grubb P. J. (1977) The maintenance of species-richness in plant communities: the importance of the regeneration niche. *Biol. Rev.* **52**, 107–145. 2.5, 4.3.4, 12.4.1

Gunnison D. & Alexander M. (1975) Resistance and susceptibility of algae to decomposition by natural microbial communities. *Limnol. Oceanogr.* **20**, 64–70. 7.3.2

Gutierrez J. (1955) Experiments on the culture and physiology of holotrichs from the bovine rumen. *Biochem. J.* **60**, 516–522. 7.2.3

Gyllenberg H. G. & Eklund E. (1974) Bacteria. In *Biology of Plant Litter Decomposition* (eds Dickinson C. H. & Pugh G. J. F.) pp. 245–268. Academic Press, London. 7.2.2

Haefner J. W. (1981) Avian community assembly rules: the foliage-gleaning guild. *Oecologia* **50**, 131–142. 6.2

Haffer J. (1969) Speciation in Amazonian forest birds. *Science* **165**, 131–137. 5.5.3

Haffer J. (1974) Avian speciation in South America. *Publ. Nutall Ornithol. Club No.* 14. Cambridge, Massachusetts. 5.5.3

Haines B. L. (1978) Elements and energy flows through colonies of the leaf-cutting ant *Atta colombica* in Panama. *Biotropica* **10**, 270–277. 7.2.2

Haines E. B. & Hanson R. B. (1979) Experimental degradation of detritus made from the salt marsh plants *Spartina alterniflora* Loisel., *Salicornia virginica* L., and *Juncus roemerianus* Scheele. *J. exp. mar. Biol. Ecol.* **40**, 27–40. 7.3.2

Hair J. D. & Holmes J. C. (1975) The usefulness of measures of diversity, niche width and niche overlap in the analysis of helminth communities in waterfowl. *Acta Parasitol. Polonica* **23**, 253–369. 9.2.1

Hairston N. G. (1964) Studies on the organization of animal communities. *J. Ecol.* **52**, (Suppl.) 227–239. 1.1

Hairston N. G. (1980a) Species packing in the salamander genus *Desmognathus*: what are the interspecific interactions involved? *Amer. Natur.* **115**, 354–366. 6.8

Hairston N. G. (1980b) The experimental test of an analysis of field distributions: competition in terrestrial salamanders. *Ecology* **61**, 817–826. 11.3.1, 11.4.1, 11.7

Hairston N. G. (1981) An experimental test of a guild: salamander competition. *Ecology* **62**, 65–72. 5.2, 5.3.1, 6.8, 11.3.1, 11.4.1, 11.7

Hairston N. G., Smith F. E. & Slobodkin L. B. (1960) Community structure, population control, and competition. *Amer. Natur.* **94**, 421–425. 6.8, 8.3.3

Haldane J. B. S. (1957) Natural selection in man. *Acta Genet. Stat. Med.* **6**, 321–332. 14.8

Hall P. R. (1934) The relation of the size of the infective dose to number of oocysts eliminated, duration of infection and immunity in *Eimeria miyairii* Ohira (?) infections in the white ant. *Iowa State College J. Sci.* **9**, 115–124. 13.2.1

Halvorsen O. (1971) Studies on the helminth fauna of Norway XVIII: On the composition of the parasite fauna of coarse fish in the River Gloma, south-eastern Norway. *Nor. J. Zool.* **19**, 181–192. 9.2.4

Hamilton W. D. (1964) The genetical evolution of social behavior I, II. *J. theoret. Biol.* **7**, 1–52. 13.2.2, 13.3.3

Hammond P. M. (1980) Speciation in the face of gene flow-sympatric parapatric speciation. In *The Evolving Biosphere* (ed. Forey P. L.) pp. 37–48. British Museum (Natural History). Cambridge University Press, Cambridge. 8.2.4

Hanlon R. D. G. (1981a) Some factors influencing microbial growth on soil animal faeces. I. Bacterial and fungal growth on particulate oak leaf litter. *Pedobiologia* **21**, 257–263. 7.2.2

Hanlon R. D. G. (1981b) Some factors influencing microbial growth on soil animal faeces. II. Bacterial and fungal growth on soil animal faeces. *Pedobiologia* **21**, 264–270. 7.2.2

Hanski I. (1982) Dynamics of regional distribution: the core and satellite species hypothesis. *Oikos* **38**, 210–221. 9.2, 9.2.1, 9.2.4, 9.3.4

Hanson R. P. (1969) Koch is dead. *Bull. Wildl. Dis. Assn.* **5**, 150–156. 9.3

Harding D. J. L. & Stuttards R. A. (1974) Microarthropods. In *Biology of Plant Litter Decomposition* (eds Dickinson C. H. & Pugh G. J. F.) pp. 489–532. Academic Press, London. 7.2.2

Harding J., Allard R. W. & Smeltzer D. G. (1966) Population studies in predominantly self-pollinated species. 9. Frequency dependent selection in *Phaseolus lunatus*. *Proc. Natl. Acad. Sci.* **56**, 99–104. 14.6

Hargrave B. T. (1970) The utilization of benthic microflora by *Hyalella azteca* (Amphipoda). *J. Anim. Ecol.* **39**, 427–437. 7.3.2

Harley J. L. (1969) *The Biology of Mycorrhiza* (2nd edn). Leonard Hill, London. 2.2

Harley J. L. (1971) Fungi in ecosystems. *J. Ecol.* **59**, 653–668. 2.2, 7.2.2

Harper J. L. (1969) The role of predation in vegetational diversity. In *Diversity and Stability in Ecological Systems* (eds Woodwell G. M. & Smith H. H.) pp. 48–62. *Brookhaven Symposia in Biology* 22. Brookhaven National Laboratory, Upton, New York. 6.4, 8.4.3

Harper J. L. (1977) *Population Biology of Plants*. Academic Press, London. 1.8, 2.2, 14.6

Harper J. L. (1982) After description. In *The Plant Community as a Working Mechanism* (ed. Newman E. I.) pp. 11–26. Blackwell Scientific Publications, Oxford. 12.5, 14.2

Harris D. (1977) Settling down: an evolutionary model for the transformation of mobile bands into sedentary communities. In *The Evolution of Social Systems* (eds Friedman J. & Rowlands M. J.) pp. 401–417. Duckworth, London. 15.4

Harris J. R. W. (1981) Competition, relatedness and efficiency. *Nature* **292**, 54–55. 13.3.3

Harris W. & Brougham R. W. (1970) The effect of grazing on the persistence of genotypes in a ryegrass population. *N. Z. J. agric. Res.* **13**, 263–278. 14.9

Hartenstein R. (1962) Soil Oribatei. I. Feeding specificity among forest soil Oribatei (Acarina). *Ann. Entomol. Soc. Amer.* **55**, 202–206. 7.2.2

Hartenstein R. (1964) Feeding, digestion, glycogen and the environmental conditions of the digestive system in *Oniscus asellus*. *J. Insect Physiol.* **10**, 611–621. 7.2.2

Hassell M. P. (1978) *The Dynamics of Arthropod Predator–Prey Systems*. Princeton University Press, Princeton. 8.3.3

Hassell M. P. (1985) Insect natural enemies as regulating factors. *J. Anim. Ecol.* **54**, 323–334. 8.3.2

Haven S. B. (1973) Competition for food between the intertidal gastropods *Acmaea scabra* and *Acmaea digitalis*. *Ecology* **54**, 143–151. 11.3.1, 11.4.1

Hawkes K., Hill K. & O'Connell J. F. (1982) Why hunters gather: optimal foraging and the Ache of eastern Paraguay. *Amer. Ethnol.* **9**, 379–398. 15.2, 15.3.1

Hayes J. L. (1981) The population ecology of a natural population of the pierid butterfly *Colias alexandre*. *Oecologia* **49**, 188–200. 8.5

Hayward G. F. & Phillipson J. (1979) Community structure and functional role of small mammals in ecosystems. In *Ecology of Small Mammals* (ed. Stoddard D. M.) pp. 135–211. Chapman & Hall, London. 7.2.1

Heald P. J. & Oxford A. E. (1953) Fermentation of soluble sugars by anaerobic holotrich ciliate protozoa of the genera *Isotricha* and *Dasytricha*. *Biochem. J.* **53**, 506–512. 7.2.3

Hegsted D. M. (1978) Protein-calorie malnutrition. *Amer. Scientist* **66**, 46–51. 15.2

Hein H. E. (1976) *Eimeria acervulina*, *E. brunetti* and *E. maxima*: pathogenic effects of single or mixed infections with low doses of oocysts in chickens. *Exp. Parasitol.* **39**, 415–421. 9.3.2, 13.2.1

Heinle D. R., Harris R. P., Ustach J. F. & Flemer D. A. (1977) Detritus as food for estuarine copepods. *Mar. Biol.* **40**, 341–353. 7.3.2

Heinrich B. (1976) Resource partitioning among some eusocial insects: bumblebees. *Ecology* **57**, 874–889. 6.4

Hendrickson J. A., Jnr. (1981) Community-wide character displacement reexamined. *Evolution* **35**, 794–810. 6.5.1, 14.6

Hertz P. E. (1983) Eurythermy and niche breadth in West Indian *Anolis* lizards: a reappraisal. In *Advances in Herpetology and Evolutionary Biology* (eds Rhodin A. & Miyata K.) pp. 472–483. Spec. Publ. Mus. Comp. Zool., Harvard University, Cambridge, Mass. 6.6.3

Hespenheid H. A. (1973) Ecological inferences from morphological data. *Ann. Rev. Ecol. Syst.* **4**, 213–230. 6.5.1

Hesse R., Allee W. C. & Schmidt K. P. (1937) *Ecological Animal Geography*. Wiley, New York. 3.1

Heyneman D. (1961) Studies on helminth immunity. III. Experimental verification of autoinfection from cysticercoids of *Hymenolepis nana* in the white mouse. *J. Infect. Dis.* **109**, 10–18. 13.3.3

Heyneman D. (1962a) Studies on helminth immunity. I. Comparison between lumenal and tissue phases of infection in the white mouse by *Hymenolepis nana* (Cestoda: Hymenolepidae). *Amer. J. trop. Med. Hyg.* **11**, 46–63. 13.3.3

Heyneman D. (1962b) Studies on helminth immunity. II. Influence of *Hymenolepis nana* (Cestoda: Hymenolepidae) in dual infections with *H. diminuta* in white mice and rats. *Exp. Parasitol.* **12**, 7–18. 13.3.3

Heyneman D. (1962c) Studies on helminth immunity. IV. Rapid onset of resistance by the white mouse against a challenging infection with eggs of *Hymenolepis nana* (Cestoda: Hymenolepidae). *J. Immunol.* **88**, 210–220. 13.3.3

Hilborn R. & Sinclair A. R. E. (1979) A simulation of the wildebeest population, other ungulates, their predators. In *Serenegeti – Dynamics of an Ecosystem* (eds Sinclair A. R. E. & Norton-Griffiths M.) pp. 287–309. University of Chicago Press, Chicago. 8.4.3

Hilborn R. & Stearns S. C. (1982) On inference in ecology and evolutionary biology: the problem of multiple causes. *Acta Biotheor.* **31**, 145–164. 9.2

Hils M. H. & Vankat J. L. (1982) Species removals from a first-year old-field plant community. *Ecology* **63**, 705–711. 11.4.2, 11.7
Hipsley E. H. & Clements F. W. (1950) Reports of the New Guinea nutrition expedition 1947. Dept. External Territories, Canberra. 15.3.4
Hipsley E. H. & Kirk N. E. (1965) Studies of dietary intake and the expenditure of energy by New Guineans. Tech. Paper No. 147, South Pacific Commission, Noumea. 15.3.4
Hixon M. A. (1980) Competitive interaction between California reef fishes of the genus *Embiotica*. *Ecology* **61**, 918–931. 11.3.1
Hobbie J. E. & Lee C. (1980) Microbial production of extracellular material: importance in benthic ecology. In *Marine Benthic Dynamics* (eds Tenore K. R. & Coull B. C.) pp. 341–346. University of South Carolina Press, Columbia, South Carolina. 7.3.3
Hobbs R. P. (1980) Interspecific interactions among gastrointestinal helminths in pikas of North America. *Amer. Midl. Natur.* **103**, 15–25. 9.2.1
Hobson P. N. (1963) Rumen bacteriology. In *Progress in Nutrition and Allied Sciences* (ed. Cuthbertson D. P.) pp. 43–55. Oliver & Boyd, Edinburgh. 7.2.3
Hoffman G. L. (1970) Intercontinental and transcontinental dissemination and transfaunation of fish parasites with emphasis on whirling disease. In *A Symposium on Diseases of Fishes and Shellfishes* (ed. Snieszko S. F.) pp. 69–91. American Fisheries Society, Washington, D.C. 9.3.2
Holbrook S. J. (1979) Habitat utilization, competitive interactions, and coexistence of three species of cricetine rodents in east-central Arizona. *Ecology* **60**, 758–769. 11.3.1
Holliman R. B., Fisher J. E. & Parker J. C. (1971) Studies on *Spirorchis parvus* (Stunkard, 1923) and its pathological effects on *Chrysemys picta picta*. *J. Parasitol.* **57**, 71–77. 9.2.1
Holling C. S. (1959) The components of predation as revealed by a study of small-mammal predation of the European pine sawfly. *Can. Entomologist* **91**, 293–320. 1.7
Holling C. S. (1973) Resilience and stability of ecological systems. *Ann. Rev. Ecol. Syst.* **4**, 1–23. 4.2.2, 10.2.2
Holmes J. C. (1961) Effects of concurrent infections of *Hymenolepis diminuta* (Cestoda) and *Moniliformis dubius* (Acanthocephala). I. General effects and comparison with crowding. *J. Parasitol.* **47**, 209–216. 6.7.1, 13.3.3
Holmes J. C. (1962) Effects of concurrent infections on *Hymenolepis diminuta* (Cestoda) and *Moniliformis dubius* (Acanthocephala). II. Effects on growth. *J. Parasitol.* **48**, 87–96. 13.3.3
Holmes J. C. (1971) Habitat segregation in sanguinicolid blood flukes (Digenea) of scorpaenid rockfishes (Perciformes) on the Pacific Coast of North America. *J. Fisheries Res. Board Canada* **28**, 903–909. 13.3.3
Holmes J. C. (1973) Site selection by parasitic helminths: interspecific interactions, site segregation, and their importance to the development of helminth communities. *Can. J. Zool.* **51**, 333–347. 9.2.1, 9.2.2
Holmes J. C. (1976) Host selection and its consequences. In *Ecological Aspects of Parasitology* (ed. Kennedy C. R.) pp. 21–39. North-Holland, Amsterdam. 9.2.1
Holmes J. C. (1979) Population regulation and parasite flow through communities. In *Host–Parasite Interfaces* (ed. Nickol B.) pp. 27–46. Academic Press, New York. 9.2.1, 9.3.2, 9.4
Holmes J. C. (1982) Impact of infectious disease agents on the population growth and geographical distribution of animals. In *Population Biology of Infectious Diseases* (eds Anderson R. M. & May R. M.) pp. 37–51. Dahlem Konferenzen. Springer-Verlag, New York. 9.3.1, 9.3.2
Holmes J. C. (1983) Evolutionary relationships between parasitic helminths and their hosts. In *Coevolution* (eds Futuyma D. J. & Slatkin M.) pp. 161–185. Sinauer, Sunderland, Mass. 9.3, 9.3.2
Holmes J. C. & Bethel W. M. (1972) Modification of intermediate host behaviour by

parasites. In *Behavioural Aspects of Parasite Transmission* (eds Canning E. U. & Wright C. A.) pp. 123–149. Academic Press, London. 9.3, 9.3.3

Holmes J. C. & Price P. W. (1980) Parasite communities: the roles of phylogeny and ecology. *Syst. Zool.* **29**, 203–213. 9.2.1

Holmes N. D., Smith D. S. & Johnston A. (1979) Effect of grazing by cattle on the abundance of grasshoppers. *J. Range Management* **23**, 310–311. 8.3.3

Holmes R. T., Bonney R. E. Jnr & Pacala S. W. (1979) Guild structure of the Hubbard Brook bird community: A multivariate approach. *Ecology* **60**, 512–520. 5.2, 5.3.2

Honigberg B. M. (1970) Protozoa associated with termites and their role in digestion. In *Biology of Termites* (eds Krishna K. & Weesner F. M.) Vol. 2, pp. 1–36. Academic Press, New York. 7.2.2

Honjo S. & Roman M. R. (1978) Marine copepod fecal pellets: production, preservation and sedimentation. *J. Mar. Res.* **36**, 45–57. 7.3.2

Hope G. S., Golson J. & Allen J. (1983) Palaeoecology and prehistory in New Guinea. *J. Human Evol.* **12**, 37–60. 15.4

Hope J. H. & Hope G. S. (1976) Palaeoenvironments for man in New Guinea. In *The Origin of the Australians* (eds Kirk R. L. & Thorne A. G.) pp. 29–54. Australian Institute of Aboriginal Studies, Canberra. 15.4

Hopkins B. (1957) The concept of minimal area. *J. Ecol.* **45**, 441–449. 4.3.1

Horn H. S. (1976) Succession. In *Theoretical Ecology* (ed. May R. M.) pp. 187–204. Blackwell Scientific Publications, Oxford. 12.2

Horn H. S. & May R. M. (1977) Limits to similarity among coexisting competitors. *Nature* **270**, 660–661. 5.5.3

Hornocker M. (1970) An analysis of mountain lion predation upon mule deer and elk in the Idaho Primitive Area. *Wildlife Monographs* **21**, 3–39. 8.4.3

Howard R. J., Christies P. R., Wakelin D., Wilson M. M. & Behnke J. M. (1978) The effect of concurrent infection with *Trichinella spiralis* on *Hymenolepis microstoma* in mice. *Parasitology* **77**, 273–279. 13.3.3

Howe H. F. (1984) Constraints on the evolution of mutualisms. *Amer. Natur.* **123**, 764–777. 14.10

Hudson H. J. (1968) The ecology of fungi on plant remains above the soil. *New Phytol.* **67**, 837–874. 7.2.2

Huey R. B. & Pianka E. R. (1974) Ecological character displacement in a lizard. *Amer. Zool.* **14**, 1127–1136. 6.6.1

Huey R. B. & Pianka E. R. (1983) Temporal separation of activity and interspecific dietary overlap. In *Lizard Ecology: Studies of a Model Organism* (eds Huey R. B., Pianka E. R. & Schoener T. W.) pp. 281–290. Harvard University Press, Cambridge, Massachusetts. 6.4

Huffaker C. B. (1958) Experimental studies on predation: dispersion factors and predator-prey oscillations. *Hilgardia* **27**, 343–383. 1.7

Huffaker C. B. (ed.) (1971) *Biological Control*. Plenum Press, New York. 8.3.3, 8.4.3

Hughes R. N. & Thomas M. L. H. (1971) The classification and ordination of shallow-water benthic samples from Prince Edward Island, Canada. *J. exp. mar. Biol. Ecol.* **7**, 1–39. 2.5

Hulscher J. B. (1973) Burying-depth and trematode infection in *Macoma balthica*. *Neth. J. Sea. Res.* **6**, 141–156. 9.3.3

Hungate R. E. (1966) *The Rumen and its Microbes*. Academic Press, New York. 7.2.3

Hurlbert S. H. (1984) Pseudoreplication and the design of ecological field experiments. *Ecol. Monogr.* **54**, 187–211. 11.1, 11.2.3, 11.3.1, 11.7

Huston M. (1979) A general hypothesis of species diversity. *Amer. Natur.* **113**, 81–101. 4.3.4

Hutchinson G. E. (1953) The concept of pattern. *Proc. Acad. Nat. Sci., (Phila.)* **105**, 1–12. 1.8

Hutchinson G. E. (1957) Concluding remarks. *Cold Spring Harbor Symp. Quant. Biol.* **22**, 415–427. 6.2, 9.2.1, 10.1

Hutchinson G. E. (1959) Homage to Santa Rosalia, or why are there so many kinds of animals? *Amer. Natur.* **93**, 145–159. 1.8, 5.5.3, 6.5.1, 6.5.2

Hutchinson G. E. (1965) *The Ecological Theater and the Evolutionary Play*. Yale University Press, New Haven 8.3.4

Hylleberg Kristensen J. (1972) Carbohydrases of some marine invertebrates with notes on their food and on the natural occurrence of the carbohydrates studied. *Mar. Biol.* **14**, 130–142. 7.3.2

Hyndman D. C. (1979) Wopkaimin subsistence: cultural ecology in the New Guinea Highland Fringe. PhD dissertation, Department of Anthropology and Sociology, University of Queensland, Brisbane. 15.3.2, 15.3.5

Hyndman D. C. (1982) Biotope gradient in a diversified New Guinea subsistence system. *Human Ecology* **10**, 219–259. 15.3.1, 15.3.2

Hynes R. A. & Chase A. K. (1982) Plants, sites and domiculture: aboriginal influence upon plant communities in Cape York Peninsula. *Archaeology in Oceania* **17**, 38–50. 15.2

Inger R. F. & Colwell R. K. (1977) Organization of contiguous communities of amphibians and reptiles in Thailand. *Ecol. Monogr.* **47**, 229–253. 5.2, 6.5.2

Inger R. F. & Greenberg B. (1966) Ecological and competitive relations among three species of frogs (genus *Rana*). *Ecology* **47**, 746–759. 11.3.1

Inouye D. W. (1978) Resource partitioning in bumblebees: experimental studies of foraging behavior. *Ecology* **59**, 672–678. 6.4

Inouye R. S. (1980) Density-dependent germination response by seeds of desert annuals. *Oecologia* **46**, 235–238. 11.7

Inouye R. S. (1981) Interactions among unrelated species: granivorous rodents, a parasitic fungus, and a shared prey species. *Oecologia* **49**, 425–427. 11.3.1, 11.7

Ionescu-Cojocaru M. & Negulescu Fl. (1976) Wheat breeding for long term leaf rust resistance in Romania. Proc. 4th Eur. Mediterr. Cereal Rusts Conf. pp. 133–135. Interlaken. 14.8

Istock C. A. (1973) Population characteristics of a species ensemble of water boatmen (Corixidae). *Ecology* **54**, 535–544. 11.2.2

Ito A. (1978) *Hymenolepis nana*: protective immunity against mouse-derived cysticercoids induced by initial inoculation with eggs. *Exp. Parasitol.* **46**, 12–19. 13.3.3

Ito A. (1980) *Hymenolepis nana*: survival in the immunized mouse. *Exp. Parasitol.* **49**, 248–257. 13.3.3

Itô Y. (1960) Ecological studies on population increase and habitat segregation among barley aphids. *Bull. National Inst. agric. Sci., Ser. C* (Japan) **11**, 45–130. 1.7

Itô Y. (1980) *Comparative Ecology* (English edn). Cambridge University Press, Cambridge. 14.4

Jackson J. B. C. (1981) Interspecific competition and species' distributions: the ghost of theories and data past. *Amer. Zool.* **21**, 889–901. 11.1

Jackson W. D. (1968) Fire, air, water and earth — an elemental ecology of Tasmania. *Proc. Ecol. Soc. Aust.* **3**, 9–16. 12.2

Jacob F. (1977) Evolution and tinkering. *Science* **196**, 1161–1166. 14.2

Jaeger R. G. (1971) Competitive exclusion as a factor influencing the distribution of two species of terrestrial salamanders. *Ecology* **52**, 632–637. 11.3.1, 11.4.1

Jaeger R. G. (1974) Interference or exploitation? A second look at competition between salamanders. *J. Herpetol.* **8**, 191–194. 6.8

James F. C. (1971) Ordinations of habitat relationships among breeding birds. *Wilson Bull.* **83**, 215–236. 5.3.2

Janzen D. H. (1966) Coevolution of mutualism between ants and acacias in Central America. *Evolution* **20**, 249–275. 14.10

Janzen D. H. (1971) Seed predation by animals. *Ann. Rev. Ecol. Syst.* **2**, 465–492. 7.2.1

Janzen D. H. (1973a) Sweep samples of tropical foliage insects: effects of seasons, vegetation types, elevation, time of day, and insularity. *Ecology* **54**, 687–708. 5.5.3

Janzen D. H. (1973b) Host plants as islands. II. Competition in evolutionary and contemporary time. *Amer. Natur.* **107**, 786–790. 13.2.2, 13.3.3

Janzen D. H. (1974) Tropical blackwater rivers, animals, and mast fruiting by the Dipterocarpaceae. *Biotropica* **6**, 69–103. 5.5.1

Janzen D. H. (1979) New horizons in the biology of plant defenses. In *Herbivores: Their Interaction with Secondary Plant Metabolites* (eds Rosenthal G. A. & Janzen D. H.) pp. 331–350. Academic Press, New York. 13.2.2

Janzen D. H. (1980a) When is it coevolution? *Evolution* **34**, 611–612. 14.3, 14.9

Janzen D. H. (1980b) Specificity of seed-attacking beetles in a Costa Rican deciduous forest. *J. Ecol.* **68**, 929–952. 1.8

Janzen D. H. (1983) Dispersal of seeds by vertebrate guts. In *Coevolution* (eds Futuyma D. J. & Slatkin M.) pp. 232–262. Sinauer, Sunderland, Mass. 14.10

Jarman P. J. (1974) The social organisation of antelope in relation to their ecology. *Behaviour* **48**, 215–267. 8.4.1, 8.5

Järvinen O. (1982) Deducing interspecific competition from community data: preface. *Ann. Zool. Fennici* **19**, 239. 9.2

Jaynes R. A., Anagnostakis S. L. & Van Alfen N. K. (1976) Chestnut research and biological control of the chestnutblight fungus. In *Perspectives in Forest Entomology* (eds Anderson J. & Kaye H.) pp. 61–70. Academic Press, New York. 14.8

Jenkins D., Watson A. & Miller G. R. (1963) Population studies on red grouse, *Lagopus lagopus scoticus* (Lath.) in north-east Scotland. *J. Anim. Ecol.* **32**, 317–376. 9.3.1

Jenkins D., Watson A. & Miller G. R. (1964) Predation and red grouse populations. *J. Appl. Ecol.* **1**, 183–195. 9.3.1

Jennings P. R. & Aquino R. C. (1968) Studies of competition in rice. III. The mechanism of competition among phenotypes. *Evolution* **22**, 529–542. 14.5

Jennings P. R. & Herrera R. M. (1968) Studies on competition in rice. II. Competition in segregating populations. *Evolution* **22**, 332–336. 14.5

Jensen L. M. (1983) Phytoplankton release of extracellular organic carbon, molecular weight composition and bacterial assimilation. *Mar. Ecol. Prog. Ser.* **11**, 39–43. 7.3.3

Jochelson W. (1925) *Archeological Investigations in the Aleutian Islands*. Publ. 367, Carnegie Institute of Washington. 10.2.2

Jochim M. A. (1981) *Strategies for Survival: Cultural Behavior in an Ecological Context*. Academic Press, New York. 15.3.5

Joern A. & Lawlor L. R. (1980) Food and microhabitat utilization by grasshoppers from arid grasslands: comparisons with neutral models. *Ecology* **61**, 591–599. 6.8

Johannes R. E. (1980) *Words of the Lagoon: Fishing and Marine Lore in Palau District of Micronesia*. University of California Press, Los Angeles. 15.3.3

Johannes R. E. & Satomi M. (1966) Composition and nutritive value of fecal pellets of a marine crustacean. *Limnol. Oceanogr.* **11**, 301–340. 7.3.2

Johnson A. (1982) Reductionism in cultural ecology: the Amazon case. *Current Anthropology* **23**, 413–428. 15.3.3

Jones D. (1962) Selective eating of the acyanogenic form of the plant *Lotus corniculatus* by various animals. *Nature* **193**, 1109–1110. 14.9

Jordan C. F. (1982) The nutrient balance of an Amazonian rain forest. *Ecology* **63**, 647–654. 7.2.2

Jorgensen C. G. (1966) *Biology of Suspension Feeding*. Pergamon Press, Oxford. 7.3.3

Joule J. & Jameson D. L. (1972) Experimental manipulation of population density in three sympatric rodents. *Ecology* **53**, 653–660. 11.3.1, 11.7

Jutsum A. R., Cherrett J. M. & Fisher M. (1981) Interactions between the fauna of citrus trees in Trinidad and the ants *Atta cephalotes* and *Azteca* sp. *J. appl. Ecol.* **18**, 187–195. 2.2

Kareiva P. (1982) Exclusion experiments and the competitive release of insects feeding on collards. *Ecology* **63**, 696–704. 6.8, 8.3.3

Kareiva P. (1983) The influence of vegetation texture on herbivore populations: resource conservation and herbivore movement. In *Variable Plants and Herbivores in Natural and Managed Ecosystems* (eds Denmo R. F. & McClure M. S.) pp. 259–289. Academic Press, New York. 8.2.1, 8.2.2

Karl B. J. & Best H. A. (1982) Feral cats on Stewart Island; their foods and their effects on kakapo. *N. Z. J. Zool.* **9**, 287–294. 10.3.1

Karl D. M., Wirsen C. O. & Jannasch H. W. (1980) Deep-sea primary production at the Galapagos hydrothermal vents. *Science* **207**, 1345–1347. 7.3.1

Karr J. R. (1971) Structure of avian communities in selected Panama and Illinois habitats. *Ecol. Monogr.* **41**, 207–233. 4.3.3, 5.5.3, 6.8

Karr J. R. (1975) Production, energy pathways, and community diversity in forest birds. In *Tropical Ecological Systems: Trends in Terrestrial and Aquatic Research* (eds Golley F. B. & Medina E.) pp. 161–176. Springer Verlag, New York. 5.5.1

Karr J. R. (1976) Within — and between — habitat avian diversity in African and neotropical lowland habitats. *Ecol. Monogr.* **46**, 457–481. 5.5.1

Karr J. R. (1980) Geographical variation in the avifaunas of tropical forest undergrowth. *Auk* **97**, 283–298. 5.5.1

Karr J. R. & James F. C. (1975) Eco-morphological configurations and convergent evolution in species and communities. In *Ecology and Evolution of Communities* (eds Cody M. L. & Diamond J. M.) pp. 258–291. Belknap Press, Cambridge Mass. 5.5.1

Karr J. R. & Roth R. R. (1971) Vegetation structure and avian diversity in several New World areas. *Amer. Natur.* **105**, 423–435. 4.3.3

Kastendiek J. (1982) Factors determining the distribution of the sea pansy, *Renilla killikeri*, in a subtidal sand-bottom habitat. *Oecologia* **52**, 340–347. 11.7

Kates K. C. & Turner J. H. (1959) An experiment on the combined pathogenic effects of *Haemonchus contortus* and *Nematodirus spathiger* on lambs. *Proc. Helminthol. Soc. Wash.* **26**, 62–67. 13.2.1

Kawanabe H., Mori S. & Mizuno N. (1959) On the food economy of "ayu" fish with relation to the production of algae. *Physiol. Ecol.* (Kyoto) **8**, 117–128 (in Japanese). 7.3.1

Kazacos K. R. & Thorson R. E. (1975) Cross-resistance between *Nippostrongylus brasiliensis and Strongyloides ratti* in rats. *J. Parasitol.* **61**, 525–529. 13.3.3

Keast A. (1968) Competitive interactions and the evolution of ecological niches as illustrated by the Australian honeyeater genus *Melithreptus* (Meliphagidae). *Evolution* **22**, 762–784. 5.5.1

Keast A. (1969) Comparisons of the contemporary mammalian faunas of the southern continents. *Quart. Rev. Biol.* **44**, 121–167. 5.5, 5.5.1

Keast A. (1970a) Food specializations and bioenergetic interrelations in the fish faunas of some small Ontario waterways. In *Marine Food Chains* (ed. Steele J. H.) pp. 377–411. Oliver & Boyd, Edinburgh. 6.4

Keast A. (1970b) Adaptive evolution and shifts in niche occupation in island birds. *Biotropica* **2**, 61–75. 6.6.2

Keast A. (1972a) Ecological opportunities and dominant families, as illustrated by the Neotropical Tyrannidae (Aves). *Evol. Biol.* **5**, 229–277. 5.5.1

Keast A. (1972b) Comparisons of contemporary mammal faunas of southern continents. In *Evolution, Mammals and Southern Continents* (eds Keast A., Erk F. C. & Glass B.) pp. 433–501. State University of New York Press, Albany. 5.5.1

Keast A. (1972c) Faunal elements and evolutionary patterns: some comparisons between the continental avifaunas of Africa, South America and Australia. *Proc. XV Intern. Ornith. Congr.* (*The Hague*) (ed. Voous, K. H.) pp. 594–622. Brill, Leiden. 5.5.1

Keeler K. H. (1977) The extrafloral nectaries of *Ipomoea cornea* (Convolvulaceae). *Amer. J. Bot.* **64**, 1182–1188. 10.2.3

Keeler K. H. (1980) Distribution of plants with extrafloral nectaries. *Amer. Midl. Nat.* **104**, 274–280. 10.2.3

Kelly R. C. (1977) *Etoro Social Structure: A Study in Structural Contradiction*. University of Michigan Press, Ann Arbor. 15.3.2, 15.3.5

Kendeigh S. C. & Pinowski J. (eds) (1973) *Productivity, Population Dynamics and Systematics of Granivorous Birds*. Polish Scientific Publishers, Warsaw. 7.2.1

Kennedy C. R. (1974) A checklist of British and Irish freshwater fish parasites with notes on their distribution. *J. Fish. Biol.* **6**, 613–644. 9.2.1

Kennedy C. R. (1978) An analysis of the metazoan parasitocoenoses of brown trout *Salmo trutta* from British lakes. *J. Fish. Biol.* **13**, 255–263. 9.2.4

Kenneth J. W. (ed.) (1963) *A Dictionary of Biological Terms* (originally edited by Henderson I. F. & Henderson W. D.) (8th edn). Oliver & Boyd, Edinburgh. 3.1

Kenyon K. W. (1969) *The Sea Otter in the Eastern Pacific*. Government Printing Office, Washington D.C. 10.2.2

Kettlewell B. (1973) *The Evolution of Melanism*. Clarendon Press, Oxford. 14.2

Key K. H. L. (1981) Species parapatry, and the morabine grasshoppers. *System. Zool.* **30**, 425–458. 8.2.4

Keymer A. E. & Anderson R. M. (1979) The dynamics of infection of *Tribolium confusum* by *Hymenolepis diminuta*: the influence of infective-stage density and spatial distribution. *Parasitology* **79**, 195–207. 9.3

Khan M. A., Putwain P. D. & Bradshaw A. D. (1975) Population interrelationships. 2. Frequency dependent fitness in *Linum*. *Heredity* **34**, 145–163. 14.6

Kikkawa J. (1966) Population distribution of land birds in temperate rainforest of southern New Zealand. *Trans. roy. Soc. N. Z. Zool.* **7**, 215–277. 4.3.3

Kikkawa J. (1968) Ecological association of bird species and habitats in eastern Australia; similarity analysis. *J. Anim. Ecol.* **37**, 143–165. 4.3.3, 4.4.1

Kikkawa J. (1974) Comparison of avian communities between wet and semi-arid habitats of eastern Australia. *Aust. Wildl. Res.* **1**, 107–116. 4.4.1, 4.4.3

Kikkawa J. (1977) Ecological paradoxes. *Aust. J. Ecol.* **2**, 121–136. Preface, 4.2.3

Kikkawa J. (1982) Ecological association of birds and vegetation structure in wet tropical forests of Australia. *Aust. J. Ecol.* **7**, 325–345. 4.4.2, 4.4.3

Kikkawa J., Ingram G. J. & Dwyer P. D. (1979) The vertebrate fauna of Australian heathlands — an evolutionary perspective. In *Heathlands and Related Shrublands. Ecosystems of the World* (ed. Specht R. L.) Vol. 9A, pp. 231–279. Elsevier, Amsterdam. 4.3.4

Kikkawa J., Lovejoy T. E., Humphrey P. S. & Humphrey S. S. (1980) Structural complexity and species clustering of birds in tropical rainforests. *Acta XVII Congr. Intern. Ornith.* (*Berlin*) (ed. Nöhring R.) pp. 962–967. Deutsche Ornithologen-Gesellschaft, Berlin. 4.4.3

Kikkawa J. & Webb L. J. (1967) Niche occupation by birds and the structural classification of forest habitats in the wet tropics, north Queensland. *Proc. XIV Congr. Int. Union For. Res. Org. Sect.* **26**, 467–482. 4.4.3

Kikkawa J. & Williams W. T. (1971) Altitudinal distribution of land birds in New Guinea. *Search* (*Syd.*) **2**, 64–65. 4.4.3

Kikuchi E. & Kurihara Y. (1977) *In vitro* studies on the effects of tubificids on the biological, chemical and physical characteristics of submerged rice-field soil and overlying water. *Oikos* **29**, 348–356. 7.3.3

Kilburn P. D. (1966) Analysis of the species-area relation. *Ecology* **47**, 831–843. 4.3.1

Kilham L. & Oliver L. (1961) The promoting effect of trichinosis on encephalomyocarditis

(EMC) virus infection in rats. *Amer. J. trop. Med.* **10**, 879–884. 13.2.1

King A. W. & Pimm S. L. (1983) Complexity, diversity, and stability: a reconciliation of theoretical and empirical results. *Amer. Natur.* **122**, 229–239. 4.1, 4.2.3

King C. M. & Moody J. E. (1982) The biology of the stoat (*Mustela erminea*) in the National Parks of New Zealand. II. Food habits. *N. Z. J. Zool.* **9**, 57–80. 10.3.1

King H. G. C. & Heath G. W. (1967) The chemical analysis of small samples of leaf material and the relationship between the disappearance and decomposition of leaves. *Pedobiologia* **7**, 192–197. 7.2.2

King K. R., Hollibaugh J. T. & Azam F. (1980) Predator-prey interactions between the larvacean *Oikopleura dioica* and bacterio-plankton in enclosed water columns. *Mar. Biol.* **56**, 49–57. 7.3.3

Kiritani K. & Kakiya N. (1975) An analysis of the predator-prey system in the paddy field. *Res. Popul. Ecol.* **17**, 29–38. 8.3.3

Kirkman H. & Reid D. D. (1979) A study of the role of the seagrass *Posidonia australis* in carbon budget of an estuary. *Aquat. Bot.*, **7**, 173–183. 7.3.3

Kitching R. L. (1971) An ecological study of water-filled tree-holes and their position in the woodland ecosystem. *J. Anim. Ecol.* **40**, 281–302. 10.2.1

Kitching R. L. (1983a) Community structure in water-filled treeholes in Europe and Australia — some comparisons and speculations. In *Phytotelmata: Terrestrial Plants as Hosts for Aquatic Insect Communities* (eds Lounibos P. & Frank H.) pp. 205–222. Plexus, Medford, New Jersey. 10.2.1

Kitching R. L. (1983b) Myrmecophilous organs of the larvae and pupae of the lycaenid butterfly, *Jamenus evagoras* (Donovan). *J. nat. Hist.* **17**, 471–481. 10.2.3

Kitching R. L. (1983c) *Systems Ecology: An Introduction to Ecological Modelling*. University of Queensland Press, Brisbane. 10.2.3

Kitching R. L. & Callaghan C. (1982) The fauna of water-filled treeholes in box forest in south-east Queensland. *Aust. ent. Mag.* **8**, 61–70. 10.2.1

Kitting C. L. (1980) Herbivore-plant interactions of individual limpets maintaining a mixed diet of intertidal marine algae. *Ecol. Monogr.* **50**, 527–550. 14.9

Klopfer P. H. & MacArthur R. H. (1960) Niche size and faunal diversity. *Amer. Natur.* **94**, 293–300. 5.5.3

Klopfer P. H. & MacArthur R. H. (1961) On the causes of tropical species diversity: niche overlap. *Amer. Natur.* **95**, 223–226. 5.5.3

Koblentz-Mishke O. J., Volkovinsky V. V. & Kabanova J. G. (1970) Plankton primary production of the world ocean. In *Scientific Exploration of the South Pacific* (ed. Wooster W. S.) pp. 183–193. Nat. Acad. Sci., Washington D.C. 7.3.1

Koch A. L. (1974) Coexistence resulting from an alternation of density dependent and density independent growth. *J. theor. Biol.* **44**, 373–386. 6.8

Kohn A. J. (1978) Ecological shift and release in an isolated population: *Conus miliaris* at Easter Island. *Ecol. Monogr.* **48**, 323–336. 6.6.3

Kohn A. J. & Nybakken J. W. (1975) Ecology of *Conus* on Eastern Indian Ocean fringing reefs: diversity of species and resource utilization. *Mar. Biol.* **29**, 211–234. 6.4, 6.5.3

Kojima K. & Yarbrough K. M. (1967) Frequency dependent selection at the esterase 6 locus in *Drosophila melanogaster*. *Proc. Natl. Acad. Sci.* **57**, 645–649. 14.6

Kontrimavichus V. L. & Atrashkevich G. I. (1982) Parasitic systems and their role in the population biology of helminths. *Parazytologia* **16**, 177–187. (In Russian). 9.2.1

Korstian C. F. & Stickel P. W. (1927) The natural replacement of blight-killed chestnut in the hardwood forests of the north east. *J. agric. Res.* **34**, 631–648. 2.5

Krebs C. J. (1978) *Ecology. The Experimental Analysis of Distribution and Abundance* (2nd edn). Harper & Row, New York. 11.5.1

Kroh G. C. & Stephenson S. N. (1980) Effects of diversity and pattern on relative yields of four Michigan first year fallow field plant species. *Oecologia* **45**, 366–371. 11.3.2, 11.5.1

Kruuk H. (1972) *The Spotted Hyena*. University of Chicago Press, Chicago. 8.4.3

Krzysik A. J. (1979) Resource allocation, coexistence, and the niche structure of a streambank salamander community. *Ecol. Monogr.* **49**, 173–194. 6.8

Kurihara Y. (1983) Study of domestic sewage waste treatment by the polychaetes, *Neanthes japonica* and *Perinereis nuntia* var. *vallata*, on an artificial tidal flat. *Int. Revue ges. Hydrobiol.* **68**, 649–670. 7.3.3

Kurihara Y., Eadie J. M., Mann S. O. & Hobson P. N. (1968) Relationship between bacteria and ciliate protozoa in the sheep rumen. *J. Gen. Microbiol* **51**, 267–288. 7.2.3

Kuris A. M. (1974) Trophic interactions: similarity of parasitic castrators to parasitoids. *Quart. Rev. Biol.* **49**, 129–148. 9.3, 9.3.2, 13.2.1, 13.3.1

Kuris A. M., Blaustein A. R. & Alio J. J. (1980) Hosts as islands. *Amer. Natur.* **116**, 570–586. 8.1.2, 8.2.4, 9.2.3

Lack D. (1947) *Darwin's Finches*. Cambridge University Press, Cambridge. 6.4, 6.6.1, 14.6

Lack D. (1971) *Ecological Isolation in Birds*. Blackwell Scientific Publications, Oxford. 6.4

Ladd H. S. (1977) Palaeoecology. In *McGraw-Hill Encyclopaedia of Science and Technology*, Vol. **9**, pp. 579–583. McGraw-Hill, New York. 3.1

Lanciani C. A. (1975) Parasite-induced alterations in host reproduction and survival. *Ecology* **56**, 689–695. 9.3

Landres P. B. & MacMahon J. A. (1980) Guilds and community organization: analysis of an oak woodland avifauna in Sonora, Mexico. *Auk* **97**, 351–365. 5.2, 5.3.2

Larsson U. & Hagstrom A. (1979) Phytoplankton exudate release as an energy source for the growth of pelagic bacteria. *Mar. Biol.* **52**, 199–206. 7.3.3

Lathrap D. (1970) *The Upper Amazon*. Praeger, New York. 15.3.3

Latter P. M. & Howson G. (1978) Studies on the microfauna of blanket bog with particular references to Enchytraeidae. II. Growth and survival of *Cognetia sphagnetorum* on various substrates. *J. Anim. Ecol.* **47**, 415–448. 7.2.2

Law R., Bradshaw A. D. & Putwain P. D. (1977) Life history variation in *Poa annua*. *Evolution* **31**, 233–246. 14.4

Lawlor L. R. & Smith J. M. (1976) The coevolution and stability of competing species. *Amer. Natur.* **110**, 79–99. 14.6

Lawrence M. J. (1984) The population genetics of quantitative characters. In *Evolutionary Ecology* (ed. Shorrocks B.) pp. 27–63. Blackwell Scientific Publications, Oxford. 14.2

Lawrence P. & Meggitt M. S. (eds) (1965) *Gods, Ghosts and Men in Melanesia*. Oxford University Press, Melbourne. 15.3.4

Laws R. M. (1970) Elephants as agents of habitat and landscape change in East Africa. *Oikos* **21**, 1–15. 8.4.3

Laws R. M., Parker I. S. C. & Johnstone R. C. B. (1975) *Elephants and Their Habitats*. Oxford University Press, Oxford. 8.4.3

Lawton J. H. (1974) Review of J. M. Smith's *Models in Ecology*. *Nature* **248**, 537. 12.1

Lawton J. H. (1978) Host plant influences of insect diversity: the effects of space and time. In *Diversity of Insect Faunas*. *Symp. Roy. Ent. Soc.* **9**. (eds Mound L. A. & Waloff N.) pp. 105–125. Blackwell Scientific Publications, Oxford. 8.2, 8.2.1

Lawton J. H. (1982) Vacant niches and unsaturated communities: a comparison of bracken herbivores at sites on two continents. *J. Anim. Ecol.* **51**, 573–595. 5.2, 8.2.3

Lawton J. H. (1983) Plant architecture and the diversity of phytophagous insects. *Ann. Rev. Ent.* **28**, 23–39. 8.2.3, 8.5

Lawton J. H. (1984a) Herbivore community organisation: general models and specific tests with phytophagous insects. In *A New Ecology: Novel Approaches to Interactive Systems* (eds Price P. W., Slabodchikoff C. N. & Gaud W. S.) pp. 329–352. Wiley, New York. 8.2.3, 8.3.2

Lawton J. H. (1984b) Non-competitive populations, non-convergent communities and vacant niches: the herbivores of bracken. In *Ecological Communities: Conceptual Issues and the Evidence* (eds Strong D. R., Jr., Simberloff D., Abele L. G. & Thistle A. B.) pp. 67–99. Princeton University Press, Princeton. 8.2.3

Lawton J. H., Cornell H., Dritschild W. & Hendrix S. D. (1981) Species as islands: comments on a paper by Kuris *et al. Amer. Natur.* **117**, 623–627. 8.2.3

Lawton J. H. & Hassell M. P. (1981) Asymmetrical competition in insects. *Nature* **289**, 793–795. 11.1, 11.5.1, 11.6.1, 11.6.2

Lawton J. H. & Price P. W. (1979) Species richness of parasites on hosts: agronyzid flies of the British Umbelliferae. *J. Anim. Ecol.* **48**, 619–637. 8.3.4

Lawton J. H. & Schroder D. (1977) Effects of plant type, size of geographic range and taxonomic isolation on number of insect species associated with British plants. *Nature* **265**, 137–140. 8.2.3

Lawton J. H. & Strong D. R. (1981) Community patterns and competition in folivorous insects. *Amer. Natur.* **118**, 317–338. 6.8, 8.3.3, 8.3.4, 11.1

Le Baron H. M. & Gressel J. (eds) (1982) *Herbicide Resistance in Plants*. Wiley, New York. 14.2

Lee K. E. & Wood T. G. (1971) *Termites and Soils*. Academic Press, New York. 7.2.2

Lee R. B. & De Vore I. (eds) (1968) *Man the Hunter*. Aldine, Chicago. 15.2

Leigh E. G. (1975) Structure and climate in tropical rain forest. *Ann. Rev. Ecol. Syst.* **6**, 67–86. 5.5.1

LeJambre L. F. (1982) Barriers to introgression in *Haemonchus*. *Abstracts Fifth Internatl. Congr. Parasitol., Toronto, 7-14 Aug., 1982*, p. 446. 9.3.2

León J. A. (1974) Selection in contexts of interspecific competition. *Amer. Natur.* **108**, 739–757. 14.6

Leonard K. J. (1977) Selection pressures and plant pathogens. *Ann. N. Y. Acad. Sci.* **287**, 207–222. 14.8

Leong T. S. & Holmes J. C. (1981) Communities of metazoan parasites in open water fishes of Cold Lake, Alberta. *J. Fish Biol.* **18**, 693–713. 9.2.1, 9.2.4

Leopold A. (1943) Deer irruptions. *Wisconsin Cons. Dept. Publ.* **321**, 3–11. 1.7

Lerner I. M. & Ho F. K. (1961) Genotype and competitive ability in *Tribolium* species. *Amer. Natur.* **95**, 329–343. 14.5

Leston D. (1973) The ant mosaic — tropical tree crops and the limiting of pests and diseases. *PANS* **19**, 311–341. 2.2, 8.3.3

Leston D. (1978) A neotropical ant mosaic. *Ann. ent. Soc. Am.* **71**, 649–653. 2.2

Levin S. A. (1974) Dispersion and population interactions. *Amer. Natur.* **108**, 207–228. 4.3.4

Levins R. (1962) Theory of fitness in a heterogeneous environment. I. The fitness set and adaptive function. *Amer. Natur.* **96**, 361–378. 9.2.2

Levins R. (1966) Strategy of model building in population biology. *Amer. Scientist* **53**, 421–431. 15.2

Levins R. (1970) Complex systems. In *Towards a Theoretical Biology* Vol. 3. *Drafts* (ed. Waddington C. H.) pp. 73–88. Edinburgh University Press, Edinburgh. 15.2

Levins R. & Culver D. (1971) Regional coexistence of species and competition between rare species. *Natl. Acad. Sci. Proc., (U.S.A.)* **68**, 1246–1248. 13.2.3

Levinton J. S. (1982) The body size — prey size hypothesis: the adequacy of body size as a vehicle for character displacement. *Ecology* **63**, 869–872. 6.6.1

Levi-Strauss C. (1966) *The Savage Mind*. Weidenfeld & Nicolson, London. 15.4

Leviten P. J. (1978) Resource partitioning by predatory gastropods of the genus *Conus* on subtidal Indo-Pacific coral reefs: the significance of prey size. *Ecology* **59**, 614–631. 6.4

Lewis B. (1977) *Gene Expression — 3: Plasmids and Phages*. Wiley, New York. 13.3.3

Lewis J. B. (1981) Coral reef ecosystems. In *Analysis of Marine Ecosystems* (ed. Longhurst R. A.) pp. 127–158. Academic Press, New York. 7.3.1

Lewis L. G. E. (1971) The life history and ecology of three paradoxosomatid millipedes (Diplopoda: Polydesmida) in northern Nigeria. *J. Zool. Lond.* **165**, 431–451. 7.2.2
Lewontin R. C. (1955) The effect of population density and composition on viability in *Drosophila melanogaster*. *Evolution*, **9**, 27–41. 14.6
Lewontin R. C. (1965) Selection for colonising ability. In *The Genetics of Colonising Species* (eds Baker H. G. & Stebbins G. L.) pp. 77–94. Academic Press, New York. 14.4
Lewontin R. C. (1969) The meaning of stability. In *Diversity and Stability in Ecological Systems* (eds Woodwell G. M. & Smith H. H.) pp. 13–24. *Brookhaven Symposia in Biology*, 22. Brookhaven National Laboratory, Upton, New York. 4.2.2
Likens G. E. (1975) Primary production of inland aquatic ecosystems. In *Primary Productivity of the Biosphere* (eds Lieth H. & Whittaker R. H.) pp. 185–202. Springer-Verlag, Berlin. 7.3.1
Lincoln R. J., Boxshall G. A. & Clark P. F. (1982) *A Dictionary of Ecology, Evolution and Systematics*. Cambridge University Press, Cambridge. 12.3.1
Lindeman R. L. (1942) The trophic-dynamic aspects of ecology. *Ecology* **23**, 399–418. 1.6.1
Linton L. R., Davies R. W. & Wrona F. J. (1981) Resource utilization indices: an assessment. *J. Anim. Ecol.* **50**, 283–292. 6.7.2
Lister B. C. (1976a) The nature of niche expansion in West Indian *Anolis* lizards. I. Ecological consequences of reduced competition. *Evolution* **30**, 659–676. 6.6.3
Lister B. C. (1976b) The nature of niche expansion in West Indian *Anolis* lizards. II. Evolutionary components. *Evolution* **30**, 677–692. 6.6.3
Lister B. C. & McMurtrie R. E. (1976) On size variation in anoline lizards. *Amer. Natur.* **110**, 311–314. 6.6.3
Lobel P. & Ogden J. C. (1981) Foraging by the herbivorous parrotfish *Sparisoma radians*. *Mar. Biol.* **64**, 173–183. 14.9
Locke L., DeWitt J., Menzie C. & Kerwin J. (1964) A merganser die off associated with larval *Eustrongylides*. *Avian Diseases* **8**, 420–427. 9.3.3
Longman K. A. & Janik J. (1974) *Tropical Forest and its Environment*. Longman, London. 12.3.4
Lotka A. J. (1925) *Elements of Physical Biology*. Williams & Wilkins, Baltimore. 1.6
Lotka A. J. (1932) The growth of mixed populations: two species competing for a common food supply. *J. Wash. Acad. Sci.* **22**, 461–469. 6.1
Louch C. D. (1962) Increased resistance to *Trichinella spiralis* in the laboratory rat following infections with *Nippostrongylus muris*. *J. Parasitol.* **48**, 24–26. 13.3.3
Lovejoy T. E. (1975) Bird diversity and abundance in Amazon forest communities. *Living Bird, Thirteenth Annual, 1974*, pp. 127–191. 4.3.3
Lovett Doust L. (1981) Population dynamics and local speciation in a clonal perennial (*Ranunculus repens*). II. The dynamics of leaves, and a reciprocal transplant experiment. *J. Ecol.* **69**, 757–768. 14.5, 14.6
Lovtrup S. (1981) Introduction to evolutionary epigenetics. In *Evolution Today* (eds Scudder G. G. E. & Reveal J. L.) pp. 139–144. *Proc. Second Intern. Cong. Syst. Evol. Bio.*, Hunt Inst. Botanical Documentation, Carnegie-Mellon University, Pittsburg. 15.4
Lowe E. F. & Lawrence J. M. (1976) Adsorption efficiencies of *Lytechinus variegatus* (Lamarck) (Echinodermata: Echinoidea) for selected marine plants. *J. exp. mar. Biol. Ecol.* **21**, 223–234. 14.9
Loyn R. H., Runnalls R. G., Forward G. Y. & Tyers J. (1983) Territorial bell miners and other birds affecting populations of insect prey. *Science* **221**, 1411–1413. 8.3.3
Lubchenco J. (1980) Algal zonation in the New England rocky intertidal community: an experimental analysis. *Ecology* **61**, 333–344. 11.3.1
Ludwig W. (1959) Die Selectionstheorie. In *Die Evolutionen der Organismen* (ed. Herberer G.) pp. 662–712. Fischer, Stuttgart. 14.6

Lugo A. E., Farnworth E. G., Pool D., Jerez P. & Kaufman G. (1973) The impact of the leaf cutter ant *Atta colombica* on the energy flow of a tropical wet forest. *Ecology* **54**, 1292–1301. 7.2.2

Lynch M. (1978) Complex interactions between natural coexploiters *Daphnia* and *Ceriodaphnia*. *Ecology* **59**, 552–564. 11.4.1, 11.4.2

MacArthur R. H. (1955) Fluctuations of animal populations, and a measure of community stability. *Ecology* **36**, 533–536. 4.2.1

MacArthur R. H. (1957) On the relative abundance of bird species. *Proc. Nat. Acad. Sci. U.S.* **43**, 293–295. 4.3.2

MacArthur R. H. (1958) Population ecology of warblers of northeastern coniferous forest. *Ecology* **39**, 599–619. 10.1

MacArthur R. H. (1969) Patterns of communities in the tropics. *Biol. Jour. Linn. Soc.* **1**, 19–30. 5.5.3

MacArthur R. H. (1972) *Geographical Ecology: Patterns in the Distribution of Species*. Harper & Row, New York. 5.4, 6.5.2

MacArthur R. & Connell J. H. (1966) *The Biology of Populations*. Wiley, New York. 15.2

MacArthur R. H. & Levins R. (1967) The limiting similarity, convergence and divergence of coexisting species. *Amer. Natur.* **101**, 377–385. 6.5.2

MacArthur R. H. & MacArthur J. (1961) On bird species diversity. *Ecology* **42**, 594–598. 5.3.2, 5.5.3

MacArthur R. & Pianka E. (1966) On optimal use of a patchy environment. *Amer. Natur.* **100**, 603–609. 5.2, 5.5.3

MacArthur R. H., Recher H. & Cody M. L. (1966) On the relation between habitat selection and species diversity. *Amer. Natur.* **100**, 319–332. 5.3.3

MacArthur R. H. & Wilson E. O. (1963) An equilibrium theory of insular geography. *Evolution* **17**, 373–387. 9.2.3

MacArthur R. H. & Wilson E. O. (1967) *The Theory of Island Biogeography*. Princeton University Press, Princeton. 4.3.1, 6.4, 8.2.1, 9.2.3, 14.4

MacGarvin M. (1982) Species-area relationships of insects on host plants: herbivores on rosebay willowherb. *J. Anim. Ecol.* **51**, 207–223. 8.2.1

MacInnis A. J. (1976) How parasites find hosts: some thoughts on the inception of host-parasite integration. In *Ecological Aspects of Parasitology* (ed. Kennedy C. R.) pp. 3–20. North-Holland, Amsterdam. 9.2.1

MacKenzie K. & Gibson D. (1970) Ecological studies of some parasites of plaice, *Pleuronectes platessa* (L.), and flounder, *Platichthys flesus* (L.). In *Aspects of Fish Parasitology* (eds Taylor A. E. R. & Muller R.) pp. 1–42. Blackwell Scientific Publications, Oxford. 9.2.1

Mackie G. L., Qadri S. V. & Reed R. M. (1978) Significance of litter size in *Musculium securis* (Bivalvia: Sphaeriidae). *Ecology* **59**, 1069–1074. 11.4.1

Maddock L. (1979) The migration and grazing succession. In *Serengeti: Dynamics of an Ecosystem* (eds Sinclair A. R. E. & Norton-Griffiths M.) pp. 104-129. Chicago University Press, Chicago. 8.4.3

Maguire B., Jr (1971) Phytotelmata: biota and community structure determination in plant-held waters. *Ann. Rev. Ecol. Syst.* **2**, 439–464. 10.4

Maguire B., Jr (1973) Niche response structure and the analytical potentials of its relationships to the habitat. *Amer. Natur.* **107**, 213–246. 6.2

Majer J. D. (1976a) The maintenance of the ant mosaic in Ghana cocoa farms. *J . appl. Ecol.* **13**, 123–144. 2.2

Majer J. D. (1976b) The ant mosaic in Ghana cocoa farms: further structural considerations. *J. appl. Ecol.* **13**, 145–155. 2.2

Majer J. D. (1976c) The influence of ants and ant manipulation on the cocoa farm fauna. *J. appl. Ecol.* **13**, 157–175. 2.2

Malik A. R., Anderson D. J. & Myerscough P. J. (1976) Studies on structure in plant communities. VII. Field and experimental analyses of *Atriplex vesicaria* populations from the Riverine Plain of New South Wales. *Aust. J. Bot.* **24**, 265–280. 12.3.3, 12.3.4

Malinowski B. (1922) *Argonauts of the Western Pacific*. Routledge, London. 15.3.3

Mann K. H. (1975) Patterns of energy flow. In *River Ecology* (ed. Whitton B. A.) pp. 248–263. Blackwell Scientific Publications, Oxford. 7.3.1

Mann K. H. (1977) Destruction of kelp-beds by sea-urchins; a cyclical phenomenon or irreversible degradation. *Helg. wiss. Meers* **30**, 455–467. 10.2.2

Mann K. H. & Breen P. A. (1972) The relationship between lobster abundance, sea urchins and kelp beds. *J. Fish. Res. Bd., Canada* **29**, 603–609. 10.2.2

Margalef R. (1963) On certain unifying principles in ecology. *Amer. Natur.* **97**, 357–374. 1.6.1

Margalef R. (1968) *Perspectives in Ecological Theory*. University of Chicago Press, Chicago. 4.1, 12.1

Margalef R. & Gutierrez E. (1983) How to introduce connectance in the frame of an expression for diversity. *Amer. Natur.* **121**, 601–607. 4.5

Marsh H., Channells P. W., Heinsohn G. E. & Morrissey J. (1982) Analysis of stomach contents of dugongs from Queensland. *Aust. Wildl. Res.* **9**, 55–67. 7.3.1

Martin D. R. (1969) Lecithodendriid trematodes from the bat, *Peropteryx kappleri* in Columbia, including discussion of allometric growth and significance of ecological isolation. *Proc. Helm. Soc. Wash.* **36**, 250–260. 9.2.2

Mather K. & Cooke P. (1962) Differences in competitive ability between genotypes of *Drosophila. Heredity* **17**, 381–407. 14.5

May R. M. (1971) Stability in a multispecies community. *Maths. Biosci.* **12**, 59–79. 4.2.3

May R. M. (1972) Will a large complex system be stable? *Nature* **238**, 413–414. 4.2.3, 9.3.2

May R. M. (1973) *Stability and Complexity in Model Ecosystems*. Princeton University Press, Princeton. 4.1, 4.2.2, 5.1, 6.5.1, 6.5.2

May R. M. (1974) On the theory of niche overlap. *Theor. Popul. Biol.* **5**, 297–332. 6.5.2

May R. M. (1975) Patterns of species abundance and diversity. In *Ecology and Evolution of Communities* (eds Cody M. L. & Diamond J. M.) Belknap Press, Cambridge, Mass. 4.3.3

May R. M. (1978) The dynamics and diversity of insect faunas. In *Diversity of Insect Faunas* (eds Mound L. A. & Waloff N.). *Symp. Roy. Ent. Soc.* Vol. **9**, pp. 188–204. Blackwell Scientific Publications, Oxford. 8.5

May R. M. (1981) Patterns in multi-species communities. In *Theoretical Ecology* (ed. May R. M.) (2nd edn) pp. 197–227. Blackwell Scientific Publications, Oxford. 4.2.2, 8.2, 8.3.1, 8.3.4

May R. M. (1983) Parasitic infections as regulators of animal populations. *Amer. Sci.* **71**, 36–45. 9.3.1

May R. M. & Anderson R. M. (1978) Regulation and stability of host-parasite population interactions. II. Destabilizing processes. *J. Anim. Ecol.* **47**, 249–267. 9.3.1

May R. M. & Anderson R. M. (1979) Population biology of infectious diseases: II. *Nature* **280**, 455–461. 9.3.1

May R. M. & MacArthur R. H. (1972) Niche overlap as a function of environmental variability. *Proc. Natl. Acad. Sci.* **69**, 1109–1113. 6.5.2

Maybury-Lewis D. (1967) *Akwe-Shavante Society*. Clarendon Press, Oxford. 15.3.3

Maynard-Smith J. (1974) *Models in Ecology*. Cambridge University Press, Cambridge. 4.1, 4.2.2, 4.2.3

Maynard-Smith J. (1976) A comment on the Red Queen. *Amer. Natur.* **110**, 325–330. 14.8

Mayr E. (1963) *Animal Species and Evolution*. Harvard University Press, Cambridge, Mass 8.2.4, 14.2

McArthur M. (1977) Nutritional research in Melanesia: a second look at the Tsembaga. In *Subsistence and Survival: Rural Ecology in the Pacific* (eds Bayliss-Smith T. & Feacham R. G.) pp. 91–128. Academic Press, London. 15.3.4

McBrien H., Harmsen R. & Crowder A. (1983) A case of insect grazing affecting plant succession. *Ecology* **64**, 1035–1039. 8.4.3

McClean G. E. & DeFries J. C. (1973) *Introduction to Behavioral Genetics*. Freeman, San Francisco. 14.9

McClure M. S. (1980) Competition between exotic species: scale insects on hemlock. *Ecology* **61**, 1391–1401. 11.2.3

McClure M. S. & Price P. W. (1975) Competition among sympatric *Erythroneura* leafhoppers (Homoptera: Cicadellidae) on American sycamore. *Ecology* **56**, 1388–1397. 5.3.1, 6.8

McCoy E. D. & Rey J. R. (1983) The biogeography of herbivorous arthropods: species accrual on tropical crops. *Ecol. Entomol.* **8**, 305–313. 8.2.3

McLay C. L. (1974) The distribution of duckweed *Lemna perpusilla* in a small southern California lake: an experimental approach. *Ecology* **55**, 262–276. 11.2.3

McNab B. K. (1971a) The structure of tropical bat faunas. *Ecology* **52**, 352–358. 5.5.1

McNab B. K. (1971b) On the ecological significance of Bergmann's Rule. *Ecology* **52**, 845–854. 6.6.1

McNaught M. L., Owen E. C., Henry K. M. & Kon S. K. (1954) The utilization of non-protein nitrogen in the bovine rumen. *Biochem. J.* **56**, 151–156. 7.2.3

McNaughton S. J. (1976) Serengeti migratory wildebeast: facilitation of energy flow by grazing. *Science* **191**, 92–94. 8.4.3

McNaughton S. J. (1977) Diversity and stability of ecological communities: a comment on the role of empiricism in ecology. *Amer. Natur.* **111**, 515–525. 4.2.3, 12.2

McNaughton S. J. (1978) Stability and diversity of ecological communities. *Nature* **274**, 252–253. 5.2

McNaughton S. J. (1979) Grassland–herbivore dynamics. In *Serengeti — Dynamics of an Ecosystem* (eds Sinclair A. R. E. & Norton-Griffiths M.) pp. 46–81. University of Chicago Press, Chicago. 8.4.3

Medawar P. B. (1969) *Induction and Intuition in Scientific Thought*. Methuen, London. 11.7

Medway W., Prier J. E. & Wilkinson J. S. (eds) (1969) *Textbook of Veterinary Clinical Pathology*. Williams & Wilkins, Baltimore. 13.3.2

Meggers B. (1971) *Amazonia: Man and Culture in a Counterfeit Paradise*. Aldin-Atherton, Chicago. 15.3.3

Menge B. A. (1972) Competition for food between two intertidal starfish species and its effect on body size and feeding. *Ecology* **53**, 635–644. 6.7.1, 11.3.2

Menge B. A. & Sutherland J. P. (1976) Species diversity gradients: synthesis of the roles of predation, competition and temporal heterogeneity. *Amer. Natur.* **110**, 351–369. 6.8

Menge J. L. & Menge B. A. (1974) Role of resource allocation, aggression and spatial heterogeneity in coexistence of two competing intertidal starfish. *Ecol. Monogr.* **44**, 189–209. 6.7.1

Mettrick D. F. & Dunkley L. C. (1969) Variation in the size and position of *Hymenolepis diminuta* (Cestoda: Cyclophyllidea) within the rat intestine. *Can. J. Zool.* **47**, 1091–1101. 9.2.1

Middleton J. (1984) Are plant toxins aimed at decomposers? *Experientia* **40**, 299–301. 7.2.2

Miles J. (1972) Experimental establishment of seedlings on a southern English heath. *J. Ecol.* **60**, 225–234. 11.4.1

Miles J. (1974) Effects of experimental interference with the stand structure on establishment of seedlings in Callunetum. *J. Ecol.* **62**, 675–687. 11.7
Miller R. I. & Harris L. D. (1977) Isolation and extirpation in wildlife reserves. *Biological Conservation* **12**, 311–315. 8.4.2
Minot E. O. (1981) Effects of interspecific competition for food in breeding blue and great tits. *J. Anim. Ecol.* **50**, 375–385. 11.2.2, 11.3.1
Mitchell C. P. & Parkinson D. (1976) Fungal feeding of oribatid mites (Acari: Cryptostigmata) in an aspen woodland soil. *Ecology* **57**, 302–312. 7.2.2
Moermond T. C. (1979) Habitat constraints on the behavior, morphology, and community structure of *Anolis* lizards. *Ecology* **60**, 152–164. 5.2
Moir R. J. (1968) Bile, digestion, ruminal physiology. In *Handbook of Physiology, Section 6: Alimentary Canal* (ed. Code C. F.) pp. 2673–2694. Am. Physiol. Soc., Washington, D.C. 7.2.3
Montgomery W. I. (1981) A removal experiment with sympatric populations of *Apodemus sylvaticus* (L.) and *A. flavicollis* (Melchoir) (Rodentia: Muridae). *Oecologia* **51**, 123–132. 11.3.1
Moore J. A. (1949) Geographic variation of adaptive characters in *Rana pipiens*. *Genetics* **31**, 304–326. 14.3
Moore J. A. (1952) Competition between *Drosophila melanogaster* and *Drosophila simulans*. II. The improvement of competitive ability through selection. *Proc. Natl. Acad. Sci.* **38**, 381–407. 14.5
Moqbel R. & Denham D. A. (1977) *Strongyloides ratti*. I. Parasitological observations on primary and secondary infections in the small intestine of rats. *J. Helminthol.* **51**, 301–308. 13.3.3
Moqbel R. & Wakelin D. (1979) *Trichinella spiralis* and *Strongyloides ratti*: immune interaction in adult rats. *Exp. Parasitol.* **47**, 65–72. 13.3.3
Moran V. C. (1980) Interactions between phytophagous insects and their *Opuntia* hosts. *Ecol. Entomol.* **5**, 153–164. 8.2.4
Moran V. C. & Southwood T. R. E. (1982) The guild composition of arthropod communities in trees. *J. Anim. Ecol.* **51**, 289–306. 8.2.4
Moreau R. E. (1983) The food of the red-billed oxpecker, *Buphagus erythrochynchus* (Stanley). *Bull. Entomol. Res.* **24**, 325–335. 9.3.2
Moriarty D. J. W. (1976) Quantitative studies on bacteria and algae in the food of the mullet *Mugil cephalus* L. and the prawn *Metapenaeus bennettae* (Racek & Dall). *J. exp. mar. Biol. Ecol.* **22**, 131–143. 7.3.3
Morisita M. (1959) Measuring of interspecific association and similarity between communities. *Mem. Fac. Sci. Kyushu Univ., Ser. E. (Biol.)* **3**, 65–80. 4.4.1
Morisita M. (1961) Figure 3.58(b), sample size and the ranges of application of various mathematical series. In *Animal Ecology* (eds Miyadi D., Kato M., Mori S., Morisita M., Shibuya T. & Kitazawa Y.) p. 260. Asakura, Tokyo (in Japanese). 4.3.2
Morisita M. (1971) Composition of Iδ-index. *Res. Popul. Ecol.* **13**, 1–27. 4.4.1
Morren G. E. B. (1974) Settlement strategies and hunting in a New Guinea society. PhD thesis, Department of Anthropology, Columbia University, New York. 15.3.2
Morren G. E. B. (1977) From hunting to herding: pigs and the control of energy in montane New Guinea. In *Subsistence and Survival: Rural Ecology in the Pacific* (eds Bayliss-Smith T. & Feacham R.) pp. 273–315. Academic Press, New York. 15.3.2, 15.3.5
Morris J. R. (1970) An ecological study of the Basommatophoran snail *Helisoma trivolvis* in central Alberta. PhD thesis, University of Alberta, Edmonton, Alberta. 9.3.2
Morris R. D. & Grant P. R. (1972) Experimental studies of competitive interaction in a two-species system. IV. *Microtus* and *Clethrionomys* species in a single enclosure. *J. Anim. Ecol.* **41**, 275–290. 11.2.3, 11.3.1
Morse D. H. (1981) Interactions among syrphid flies and bumblebees on flowers. *Ecology* **62**, 81–88. 11.7

Motomura I. (1932) On the statistical treatment of communities. *Jap. J. Zool.* **44**, 379–398 (in Japanese). 4.3.2

Mould D. L. & Thomas G. J. (1958) The enzyme degradation of starch by holotrich protozoa from sheep rumen. *Biochem. J.* **69**, 327–337. 7.2.3

Mueller-Dombois D. & Ellenberg H. (1974) *Aims and Methods of Vegetation Ecology.* Wiley, New York. 1.1

Mungomery R. W. (1936) A survey of the feeding habits of the giant toad (*Bufo marinus* L.) and notes on its progress since its introduction into Queensland. *Proc. Qld. Soc. Sugar Cane Tech.* **1936**, 63–74. 10.3.2

Murdoch W. W. (1969) Switching in general predators: experiments on predator specificity and stability of prey populations. *Ecol. Monogr.* **39**, 335–354. 10.1

Murdoch W. W., Avery S. & Smyth M. E. B. (1975) Switching in predatory fish. *Ecology* **56**, 1094–1105. 10.1

Murdoch W. W., Evans F. C. & Peterson C. H. (1972) Diversity and pattern in plants and insects. *Ecology* **53**, 819–829. 8.2.2

Murdoch W. W. & Oaten A. (1975) Predation and population stability. *Adv. Ecol. Res.* **9**, 2–131. 10.1, 14.9

Murray R. M., Marsh H., Heinsohn G. E. & Spain A. V. (1977) The role of the midgut caecum and large intestine in the digestion of sea grasses by the dugong (Mammalia: Sirenia). *Comp. Biochem. Physiol.* **56A**, 7–10. 7.3.1

Nalewajko C. & Marin L. (1969) Extracellular production in relation to growth of four planktonic algae and of phytoplankton populations from Lake Ontario. *Can. J. Bot.* **47**, 405–413. 7.3.3

Nayak D. P. & Kelly G. W. (1969) Synergistic effect of *Ascaris* migration and influenza infection in mice. *J. Parasitol.* **51**, 297–298. 13.2.1

Nelson R. R. (1973) Pathogen variation and host resistance. In *Breeding Plants for Disease Resistance* (ed. Nelson R. R.) pp. 40–48. Pennsylvania State University Press, Pennsylvania. 14.8

Neraasen T. A. & Holmes J. C. (1975) The circulation of cestodes among three species of geese nesting on the Anderson River Delta, Canada. *Acta Parasit. Polonica* **23**, 277–289. 9.2.1

Neuhauser E. & Hartenstein R. (1978) Phenolic content and palatability of leaves and wood to soil isopods and diplopods. *Pedobiologia* **18**, 99–109. 7.2.2

New T. R. (1979) Phenology and relative abundance of Coleoptera on some Australian wattles. *Aust. J. Zool.* **27**, 9–16. 10.2.3

New T. R. (1983) Colonisation of seedling acacias by arthropods in southern Victoria. *Aust. ent. Mag.* **10**, 13–18. 10.2.3

Newell R. (1965) The role of detritus in the nutrition of two marine deposit feeders, the prosobranch *Hydrobia ulvae* and the bivalve *Macoma balthica*. *Proc. Zool. Soc. London* **144**, 15–45. 7.3.2

Newman E. I. (1982) Niche separation and species diversity in terrestrial vegetation. In *The Plant Community as a Working Mechanism* (ed. Newman E. I.) pp. 61–77. Blackwell Scientific Publications, Oxford. 1.8

Newman W. A. & Stanley S. M. (1981) Competition wins out overall: reply to Paine. *Paleobiology* **7**, 561–569. 6.8

Nicholson A. J. (1933) The balance of animal populations. *J. Anim. Ecol.* **2**, (Suppl.) 132–178. 1.7

Nicholson P. B., Bocock K. L. & Heal O. W. (1966) Studies on the decomposition of the faecal pellets of a millipede (*Glomeris marginata* (Villers)). *J. Ecol.* **54**, 755–767. 7.2.2

Niering W. A. & Goodwin B. H. (1974) Creation of relatively stable shrublands with herbicides: arresting "succession" on rights-of-way and pastureland. *Ecology* **55**, 784–795. 12.2

Nietschmann B. (1973) *Between Land and Water: The Subsistence Ecology of the Miskito Indians, Eastern Nicaragua*. Seminar Press, New York. 15.3.3

Niven B. S. & Stewart M. G. (1982) The precise environment of some well-known animals. VII. The cane toad (*Bufo marinus*) AES Working Paper 3/82, Griffith University, Brisbane. 10.3.2

Noble I. R. & Slatyer R. O. (1978) The effect of disturbances on plant succession. *Proc. Ecol. Soc. Aust.* **10**, 135–145. 12.2

Noble I. R. & Slatyer R. O. (1980) The use of vital attributes to predict successional changes in plant communities subject to recurrent disturbances. *Vegetatio* **43**, 5–21. 12.2

Nolan J. V. (1975) Quantitative models of nitrogen metabolism in sheep. In *Digestion and Metabolism in the Ruminant* (*Proc. IV Intern. Symp. Ruminant Physiol.*) (eds McDonald I. W. & Warner A. C. I.) pp. 416–431. University of New England Publ. Unit, Armidale, NSW. 7.2.3

Noon B. R. (1981) The distribution of an avian guild along a temperate elevational gradient: the importance and expression of competition. *Ecol. Monogr.* **51**, 105–124. 5.3.1

Norgan N. G., Ferro-Luzzi A. & Durnin J. V. G. A. (1974) The energy and nutrient intake and nutrient expenditure of 204 New Guinean adults. *Phil. Trans. roy. Soc., Ser. B* **268**, 309–348. 15.3.4

Norton-Griffiths M. (1968) The feeding behaviour of the oystercatcher (*Haematopus ostralegus*). D. Phil. thesis, University of Oxford [cited in Curio (1976)]. 6.6.3

Noy-Meir I. & Anderson D. J. (1970) Multiple pattern analysis, or multiscale ordination: pathway to a vegetation hologram? *Proc. Int. Symp. Stat. Ecol.* III., pp. 207–225. 12.3.2

Nye P. H. & Tinker P. B. (1977) *Solute Movement in the Soil-Root System*. Blackwell Scientific Publications, Oxford. 2.2

O'Donald P. & Pilecki C. (1970) Polymorphic mimicry and natural selection. *Evolution* **24**, 395–401. 14.9

O'Dowd D. (1979) Foliar nectar production and ant activity on a neotropical tree, *Ochroma pyramidale*. *Oecologia* **43**, 233–248. 10.2.3

Odum E. P. (1969) The strategy of ecosystem development. *Science* **164**, 262–270. 1.6.1, 12.2

Odum E. P. (1971) *Fundamentals of Ecology* (3rd edn). W. B. Saunders, Philadelphia. 1.6.1

Odum H. T. (1971) *Environment, Power and Society*. Wiley–Interscience, New York. 1.6

Ogden J. C. & Ebersole J. P. (1981) Scale and community structure of coral reef fishes: a long-term study of a large artificial reef. *Mar. Ecol. Prog., Ser.* **4**, 97–103. 2.5

Ogden J. C. & Lobel P. S. (1978) The role of herbivorous fishes and urchins in coral reef communities. *Env. Biol. Fish.* **3**, 49–63. 7.3.1

Ogilvie B. M. & Jones V. E. (1971) *Nippostrongylus brasiliensis*: a review of immunity and the host/parasite relationship in the rat. *Exp. Parasitol.* **29**, 138–177. 13.3.3

Ohtsuka R. (1977) The sago eaters. In *Sunda and Sahul: Prehistoric Studies in southeast Asia, Melanesia and Australia* (eds Allen J., Golson J. & Jones R.) pp. 465–492. Academic Press, London. 15.3.3

Okhotina M. V. & Nadtochy E. V. (1970) Effect of *Mammanidula asperocutis* Sadovskaja in Skrjabin, Sihobalova et Sulc, 1954 (Nematoda), on the population size of shrews of the genus *Sorex*. *Acta Parasit. Polonica* **18**, 81–84. 9.3

O'Neill J. P. (1974) The birds of Balta, a Peruvian dry tropical forest locality, with an analysis of their origins and ecological relationships. PhD thesis, Louisiana State University, Baton Rouge, La. 5.4

O'Neill J. P. & Pearson D. L. (1974) Estudio priliminar de las aves de Yarinacocha, Departmento de Loreto, Peru. *Publ. Mus. Nat. Javier Prado, Ser. A Zool.* **25**, 1–13. 5.4

Oomen H. A. P. C. (1961) The nutrition situation in western New Guinea. *Trop. geogr. Med.* **13**, 321–335. 15.3.4

Oomen H. A. P. C. (1970) Interrelationship of the human intestinal flora and protein utilization. *Proc. Nutr. Soc.* **29**, 197–206. 15.3.4

Orians G. H. (1969) The number of bird species in some tropical forests. *Ecology* **50**, 783–801. 5.5.3, 6.8

Osman R. W. (1977) The establishment and development of a marine epifaunal community. *Ecol. Monogr.* **47**, 37–63. 9.3.4

Osmond C. B., Björkman O. & Anderson D. J. (1981) *Physiological Processes in Plant Ecology*. Springer-Verlag, Berlin. 12.3.3

Ostfeld R. S. (1982) Foraging strategies and prey switching in the California sea otter. *Oecologia* **53**, 170–178. 10.2.2

Otsuki A. & Wetzel R. G. (1974) Release of dissolved organic matter by autolysis of a submerged macrophyte: *Scirpus subterminalis*. *Limnol. Oceanogr.* **19**, 842–845. 7.3.3

Ould P. & Welch H. E. (1980) The effect of stress on the parasitism of mallard ducklings by *Echinuria uncinata* (Nematoda: Spirurida). *Can. J. Zool.* **58**, 228–234. 9.3

Oxford A. E. (1951) The conversion of certain soluble sugars to a glucosan by holotrich ciliates in the rumen of sheep. *J. gen. Microbiol.* **5**, 83–90. 7.2.3

Pacala S. & Roughgarden J. (1982a) The evolution of resource partitioning in a multidimensional resource space. *Theor. Popul. Biol.* **22**, 127–145. 6.5.3

Pacala S. & Roughgarden J. (1982b) Resource partitioning and interspecific competition in two two-species insular *Anolis* lizard communities. *Science* **217**, 444–446. 14.6

Pacala S. W. & Roughgarden J. (1985) Population experiments with the *Anolis* lizards of St Maaten and St Eustatius. *Ecology,* **66**, 129–141. 6.7.1

Padoch C. & Vayda A. P. (1983) Patterns of resource use and human settlement in tropical forests. In *Ecosystems of the World, Vol.* **14***A*. *Tropical Rain Forest Ecosystems: Structure and Function* (ed. Golley F. B.) pp. 301–313. Elsevier, Amsterdam. 15.1

Paffenhöffer G. A. & Strickland J. D. H. (1970) A note on the feeding of *Calanus heligolandicus* on detritus. *Mar. Biol.* **5**, 97–99. 7.3.2

Paine R. T. (1966) Food web complexity and species diversity. *Amer. Natur.* **100**, 65–75. 10.1

Paine R. T. (1969) A note on trophic complexity and community stability. *Amer. Natur.* **103**, 91–93. 10.1, 10.2.2

Paine R. T. (1980) Food webs: interaction strength, linkage and community infrastructure. *J. Anim. Ecol.* **49**, 667–685. 2.2, 4.1, 11.3.1

Paine R. T. (1981) Barnacle ecology: is competition important? The forgotten roles of disturbance and predation. *Paleobiology* **7**, 553–560. 6.8

Paine R. T. & Levin S. A. (1981) Intertidal landscapes: disturbance and the dynamics of pattern. *Ecol. Monogr.* **51**, 145–178. 6.8

Pajunen V. I. (1982) Replacement analysis of non-equilibrium competition between rock pool corixids (Hemiptera, Corixidae). *Oecologia* **52**, 153–155. 11.4.2

Papavizas G. C. (ed.) (1981) *Biological Control in Crop Production*. Allanheld & Osmun, New York. 8.3.3

Park T. (1948) Experimental studies of interspecies competition. I. Competition between populations of the flour beetles, *Tribolium confusum* Duval and *Tribolium castaneum* Herbst. *Ecol Monogr.* **18**, 265–307. 1.7, 9.3.2

Parker T. A., III. (1980) *Birds of the Tambopata Reserve, Peru*. Privately printed. 5.4

Parrish J. A. D. & Bazzaz F. A. (1976) Underground niche separation in successional plants. *Ecology* **57**, 1281–1288. 6.4

Paterson S. S. (1956) *The Forest Area of the World and its Potential Productivity*. Special publication, Department of Geography, Royal University of Goteborg, Sweden. 12.3.2

Pavlovsky E. N. (1966). *The Natural Nidality of Transmissible Diseases*. University of Illinois Press, Urbana. 9.3, 9.3.4

Pearre S., Jr (1979) Niche modification in Chaetognatha infected with larval trematodes (Digenea). *Int. Rev. ges. Hydrobiol.* **64**, 193–206. 9.3.3

Pearsall W. H. (1964) The development of ecology in Britain. *J. Ecol.* **52** *(Suppl.)* 1–12. 1.1, 1.6

Pearse J. S. & Hines A. H. (1979) Expansion of a central California kelp forest following the mass mortality of sea urchins. *Mar. Biol.* **51**, 83–91. 10.2.2

Pearson D. L. (1975) The relation of foliage complexity to ecological diversity of three Amazonian bird communities. *Condor* **77**, 453–466. 5.3.2

Pearson D. L. (1977) A pantropical comparison of bird community structure on six lowland forest sites. *Condor* **79**, 232–244. 5.3.2, 5.5.1

Pearson D. L. (1982) Historical factors and bird species richness. In *Biological Diversification in the Tropics* (ed. Prance G. T.) pp. 441–452. Columbia University Press, New York. 4.3.4

Pearson D. L. & Mury E. J. (1979) Character divergence and convergence among tiger beetles (Coleoptera: Cicindelidae). *Ecology* **60**, 557–566. 6.5.1

Pearson D. L., Tallman D. & Tallman E. (1977) *The Birds of Limoncocha, Napo Province, Ecuador*. Instituto Linginstico de Verano, Suito, Ecuador. 5.4

Peckarsky B. L. & Dodson S. I. (1980) An experimental analysis of biological factors contributing to stream community structure. *Ecology* **61**, 1283–1290. 11.2.3

Peet R. K. (1975) Relative diversity indices. *Ecology* **56**, 496–498. 4.3.3

Penhale P. A. & Smith W. O., Jr (1977) Excretion of dissolved organic carbon by eelgrass (*Zostera marina*) and its epiphytes. *Limnol. Oceanogr.* **22**, 400–407. 7.3.3

Perrin W. F. & Powers J. E. (1980) Role of a nematode in natural mortality of spotted dolphins. *J. Wildl. Manag.* **44**, 960–963. 9.3

Person C., Groth J. V. & Mylyk O. M. (1976) Genetic change in host-parasite populations. *Ann. Rev. Phytopath.* **14**, 177–188. 14.8

Peters R. H. (1983) *The Ecological Implications of Body Size*. Cambridge University Press, Cambridge. 8.4.1

Peters R. H. & Wassenberg K. (1983) The effect of body size on animal abundance. *Oecologia* **60**, 89–96. 8.4.1

Peterson B. J., Hobbie J. E. & Haney J. F. (1978) *Daphnia* grazing on natural bacteria. *Limnol. Oceanogr.* **23**, 1039–1044. 7.3.3

Peterson C. H. (1979) The importance of predation and competition in organizing the intertidal epifaunal communities of Barnegate Inlet, New Jersey. *Oecologia* **39**, 1–24. 11.3.1

Peterson C. H. & Andre S. V. (1980) An experimental analysis of interspecific competition among marine filter feeders in a soft-sediment environment. *Ecology* **61**, 129–139. 11.4.1, 11.6.2, 11.7

Petranka J. W. & McPherson J. K. (1979) The role of *Rhus copallina* in the dynamics of the forest-prairie ecotone in north-central Oklahoma. *Ecology* **60**, 956–965. 11.7

Petras M. L. (1967a) Studies on natural populations of *Mus*. I. Biochemical polymorphisms and their bearing on breeding structure. *Evolution* **21**, 259–260. 13.3.3

Petras M. L. (1967b) Studies on natural populations of *Mus*. II. Polymorphism at the T locus. *Evolution* **21**, 466–478. 13.3.3

Petrusewicz K. & Macfadyen A. (1970) *Productivity of Terrestrial Animals: Principles and Methods*. IBP Handbook No. 13. Blackwell Scientific Publications, Oxford. 7.2.1

Petrushevski G. K. & Shulman S. S. (1961) The parasitic diseases of fishes in the natural waters of the USSR. In *Parasitology of Fishes* (eds Dogiel V. A., Petrushevski G. K. &

Polyanski Yu. I.) (English edn) pp. 299–319. Oliver & Boyd, Edinburgh. 9.3
Pfeiffer W. J. & Wiegert R. G. (1981) Grazers on *Spartina* and their predators. In *The Ecology of a Salt Marsh* (eds Pomeroy L. R. & Wiegert R. G.) pp. 87–112. Springer-Verlag, New York. 7.2.1
Pianka E. R. (1966a) Latitudinal gradients in species diversity: a review of concepts. *Amer. Natur.* **100**, 33–46. 4.3.4, 5.5.3
Pianka E. R. (1966b) Convexity, desert lizards and spatial heterogeneity. *Ecology* **47**, 1055–1059. 5.5.2
Pianka E. R. (1967) On lizard species diversity: North American flatland deserts. *Ecology* **48**, 333–351. 5.5.2
Pianka E. R. (1969a) Habitat specificity, speciation and species density in Australian desert lizards. *Ecology* **50**, 498–502. 5.5.2
Pianka E. R. (1969b) Sympatry of desert lizards (*Ctenotus*) in Western Australia. *Ecology* **50**, 1012–1030. 5.5.2, 6.4, 6.5.1
Pianka E. R. (1970) On *r*- and *K*- selection. *Amer. Natur.* **104**, 592–597. 8.4.3
Pianka E. R. (1971) Lizard species density in the Kalahari desert. *Ecology* **52**, 1024–1029. 5.5.2
Pianka E. R. (1973) The structure of lizard communities. *Ann. Rev. Ecol. Syst.* **4**, 53–74. 5.5.2, 6.4
Pianka E. R. (1974) Niche overlap and difference competition. *Proc. Natl. Acad. Sci., USA* **71**, 2141–2145. 6.5.2
Pianka E. R. (1980) Guild structure in desert lizards. *Oikos* **35**, 194–201. 4.3.4
Pianka E. R. (1981) Competition and niche theory. In *Theoretical Ecology* (ed. May R. M.) pp. 167–196. Blackwell Scientific Publications, Oxford. 6.2, 6.5.2
Pielou E. C. (1961) Segregation and symmetry in two-species populations as studied by nearest neighbour relations. *J. Ecol.* **49**, 255–269. 12.4.1
Pielou E. C. (1975) *Ecological Diversity*. Wiley–Interscience, London. 10.2.1
Pielou E. C. (1977) *Mathematical Ecology*. Wiley–Interscience, New York. 4.4.2
Pierce N. E. (1983) The ecology and evolution of symbioses between lycaenid butterfly larvae and ants. PhD thesis, Harvard University. 10.2.3
Pierce N. E. & Mead P. S. (1981) Parasitoids as selective agents in the symbiosis between lycaenid butterfly larvae and ants. *Science* **211**, 1185–1187. 9.3.2
Pigott C. D. (1970) Soil formation and development on the Carboniferous Limestone of Derbyshire. II. The relation of soil development to vegetation on the plateau near Coombs Dale. *J. Ecol.* **58**, 529–541. 2.2
Pimentel D., Feinberg E. H., Wood D. W. & Hayes J. T. (1965) Selection, spatial distribution and the co-existence of competing fly species. *Amer. Natur.* **99**, 97–108. 14.5, 14.6
Pimlott D. H. (1975) The ecology of the wolf in North America. In *The Wild Canids — Their Systematics, Behavioural Ecology and Evolution* (ed. Fox M. W.) pp. 280–285. Van Nostrand Reinhold, New York. 8.4.3
Pimm S. L. (1982) *Food Webs*. Chapman & Hall, London. 4.2, 4.2.2, 4.2.3, 10.4
Pimm S. L. (1984) The complexity and stability of ecosystems. *Nature* **307**, 321–326. 4.2.2
Pimm S. L. & Lawton J. H. (1980) Are food webs divided into compartments? *J. Anim. Ecol.* **49**, 879–898. 4.2.3
Pinder J. E. (1975) Effects of species removal on an old-field plant community. *Ecology* **56**, 747–751. 11.7
Pippet J. (1975) The marine toad, *Bufo marinus*, in Papua New Guinea. *P. N. G. Agric. Journ.* **26**, 23–30. 10.3.2
Pirozynski K. A. (1976) Fossil fungi. *Ann. Rev. Phytopath.* **14**, 237–308. 14.8
Platt W. J. & Weis I. M. (1977) Resource partitioning and competition within a guild of fugitive prairie plants. *Amer. Natur.* **111**, 479–513. 5.2
Pleasants J. M. (1980) Competition for bumblebee pollinators in Rocky Mountain plant

communities. *Ecology* **61**, 1446–1460. 6.4

Plowman K. P. (1979) Litter and soil fauna of two Australian subtropical forests. *Aust. J. Ecol.* **4**, 87–104. 7.2.2

Plowman K. P. (1981) Distribution of Cryptostigmata and Mesostigmata (Acari) within the litter and soil layers of two subtropical forests. *Aust. J. Ecol.* **6**, 365–374. 7.2.2

Pomeroy L. R. & Deibel D. (1980) Aggregation of organic matter by pelagic tunicates. *Limnol. Oceanogr.* **25**, 643–652. 7.3.2

Pomeroy L. R. & Johannes R. E. (1968) Occurrence and respiration of ultraplankton in the upper 500 meters of the ocean. *Deep-Sea Res.* **15**, 381–391. 7.3.2

Pontin A. J. (1960) Population stabilization and competition between the ants *Lasius flavus* (F.) and *L. niger* (L.). *J. Anim. Ecol.* **30**, 47–54. 11.3.1, 11.3.2

Pontin A. J. (1969) Experimental transplantation of nest-mounds of the ant *Lasius flavus* (F.) in habitat containing also *L. niger* (L.) and *Myrmica scabrinodis* Nyl. *J. Anim. Ecol.* **38**, 747–754. 11.3.2, 11.5.1, 11.5.2

Poole R. W. (1974). *An Introduction to Quantitative Ecology*. McGraw-Hill, New York. 4.4.2

Poore M. E. D. (1964) Integration in the plant community. *J. Ecol.* **52** (*Suppl.*), 213–226. 1.1

Porter J. W. (1976) Autotrophy, heterotrophy, and resource partitioning in Caribbean reef-building corals. *Amer. Natur.* **110**, 731–742. 6.4

Pound R. & Clements F. E. (1898) *The Phytogeography of Nebraska*. University of Nebraska Publ. Lincoln. 1.5

Pounden W. D. & Hibbs J. W. (1950) The development of calves raised without protozoa and certain other characteristic rumen microorganisms. *J. Dairy Sci.* **33**, 639–644. 7.2.3

Preston F. W. (1948) The commonness and rarity of species. *Ecology* **29**, 254–283. 1.8, 4.3.2

Preston F. W. (1962) The canonical distribution of commonness and rarity. *Ecology* **43**, 185–215; 410–432. 4.3.1, 4.3.2, 8.2.1

Price M. V. (1978) The role of microhabitat in structuring desert rodent communities. *Ecology* **59**, 910–921. 11.3.2

Price P. W. (1980) *Evolutionary Biology of Parasites*. Princeton University Press, Princeton. 9.1, 9.2.1, 9.2.2, 9.2.3, 9.2.5

Price P. W. (1984a) Communities of specialists: vacant niches in ecological and evolutionary time. In *Ecological Communities: Conceptual Issues and the Evidence* (eds Strong D., Simberloff D., Abele L. G. & Thistle A. B.) pp. 510–523. Princeton University Press, Princeton. 9.2.1

Price P. W. (1984b) Alternative paradigms in community ecology. In *A New Ecology: Novel Approaches to Interactive Systems* (eds Price P. W., Slobodchikoff C. N. & Gaud W. S.) pp. 449–454. Wiley, New York. 9.2.2, 9.4

Price P. W. & Clancy K. M. (1983) Patterns in number of helminth parasite species in freshwater fishes. *J. Parasitol.* **69**, 449–454. 9.2.3

Price P. W., Slobodchikoff C. N. & Gaud W. S. (eds) (1984) *A New Ecology: Novel Approaches to Interactive Systems*. Wiley, New York. 9.2

Pulliam H. R. (1975) Coexistence of sparrows: a test of community theory. *Science* **189**, 474–476. 5.1

Putwain P. D. & Harper J. L. (1970) Studies in the dynamics of plant populations. III. The influences of associated species on populations of *Rumex acetosa* L. and *R. acetosella* L. in grassland. *J. Ecol.* **58**, 251–264. 11.4.1, 11.7

Putwain P. D. & Harper J. L. (1972) Studies in the dynamics of plant populations. V. Mechanisms governing sex ratio in *Rumex acetosa* and *R. acetosella*. *J. Ecol.* **60**, 113–129. 14.7

Pyke G. H., Pulliam H. R. & Charnov E. L. (1977) Optimal foraging: a selective review of theory and tests. *Quart. Rev. Biol.* **52**, 137–154. 6.5.5, 15.2

Quinn J. F. & Dunham A. E. (1983) On hypothesis testing in ecology and evolution. *Amer. Natur.* **122**, 602–617. 6.3, 6.8, 14.1

Rabenold K. N. (1978) Foraging strategies, diversity, and seasonality in bird communities of Appalachian spruce-fir forests. *Ecol. Monogr.* **48**, 397–424. 6.6.3

Rachootin S. P. & Thomson K. S. (1981) Epigenetics, paleontology, and evolution. In *Evolution Today* (eds Scudder G. G. E. & Reveal J. L.) pp. 181–193, Proceedings of the Second International Congress of Systematic and Evolutionary Biology. Hunt Institute for Botanical Documentation, Carnegie-Mellon University, Pittsburgh. 15.4

Radcliffe-Brown A. R. (1964) *The Andaman Islanders.* The Free Press, New York. 15.3.1

Ramensky L. G. (1926) Die Grundgesetzmassigkeiten in Aufbau der Vegetationsdecke. *Bot. Centralblatt. N. F.* **7**, 453–455. 9.2.2

Ramensky L. G. (1930) Zur Methodik der vergleichenden Bearbeitung und Ordnung von Pflanzenlisten und anderen Objekten, die durch mehrere, verschiedenartig wirkende Faktorem bestimmt werden. *Beitr. Biol. Pflanz.* **18**, 269–304. 12.2

Rand A. S. (1967) Predator-prey interactions and the evolution of aspect diversity. *Atas Symp. Biota amazonica* (*Zool.*) **5**, 73–83. 6.8

Rappaport R. A. (1968) *Pigs for the Ancestors*: *Ritual in the Ecology of a New Guinea People*. Yale University Press, New Haven. 15.3.4

Rathcke B. J. (1976a) Insect–plant patterns and relationships in the stem-boring guild. *Amer. Midl. Natur.* **96**, 98–117. 5.2, 5.3.1

Rathcke B. J. (1976b) Competition and coexistence within a guild of herbivorous insects. *Ecology* **57**, 76–87. 8.3.3

Rau M. E. (1983) Establishment and maintenance of behavioural dominance in male mice infected with *Trichinella spiralis*. *Parasitology* **86**, 319–322. 9.3

Raunkiaer C. (1934) *The Life Forms of Plants and Statistical Plant Geography*. Clarendon Press, Oxford. 1.3, 5.5

Rausher M. C. (1981) The effect of native vegetation on the susceptibility of *Aristolochia reticulata* (Aristolochiaceae) to herbivore attack. *Ecology* **62**, 1187–1195. 8.2.2

Raynal D. J. & Bazzaz F. A. (1975) Interference of winter annuals with *Ambrosia artemisifolia* in early successional fields. *Ecology* **56**, 35–49. 11.7

Read C. P. (1970) *Parasitism and Symbiology*. Ronald Press, New York. 13.3.2, 13.3.3

Recher H. F. (1974) Colonisation and extinction: the birds of Lord Howe Island. *Aust. nat. Hist.* **18**, 64–69. 10.3.1

Redfield J. A., Krebs C. J. & Taitt M. J. (1977) Competition between *Peromyscus maniculatus* and *Microtus townsendii* in grasslands of coastal British Columbia. *J. Anim. Ecol.* **46**, 607–616. 11.3.1, 11.7

Rees C. J. C. (1983) Microclimate and the flying hemiptera fauna of a primary lowland rain forest in Sulawesi. In *Tropical Rain Forest*: *Ecology and Management* (eds Sutton S. L., Whitmore T. C. & Chadwick A. C.) pp. 121–136. Blackwell Scientific Publications, Oxford. 8.3.3

Reese J. C. (1979) Interactions of allelochemicals with nutrients in herbivore food. In *Herbivores, their Interactions with Secondary Plant Metabolites* (eds Rosenthal G. A. & Janzen D. H.) pp. 309–330. Academic Press, New York. 7.2.1

Reichman O. J., Woodin S. A., Jackson J. B. C., Hayward T. L., McGowan J. A., Carroll C. R., Smith C. C., Balda R. P., Russert Kraemer L., Levins R., Carpenter F. L., Kodric-Brown A., Brown J. H., Davidson D. W., Wright S. J., Fleming T. H. (1979) Competition between distantly related taxa. *Amer. Zool.* **19**, 1027–1175. 5.2

Reighard J. (1908) An experimental field study of warning coloration in coral-reef fishes. *Publ. Carnegie. Instn.* **103**, 257–325. 14.9

Reiswig H. M. (1975) Bacteria as food for temperate-water marine sponges. *Can. J. Zool.* **53**, 582–589. 7.3.3

Rejmanek M. & Stary P. (1979) Connectance in real biotic communities and critical values for stability of model ecosystems. *Nature* **280**, 311–313. 4.2.2

Rey J. R., McCoy E. D. & Strong D. R. Jr (1981) Herbivore pests, habitat islands, and the species-area relationship. *Amer. Natur.* **117**, 611–622. 8.2.3, 8.2.4, 9.2.3

Reyes V. G. & Tiedje J. M. (1976a) Ecology of the gut microbiota of *Tracheoniscus rathkei* (Crustacea: Isopoda). *Pedobiologia* **16**, 67–74. 7.2.2

Reyes V. G. & Tiedje J. M. (1976b) Metabolism of C-labeled plant materials by woodlice (*Tracheoniscus rathkei* Blandt) and soil microorganisms. *Soil Biol. Biochem.* **8**, 103–108. 7.2.2

Reynoldson T. B. (1964) Evidence for intraspecific competition in field populations of triclads. *J. Anim. Ecol.* **33**, 187–207. 1.7

Reynoldson T. B. & Bellamy L. S. (1970) The establishment of interspecific competition in field populations, with an example of competition in action between *Polyceli nigra* (Mull.) and *P. teguis* (Ijima) (Turbellaria, Tricladida). In *Proceedings of the Advanced Study Institute on Dynamics of Numbers in Populations, Oosterbeck* (eds den Boer P. J. & Gradwell G. R.) pp. 282–297. Centre for Agricultural Publication and Documentation, Wageningen. 11.5.1, 11.5.2

Reynoldson T. B. & Davies R. W. (1970) Food niche and co-existence in lake-dwelling triclads. *J. Anim. Ecol.* **39**, 599–617. 6.4

Rhoades D. F. (1979) Evolution of plant chemical defense against herbivores. In *Herbivores: Their Interaction with Secondary Plant Metabolites* (eds Rosenthal G. A. & Janzen D. H.) pp. 4–55. Academic Press, New York. 13.2.1, 13.2.2

Rhoads J. W. (1981) Variation in land-use strategies among Melanesian sago eaters. *Canberra Anthropology* **4**(2), 45–73. 15.3.3

Rhoads J. W. (1982) Sago palm management in Melanesia: an alternative perspective. *Archaeol. Oceania* **17**, 20–27. 15.4

Rhodes I. (1968) Yield of contrasting ryegrass varieties in monoculture and mixed culture. *J. Br. Grassl. Soc.* **23**, 156–158. 14.6

Rice E. L. (1974) *Allelopathy*. Academic Press, New York. 2.2

Richards P. W. (1964) *The Tropical Rain Forest* (2nd edn). Cambridge University Press, Cambridge. 5.5, 5.5.1

Ricklefs R. E., Cochran D. & Pianka E. R. (1981) A morphological analysis of the structure of communities of lizards in desert habitats. *Ecology* **62**, 1474–1483. 5.2

Ricklefs R. E. & O'Rourke K. (1975) Aspect diversity in moths: a temperate-tropical comparison. *Evolution* **29**, 313–324. 6.8

Ricklefs R. E. & Travis J. (1980) A morphological approach to the study of avian community organization. *Auk* **97**, 321–338. 5.2, 5.5.1

Riedl R. (1978) *Order in Living Organisms: A Systems Analysis of Evolution*. John Wiley, New York. 15.4

Rigby C. & Lawton J. H. (1981) Species-area relationships of arthropods on host plants: herbivores on bracken. *J. Biogeogr.* **8**, 125–133. 8.2.1

Rigby D. W. & Chobota B. (1966) The effects of *Trypanosoma lewisi* on the development of *Hymenolepis diminuta* in concurrently infected white rats. *J. Parasitol.* **52**, 389–394. 13.3.3

Riley G. A. (1963) Organic aggregates in the sea water and the dynamics of their formation and utilization. *Limnol. Oceanogr.* **8**, 372–381. 7.3.2

Riley G. A. (1970) Particulate organic matter in the sea. *Adv. Mar. Biol.* **8**, 1–118. 7.3.2

Risch S. (1980) The population dynamics of several herbivorous beetles in a tropical agroecosystem: the effect of intercropping corn, beans and squash in Costa Rica. *J. appl. Ecol.* **17**, 593–612. 8.2.2

Robertson D. R. (1982) Fish feces as fish food on a Pacific coral reef. *Mar. Ecol. Prog. Ser.* **7**, 253–265. 7.3.2

Robertson M. L., Mills A. L. & Zieman J. C. (1982) Microbial synthesis of detritus-like

particulates from dissolved organic carbon released by tropical seagrasses. *Mar. Ecol. Prog. Ser.* **7**, 279–285. 7.3.3
Robinson S. K. & Holmes R. T. (1982) Foraging behavior of forest birds: the relationships among search tactics, diet, and habitat structure. *Ecology* **63**, 1918–1931. 5.2
Rodin L. E. & Bazilevich N. I. (1967) *Production and Mineral Cycling in Terrestrial Vegetation* (English edn). Oliver & Boyd, Edinburgh. 7.2.2
Rogers W. P. (1963) Physiology of infection with nematodes: some effects of the host stimulus on infective stages. *Ann. N. Y. Acad. Sci.* **113**, 208–216. 13.2.2
Rohde K. (1978a) Latitudinal differences in host-specificity of marine Monogenea and Digenea. *Mar. Biol.* **47**, 125–134. 9.2.3
Rohde K. (1978b) Latitudinal gradients in species diversity and their causes. II. Marine parasitological evidence for a time hypothesis. *Biol. Zentralbl.* **97**, 405–418. 9.2.3
Rohde K. (1979) A critical evaluation of intrinsic and extrinsic factors responsible for niche restriction in parasites. *Amer. Natur.* **114**, 648–671. 9.2.1, 9.2.2, 9.2.3
Rohde K. (1982) *Ecology of Marine Parasites*. University of Queensland Press, Brisbane. 9.2.2, 9.3.2
Röhnert U. (1950) Wasserfüllte Baumhöhlen und ihre Besiedlung. Ein Beitrag zu Fauna dendrolimnetica. *Arch. Hydrobiol.* **44**, 472–514. 10.2.1
Rolls E. C. (1969) *They All Ran Wild: the Story of Pests on the Land in Australia*. Angus & Robertson, Sydney. 10.3.3
Root R. B. (1967) The niche exploitation pattern of the blue-gray gnatcatcher. *Ecol. Monogr.* **37**, 317–350. 5.1, 5.2, 5.3.1, 6.4, 9.2.1
Root R. B. (1973) Organization of a plant-arthropod association in simple and diverse habitats: the fauna of collards (*Brassica oleracea*). *Ecol. Monogr.* **43**, 95–124. 5.2, 8.2.2
Root R. B. & Chaplin S. J. (1976) The life-styles of tropical milkweed bugs, *Oncopeltus* (Hemiptera: Lygaeidae), utilizing the same hosts. *Ecology* **57**, 132–140. 5.2
Root R. B. & Tahvanainen J. O. (1969) Role of winter corn, *Barbarea vulgaris* as a temporal host in the seasonal development of the crucifer fauna. *Ann. Ent. Soc. Amer.* **62**, 852–855. 8.2.4
Rosen D. E. (1982) Do current theories of evolution satisfy the basic requirements of explanation? *Syst. Zool.* **31**, 76–85. 15.4
Rosenzweig M. L. (1973) Evolution of the predator isocline. *Evolution* **27**, 84–94. 14.8
Ross E. B. (1978) Food taboos, diet and hunting strategy: the adaptation to animals in Amazonian cultural ecology. *Current Anthropology* **19**, 1–36. 15.3.3
Ross H. H. (1957) Principles of natural coexistence indicated by leafhopper populations. *Evolution* **11**, 113–129. 6.8
Rotenberry J. T. (1980) Dietary relationships among shrub-steppe passerine birds: competition or opportunism in a variable environment? *Ecol. Monogr.* **50**, 93–110. 6.5.1
Rotenberry J. T. & Wiens J. A. (1980) Habitat structure, patchiness and avian communities in North American steppe vegetation: A multivariate analysis. *Ecology* **61**, 1228–1250. 5.3.1
Rotenberry J. T. & Wiens J. A. (1981) Morphological size ratios and competition in ecological communities. *Amer. Natur.* **117**, 592–599. 6.8
Roth V. L. (1981) Constancy in the size ratios of sympatric species. *Amer. Natur.* **118**, 394–404. 6.5.1
Rothman A. H. (1959) The role of bile salts in the biology of tapeworms. II. Further observations on the effects of bile salts on metabolism. *J. Parasitol.* **45**, 379–383. 13.2.2
Rothstein S. I. (1973) The niche-variation model — is it valid? *Amer. Natur.* **107**, 598–620. 6.6.3
Roughgarden J. (1972) Evolution of niche width. *Amer. Natur.* **106**, 683–719. 6.6.3
Roughgarden J. (1974a) Species packing and the competition function with illustrations

from coral reef fish. *Theor. Popul. Biol.* **5**, 1–24. 6.5.2
Roughgarden J. (1974b) Niche width: biogeographic patterns among *Anolis* lizard populations. *Amer. Natur.* **108**, 429–442. 6.6.3
Roughgarden J. (1976) Resource partitioning among competing species — a coevolutionary approach. *Theor. Popul. Biol.* **9**, 388–424. 6.5.1, 6.5.2, 14.6
Roughgarden J. (1983) Competition and theory in community ecology. *Amer. Natur.* **122**, 583–601. 4.3.4, 6.3
Roughgarden J. & Feldman M. (1975) Species packing and predation pressure. *Ecology* **56**, 489–492. 6.5.2
Roughgarden J., Heckel D. & Fuentes E. (1983) Co-evolutionary theory and the biogeography and community structure of *Anolis*. In *Lizard Ecology: Studies of a Model Organism* (eds Huey R. B., Pianka E. R. & Schoener T. W.) pp. 371–410. Harvard University Press, Cambridge, Mass. 6.5.1, 6.5.3
Royama T. (1970) Factors governing the hunting behaviour and selection of food by the great tit (*Parus major* L.). *J. Anim. Ecol.* **39**, 619–668. 10.1
Sabo S. R. (1980) Niche and habitat relations in subalpine bird communities of the White Mountains of New Hampshire. *Ecol. Monogr.* **50**, 241–259. 5.2, 5.3.2
Sahlins M. (1976) *The Use and Abuse of Biology*. University of Michigan Press, Ann Arbor. 15.2
Saito S. (1969) Energetics of isopod populations in a forest of central Japan. *Res. Popul. Ecol.* **11**, 229–258. 7.2.2
Saitô T. (1965) Coaction between litter-decomposing by hymenomycetes and their associated microorganisms during decomposition of beech litter. *Sci. Rep. Tôhoku Univ. 4th Ser.* **31**, 155–173. 7.2.2
Saitô T. (1966) Sequential pattern of decomposition of beech litter with special reference to microbial succession. *Ecol. Rev.* **16**, 245–254. 7.2.2
Sakai K. I. (1955) Competition in plants and its relation to selection. *Cold Spring Harbor Symp. Quant. Biol.* **20**, 137–157. 14.5
Sakai K. I. (1961) Competitive ability in plants: its inheritance and some related problems. *Symp. Soc. exp. Biol.* **15**, 245–263. 14.5
Sale P. F. (1977) Maintenance of high diversity in coral reef fish communities. *Amer. Natur.* **111**, 337–359. 6.8
Sale P. F. (1980) The ecology of fishes on coral reefs. *Oceanogr. mar. Biol. Ann. Rev.* **18**, 367–421. 2.5
Sale P. F. & Williams D. M. (1982) Community structure of coral reef fishes: are the patterns more than those expected by chance? *Amer. Natur.* **120**, 121–127. 6.8
Salmon J. T. (1975) The influence of man on the biota. In *Biogeography and Ecology in New Zealand* (ed. Kuschel G.) pp. 643–661. W. Junk, The Hague. 10.3.1
Samuel W. M. & Barker M. J. (1979) The winter tick, *Dermacventor albipictus* (Packard, 1869) on moose, *Alces alces* (L.) of central Alberta. *Proc. No. Amer. Moose Conf. Workshop* **15**, 303–348. 9.3
Sanders H. L. (1968) Marine benthic diversity: a comparative study. *Amer. Natur.* **102**, 243–282. 5.5.3
Sands W. A. (1969) The association of termites and fungi. In *Biology of Termites* (eds Krishna K. & Weesner F. M.) Vol. 1, pp. 495–524. Academic Press, London. 7.2.2
Sanford W. W. (1974) The ecology of orchids. In: *The Orchids: Scientific Studies* (ed. Withner C. L.) pp. 1–100. Wiley, New York. 2.2
Sankurathri C. S. & Holmes J. C. (1976) Effects of thermal effluents on parasites and commensals of *Physa gyrina* Say (Mollusca: Gastropoda) and their interactions at Lake Wabamun, Alberta. *Can. J. Zool.* **54**, 1742–1753. 9.3.2
Satchell J. E. (1974) Introduction. Litter interface of animate/inanimate matter. In *Biology of Plant Litter Decomposition* (eds Dickinson G. H. & Pugh G. J. F.) pp. xiii-xliv. Academic Press, London. 7.2.1

Satchell J. E. & Lowe D. G. (1967) Selection on leaf litter by *Lumbricus terrestris*. In *Progress in Soil Biology* (eds Graff O. & Satchell J. E.) pp. 102–119. New-Holland, Amsterdam. 7.2.2

Sauer C. O. (1969) *Seeds, Spades, Hearths and Herds: The Domestication of Animals and Foodstuffs*. MIT Press, Cambridge, Massachusetts. 15.4

Schad G. A. (1966) Immunity, competition and natural regulation of helminth populations. *Amer. Natur.* **100**, 359–364. 13.3.3

Sehad G. A., Nawalinski T. A. & Kochar V. (1983) Human ecology and the distribution of hookworm populations. In *Human Ecology and Infectious Diseases* (eds Croll N. A. & Cross J. H.) pp. 188–223. Academic Press, New York. 9.3.4

Schaller G. B. (1967) *The Deer and the Tiger: a Study of Wildlife in India*. University of Chicago Press, Chicago. 8.4.3

Schaller G. B. (1972) *The Serengeti Lion — a Study of Predator–Prey Relations*. University of Chicago Press, Chicago. 8.4.3

Schalm O. W. (1961) *Veterinary Hematology*. Lea & Febiger, Philadelphia. 13.3.2

Schmidt R. A. (1978) Diseases in forest ecosystems. In *Plant Disease: an Advanced Treatise* (eds Horsfall J. G. & Cowling E. B.) Vol. 2, pp. 287–315. Academic Press, New York. 14.8

Schoener T. W. (1965) The evolution of bill size differences among sympatric congeneric species of birds. *Evolution* **19**, 189–213. 5.5.3, 6.5.1

Schoener T. W. (1968) The *Anolis* lizards of Bimini: resource partitioning in a complex fauna. *Ecology* **49**, 704–726. 6.5.3

Schoener T. W. (1969a) Models of optimal size for solitary predators. *Amer. Natur.* **103**, 277–313. 6.4, 6.6.1

Schoener T. W. (1969b) Size pattern in West Indian *Anolis* lizards. I. Size and species diversity. *Syst. Zool.* **18**, 386–401. 6.5.1, 14.6

Schoener T. W. (1970) Size patterns in West Indian *Anolis* lizards. II. Correlations with the sizes of particular sympatric species — displacement and convergence. *Amer. Natur.* **104**, 155–174. 6.5.1, 6.5.4

Schoener T. W. (1971) Large-billed insectivorous birds: a precipitous diversity gradient. *Condor* **73**, 154–161. 5.5.3, 6.8

Schoener T. W. (1974a) Resource partitioning in ecological communities. *Science* **185**, 27–39. 6.3, 6.4, 6.5.1, 6.5.3

Schoener T. W. (1974b) Temporal resource partitioning and the compression hypothesis. *Proc. Natl. Acad. Sci. USA* **71**, 4169–4172. 6.4, 6.5.5

Schoener T. W. (1974c) Some methods for calculating competition coefficients from resource-utilization spectra. *Amer. Natur.* **108**, 332–340. 6.5.5, 6.7.2

Schoener T. W. (1974d) Competition and the form of habitat shift. *Theor. Popul. Biol.* **6**, 265–307. 6.5.2, 6.6.2, 6.7.2

Schoener T. W. (1975) Presence and absence of habitat shift in some widespread lizard species. *Ecol. Monogr.* **45**, 233–258. 6.6.2

Schoener T. W. (1976) Alternatives to Lotka–Volterra competition: models of intermediate complexity. *Theor. Popul. Biol.* **10**, 309–333. 6.5.2, 6.7.2

Schoener T. W. (1977) Competition and the niche. In *Biology of the Reptilia*. (eds Gans C. & Tinkle D. W.) Vol. 7, pp. 35–136. Academic Press, London. 6.4, 6.5.3, 6.6.3

Schoener T. W. (1978) Effect of limited food search on the outcome of single-level competition. *Theor. Popul. Biol.* **13**, 365–381. 6.5.2, 6.7.2

Schoener T. W. (1982) The controversy over interspecific competition. *Amer. Sci.* **70**, 586–595. 6.5.5, 6.8, 9.2

Schoener T. W. (1983a) Overview. In *Lizard Ecology: Studies of a Model Organism* (eds Huey R. B., Pianka E. R. & Schoener T. W.) pp. 233–240. Harvard University Press, Cambridge, Massachusetts. 6.4

Schoener T. W. (1983b) Field experiments on interspecific competition. *Amer. Natur.* **122**,

240–285. 6.7.1, 6.7.2, 6.8, 8.4.3, 11.1, 11.1.1, 11.2.2, 11.5.1, 11.5.2, 11.6.1, 11.6.2, 11.7

Schoener T. W. (1984) Size differences among sympatric, bird-eating hawks: a worldwide survey. In *Ecological Communities: Conceptual Issues and the Evidence* (eds Strong D. R., Simberloff D., Abele L. G. & Thistle A. B.) pp. 254–281. Princeton University Press, Princeton. 6.5.1, 6.5.2

Schoener T. W. Choice of species pool affects conclusions from null models: an illustration with lizard sizes and habitats (in manuscript). 6.5.1

Schoener T. W. & Janzen D. H. (1968) Notes on environmental determinants of tropical versus temperate insect size patterns. *Amer. Natur.* **102**, 207–224. 5.5.3, 6.5.1

Schoonhaven L. M. & Jermy T. (1977) A behavioral and electrophysiological analysis of insect feeding deterrents. In *Crop Protection Agents — Their Biological Evaluation* (ed. McFarlane N. R.) pp. 133–146. Academic Press, New York. 13.2.2

Schroder G. D. & Rosenzweig M. L. (1975) Perturbation analysis of competition and overlap in habitat utilization between *Dipodomys ordii* and *Dipodomys merriami*. *Oecologia* **19**, 9–28. 11.1, 11.4.2

Schröter C. & Kirchner O. (1896, 1902) *Die Vegetation der Bodensees*, *Kommissionverlag der Schriften des Vereins fur Geschicte des Bodensees u. Seiner Umgebung* (von T. J. Stettner). Lindair. I, 122 pp. II, 865 pp. 1.1

Schwartz A. & Garrido O. H. (1972) The lizards of the *Anolis equestris* complex in Cuba. *Stud. Fauna Curacao, Other Caribb. Isl.* **39**, 1–86. 6.5.4

Scott P. R., Johnson R., Wolfe M. S., Lowe H. J. B. & Bennett F. G. A. (1980) Host-specificity in cereal parasites in relation to their control. *Appl. Biol.* **5**, 349–393. 1.8

Seaton A. J. P. & Antonovics J. (1967) Population interrelationships. I. Evolution in mixtures of *Drosophila* mutants. *Heredity* **22**, 19–33. 14.6

Segal A., Manisterski J., Fischbeck G. & Wahl I. (1980) How plant populations defend themselves in natural ecosystems. In *Plant Disease*: *An Advanced Treatise* (eds Horsfell J. G. & Cowling E. B.) Vol. 5, pp. 75–102. Academic Press, New York. 14.8

Seifert R. P. (1982) Neotropical *Heliconia* insect communities. *Quart. Rev. Biol.* **57**, 1–28. 8.3.3

Seifert R. P. & Seifert F. H. (1976) A community matrix analysis of *Heliconia* insect communities. *Amer. Natur.* **110**, 461–483. 8.3.3

Seifert R. P. & Seifert F. H. (1979) A *Heliconia* insect community in a Venezuelan cloud forest. *Ecology* **60**, 462–467. 8.3.3, 11.5.1, 11.5.2

Seki H. (1972) The role of microorganisms in the marine food chain with reference to organic aggregates. *Mem. 1st. Ital. Idrobiol.* **29**, (*Suppl.*), 245–254. 7.3.3

Selander R. K. (1966) Sexual dimorphism and differential niche utilization in birds. *Condor* **68**, 113–151. 6.6.3

Selander R. K. (1970) Behavior and genetic variations in natural populations. *Amer. Zool.* **10**, 53–66. 13.3.3

Shapiro A. M. (1976) Beau geste? *Amer. Natur.* **110**, 900–902. 9.3.3

Shaw G. L. & Quadagna D. (1975) *Trypanosoma lewisi* and *T. cruzi*. Effect of infection on gestation in the rat. *Exp. Parasitol.* **37**, 211–217. 13.2.1

Shelford V. E. (1913) *Animal Communities in Temperate America*. Chicago University Press, Chicago. 1.4

Shorrocks B., Atkinson W. & Charlesworth P. (1979) Competition on a divided and ephemeral resource. *J. Anim. Ecol.* **48**, 899–908. 8.3.1

Sillitoe P. (1983) *Roots of the Earth*: *Crops in the highlands of Papua New Guinea*. New South Wales University Press, Kensington. 15.3.4

Simberloff D. S. (1978) Using island biogeographic distributions to determine if colonization is stochastic. *Amer. Natur.* **112**, 713–726. 11.1, 14.8

Simberloff D. S. (1982) The status of competition theory in ecology. *Ann. Zool. Fennici* **19**, 241–254. 9.2

Simberloff D. & Boecklen W. (1981) Santa Rosalia reconsidered. *Evolution* **35**, 1206–1228. 6.3, 6.5.1, 6.5.2, 9.2.2

Simenstad C. A., Estes J. A. & Kenyon K. W. (1978) Aleuts, sea otters, and alternate stable-state communities. *Science* **200**, 403–411. 10.2.2

Simons M. D. (1972) Crown rust tolerance of *Avena sativa*-type oats derived from wild *Avena sterilis*. *Phytopath.* **62**, 1444–1446. 14.8

Simons M. D., Martens J. W., KcKenzie R. I. H., Nishiyama I., Sadanaga K., Sebesta J. & Thomas H. (1978) Oats: a standardised system of nomenclature for genes and chromosomes and catalog of genes governing characters. *U.S. Dep. Agric., Agric. Handb.* **509**. 14.8

Simpson E. H. (1949) Measurement of diversity. *Nature* **163**, 688. 4.3.3, 4.4.1

Sinclair A. R. E. (1975) The resource limitation of trophic levels in tropical grassland ecosystems. *J. Anim. Ecol.* **44**, 497–520. 8.5

Sinclair A. R. E. (1979a) Dynamics of the Serengeti ecosystem: process and pattern. In *Serengeti — Dynamics of an Ecosystem* (eds Sinclair A. R. E. & Norton-Griffiths M.) pp. 1–30. University of Chicago Press, Chicago. 8.4.3

Sinclair A. R. E. (1979b) The eruption of the ruminants. In *Serengeti — the Dynamics of an Ecosystem* (eds Sinclair A. R. E. & Norton-Griffiths M.) pp. 82–103. University of Chicago Press, Chicago. 8.4.3

Sinclair A. R. E. (1985) Does interspecific competition or predation shape the African ungulate community? *J. Anim. Ecol.* **54**, 899–918. 8.4.3

Sinclair A. R. E., Dublin H. & Borner M. (1985) Population regulation of Serengeti wildebeest: A test of the food hypothesis. *Oecologia* **65**, 266–268. 8.4.3

Sinclair A. R. E. & Norton-Griffiths M. (eds.) (1979) *Serengeti — Dynamics of an Ecosystem*. University of Chicago Press, Chicago. 8.4.3

Sinclair A. R. E. & Norton-Griffiths M. (1982) Does competition or facilitation regulate migrant ungulate populations in the Serengeti? A test of hypotheses. *Oecologia* **53**, 364–369. 8.4.3

Sinskaia E. N. (1931) The study of species in their dynamics and inter-relation with different types of vegetation. *Bull. appl. Bot., Gen. & Plant Breeding* **25**(2), 1–92. 14.3, 14.4

Sinskaia E. N. & Beztuzcheva A. A. (1931) The forms of *Camelina sativa* in connection with climate, flax and man. *Bull. appl. Bot., Gen. & Plant Breeding* **25**(2), 98–200. 14.4

Skinner G. H. & Whittaker J. B. (1981) An experimental investigation of interrelationships between the wood-ant (*Formica rufa*) and some tree-canopy herbivores. *J. Anim. Ecol.* **50**, 313–326. 8.3.3

Slansky F. (1973) Latitudinal gradients in species diversity of the New World swallowtail butterflies. *J. Res. Lepid.* **11**, 201–217. 8.3.4

Slatkin M. W. (1980) Ecological character displacement. *Ecology* **61**, 163–177. 6.5.1, 6.6.1, 14.6

Slobodchikoff C. N. & Schulz W. C. (1980) Measures of niche overlap. *Ecology* **61**, 1051–1055. 6.7.2

Slobodkin L. B. (1968) How to be a predator. *Amer. Zool.* **8**, 43–51. 9.3.3

Smith C. C. (1975) The coevolution of plants and seed predators. In *Coevolution of Animals and Plants* (eds Gilbert L. E. & Raven P. H.) pp. 53–77. University of Texas Press, Austin. 7.2.1

Smith C. C. (1981) The indivisible niche of *Tamiasciurus*: an example of non-partitioning of resources. *Ecol. Monogr.* **51**, 343–363. 6.4

Smith D. C. (1981) Competitive interaction of the striped plateau lizard (*Sceloporus virgatus*) and the tree lizard (*Urosaurus ornatus*). *Ecology* **62**, 679–687. 11.3.1, 11.7

Smith D. W. & Cooper S. D. (1982) Competition among Cladocera. *Ecology* **63**, 1004–1015. 11.5, 11.6.2

Smith E. A. (1979) Human adaptation and energetic efficiency. *Human Ecology* **7**, 53–74. 15.2

Smith H. (1976) *Modern Views on Microbial Pathogenicity*. Meadowfield Press, Shildon. 13.3.3

Smith J. N. M., Grant P. R., Grant B. R., Abbott I. J. & Abbott L. K. (1978) Seasonal variation in feeding habits of Darwin's ground finches. *Ecology* **59**, 1137–1150. 6.5.5

Smith N. G. (1968) The advantage of being parasitized. *Nature* **219**, 690–694. 9.3.2, 14.10

Smith N. G. (1979) Alternate responses by hosts to parasites which may be helpful or harmful. In *Host–Parasite Interfaces* (ed. Nickol B. B.) pp. 7–15. Academic Press, New York. 9.3.2, 14.10

Smith V. S. (1946) Studies on reactions of rat serum to eggs of *Trichosomoides crassicauda*: a nematode of the urinary bladder. *J. Parasitol.* **32**, 136–141. 13.3.2

Smithers C., McAlpine D., Colman P. & Grey M. (1974) Island invertebrates. *Aust. nat. Hist.* **18**, 60–63. 10.3.1

Smith-Trail D. R. (1980) Behavioral interactions between parasites and hosts: host suicide and the evolution of complex life cycles. *Amer. Natur.* **116**, 77–91. 9.3.3

Smyth J. D. & Haslewood G. A. D. (1963) The biochemistry of bile as a factor determining host specificity in intestinal parasites, with particular reference to *Echinococcus granulosus*. *Ann. N.Y. Acad. Sci.* **113**, 234–260. 13.2.2

Snow W. E. (1958) Stratification of arthropods in a wet stump cavity. *Ecology* **39**, 83–88. 10.2.1

Sobey D. G. & Kenworthy J. B. (1979) The relationship between herring gulls and the vegetation of their breeding colonies. *J. Ecol.* **67**, 469–496. 2.2

Sobey W. R. (1960) Myxomatosis: the virulence of the virus and its relation to genetic resistance in the rabbit. *Aust. J. Sci.* **23**, 53–55. 14.8

Sogandares-Bernal F. (1959) Digenetic trematodes of marine fishes from the Gulf of Panama and Bimini, British West Indies. *Tulane Stud. Zool.* **7**, 69–117. 9.2.2

Solbrig O. T. & Simpson B. B. (1974) Components of regulation of population of dandelions in Michigan. *J. Ecol.* **62**, 473–486. 14.4

Soma K. & Saitô T. (1979) Ecological studies of soil organisms with reference to the decomposition of pine needles. I. Soil macrofaunal and mycofloral surveys in coastal pine plantations. *Rev. Ecol. Biol. Sol.* **16**, 337–354. 7.2.2

Soma K. & Saitô T. (1983) Ecological studies of soil organisms with reference to the decomposition of pine needles. II. Litter feeding and breakdown by the woodlouse *Porcellio scaber*. *Plant & Soil* **75**, 139–151. 7.2.2

Sorokin Yu. I. (1978) Decomposition of organic matter and nutrient regeneration. In *Marine Ecology* (ed. Kinne O.) Vol. 4, pp. 501–616. Wiley, New York. 7.3.3

Sorokin Yu. I. (1981) Microheterotrophic organisms in marine ecosystems. In *Analysis of Marine Ecosystems* (ed. Longhurst R. A.) pp. 293–342. Academic Press, London. 7.3.1

Soulé M. E., Wilcox B. A. & Holtby C. (1979) Benign neglect: A model of fauna collapse in the game reserves of East Africa. *Biological Conservation* **15**, 259–272. 8.4.2

Soulsby E. J. (1962) Antigen-antibody reactions in helminth infections. *Adv. Immunol.* **2**, 265–308. 13.2.2

Southern H. N. (1970) Ecology at the cross-roads. *J. Ecol.* **58**, 1–11. 2.2

Southwood T. R. E. (1973) The insect/plant relationship — an evolutionary perspective. *Symp. Roy. Ent. Soc. Lond.* **6**, 3–30. 8.3.3

Southwood T. R. E. (1977) Habitat, the templet for ecological strategies. Presidential address to the British Ecological Society, 5 January 1977. *J. Anim. Ecol.* **46**, 337–365. 8.3.1, 10.3.3

Southwood T. R. E. (1978) The components of diversity. In *Diversity of Insect Fauna* (eds Mound L. A. & Waloff N.) pp. 19–40. *Symp. roy. ent. Soc. Lond.*, Vol. 9. Blackwell Scientific Publications, Oxford. 8.5

Southwood T. R. E. (1981) Bionomic strategies and population parameters. In *Theoretical Ecology* (ed. May R. M.) (2nd edn) pp. 30–52. Blackwell Scientific Publications, Oxford. 8.4.3, 8.5

Southwood T. R. E., Brown V. K. & Reader P. M. (1979) The relationship of plant and insect diversities in succession. *Biol. J. Linn. Soc.* **12**, 327–348. 8.2.2

Southwood T. R. E., Moran V. C. & Kennedy C. E. J. (1982) The assessment of arborical insect fauna: comparisons of knockdown sampling and faunal lists. *Ecol. Entomol.* **7**, 331–340. 8.2.4

Specht R. L. (1979) The sclerophyllous (heath) vegetation of Australia: the eastern and central States. In *Heathlands and Related Shrublands. Ecosystems of the World* Vol. 9A. (ed. Specht R. L.) pp. 125–210. Elsevier, Amsterdam. 4.3.1, 4.3.4

Spiess E. B. (1957) Relation between frequencies and adaptive values of chromosome arrangements in *Drosophila persimilis. Evolution* **11**, 84–93. 14.6

Spinage C. A. (1962) Rinderpest and faunal distribution patterns. *African Wildl.* **16**, 55–60. 9.3.2

Sprent J. F. A. (1962) A study of adaption tolerance — the growth of ascaridoid larvae in indigenous and non-indigenous intermediate hosts. *Proc. First Regional Symp. Scient. Knowl. Trop. Parasit., Univ. Singapore*, Nov. 1962, pp. 261–266. 13.2.2

Spriggs M. (1978) Taro irrigation in Oceania. Dept. of Prehistory Research School of Pacific Studies, Australian National University, Canberra (mimeograph). 15.3.4

Spriggs M. (1982) Traditional uses of fresh water in Papua New Guinea: past neglect and future possibilities. In *Traditional Conservation in Papua New Guinea: Implications for Today* (eds Morauta L., Pernetta J. & Heaney W.) pp. 257–271. Monograph 16, Institute of Applied Social and Economic Research, Boroko, Papua New Guinea. 15.3.4

Stakman E. C. & Piemeisel F. J. (1917) Biologic forms of *Puccinia graminis* on cereals and grasses. *J. agric. Res.* **10**, 429–495. 14.8

Stamps N. E. (1981) Behavior of parasitized aposematic caterpillars: advantageous to the parasitoid or the host? *Amer. Natur.* **118**, 715–725. 9.3.3

Stamps J. A. (1983) Sexual selection, sexual dimorphism and territoriality in lizards. In *Lizard Ecology: Studies of a Model Organism* (eds Huey R. B., Pianka E. R. & Schoener T. W.) pp. 169–204. Harvard University Press, Cambridge, Mass. 6.6.3

Stanton M. L. (1983) Spatial patterns in the plant community and their effects upon insect search. In *Herbivorous Insects. Host Seeking Behaviour and Mechanisms* (ed. Ahmad S.) pp. 125–157. Academic Press, New York. 8.2.2

Stark N. (1972) Nutrient cycling pathways and litter fungi. *Bioscience* **22**, 355–360. 7.2.2

Stark N. (1973) *Nutrient Cycling in a Jeffrey Pine Ecosystem*. University of Montana Press, Missoula. 7.2.2

Stearns S. C. (1977) The evolution of life history traits. *Ann. Rev. Ecol. Syst.* **8**, 145–171. 9.3.3, 14.4

Stebaev I. V. & Reznikova J. I. (1972) Two interaction types of ants living in steppe ecosystem in south Siberia, USSR. *Ekol. Pol.* **20**, 103–109. 6.7.1

Steele F. & Bourne A. (eds) (1975) *The Man/Food Equation*. Academic Press, New York. 15.2

Stephenson W. & Searles R. B. (1960) Experimental studies on the ecology of intertidal environments at Heron Island. I. Exclusion of reef fish from beach rock. *Aust. J. mar. Freshwater Res.* **2**, 241–267. 7.3.1

Stewart G. L., Reddington J. J. & Hamilton M. (1980) *Eimeria nieschulzi* and *Trichinella spiralis*: analysis of concurrent infection in the rat. *Exp. Parasitol.* **50**, 115–123. 13.3.3

Stiles E. W. (1978) Avian communities in temperate and tropical alder forests. *Condor* **80**, 276–284. 5.5.3

Stimson J. (1970) Territorial behavior of the owl limpet, *Lottia gigantea. Ecology* **51**, 113–118. 11.3.1

Strong D. R., Jr. (1974) Rapid asymptotic species accumulation in phytophagous insect communities; the pests of cacao. *Science* **185**, 1064–1066. 8.2.4

Strong D. R., Jr (1977) Microbial herbivores. *Science* **197**, 1071. 14.8

Strong D. R., Jr (1979) Biogeographic dynamics of insect-host plant communities. *Ann. Rev. Entomol.* **24**, 89–119. 8.2, 8.2.3

Strong D. R., Jr (1982) Harmonious coexistence of hispine beetles on *Heliconia* in experimental and natural communities. *Ecology* **63**, 1039–1049. 6.8

Strong D. R. Jr, Lawton J. H. & Southwood T. R. E. (1984a) *Insects on Plants: Community Patterns and Mechanisms*. Blackwell Scientific Publications, Oxford. 8.1, 8.2, 8.2.1, 8.2.2, 8.2.3, 8.3.2, 8.3.3, 8.3.4, 8.5

Strong D. R. Jr & Levin D. A. (1979) Species richness of plant parasites and growth forms of their hosts. *Amer. Natur.* **114**, 1–22. 8.2.3

Strong D. R. Jr, McCoy E. D. & Rey J. R. (1977) Time and the number of herbivore species: the pests of sugarcane. *Ecology* **58**, 167–175. 8.2.4

Strong D. R. Jr, Simberloff D. S., Abele L. G. & Thistle A. B. (eds) (1984b) *Ecological Communities: Conceptual Issues and the Evidence*. Princeton University Press, Princeton. 6.5.4, 9.2, 11.1

Strong D. R. Jr, Szyska L. A. & Simberloff D. S. (1979) Tests of community-wide character displacement against null hypotheses. *Evolution* **33**, 897–913. 6.3, 6.5.1, 11.1, 14.6

Sturkie P. D. & Newman H. J. (1951) Plasma proteins of chickens as influenced by time of laying, ovulation, number of blood samples taken and plasma volume. *Poultry Sci.* **30**, 240–248. 13.3.2

Sugihara G. (1980) Minimal community structure: an explanation of species abundance patterns. *Amer. Natur.* **116**, 770–787. 4.3.1, 4.3.2

Sulgostowska T. (1963) Trematodes of birds in biocoenosis of the Lakes Druzno, Godapiwo, Mamry Pónocne and Świeçajty. *Acta Parasitol. Polon.* **11**, 239–264. 9.2.4

Sutherland J. P. (1978) Functional roles of *Schizoporella* and *Styela* in the fouling community at Beaufort, North Carolina. *Ecology* **59**, 257–264. 11.7

Swift M. J., Heal O. W. & Anderson J. M. (1979) *Decomposition in Terrestrial Ecosystems*. Blackwell Scientific Publications, Oxford. 2.4

Symons L. E. A. & Steel J. W. (1978) Pathogenesis of the loss of production in gastro-intestinal parasitism. In *The Epidemiology and Control of Gastro-Intestinal Parasites of Sheep in Australia* (eds Donald A. D., Southcott W. H. & Dineen J. K.). CSIRO Division of Animal Health, Melbourne. 9.3

Sztejnberg A. & Wahl I. (1969) Comparative resistance of *Avena sterilis* selections at seedling and more advanced stages of growth to *Puccinia graminis avenae* (stem rust). *Res. Rep. Sci. Agric.*, Hebrew University of Jerusalem. 14.8

Tahvanainen J. O. & Root R. B. (1972) The influence of vegetational diversity on the population ecology of a specialised herbivore *Phyllotreta cruciferae* (Coleopotera: Chrysomelidae). *Oecologia* **10**, 321–346. 8.2.2

Tanno T. (1976) The Mbuti nethunters in the Ituri forest, Eastern Zaire: their hunting activities and band composition. *Kyoto African Studies* **10**, 101–135. 15.3.5

Tansley A. G. (1920) The classification of vegetation and the concepts of development. *J. Ecol.* **8**, 118–149. 12.1, 12.3.1

Tansley A. G. (1935) The use and abuse of vegetational concepts and terms. *Ecology*. **16**, 284–307. 1.5, 1.6

Taylor L. R., Woiwod I. P. & Perry J. N. (1978) The density-dependence of spatial behaviour and the rarity of randomness. *J. Anim. Ecol.* **47**, 383–406. 12.3.4

Taylor P. R. & Littler M. M. (1982) The roles of compensatory mortality, physical disturbance and substrate retention in the development and organization of a sand-influenced rocky-intertidal community. *Ecology* **63**, 135–146. 11.3.1, 11.7

Taylor R. H. (1979a) Predation of sooty terns at Raoul Island by rats and cats. *Notornis* **26**, 199–202. 10.3.1

Taylor R. H. (1979b) How the Macquarie Island parrakeet became extinct. *N.Z. J. Ecol.* **2**, 42–45. 10.3.1

Tenore K. R. (1975) Detrital utilization by the polychaete, *Capitella capitata*. *J. mar. Res.* **33**, 261–274. 7.3.2

Tenore K. R. (1977) Growth of *Capitella capitata* cultured on various levels of detritus derived from different sources. *Limnol. Oceanogr.* **22**, 936–941. 7.3.2

Tenore K. R. (1980) Organic nitrogen and caloric content of detritus. *Estuar. Coast. Shelf Sci.* **12**, 39–47. 7.3.2

Tenore K. R., Hanson R. B., Dornseif B. E. & Wiederhold C. N. (1979) The effect of organic nitrogen supplement on the utilization of different sources of detritus. *Limnol. Oceanogr.* **24**, 350–355. 7.3.3

Terada M. & Oshima Y. (1970) A preliminary report on the analysis of growth and nitrogen budgets of experimental population of *Armadillidium vulgare*. *Sci. Res. Sch. Educ. Waseda Univ. Ser. Biol. Geol.* **19**, 17–34. 7.2.2

Terborgh J. (1971) Distribution on environmental gradients: theory and a preliminary interpretation of distributional patterns in the avifauna of the Cordillera Vilcabamba, Peru. *Ecology* **52**, 23–40. 5.3.3

Terborgh J. (1973) On the notion of favourableness in plant ecology. *Amer. Natur.* **107**, 481–501. 5.5.3

Terborgh J. (1977) Bird species diversity on an Andean elevational gradient. *Ecology* **58**, 1007–1019. 5.5.3

Terborgh J. (1980a) Causes of tropical species diversity. In *Acta. XVII Congr. Intern. Ornith.* (*Berlin*) (ed. Nöhring R.) pp. 955–961. Deutsche Ornithologen-Gesellschaft, Berlin. 5.5.3

Terborgh J. (1980b) Vertical stratification of neotropical forest bird community. In *Acta XVII Congr. Intern. Ornith. (Berlin)* (ed. Nöhring R.) pp. 1005–1012. Deutsche Ornithologen-Gesellschaft, Berlin. 5.3.2

Terborgh J. & Diamond J. M. (1970) Niche overlap in feeding assemblages of New Guinea birds. *Wilson Bull.* **82**, 29–52. 5.5.1

Terborgh J., Fitzpatrick J. & Emmons L. (1984) Annotated checklist of the bird and mammal species of Cocha Cashu Biological Station, Manu National Park, Peru. *Publ. Field Mus. Zool. Ser.* **21**, 1–29. 5.4

Terborgh J. & Weske J. S. (1975) The role of competition in the distribution of Andean birds. *Ecology* **56**, 562–576. 5.1

Theodor O. & Costa M. (1967) A survey of the parasites of wild mammals and birds in Israel. Part I. Ectoparasites. *Israel Acad. Sci. Human. Sect. Sci.* **1967**, 5–117. 9.2.1

Thompson J. N. (1982) *Interaction and Coevolution*. Wiley, New York. 14.9

Tilly L. J. (1968) The structure and dynamics of Cone Spring. *Ecol. Monogr.* **38**, 169–197. 7.3.1

Tilman D. (1978) Cherries, ants and tent caterpillars: timing of nectar production in relation to susceptibility of caterpillars to ant predation. *Ecology* **59**, 686–692. 10.2.3

Tilman D. (1980) Resources: a graphical — mechanistic approach to competition and predation. *Amer. Natur.* **116**, 362–393. 1.8

Tilman D. (1982) *Resource Competition and Community Structure*. Princeton University Press, Princeton. 9.3.4

Tinbergen L. (1960) The natural control of insects in pine woods. I. Factors affecting the intensity of predation by song birds. *Arch. Neerl. Zool.* **13**, 265–336. 14.9

Tinkle D. W. (1982) Results of an experimental density manipulation in an Arizona lizard community. *Ecology* **63**, 57–65. 11.3.1, 11.4.2

Titman D. (1976) Ecological competition between algae: experimental confirmation of resource-based competition theory. *Science* **192**, 463–465. 6.4, 6.7.2

Todd A. W., Gunson J. R. & Samuel W. M. (1981) Sarcoptic mange: an important disease of coyotes and wolves of Alberta. In *Proc. First Worldwide Furbearer Conference* (eds Chapman J. A. & Pursley D.) pp. 706–729. Worldwide Furbearer Conference, Frostburg, Maryland. 9.3

Toft C. A. (1980) Feeding ecology of thirteen synotopic species of anurans in a seasonal tropical environment. *Oecologia* **45**, 131–141. 6.4, 6.5.5

Toft C. A. (1981) Feeding ecology of Panamanian litter anurans: patterns in diet and foraging mode. *J. Herpetol.* **15**, 139–144. 6.5.1

Toft C. A. (1985) Resource partitioning in amphibians and reptiles. *Copeia* **1985**, 1–21. 6.4, 6.8

Toft C. A. & Shea P. J. (1983) Detecting community-wide patterns: estimating power strengthens statistical inference. *Amer. Natur.* **122**, 618–625. 6.3

Toft C. A., Trauger D. L. & Murdy H. W. (1982) Test for species interactions: breeding phenology and habitat use in subarctic ducks. *Amer. Natur.* **120**, 586–613. 6.5.4

Tomiyama K., Sakai R., Sakuma T. & Ishizaka N. (1967) The role of polyphenols in the defense reaction of plants induced in infection. In *The Dynamic Role of Molecular Constituents in Plant–Parasite Interaction* (eds Mirocha L. J. & Uritani I.) pp. 165–182. Amer. Phytopathol. Soc., St Paul, Minnesota. 13.3.3

Townsend P. K. (1974) Sago production in a New Guinea economy. *Human Ecology* **2**, 217–236. 15.3.2

Trenbath B. R. (1974) Biomass productivity of mixtures. *Adv. Agronomy* **26**, 177–210. 11.4.1

Troyer K. (1984) Microbes, herbivory and the evolution of social behavior. *J. theor. Biol.* **106**, 157–169. 7.2.3

Tsuchiya M. & Kurihara Y. (1979) Feeding habits and food sources of deposit-feeding polychaete *Neanthes japonica* (Izuka). *J. exp. mar. Biol. Ecol.* **36**, 79–89. 7.3.2

Turelli M. (1981) Niche overlap and invasion of competitors in random environments. I. Models without demographic stochasticity. *Theor. Popul. Biol.* **20**, 1–56. 6.5.2

Turesson G. (1925) The plant species in relation to habitat and climate. *Hereditas* **6**, 147–236. 14.4

Turkington R. & Harper J. L. (1979) The growth, distribution and neighbour relationships of *Trifolium repens* in a permanent pasture. IV. Fine scale biotic differentiation. *J. Ecol.* **67**, 245–254. 14.6

Turnbull C. M. (1962) *The Forest People*. Simon & Schuster, New York. 15.3.3

Turnbull C. M. (1965) *Wayward Servants: the Two Worlds of the African Pygmies*. Eyre & Spottiswoode, London. 15.3.5

Turner J. R. G. (1984) Darwin's coffin and Doctor Pangloss — do adaptationist models explain mimicry? In *Evolutionary Ecology* (ed. Shorrocks B.) pp. 313–361. Blackwell Scientific Publications, Oxford. 14.9

Turner T. S. (1979) The Ge and Bororo societies as dialectical systems: a general model. In *Dialectical Societies* (ed. Maybury-Lewis D.) pp. 147–178. Harvard University Press, Cambridge, Massachusetts. 15.3.3

Turnipseed S. G. & Kogan M. (1976) Soybean entomology. *Ann. Rev. Ent.* **21**, 247–282. 8.2.4

Tutin T. G. (1942) Biological flora of the British Isles. *Zostera* L. *J. Ecol.* **30**, 217–226. 2.5

Tyler M. J. (1975) The cane toad, *Bufo marinus*. An historical account and modern assessment. A Report to the Vermin and Noxious Weeds Destruction Board, Victoria and the Agricultural Protection Board, WA. 10.3.2

Ucko P. J. & Dimbleby G. W. (eds) (1969) *The Domestication and Exploitation of Plants and Animals*. Duckworth, London. 15.4

Uetz G. W. (1977) Coexistence in a guild of wandering spiders. *J. Anim. Ecol.* **46**, 531–542. 6.4

Ullstrup A. J. (1972) The impacts of the southern corn leaf blight epidemics of 1970–1971. *Ann. Rev. Phytopath.* **10**, 37–50. 14.8

Ullyett G. C. (1950) Competition for food and allied phenomena in sheep-blowfly populations. *Phil. Trans. Roy. Soc. Lond.* **234**, 77–174. 10.1

Underwood A. J. (1976) Food competition between age-classes in the intertidal neritacean *Nerita atramentosa* Reeve (Gastropoda: Prosobranchia). *J. exp. mar. Biol. Ecol.* **23**, 145–154. 11.4.1, 11.6.1

Underwood A. J. (1978) An experimental evaluation of competition between three species of intertidal prosobranch gastropods. *Oecologia* **33**, 185–202. 11.2.2, 11.3.1, 11.5.1, 11.5.2, 11.6.1, 11.7

Underwood A. J. (1980) The effects of grazing by gastropods and physical factors on the upper limits of distribution intertidal macroalgae. *Oecologia* **46**, 201–213. 7.3.1

Underwood A. J. (1981) Techniques of analysis of variance in experimental marine biology and ecology. *Ann. Rev. Oceanogr. Mar. Biol.* 513–605. 11.1, 11.7

Underwood A. J. & Denley E. J. (1984) Paradigms, explanations and generalizations in models for the structure of intertidal communities on rocky shores. In *Ecological Communities: Conceptual Issues and the Evidence* (eds Strong D. R., Simberloff D. S., Abele L. G. & Thistle A. B.) pp. 151–180. Princeton University Press, Princeton. 11.1, 11.2.2

Underwood A. J., Denley E. J. & Moran M. J. (1983) Experimental analyses of the structure and dynamics of mid-shore rocky intertidal communities in New South Wales. *Oecologia* **56**, 202–219. 11.4.2

Usinger R. L. (1962) Foreword to H.W. Bates. *The Naturalist in the River Amazon*, pp. v-viii. University of California Press, Berkeley. 6.1

Van Beurden E. (1978) Report on Stage 1 of an ecological and physiological study of the Queensland cane toad *Bufo marinus*. Report to the Australian National Parks and Wildlife Service, Canberra. 10.3.2

Vance R. R. (1972) Competition and mechanism of coexistence in three sympatric species of intertidal hermit crabs. *Ecology* **53**, 1062–1074. 6.8

Vance R. R. (1978) Predation and resource-partitioning in one-predator two-prey model communities. *Amer. Natur.* **112**, 797–813. 6.8

van der Drift J. & Witkamp M. (1960) The significance of the break-down of oak litter by *Enoicyla pusilla* Burm. *Arch. Neerland. Zool.* **13**, 486–492. 7.2.2

Van der Gulden W. J. I. (1967) Diurnal rhythm in egg production by *Sypacia muris*. *Exp. Parasitol.* **21**, 344–347. 13.3.2

Vandermeer J. H. (1972) Niche theory. *Ann. Rev. Ecol. Syst.* **3**, 107–132. 6.7.2

Van der Plank J. E. (1963) *Plant Diseases: Epidemics and Control*. Academic Press, New York. 13.2.2

Van der Plank J. E. (1968) *Disease Resistance in Plants*. Academic Press, New York. 14.8

van Emden H. F. & Way M. J. (1973) Host plants in the population dynamics of insects. *Symp. Roy. Ent. Soc.* **6**, 181–199. 8.3.3

Van Horne B. (1982) Niches of adult and juvenile deer mice (*Peromyscus maniculatus*) in seral stages of coniferous forest. *Ecology* **63**, 992–1003. 6.6.3

van Soest P. J. (1982) *Nutritional Ecology of the Ruminant*. O. & B. Books, Corvallis, Oregon. 7.2.3

Van Valen L. (1965) Morphological variation and width of ecological niche. *Amer. Natur.* **99**, 377–390. 6.6.3, 14.7

Van Valen L. (1973) A new evolutionary law. *Evolutionary Theory* **1**, 1–30. 14.3, 14.8

Varley G. C., Gradwell G. R. & Hassell M. P. (1973) *Insect Population Ecology: an Analytical Approach*. Blackwell Scientific Publications, Oxford. 8.3.2, 8.3.3

Vesey-Fitzgerald D. F. (1965) The utilization of natural pastures by wild animals in the Rukwa Valley, Tanganyika. *E. Afr. Wildlife J.* **3**, 38–48. 12.2

Vesey-Fitzgerald D. F. (1960) Grazing succession among East African game animals. *Journal Mammal* **41**, 161–172. 8.4.3

Vinson S. B. & Iwantsch G. F. (1980) Host regulation by insect parasitoids. *Quart. Rev.*

Biol. **55**, 143–165. 13.3.1

Vogl R. J. (1974) Effects of fire on grasslands. In *Fire and Ecosystems* (eds Kozlowski T. T. & Ahlgren C. E.) pp. 139–194. Academic Press, New York. 12.2

Vogt K. A., Grier C. C., Meier G. E. & Keyes M. R. (1983) Organic matter and nutrient dynamics in forest floors of young and mature *Abies amabilis* stands in western Washington, as affected by fine-root input. *Ecol. Monogr.* **53**, 139–157. 7.2.2

Volterra V. (1926) Variatzioni e fluttuazioni del numero d'individui in specie animali conviventi. *Mem. R. Acad. Linei Ser.* **6**, II. 1.7, 6.1

von Humboldt A. (1850) *Essay sur la Géographie des Plantes*. Levrault, Schoell et Cie, Paris. 1.3

Waddell E. W. (1972) *The Mound Builders: Agricultural Practices, Environment and Society in the Central Highlands of New Guinea*. University of Washington Press, Seattle. 15.3.4

Waddell E. (1975) How the Enga cope with frost: responses to climatic perturbation in the Central Highlands of New Guinea. *Human Ecology* **3**, 249–274. 15.3.4

Wahl I., Eshed N., Segal A. & Sobel Z. (1978) Significance of wild relatives of small grains and other wild grasses in cereal powdery mildews. In *The Powdery Mildews* (ed. Spencer D. M.) pp. 83–100. Academic Press, London. 14.8

Waiko J. & Jiregari K. (1982) Conservation in Papua, New Guinea: custom and tradition. In *Traditional Conservation in Papua, New Guinea: implications for today* (eds Morauta L., Pernetta J. & Heaney W.) pp. 21–43, Monograph 16, Institute of Applied Social and Economic Research, Boroko, Papua, New Guinea. 15.3.5

Walker D. (1970) Direction and rate in some British post-glacial hydroseres. In *Studies in the Vegetational History of the British Isles* (eds Walker D. & West R.) pp. 117–137. Cambridge University Press, Cambridge. 1.5, 12.2

Walker D. (1982) The development of resilience in burned vegetation. In *The Plant Community as a Working Mechanism* (ed. Newman E. I.) pp. 27–43. Blackwell Scientific Publications, Oxford. 12.5

Waloff N. & Thompson P. (1980) Census data of populations of some leafhoppers (Auckenorrhyncha, Homoptera) of acid grassland. *J. Anim. Ecol.* **49**, 395–416. 8.3.2

Walsh J. J. (1981) Shelf-sea ecosystems. In *Analysis of Marine Ecosystems* (ed. Longhurst R. A.) pp. 159–196. Academic Press, London. 7.3.1

Walter H. (1973) *Vegetation of the Earth*. Springer Verlag, New York. 5.5, 5.5.1

Walters C. J., Stocker M. & Haber G. C. (1981) Simulation and optimisation models for wolf-ungulate system. In *Dynamics of Large Mammal Populations* (eds Foster C. W. & Smith T. D.) pp. 317–338. Wiley-Interscience, New York. 8.4.3

Ward L. K. & Lakhani K. W. (1977) The conservation of juniper: the fauna of food-plant island sites in southern England. *J. appl. Ecol.* **14**, 121–135. 8.2.1

Warner R. E. (1968) The role of introduced diseases in the extinction of the endemic Hawaiian avifauna. *Condor* **70**, 101–120. 9.3.2

Waser N. M. (1978) Competition for hummingbird pollination and sequential flowering in two Colorado wildflowers. *Ecology* **59**, 934–944. 11.2.3, 11.5.1

Watanabe H. (1967) Amount of ingestion of mature woodlice (*Armadillidium vulgare*) under various temperatures. *Jap. J. Ecol.* **17**, 134–135 (in Japanese). 7.2.2

Watson P. (1969) Evolution in closely adjacent populations. VI. An entomophilous species, *Potentilla erecta*, in two contrasting habitats. *Heredity* **24**. 407–422. 14.4

Watt A. S. (1919) On the causes of failure of natural regeneration in British oakwoods. *J. Ecol.* **7**, 173–203. 12.2

Watt A. S. (1947) Pattern and process in the plant community. *J. Ecol* **35**, 1–22. 1.8, 2.2, 12.5

Watt A. S. (1957) The effects of excluding rabbits from grassland B (Mesobrometum) in Breckland. *J. Ecol.* **45**, 861–878. 1.8

Watt W. D. (1966) Release of dissolved organic material from the cells of phytoplankton populations. *Proc. roy. Soc. Lond. Ser. B* **164**, 521–551. 7.3.3

Wavre M. & Brinkhurst R. O. (1971) Interactions between some tubificid oligochaetes and bacteria found in the sediments of Toronto Harbour, Ontario. *J. Fish. Res. Bd. Canada* **28**, 335–341. 7.3.3

Weatherly N. F. (1971) Effects on litter size and litter survival in Swiss mice infected with *Trichinella spiralis* during gestation. *J. Parasitol.* **57**, 298–301. 13.2.1

Webb L. J. (1977) Ecological considerations and safeguards in the modern use of tropical rainforests as a source of pulpwood: example, the Madang area, Papua, New Guinea. Office of Environment and Conservation, Dept. of Natural Resources, Waigani, Papua New Guinea. 15.1

Webb L. J., Tracey J. G. & Williams W. T. (1972) Regeneration and pattern in the subtropical rain forest. *J. Ecol.* **60**, 675–695. 2.5

Weinack O. M., Snoeyenbos G. H., Smyser C. F. & Soerjadi A. S. (1981) Competitive exclusion of intestinal colonization of *Escherichia coli* in chicks. *Avian Disease* **25**, 696–705. 9.3.2

Weller R. A. (1957) The amino acid composition of hydrolysates of microbial preparations from the rumen of sheep. *Aust. J. biol. Sci.* **10**, 384–389. 7.2.3

Werner E. E. & Hall D. J. (1977) Competition and habitat shift in two sunfishes (Centrarchidae). *Ecology* **58**, 869–876. 11.2.2, 11.3.1, 11.4.1

Werner E. E. & Hall D. J. (1979) Foraging efficiency and habitat switching in competing sunfish. *Ecology* **60**, 256–264. 6.7.1, 11.2.2, 11.3.1

Werner P. A. (1977) Colonization success of a "biennial" plant species: experimental field studies of species cohabitation and replacement. *Ecology* **58**, 840–849. 11.2.3

Werner P. A. (1979) Competition and coexistence of similar species. In *Topics in Plant Population Biology* (eds Solbrig O. T., Jain S., Johnson G. B. & Raven P. H.) pp. 287–312. Columbia University Press, New York. 6.4

Werner P. A. & Platt W. J. (1976) Ecological relationships of co-occurring golden rods (*Solidago*: Compositae). *Amer. Natur.* **110**, 959–971. 1.8

Wetzel R. G. & Penhale P. A. (1979) Transport of carbon and excretion of dissolved organic carbon by leaves and roots/rhizomes in seagrasses and their epiphytes. *Aquat. Bot.* **6**, 149–158. 7.3.3

Wheeler A. (1980) Fish-algal relations in temperate waters. In *The Shore Environment* (eds Price J. H., Irvine D. E. G. & Farnham W. F.) pp. 677–698. Academic Press, London. 7.3.1

Wheeler A. G. Jr. (1974) Phytophagous arthropod fauna of crownvetch in Pennysylvania. *Canadian Entomologist* **106**, 897–908. 8.2.4

White M. J. D. (1978) *Modes of Speciation*. W. H. Freeman, San Francisco. 8.2.4

White P. S. (1979) Pattern, process and natural disturbance in vegetation. *Bot. Rev.* **45**, 229–299. 12.2, 12.4

Whitlock J. H. & Georgi J. R. (1976) Biological controls in mixed trichostrongyle infection. *Parasitology* **72**, 207–224. 13.2.1

Whitmore T. C. (1982) On pattern and process in forests. In *The Plant Community as a Working Mechanism* (ed. Newman E. I.) pp. 45–59. Blackwell Scientific Publications, Oxford. 12.2, 12.3.4

Whittaker R. H. (1952) A study of summer foliage insect communities in the Great Smoky Mountains. *Ecol Monogr.* **22**, 1–44. 1.5

Whittaker R. H. (1956) Vegetation of the Great Smoky Mountains. *Ecol. Monogr.* **26**, 1–80. 1.5

Whittaker R. H. (1962) Classification of natural communities. *Bot. Rev.* **28**, 1–239. 1.4

Whittaker R. H. (1965) Dominance and diversity in land plant communities. *Science* **147**, 250–260. 1.5

Whittaker R. H. (1967) Gradient analysis of vegetation. *Biol Rev.* **42**, 207–264. 1.5, 5.5.3

Whittaker R. H. (ed.) (1973) *Ordination and Classification of Communities. Handbook of*

Vegetation Science. **5**. W. Junk, The Hague. 4.4.2
Whittaker R. H. (1975) *Communities and Ecosystems* (2nd edn) Macmillan, New York. 1.1, 1.6, 4.4, 5.1, 5.5, 7.3.1, 12.1
Wiens J. A. (1977) On competition and variable environments. *Amer. Natur* **65**, 590–597. 5.2, 6.2
Wiens J. A. & Rotenberry J. T. (1979) Diet niche relationships among North American grassland and shrubsteppe birds. *Oecologia* **42**, 253–292. 5.2
Wiens J. A. & Rotenberry J. T. (1980) Bird community structure in cold shrub deserts: competition or chaos? *Acta XVII Congr. Intern. Ornith.* (*Berlin*) (ed. Nöhring R.) pp. 1063–1070. Deutsche Ornithologen-Gesellschaft, Berlin. 6.6.3, 6.8
Wilbur H. W. (1972) Competition, predation, and the structure of the *Ambystoma–Rana sylvatica* community. *Ecology* **53**, 3–21. 11.3.1, 11.4.1, 11.5.1, 11.5.2, 11.7
Wilden A. (1972) *System and Structure: Essays in Communication and Exchange*. Tavistock, London. 15.4
Wiley E. O. & Brooks D. R. (1982) Victims of history — a non-equilibrium approach to evolution. *Syst. Zool.* **31**, 1–24. 15.4
Wilhm J. M. (1968) Use of biomass units in Shannon's formula. *Ecology* **49**, 153–156. 4.3.3
Willet K. C. (1956) An experiment on dosage in human trypanosomiasis. *Ann. trop. Med. Parasitol.* **50**, 75–80. 13.2.1
Williams A. H. (1981) An analysis of competitive interactions in a patchy back-reef environment. *Ecology* **62**, 1107–1120. 11.3.1, 11.6.2, 11.7
Williams C. B. (1947) The logarithmic series and its application to biological problems. *J. Ecol.* **34**, 253–272. 4.3.2
Williams C. B. (1964) *Patterns in the Balance of Nature*. Academic Press, London. 9.2.3
Williams E. E. (1972) The origin of faunas. Evolution of lizard congeners in a complex island fauna: a trial analysis. *Evol. Biol.* **4**, 47–89. 6.5.1
Williams F. E. (1936) *Papuans of the Trans-Fly*. Clarendon Press, Oxford. 15.3.3
Williams G. R. (1973) Birds. In *The Natural History of New Zealand* (ed. Williams G. R.) pp. 304–333. A. H. & A. W. Reed, Wellington. 10.3.1
Williams J. C., Knox J. W., Baumann B. A., Snider T. G., Kimball M. D. & Hoerner T. J. (1983) Seasonal changes of gastrointestinal nematode populations in yearling beef cattle in Louisiana with emphasis on prevalence of inhibition in *Ostertagia ostertagi*. *Intern. J. Parasitol.* **13**, 133–143. 9.2.5
Williams W. T. (ed.) (1976) *Pattern Analysis in Agricultural Science*. Elsevier, Amsterdam. 4.4.2, 4.4.3
Williamson M. (1972) *The Analysis of Biological Populations*. Edward Arnold, London. 11.5.1
Williamson M. H. (1981) *Island Populations*. Oxford University Press, Oxford. 8.2.4
Williamson P. (1971) Feeding ecology of the red-eyed vireo (*Vireo olivaceus*) and associated foliage-gleaning birds. *Ecol. Monogr.* **41**, 129–152. 5.2
Willson M. F. (1969) Avian niche size and morphological variation. *Amer. Natur.* **103**, 531–544. 6.6.3
Willson M. F. (1974) Avian community organization and habitat structure. *Ecology* **55**, 1017–1029. 5.2, 5.3.2
Wilson D. E. (1973) Bat faunas: a trophic comparison. *Syst. Zool.* **22**, 14–29. 5.5.1
Wilson D. S. (1975) The adequacy of body size as a niche difference. *Amer. Natur.* **109**, 769–784. 6.4, 6.5.1
Wilson E. O. (1961) The nature of the taxon cycle in the Melanesian ant fauna. *Amer. Natur.* **95**, 169–193. 6.6.2
Wilson E. O. (1969) The species equilibrium. In *Diversity and Stability in Ecological Systems* (eds Woodwell G. M. & Smith H. H.) pp. 38–47. Brookhaven Symposia in Biology, No. **22**. Brookhaven National Laboratory, Upton, New York. 9.2.3

Winterhalder B. (1980) Environmental analysis in human evolution and adaptation research. *Human Ecology* **8**, 135–170. 15.2

Winterhalder B & Smith E. A. (1981) *Hunter-Gatherer Foraging Strategies: Ethnographic and Archaeological Analyses*. University of Chicago Press, Chicago. 15.2

Wise D. H. (1981a) A removal experiment with darkling beetles: lack of evidence for interspecific competition. *Ecology* **62**, 727–738. 11.4.1, 11.7

Wise D. H. (1981b) Inter- and intra-specific effects of density manipulations upon females of two orb-weaving spiders (Araneae: Araneidae). *Oecologia* **48**, 252–256. 11.4.1, 11.7

Wisniewski W. L. (1958) Characterization of the parasitofauna of a eutrophic lake. *Acta Parasitol. Polon.* **6**, 1–64. 9.2.4

Wodzicki K. A. (1950) Introduced mammals of New Zealand. *D. S. I. R. Bull, Wellington* **98**, 1–255. 10.3.1

Wood A. & Humphreys G. S. (1982) Traditional soil conservation in Papua, New Guinea. In *Traditional Conservation in Papua, New Guinea: Implications for Today* (eds Morauta L., Pernetta J. & Heaney W.) pp. 93–114. Monograph **16**, Institute of Applied Social and Economic Research, Boroko, Papua New Guinea. 15.3.4

Wood T. K. & Guttman S. I. (1983) *Enchenopa binotata* complex: sympatric speciation? *Science* **220**, 310–312. 8.5

Wootten R. (1973) The metazoan parasite-fauna of fish from Hanningfield Reservoir, Essex, in relation to features of the habitat and host populations. *J. Zool. Lond.* **171**, 323–331. 9.2.4

Wrobel D. J., Gergitz W. F. & Jaeger R. G. (1980) An experimental study of interference competition among terrestrial salamanders. *Ecology* **61**, 1034–1039. 11.2.2

Yarbrough K. M. & Kojima K. (1967) The mode of selection at the polymorphic esterase 6 locus in cage populations of *Drosophila melanogaster*. *Genetics* **57**, 677–686. 14.6

Yen D. E. (1976a) The ethnobotany of the Tasaday: II. Plant names of the Tasaday, Manobo Blit and Kemato Tboli. In *Further Studies on the Tasaday* (eds Yen D. E. & Nance J.) pp. 137–158. Panamin Foundation Research Series No. 2, Makati, Rizal, Philippines. 15.3.1

Yen D. E. (1976b) The ethnobotany of the Tasaday: III. Notes on the subsistence system. In *Further Studies on the Tasaday* (eds Yen D. E. & Nance J.) pp. 159–183. Panamin Foundation Research Series No. **2**, Makati, Rizal, Philippines. 15.3.1

Yen D. E. & Nance J. (eds.) (1976) *Further Studies on the Tasaday*. Panamin Foundation Research Series No. **2**, Makati, Rizal, Philippines. 15.3.1

Yodzis P. (1980) The connectance of real ecosystems. *Nature* **284**, 544–545. 4.2.2

Yodzis P. (1981) The stability of real ecosystems. *Nature* **289**, 674–676. 9.3.1

Zemenhof P. J. & Eichon H. H. (1967) Study of microbial evolution through loss of biosynthetic functions: establishment of defective mutants. *Nature* **216**, 456. 13.3.2

Zieman J. C. (1975) Quantitative and dynamic aspects of the ecology of turtle grass *Thalassia testudinum*. In *Estuarine Research* (ed. Cronin L. E.) Vol. 1, pp. 541–562. Academic Press, New York. 7.3.3

Zlotin R. I. & Khodashova K. S. (1980) *The Role of Animals in Biological Cycling of Forest-Steppe Ecosystem* (English edn). Dowden, Hutchinson & Ross, Stroudsburg, Pennsylvania. 7.2.2

ZoBell C. E. (1963) Domain of the marine microbiologist. In *Symposium on Marine Microbiology* (ed. Oppenheimer C. H.) pp. 3–24. Charles C. Thomas, Springfield, Illinois. 7.3.3

ZoBell C. E. & Feltham C. B. (1938) Bacteria as food of certain marine invertebrates. *J. mar. Res.* **4**, 312–327. 7.3.3

Winterhalder B. (1980) Environmental analysis in human evolution and adaptation research. *Human Ecology* 8, 135–170. [illegible]

Winterhalder B. & Smith E. A. (1981) *Hunter-Gatherer Foraging Strategies. Ethnographic and Archaeological Analyses.* University of Chicago Press, Chicago. [illegible]

Wise D. H. (1981a) A removal experiment with darkling beetles: lack of evidence for interspecific competition. *Ecology* 62, 727–738. [illegible]

Wise D. H. (1981b) Inter- and intraspecific effects of density manipulations upon females of two orb-weaving spiders (Araneae: Araneidae). *Oecologia* 48, 252–256. [illegible]

Wisniewski W. L. (1958) Characterization of the parasitofauna of a eutrophic lake. *Acta Parasitol. Polon.* 6, 1–64. [illegible]

Wodzicki K. A. (1950) Introduced mammals of New Zealand. *D.S.I.R. Bull. Wellington* 98, 1–255. [illegible]

Wood A. & Humphreys G. S. (1982) Traditional soil conservation in Papua New Guinea. In *Traditional Conservation in Papua New Guinea: Implications for Today* (eds L. Morauta, J. Pernetta & W. Heaney), pp. 93–114. Monograph 16, Institute of Applied Social and Economic Research, Boroko, Papua New Guinea. [illegible]

Wood T. K. & Guttman S. I. (1983) *Enchenopa binotata* complex: sympatric speciation? *Science* 220, 310–312. [illegible]

Wootten R. (1973) The metazoan parasite-fauna of fish from Hanningfield Reservoir, Essex, in relation to features of the habitat and host populations. *J. Zool., Lond.* 171, 323–331. [illegible]

Wrobel D. J., Gergits W. F. & Jaeger R. G. (1980) An experimental study of interference competition among terrestrial salamanders. *Ecology* 61, 1034–1039. [illegible]

Yarbrough K. M. & Kojima K. (1967) The mode of selection at the polymorphic esterase 6 locus in cage populations of *Drosophila melanogaster*. *Genetics* 57, 677–686. [illegible]

Yen D. E. (1976a) The ethnobotany of the Tasaday. II. Plant names of the Tasaday, Manobo Blit and Kemato Tboli. In *Further Studies on the Tasaday* (eds D. E. Yen & J. Nance) pp. 137–158. Panamin Foundation Research Series No. 2, Makati, Rizal, Philippines. [illegible]

Yen D. E. (1976b) The ethnobotany of the Tasaday. III. Notes on the subsistence system. In *Further Studies on the Tasaday* (eds D. E. Yen & J. Nance) pp. 159–183. Panamin Foundation Research Series No. 2, Makati, Rizal, Philippines. [illegible]

Yen D. E. & Nance J. (eds) (1976) *Further Studies on the Tasaday*. Panamin Foundation Research Series No. 2, Makati, Rizal, Philippines. [illegible]

Yodzis P. (1980) The connectance of real ecosystems. *Nature* 284, 544–545. [illegible]

Yodzis P. (1981) The stability of real ecosystems. *Nature* 289, 674–676. [illegible]

Zamenhof P. J. & Eichhorn H. H. (1967) Study of microbial evolution through loss of biosynthetic functions: establishment of defective mutants. *Nature* 216, 456–458. [illegible]

Zieman J. C. (1975) Quantitative and dynamic aspects of the ecology of turtle grass (*Thalassia testudinum*). In *Estuarine Research* (ed. L. E. Cronin) Vol. 1, pp. 541–562. Academic Press, New York. [illegible]

Zlotin R. I. & Khodashova K. S. (1980) *The Role of Animals in Biological Cycling of Forest-Steppe Ecosystems*. (English trans.) Dowden, Hutchinson & Ross, Stroudsburg, Pennsylvania. [illegible]

ZoBell C. E. (1963) Domain of the marine microbiologist. In *Symposium on Marine Microbiology* (ed. C. H. Oppenheimer) pp. 3–24. Charles C. Thomas, Springfield, Illinois. [illegible]

ZoBell C. E. & Feltham C. B. (1938) Bacteria as food for certain marine invertebrates.

Subject Index